TASER® Conducted Electrical Weapons: Physiology, Pathology, and Law

The below-listed authors have a material, financial, or other relationship with a related business or other entity whose products or services may be discussed in, or directly affected in the marketplace by, this manuscript. This relationship is specified in the author's chapter.

Jeffrey D. Ho
Mark W. Kroll
Ronald Moscati
Robert F. Reardon
Donald M. Dawes
James D. Sweeney
Robert A. Stratbucker

Mark W. Kroll · Jeffrey D. Ho
Editors

TASER® Conducted Electrical Weapons: Physiology, Pathology, and Law

Springer

Editors
Mark W. Kroll
University of Minnesota
Biomedical Engineering
Crystal Bay, Minnesota, USA
mark@kroll.name

Jeffrey D. Ho
Department of Emergency Medicine
Hennepin County Medical Center
Minneapolis, Minnesota, USA
hoxxx010@umn.edu

ISBN 978-0-387-85474-8 e-ISBN 978-0-387-85475-5
DOI 10.1007/978-0-387-85475-5

Library of Congress Control Number: 2008933389

Printed on acid-free paper

springer.com

Contents

Contributors

Geoffrey P. Alpert University of South Carolina, Department of Criminal Justice and Sociology

Loren Berenbom Clinical Assistant Professor; Director, Electrophysiology, Mid America Cardiology, University of Kansas Hospital

Greg Bingham Oakland, CA Police Department (Ret.), gkcglb@aol.com

Michael A. Brave LAAW International, Inc., TASER International, Inc., brave@laaw.com

James E. Brewer jamesbrewer@wcta.net

Theodore C. Chan Department of Emergency Medicine, University of California, San Diego Medical Center, tcchan@ucsd.edu

Samuel Cloud Department of Emergency Medicine, SUNY at Buffalo, Erie County Medical Center, New York, USA

Peter J. Cuenca Department of Combat Medic Training, Department of Emergency Medicine, Brooke Army Medical Center

Donald M. Dawes Department of Emergency Medicine, Lompoc District Hospital, Lompoc, CA, donalddawes@aol.com

Vincent J.M. Di Maio Chief Medical Examiner (retired), Bexar County (San Antonio), Texas, vincent_dimaio@yahoo.com

Theresa G. Di Maio

Derek J. Dosdall Department of Biomedical Engineering, University of Alabama, Birmingham, Alabama, USA, djd@crml.uab.edu

Igor R. Efimov Department of Biomedical Engineering, Washington University, igor@wustl.edu

Michael A. Evans AIT Laboratories, maevans@aitlabs.com

Vadim V. Fedorov Department of Biomedical Engineering, Washington University, igor@wustl.edu

Joshua Gunn AIT Laboratories, JGunn@aitlabs.com

Mark Henych Advanced Research Solutions, Estero, FL

Jeffrey D. Ho Department of Emergency Medicine, Hennepin County Medical Center, University of Minnesota Medical School, Minneapolis, MN, USA, hoxxx010@umn.edu

Raymond E. Ideker Departments of Medicine, Biomedical Engineering, and Physiology, University of Alabama-Birmingham, Volker Hall B140, 1670 University Blvd, Birmingham, AL 35294-0019, USA, rei@crml.uab.edu

Robert J. Kaminski University of South Carolina, Department of Criminal Justice and Sociology

Steven B. Karch Consultant Pathologist and Toxicologist, Berkeley, California, skarch@sonic.net

Geran Kostecki Department of Biomedical Engineering, Washington University in St. Louis

M. Scott Kriger AIT Laboratories

Mark W. Kroll University of Minnesota, Biomedical Engineering, mark@kroll.name

Dhanunjaya R. Lakkireddy Clinical Assistant Professor and Staff Electrophysiologist, Mid America Cardiology, University of Kansas Hospitals, Suite G600, 3901 Rainbow Blvd, Kansas City, KS 66160, USA, dlakkireddy@mac.md

Andreas K. Lauer Casey Eye Institute, Department of Ophthalmology, Oregon Health & Science University, Legacy Hospitals, lauera@ohsu.edu

Leonid Livshitz Department of Biomedical Engineering, Washington University in St. Louis

Jonathan L. Marinaro Trauma Critical Care, Departments of Surgery and Emergency Medicine University of New Mexico Health Sciences

Deborah C. Mash Department of Neurology, Miller School of Medicine, University of Miami, dmash@med.miami.edu

Jason Mathis University of South Carolina, Department of Criminal Justice and Sociology

John G. McManus Department of Emergency Medicine, Brooke Army Medical Center, john.mcmanus@amedd.army.mil

Charlie Mesloh Florida Gulf Coast University, cmesloh@fgcu.edu

Greg Meyer Captain, Los Angeles Police Department (Ret.), gregmeyer@earthlink.net

Ronald Moscati Department of Emergency Medicine, SUNY at Buffalo, Erie County Medical Center, moscati@buffalo.edu

Dorin Panescu Chief Technical Officer with NewCardio, Inc., Santa Clara, CA, dpanescu@NewCardio.com

John G. Peters President and Founder, Institute for the Prevention of In-Custody Deaths, Inc., john@ipicd.com

Justin Ready Department of Sociology, John Jay College of Criminal Justice, 899 Tenth Ave, New York, NY 10019, jready@jjay.cuny.edu

Robert Reardon Department of Emergency Medicine, Hennepin County Medical Center, rfreardon@gmail.com

Jeffrey Rojek University of South Carolina, Department of Criminal Justice and Sociology

Michael R. Smith University of South Carolina, mrsmith@gwm.sc.edu

Robert A. Stratbucker Faculty, University of Nebraska, Colleges of Medicine and Engineering (Ret.), rstratbucker@cox.net

Samuel J. Stratton Professor, University of California, Los Angeles School of Public Health and the David Geffen School of Medicine at the University of California, Los Angeles; Medical Director, Health Disaster Management/ Emergency Medical Services Orange County Health Care Agency, strattos@ucla.edu

James D. Sweeney Department of Bioengineering, Florida Gulf Coast University, jsweeney@fgcu.com

Patrick Tchou Cleveland Clinic, Section of Cardiac Electrophysiology and Pacing, Department of Cardiovascular Medicine, The Heart and Vascular Institute, tchoup@ccf.org

James L. Vacek Director of Cardiovascular Research, The Center for Cardiovascular Scholarship, Mid America Cardiology, The University of Kansas Hospital

Subba Reddy Vanga St. Luke's Hospital, Chesterfield, MO 63017, USA

Gary M. Vilke Department of Emergency Medicine, University of California, San Diego Medical Center, CA, USA, gmvilke@ucsd.edu

John G. Webster Department of Biomedical Engineering, University of Wisconsin, webster@engr.wisc.edu

Charles V. Wetli Chief Medical Examiner and Director of Forensic Sciences, Suffolk, County, NY (retired); Clinical Professor of Pathology, SUNY at Stony Brook (retired), thanatopsis888@yahoo.com

Michael D. White School of Criminology and Criminal Justice, Arizona State University, mdwhite1@asu.edu

S. Robert Witherspoon Casey Eye Institute, Department of Ophthalmology, Oregon Health & Science University, Devers Eye Institute, Legacy Hospitals, witherss@ohsu.edu

Ross Wolf University of Central Florida, American Military University

Introduction

Seldom has a technology had such rapid adoption as the TASER® conducted electrical weapon (CEW). We all think of mobile phones and their rapid adoption but the compound average growth rate of mobile phone penetration over the past 10 years was 5.2% while that of the CEW was an astounding 84%. It is now the standard nonlethal tool for the majority of the law enforcement agencies in all of the English speaking countries and has rapidly spread to 45 countries worldwide.

Along the way there was an enormous collision of confusions. The interaction with law enforcement for the average citizen is the meek acceptance of a traffic ticket. Thus, we have no personal experience with the difficult challenge facing a police officer when taking a resistant subject into custody. Or, when controlling a psychotic person needing help for a medical emergency. For centuries before the CEW the primary tools were the clubs which are now euphemistically referred to as batons. Secondly, the total understanding of electricity for the average citizen is that it is something useful that comes out of sockets in the wall but that it is very dangerous and it should not contact the body. Finally, there is no appreciation of the difficult problem of arrest-related deaths which now take 800 lives annually in North America alone.

This collision of confusions is now being confronted by a rapidly growing body of science. Nearly a dozen animal and human studies of CEWs are now performed per year. Unfortunately, all of this scientific data has never been put together nor has it been combined within the context of the law enforcement use-of-force requirements and the difficult problems of arrest-related deaths. Thus, we recognized that there was a need for a standard reference to deal with these issues.

The book begins with the background of the history of CEWs both from a user and a scientific perspective. Captain Greg Meyer was a senior researcher for the innovative Los Angeles Police Department in use-of-force techniques and writes the user history. Dr. Stratbucker was well known for early research on defibrillation and was a medical pioneer in the study of CEWs. Dr. Mesloh

---

TASER® is a registered trademark of TASER International, Inc.

follows up with the background of police use-of-force and Dr. McManus covers the broader context of nonlethal weapons and the directions for the future.

Dr. James Sweeney – who is well known for his work in electrical transcutaneous muscle stimulation to help paralysis victims – explains transcutaneous muscle stimulation. Dr. Panescu – a recognized authority in the flow of electrical current through the human body – explains exactly where the CEW current does and does not flow.

The central and critical portion of the book covers the issue of possible side effects. We begin with the chapter of Dr. Webster, a prominent authority in biomedical engineering, who writes on animal models that have been used to study these devices. Dr. Chan has participated in about a dozen human studies of CEWs and other police tools. He wrote the important chapter on bridging the human and animal data as there are areas in which the animal and human data disagree. Dr. Ideker is the internationally recognized expert on lethal cardiac arrhythmias and writes on that topic. Dr. Ho follows up with chapters on the electrocardiographic and blood serum effects. Dr. Reardon covers the important topic of echocardiographic monitoring of the heart during CEW applications. The possibility of kidney damage from rhabdomyolysis is covered in the chapter by Dr. Moscati. Dr. Dawes discusses the effects on human respiration in one chapter and neuroendocrine effects in another. Dr. Moscati covers the effects with alcohol intoxication

Dr. Efimov – world recognized authority on electroporation (direct electrical damage to cells) – contributed the chapter on electroporation and possible nerve damage while Dr. Lauer covers the issue of facial and head injuries including the eye. A common concern is the possible dysynergy with illegal drugs and that is covered by Dr. Tchou who is a senior cardiac electrophysiologist with the Cleveland Clinic. A common concern with the public is the possible interaction with a pacemaker or implantable defibrillator and Dr. Lakireddy – who has researched this issue – presents the facts in his chapter.

The law enforcement need and the field experience are covered in three chapters by experienced police officers and law enforcement researchers. The chapter by Sgt. Bingham deals with risk management issues in a large police department. The field results in New York City are covered by Dr. White and the data in two other cities is covered by Dr. Smith. Statistician Brewer then presents the statistical evidence surrounding other common questions.

The important issues of the in-custody death are covered in the chapter by Dr. Stratton. Dr. Karch explains the cardiac effects of illegal stimulants. Dr. DiMaio covers the issue of excited delirium. The in-custody death creates great challenges for a medical examiner and the issue of toxicology is also covered by Dr. Evans' chapter while the brain analysis is covered by Dr. Mash who pioneered the techniques of brain analysis for excited delirium deaths. Dr. Wetli – who coined the modern term of "excited delirium" – provides a pathology checklist.

Finally, the legal issues loom very large in this case since the CEW is an electrical device that interacts daily with law enforcement. Michael Brave penned the chapter on CEW law and Dr. Peters covers the complicated issue of the meaning of science and logic in the courtroom.

Because of his scientific background in the use of electrical stimulation for medical devices Mark was invited to join the TASER International board in 2003 and there got his first exposure to the challenges faced by police. Jeff recognized a need for more scientific studies in this area and was able to secure grants from TASER to do the first formal human studies.

As opposed to the pacemaker and implantable defibrillator industry we have an unusual situation where one manufacturer dominates an area of technology. Thus we had a special challenge to find authors with the maximum amount of independence and minimal connections to TASER in order to keep the scientific credibility at its highest level. That had to be balanced with the fact that many of the top authorities in this area became so by working cooperatively in some cases with the primary manufacturer. As it is often put, we could not cross the sometimes thin line between independence and ignorance. We think we have achieved the right balance between knowledge and objectivity. Only one of the 50 authors is an employee of TASER International, and the majority have derived no consulting income from them. Dr. Sweeney, Dr. Stratbucker, and Dr. Kroll are on the Scientific and Medical Advisory Board.

To keep the scientific rigor at its highest level we also took the unusual step of having the chapters peer reviewed. Almost every chapter was anonymously reviewed by one to three experts in the field. This was done in addition to careful review by the coeditors.

We have dramatically different backgrounds. Jeff is a board certified emergency physician and a licensed law enforcement officer. Mark is a scientist who studied the use of electrical stimulation for therapy. What we have in common is a career dedication to saving lives especially those in emergency situations. Neither of us expected to be involved with TASER CEWs but our involvement has now reached the ultimate level of commitment as we have both experienced the amazing effects of this weapon. Thus, we joined that special club of over 1.6 million people that know what it feels like to have your body briefly controlled by special electrical waveforms. We hope you enjoy reading this book as much as we have enjoyed putting it together.

Chapter 1
Conducted Electrical Weapons: A User's Perspective

Greg Meyer

1.1 Emergence of Conducted Electrical Weapons

Since the mid-1970s, law enforcement officials have used conducted electrical weapons (CEWs) with varying degrees of success to control violent individuals who resist arrest or help. The TASER® devices have been the most widely used CEWs, ranging from the original 7 watt models (minimally effective as pain-compliance tools) to the 26-watt M-26 model (very effective as neuromuscular incapacitation tools) and now back to a new 7-watt model with the more effective waveform of the X-26. The actual power delivered to the suspect is far less than the power generated internally at the transformer input. Since delivered power is more complicated to measure, a practice was developed to rate CEWs by their internal power. So, the "7-watt" TASER X-26 delivers only a typical 1.3 watts to the suspect—not 7 watts—as its internal power might imply.

During his service with the Los Angeles Police Department, the author personally used or directed 11-watt TASER devices on the street eight times in the mid-1980s. These included the following: two knife-point threatened suicides, one standoff between the police and a resisting suspect armed with a knife, an enraged man (handcuffed in the backseat of a police car, repeatedly bashing his head against the door frame), and several people experiencing "excited delirium" influenced by use of drugs—in these cases, phencyclidine (PCP or angel dust)—that caused then to exhibit superhuman strength. The risk of serious injury to the suspects (as well as to the officers attempting to subdue them) was far greater using conventional methods than a CEW. It is also noteworthy that no device failure or sudden in-custody death occurred in these 8 confrontations.

Since 1999 there has been a dramatic proliferation of modern, more effective TASER CEWs and an increase in their use by law enforcement throughout the United States, Canada, the United Kingdom, and at least 43 other countries. Presently, over 12,000 law enforcement and corrections agencies use the devices.

G. Meyer (✉)
Captain, Los Angeles Police Department (Ret.)
e-mail: gregmeyer@earthlink.net

M.W. Kroll, J.D. Ho (eds.), *TASER® Conducted Electrical Weapons: Physiology, Pathology, and Law*, DOI 10.1007/978-0-387-85475-5_1,

Today's TASER CEW is a handheld device using compressed nitrogen to launch two tiny barbed darts toward its target. The darts are tethered to the device (which contains the power source) by thin, insulated wires that project outward to maximum distances of 15–35 feet, depending on the cartridge selected. In fast-breaking police use-of-force situations, experience proves that accuracy is greatest in the 12–15 foot range. When the two darts connect to the target (by attaching to clothing or sticking into the skin), an electronic circuit is completed. The TASER CEW delivers short pulses with very low average current. When successfully deployed (ideally a spread of 1–2 feet between the darts), the TASER CEW current interrupts the electrical signals from the central nervous system to the peripheral body. This interruption overwhelms the motor nervous system and causes the body to experience sudden shaking and rigidity, typically leading to a loss of balance and a fall to the ground. The default cycle is 5 seconds in length, which may be overridden if the trigger is held down. The cycle may be repeated if needed. Officers may safely touch and handcuff the suspect during the cycle if they are careful not to interject themselves into the path of the circuit. Ordinarily, subjects quickly recover from the temporary incapacitation effects of the device. Some subjects renew their resistance resulting in multiple applications of the device to subdue them. In addition to the "dart mode," the device may be used in "drive-stun" mode, i.e., touching the subject with the device without darts. This mode generally affects only the sensory nervous system, providing pain but not incapacitation. The dart mode is preferred by law enforcement because it is more helpful in the control of the suspect.

The interesting history of CEWs begins with John H. "Jack" Cover, an aerospace engineer, who invented a nonlethal CEW weapon to safely subdue the airline "skyjackers" of the 1970s. The government was deploying sky marshals armed with .38 caliber revolvers in response to these threats and the potential dangers of lethal weapons to passengers and aircraft were obvious. Still, the CEWs did not begin to gain acceptance until the 1980s.

1.2 Resisting Arrest and Excited Delirium

One of the most difficult and dangerous police responsibilities is to overcome the resistance of a hyperagitated person in order to bring about an arrest or to assist a suicidal subject—without inflicting or receiving significant injuries.[1]

Unfortunately, years of fictional depictions of police work give an unrealistic idea of the level of force it takes to control an actively resisting person. Police officers, paramedics, emergency room staff, and psychiatric facility staffs know the unpleasant realities. They also know it is especially difficult to control an alcohol-intoxicated or drug-abusing subject, especially drugs such as methamphetamine, cocaine, LSD, or PCP. In some cases it may take six or more officers to restrain them. Likewise, psychotic individuals may exhibit superhuman strength and extreme agitation with bizarre behavior. In the past this collection of symptoms has been called by various names, such as Bell's Mania, agitated delirium, acute exhaustive mania, and other terms. It is now known as "excited delirium" and will be discussed in great detail in later chapters.

1.3 Tragedy Leads to Change

Until recently, the only tools available for law enforcement to control those resisting arrest have been: firearms (revolvers, carbines, shotguns), batons (billy clubs, nightsticks), saps (leather-bound, lead-filled striking devices), neck restraints (bar-arm holds across the trachea and carotid holds), and handcuffs. Except the firearms all require close bodily contact. Scores (if not hundreds) of officers have suffered major injuries and death attempting to subdue or take into custody combative suspects under the influence of drugs like PCP, cocaine, LSD, and methamphetamines; as well as schizophrenius and bi-polar subjects off their medications (all known to be associated with superhuman strength).

A series of tragic events in the 1970s caused several major law enforcement agencies to give less-injurious alternative devices serious consideration. In 1977, officer Roger Scott was disarmed and fatally shot in the face by a naked man on PCP. That same year, four police officers responding to a single incident suffered broken bones and concussions subduing another naked PCP user. Again, in an unrelated incident, sergeant Kurt Barz, shot and killed a naked man on PHP (a PCP analog); the man had twice taken the sergeant baton and was about to overpower him. In 1978, a deputy sheriff was disarmed and shot to death in a struggle with a PCP suspect. These events occurred in the greater Los Angeles area, and similar events were occurring elsewhere as well, and all of them involved suspects in the hyperagitated state of "excited delirium."[1]

A significant turning point occurred in 1979. On January 3, an emotionally distraught Eulia Love was shot and killed when she attacked two Los Angeles Police Department officers with an 11-inch butcher knife. The officers shot her after repeated verbal efforts and the use of a police baton failed to bring her under control. In the wake of that incident, the Board of Police Commissioners directed "continued research into the use of intermediate weapons and/or control devices which have the potential to significantly reduce reliance upon deadly force."[2]

On April 30, 1981, following a 1-year field test, the LAPD adopted the TASER CEW and chemical irritant spray as authorized nonlethal weapons. The enabling document, signed by the Chief of Police, stated the purpose of these devices:

> *Consistent with the Department's philosophy of using the minimum amount of force necessary to control violent suspects, the Department has tested and approved the use of certain nonlethal control devices. It is anticipated that the use of these devices will result in fewer altercation-related injuries to officers and suspects.*[3]

The original 7-watt TASER model proved ineffective on PCP suspects experiencing excited delirium. Its inventor tuned it up and field tested the 11-watt model which was then used for some years by the LAPD and proved to be very effective at subduing violent suspects, including those under the influence of PCP. In 1986 alone, the device was successfully used 600 times by the LAPD.[4]

However, in the late 1980s, the Tasertron CEW company rolled back the LAPD devices to 7 watts to match the power output of the rest of the company's law enforcement distribution. Unfortunately, this led to numerous failures of the device on the street. Officers lost confidence in the device, and its use was dramatically curtailed. The Rodney King incident in 1991 is perhaps the most famous failure of the device. King was twice taken to the ground by a 7-watt CEW but he was able to overcome its effects, leading officers to use batons and kicks in attempt to subdue him. This resulted in two officers being convicted in federal court and sent to prison, costing the City of Los Angeles $3.8 million in damages.

Today cocaine and methamphetamines are more prevalent on American streets than PCP (still a major problem in some cities). The effects are similar to PCP in terms of the level of violence and agitation experienced by some abusers of these drugs.

In response, TASER International, Scottsdale, Arizona, began marketing the "M-26," a 26-watt TASER CEW to law enforcement in 1999. Then, in 2003, the company introduced the "X-26," an improved 7-watt TASER CEW that has been widely distributed. Properly deployed modern TASER CEWs have proven to be very effective at controlling heavy aggression and resistance.

Several major police agencies (Phoenix, Houston, Miami, and Cincinnati among them) have opted for "full deployment" (i.e., a TASER CEW on every field officer's equipment belt).

1.4 Use of Force and Nonlethal Weapons

By law, police force may be used to make an arrest, prevent escape, or overcome resistance.[5] Many police use-of-force situations are sudden close-contact situations requiring immediate, instinctive response. Some begin as "standoffs" (with time for planning and maneuvering) but change to "immediate-response" situations if the suspect increases resistance. This escalation of events during a standoff is more likely when officers approach the suspect without formulating a plan, or if they fail to take aggressive actions to control the suspect before the standoff situation deteriorates.[6] Ideally, an officer adjusts the level of force in response to the changing levels of the suspect's resistance, in an effort to quickly overcome that resistance in a manner calculated to minimize injuries to all parties.

The term "nonlethal weapons" is used generically to identify innovative alternatives to traditional nonfirearm weapons and tactics (such as batons, flashlights, martial arts techniques, and miscellaneous bodily force) that police have used for more than 150 years. Three interchangeable terms are used to describe law enforcement tools: nonlethal weapons, less than lethal weapons, and less-lethal weapons. The National Criminal Justice Reference Thesaurus defines nonlethal weapons as "coercive weapons intended to prevent substantial risk of serious injury or deadly harm."[7] Researcher Peak defined them in terms of their effects:

> *Any weapon where there is only a temporary effect and minimal medical implications to normally healthy subjects; there is a high probability of instantaneous control over a highly motivated suspect; and there are observable effects, with a high probability of affecting only the intended targets.*[8]

Nonlethal weapons have also been defined in terms of the tactics and situations for which they are appropriate:

> *Devices which may be used to aggressively take control of a deteriorating tactical situation prior to that point in time when control holds, batons, or deadly force may become necessary; and when it is unsafe for an officer to move to within contact range of the suspect; and when attempts by officers to control the suspect by conventional means will likely result in serious injury to officers, suspects, or both.*[9]

The United States Department of Defense defines nonlethal weapons as

> *Weapons that are explicitly designed and primarily employed so as to incapacitate personnel or materiel, while minimizing fatalities, permanent injury to personnel, and undesired damage to property and the environment.*[10]

Some consider nonlethal weapons like the TASER CEW as "shooting-avoidance" tools. The concept of using nonlethal weapons to reduce the number of shootings by police is grounded in the belief that, in some situations, nonlethal weapons could control a suspect early in the confrontation, before an unarmed but resisting suspect has the opportunity to become armed and attack the officer. Also, a suspect armed with less than a firearm could be controlled before using it on himself or anyone else. In recent years, there have been an increasing number of so-called "saves" (i.e., safe control of a person who would likely otherwise have been shot with a firearm).

The factors of time (sudden attack vs. a standoff situation) and distance (between the officer and the suspect) are crucial to determine whether nonlethal weapons are appropriate for the situation. In standoff situations, nonlethal weapons should be used early and aggressively to bring the situation to a conclusion as quickly as possible, before it deteriorates into a confrontation requiring a greater level of force.[9]

During more than 28 years of LAPD use, the TASER CEW and chemical irritant spray have been used in thousands of standoff situations. Such cases occasionally deteriorate into immediate-response, deadly force situations if the suspect attacks with a knife, club, or other dangerous implement (less than a firearm) and is not quickly controlled. Even in its early years on the street there were instances where standoff situations involving the use of knives[11] and other weapons, as well as a number of suicide threats[12] were brought to swift conclusion through aggressive use of the TASER CEW.

The vast majority of nonlethal weapon incidents involve unarmed suspects who exhibit resistive, violent, or bizarre behavior, thus presenting a significant safety threat to themselves, others and the officers whose job it is to intervene.

Today, controversies surround CEW use, especially in incidents of in-custody death, or when a CEW is used on very young, very old, or restrained people. The headlines "*Man Dies After Police Use TASER*" are reminiscent of similar headlines in years past when there were deaths reported after use of pepper spray, neck restraint, and other tactics. The question of best police procedure remains, however, when dealing with a violent elderly, very young, or restrained person who may be armed. How are the police to take such a person into custody if not by TASER CEW which often quickly ends the incident and causes fewer and less severe injuries than many other police tools and tactics?

From a tactics training and policy point of view, there is continuing controversy about whether to restrict the number of times an individual may receive the CEW shock during an incident. (Several medical studies on this point will be referenced elsewhere.) It is a common misunderstanding that CEWs can cause electrocution, especially after multiple exposures. Since electrocution occurs within 1 or 2 seconds after exposure to lethal amounts of electrical current in the body, and all known in-custody deaths occur after that interval, a number of medical research experts eliminated CEWs as electrocuting agents, so the common suggestion to limit application time may have no scientific basis.

Table 1.1 Injuries to suspects by effective force type[13]

Effective force type	No injury	Temporary effects only	Minor injury	Moderate injury	Major injury	Total cases	Injury rate (%)
Baton	24	0	24	66	7	121	80
Kick	20	0	9	12	0	41	51
Punch	6	0	5	15	1	27	78
Misc. bodily force	51	0	20	58	6	135	62
Flashlight	4	0	0	14	6	24	83
Swarm	33	0	3	10	1	47	30
Chem CS/CN	0	18	1	0	0	19	5
TASER® CEW	0	88	0	0	0	88	0
Total	138	106	62	175	21	502	n/a

Table 1.2 Injuries to officers by effective force type[13]

Effective force type	No injury	Temporary effects only	Minor injury	Moderate injury	Major injury	Total cases	Injury rate (%)
Baton	99	0	4	10	8	121	18
Kick	36	0	0	3	2	41	12
Punch	19	0	0	5	3	27	30
Misc. bodily force	109	0	5	13	8	135	19
Flashlight	20	0	3	1	0	24	17
Swarm	39	0	1	6	1	47	17
Chem CS/CN	14	5	0	0	0	19	0
TASER® CEW	88	0	0	0	0	88	0
Total	424	5	13	38	22	502	n/a

Tables 1.1 and 1.2 give the statistics for a stratified random sample of suspect and officer injuries from the first half of 1989. Figures 1.1 and 1.2 depict the injury rates. (Note that the LAPD used a mixture of 11-watt and 7-watt TASER CEW devices at the time.)

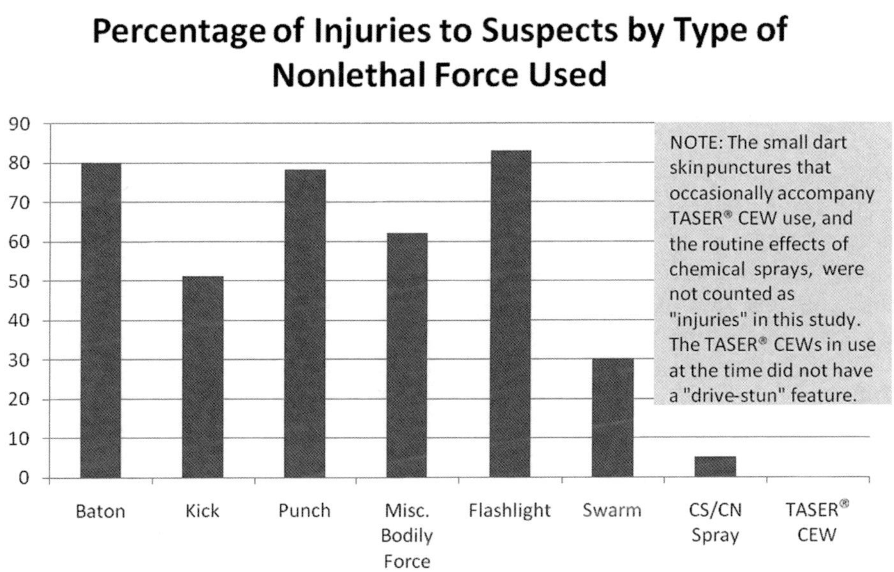

Fig. 1.1 Percentage of injuries to suspects by type of nonlethal force used (note: the small dart skin punctures that occasionally accompany TASER® CEW use, and the routine effects of chemical sprays, were not counted as "injuries" in this study. The TASER® CEWs in use at the time did not have a "drive-stun" feature)

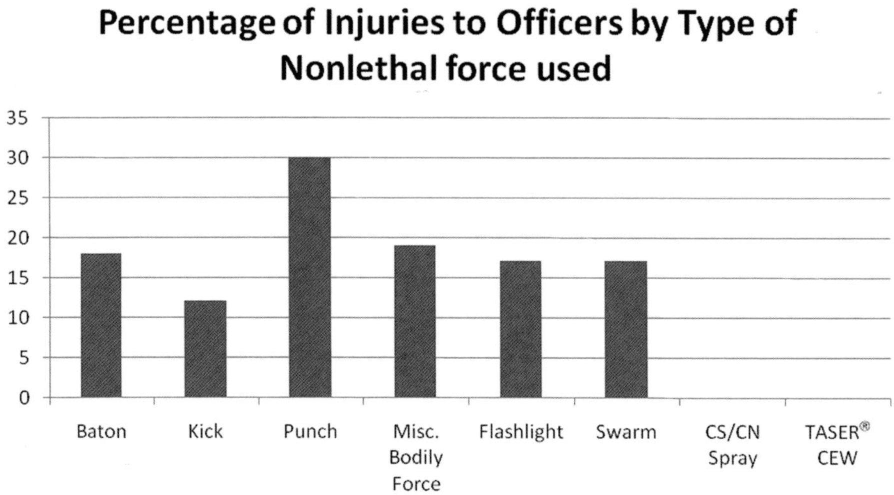

Fig. 1.2 The percentage of injuries to officers by type of nonlethal force used (statistics from 502 nonlethal use of force cases[13])

1.5 Conclusion

The aggressive use of available nonlethal weapons early in standoff confrontations predictably results in fewer and less severe injuries to suspects and officers. The usual tactical alternative of prolonging the standoff frequently leads to the use of more injurious degrees of force, including deadly force.

There will always be potentially violent confrontations between police officers and resistive suspects. The challenge for law enforcement is to provide the leadership to educate officers, the public, politicians, and the media to accept a counterintuitive, yet fundamental, change in the way law enforcement deals with these confrontations. The police must acquire effective nonlethal weapons and must use them aggressively in appropriate situations to reduce injuries to suspects and to themselves.

Police officers, supervisors, and managers must systematically search for effective tactics to reduce the number of violent confrontations. For those confrontations that are unavoidable, the number and severity of injuries can be reduced significantly through the development and use of effective nonlethal weapons such as the TASER CEW.

Nonlethal weapons technology allows law enforcement to have policy, training, equipment, tactics, and review practices that will lead to safer outcomes in many dangerous incidents. Law enforcement and corrections agencies that use nonlethal weapons see a tremendous decrease in injuries to officers and suspects, reduced personnel complaints, reduced liability in lawsuits, and an improved public image for the agency.

Notes

1. Meyer, G. Your nonlethal weapons alternatives, Journal of California Law Enforcement 1981;15(1):125–127.
2. Los Angeles Police Department, The Report of the Board of Police Commissioners Concerning the Shooting of Eulia Love and the Use of Deadly Force: Part III — Training and Community Relations, 1980.
3. Los Angeles Police Department, Office of the Chief of Police, Nonlethal Control Devices, Special Order No. 16, 30 April 1981.
4. Los Angeles Police Department, Statistical Digest, 1989.
5. Section 835a, California Penal Code.
6. Meyer, G. Nonlethal Weapons.
7. U.S., National Institute of Justice, National Criminal Justice Thesaurus, Department of Justice, January 1987, p. 174.
8. Peak, K. The quest for alternatives to lethal force: a heuristic view, Journal of Contemporary Criminal Justice 1990;6(1):9.
9. Meyer, G. Nonlethal Weapons.
10. Department of Defense Directive (DODD) 3000.3. *Policy for Nonlethal Weapons*, July 1996.
11. Rippee, G. Knife-wielder's life saved by use of Taser dart gun, Daily News, Van Nuys, California 1983;6 20(1):2.
12. Meyer, G. Los Angeles Police Department, Sergeant's Daily Report, 17 August 1984; and personal knowledge of other similar incidents in which the TASER CEW was used aggressively to control people who were threatening suicide.
13. Meyer, G. Nonlethal Weapons vs. Conventional Police Tactics: The Los Angeles Police Department Experience, Master's Thesis, California State University, Los Angeles 1991.

Chapter 2
The Scientific History

Robert A. Stratbucker*

We begin with the man who started it all—still among us—master "bioengineer" John H. Cover of Chula Vista California, and his revolutionary 7-watt TASER CEW.

In the mid-1960s—as a NASA aerospace physicist—he came to the conclusion that he had all of the critical components required to emulate the fictional less-lethal "electric rifle" popularized by its equally fictional creator "Thomas A. Swift" [1,2].

The USA Heart, Cancer, and Stroke initiative emerged at that time, diverting much defense and aerospace spending to life science and also bioengineering initiatives such as The Artificial Heart Program. The timing was perfect for an electronic flip-flop from aerospace into health systems and it may have carried Jack Cover with it. Edwin Meese, 70th Attorney General under then President Reagan, was an ardent proponent of nonlethal weapon research. Back then the pop-phrase was "less than lethal" [3].

Contract research, through the National Institute of Justice (NIJ), supported hardware development for the first time. The NIJ also sponsored a series of futuristic brainstorming symposia, sometimes touting Cover's work as feats accomplished without government support. Cover's earliest patent application was filed in 1970 but resubmitted in July 1972. It issued as US Patent #3,803,463 on April 9, 1974. The patent's specification disclosed all of the electronic as well as the ballistic details of a handheld, battery-powered, less-lethal "electric rifle." The circuitry disclosed in this patent was soon imitated by offshore manufacturers in the 1980s who then flooded the USA with shocking-only, handheld clones known

---

**Robert A. Stratbucker reports serving as a full time employee of TASER International, Inc. and as a medical and scientific consultant to TASER International, Inc. Dr. Stratbucker reports as a stockholder of shares of TASER International, Inc. and with patents assigned to TASER International, Inc. No other potential conflict of interest relevant to this chapter was reported.*

R.A. Stratbucker (✉)
Faculty, University of Nebraska, Colleges of Medicine and Engineering (Ret.)
e-mail: rstratbucker@cox.net

M.W. Kroll, J.D. Ho (eds.), *TASER® Conducted Electrical Weapons: Physiology, Pathology, and Law*, DOI 10.1007/978-0-387-85475-5_2,

as "stun guns." These were generally ineffective gadgets that sold for $50 or less. A "stun gun" is a pain-compliance device without the probes and connecting wires required to allow effective control such as seen with the TASER CEW.

Cover had hoped that the local Los Angeles Police Department might eventually serve as an enthusiastic and technically competent test market for his invention. Ultimately they did—despite the fact that electrical safety was a foreign concept with little or no historical basis in police work. His scientific references on health and safety matters included the famous human research on electrical hazards from Dalziel [4], and the animal studies of Ferris [5] along with an epidemiologic-engineering report on electric fences done by the technical staff of the Underwriters Laboratories [6].

Now to the past.

2.1 The Giants of Electricity and Bioelectricity

2.1.1 von Guericke

German scientist Otto von Guericke (1602–1686)—best known for his studies of vacuums—also made important contributions to the field of electricity. From fused sulfur scraps he fabricated a 2-foot rotating spherical ball of the nonconducting sulfur and mounted it to spin on an axle. He could easily harvest electric shock-producing quantities of frictional electric charge just by pressing his hand lightly on the spinning ball so long as he stood on dry wood. But, he had no way to store the static charge he produced. His accomplishments—as well as his problems—gained the attention of Dutch scientists in Leiden, a few hundred kilometers to the west. They had earlier demonstrated the ability to store electric charge in foil-wrapped glass vessels, remarkably simple devices soon known throughout Europe (and now, thanks to Ben Franklin, in the new world) as "Leiden Jars," the first electrical capacitors.

2.1.2 Galvani

Over a century later, practicing physician–physiologist Luigi Galvani (1737–1798) was engaged in bioelectricity research using an "in vivo" preparation, later to become widely known as the "rheoscopic" frog. Galvani built and used electrostatic generators of the von Guericke type to make frog legs twitch. Galvani also developed a stimulator (or "battery") using two dissimilar metals and thus may have discovered the electrochemical cell. His physicist-neighbor Volta quietly recognized the result and is usually credited with the discovery of such cells, their chemical basis, and their voltage addition in series to create a true battery. Both the electrical parameter "voltage" and the unit—the "volt"—are named in his honor. There could be no CEW without Galvani's discovery that electric current could control muscles. Remarkably, his name remains attached to the non-pulsatile form of electrical muscle stimulation that is called "Galvanic" stimulation.

2.1.3 Benjamin Franklin

Born in colonial Boston and unhappily apprenticed to his printer brother, Franklin (1706–1790) escaped to Philadelphia and was self-educated in the sciences through his love of books. Although he never freed himself completely from print and paper he used its economic value to support his fruitful avocations, most notably the science of electricity and specifically atmospheric electricity. As a teenaged entrepreneur he supported himself and relocated to London for several years. Later, back in Philadelphia, in parallel with turning a few nascent print shops into profitable enterprises, he steadfastly elevated both his political and scientific stature. In portraits, he is frequently within arm's length of new electrical apparatus, particularly the Leiden Jar Many of his discoveries were related to the "Magic Jar" including the consequences of serial and parallel connections that were basic principles of electrical capacitance not previously appreciated. An astonishing desktop discovery was that the electrical energy stored in a jar remained in the glass upon careful disassembly. Thus, he showed that the strain resided only in the glass and not at all in the metallic constituents inside or out.

During his voyages to the European continent, Franklin observed atmospheric electricity in several shipboard forms, most notably from lightning and the coronal emissions from elevated metallic projections that was known as St. Elmo's fire. Through these experiments and observations he discovered the most important and fundamental fact of electricity. This was the fact that "vitreous" and "resinous" "electricity" were simply positive and negative charges of the same electrical force.

Franklin later submitted a paper to the British Royal Society describing how one could detect electricity in clouds. The "key" on the conductive, damp, hemp, kite string was a handy metallic point that could be coaxed safely into alignment with his grounded finger to catalog the strength of discharges created by each threatening cloud deck. The pain was the indicator of tolerable intensity. Critically important in all of this was the rainproof wooden shed, the dry wooden stool with the seated operator well isolated from earth ground and, finally, a wax stick for safe handling of the much used but never touchable by a body part, grounding wire. Franklin did not say he had actually performed this experiment. Indeed, many historians believe he was merely suggesting this to punish the British correspondent who was repeating his experiments and then publishing them as his own.

Unfortunately, one Professor Rickmann in St. Petersburg performed the experiment and was killed in front of his wife and children thus becoming the first human to be electrocuted by current delivered through a wire. A remorseful Franklin mailed a large check to the widow. The significance of Franklin's suggested experiment and Professor Rickmann's unintended sacrifice is that it established that electrical current—at a certain level—can be lethal to man.

2.1.4 Coulomb

Volta, in Italy, and Franklin, in the Colonies, were aided significantly in electrical matters by their close association with influential French scientists and politicians including Napoleon himself. Lt. Charles Augustin Coulomb—a contemporary French military engineer (1736–1806) was making scientific history by adapting field-grade surveying compasses to the delicate task of measuring extremely small mechanical forces, such as those associated with hitherto undetectable static electrical charges. Such improvement was attained by replacing friction-prone pivots with fine-wire torsion elements. The resulting instrument was an "electrometer" capable of measuring the tiny forces between charged bits of paper. The international unit for electrical charge is the "coulomb" named in his honor.

Another major breakthrough was the discovery of the connection between electricity and magnetism. Coulomb remarked in his writings of the predictable "flicks" in compass needles whenever he touched the metallic eyepiece on his sensitive electrometer. So annoyed was Coulomb that he went to technical extremes to suppress these flicks as confounding artifacts, apparently without the slightest suspicion they could have been electromagnetically created. Unfortunately, he did not recognize the significance of this observation and that was left to Oersted.

2.1.5 Oersted

During a demonstration to students at the University of Copenhagen in 1820, Hans Christian Oersted (1777–1851) discovered that a compass (left by the previous lecturer) was disturbed by an electric current. What he observed and promptly reported was steady magnetic attraction, not a transient disturbance. Fortunately, he submitted a paper to the French Academy of Science. The unit of magnetic field strength is now the "oersted."

2.1.6 Ampere

Andre-Marie Ampere (1775–1836) at that dangerous time in history was recovering from the politically motivated guillotining of his father. The self-taught son was not only a superb mathematician but an accomplished natural scientist. Within 3 weeks of the announcement of Oersted's discovery, Ampere had developed all the mathematical formulas describing exactly how electrical current generates a magnetic field, and performed all the required experiments to prove it. Amazingly, in the near two centuries since, no one has been able to improve on "Amperes" law and the unit of electrical current is the "ampere" in his honor. The ubiquitous electric motor would not be possible without the discovery by Oersted of the ability of current to generate a magnetic field and the quantification of this effect by Ampere.

2.1.7 Faraday

Michael Faraday (1791–1867) was a butcher's son and a self-taught British chemist who apprenticed in his early years with Sir Humphrey Davy assisting with electrolytic decompositions of metals. Faraday experimented with the process of repeatedly extracting charges from capacitors without their becoming depleted. He cited Franklin frequently, particularly as the first scientist to point out that fixed charges cannot exist in metals. Such "strains," said Faraday, can only remain fixed in solid dielectrics—the critical component in capacitors. It is of no surprise then that the international unit of electrical capacitance is the "farad."

Faraday was intrigued with the new electrical concept of electromagnetism. He visited with contemporaries Andre-Marie Ampere and Georg Simon Ohm. Ohm (after whom the unit of resistance is named), at his laboratory in Bavaria, was engaged in the mathematical characterization of electrical conductivity, painstaking work directed not only to categorizing hundreds of materials in terms of their basic property of resistivity, but to developing the mathematics to deal with complex networks of such resistance elements.

Upon returning to England in 1832, Faraday refined his concepts of electromagnetic induction, the generation of electric current from changing magnetic fields. Ampere and Faraday, with capacitor banks for current sources, had repeatedly observed evidence for induction currents produced by the sudden creation or collapse of a magnetic field.

Yet another of Faraday's contributions to electromagnetism was the induction of sine-wave currents by the simple rotation of magnets around coils or—conversely—rotation of coils around magnets. This led to his "dynamo" or the first electric generator. This in turn, led to what was probably the first conducted electrical weapon—the electric whale harpoon as seen in Fig. 2.1 [7]. Note that this TASER CEW predecessor has only one wire and one barb as the ocean's salt water provides the return path.

2.1.8 du Bois-Reymond

As noted earlier, electromagnetic induction transformed electricity from a curiosity to a commodity by allowing the construction of generators and transformers. Bioelectrical knowledge lagged industrial know-how for many reasons, chief among them being the lack of adequate instrumentation.

German-born Emil du Boise-Reymond (1818–1896) is considered by many the founder of the discipline now called electrophysiology. He began with a fervent interest in electric fishes, species which possess a built-in conducted electrical weapon for stunning prey. Earlier experimenters used electric fishes and "rheoscopic" frogs as a proof of the existence of "animal" electricity. Du Boise-Reymond used an adjustable coil, much like a Faraday induction transformer, with a spring-loaded vibrating armature that could quickly manufacture countless combinations of stimuli reminiscent of a music box or mechanical piano

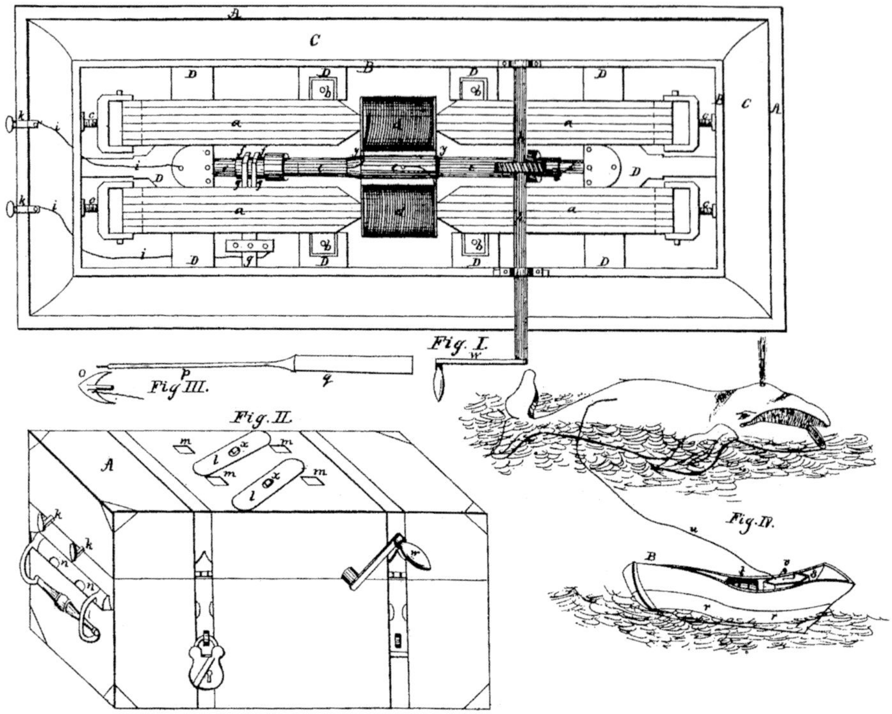

Fig. 2.1 1852 whaling patent No. 8,843

controller. Using the frog leg as a calibrated strength-of-contraction force monitor, he was able to identify and characterize muscle parameters such as twitch, tetanus, and fatigue. This research set the stage for the setting of the TASER X26 pulse rate at 19 pulses per second to gain control without causing tetanus. He also developed the first instrument capable of recording the small voltage (<0.01 volt) coming from the body. This led eventually to the electrocardiogram.

2.2 D'Arsonval

Paris physician and physiologist Jaques-Arsene D'Arsonval (1851–1940) is much better known for his electrical engineering contributions than those in electrophysiology. He used the new electromagnetic needle with small currents thus creating the first direct reading current and voltage meters.

Using super thin varnish insulated wire for sensitive windings, his rugged "galvanometer" has survived well over a century without major change. The design is still referred to as the D'Arsonval movement and is the heart of any needle-indicating meter.

While improving galvanometers, he was actively engaged in health-related electrical research. He is credited with being first to describe an alternating current phenomenon in humans and other animals wherein the electrical stimulation capability declines with increasing frequency. Higher frequencies deliver heating but not membrane activation (stimulation). It is the technical basis of clinical diathermy.

D'Arsonval would publicly subject himself to otherwise lethal high current shocks, but safe due to the high frequencies. Higher frequencies do not effect the slower reacting cardiac cells and this explains D'Arsonval's survival of these high currents. A similar phenomenon applies to TASER CEW waveforms. Medical industry safety standards now recognize this, and safety limits are adjusted according to the dominant frequency of the electrical stimulation.

2.2.1 The Edison Versus Tesla Arena

Thomas Edison (1847–1931) built and sold the first incandescent bulb. Contrary to legend, Edison had nothing to do with the invention; Swan first published the idea of a glowing wire in a vacuum and Sawyer invented the idea of replacing the vacuum with nitrogen. Edison was pushing the city of New York to become illuminated by his bulbs and powered by his direct current dynamos. Industrialist George Westinghouse was pushing for an alternating current (AC) standard—already popular in Europe—which would have permanently excluded Edison's already obsolete (DC) plan.

Pseudo-scientific arguments dealing with this controversy were frequently voiced in the press reminiscent of the many urban myths today surrounding CEWs. In the height of the industrial AC–DC war, Edison bribed enough state legislators in New York to make AC "electrocution" the only legal means to dispense capital punishment. Edison was convinced that only this desperate marketing approach could save his DC systems by suggesting that AC was far too dangerous. A German immigrant Kemmler then received a lethal dose of AC after his conviction for an axe murder. The press observed that the condemned had been successfully "westinghoused" thanks to Edison press kits. This macabre affair provides a small window into the finality if not the lethality of accidental or even intentional electric shock exposure. Nicola Tesla (1856–1943), Croatian-born electrical engineer, immigrated to the United States in the midst of the electrical war and became a dominant force in US electrification. He correctly forecasted that DC could never survive in economic competition with AC because transformers will not work on DC and hence voltages cannot be converted up and down. Tesla was correct and Edison's idea of DC electrification is now relegated to portable devices such as flashlights and cars.

2.2.2 20th Century Electrophysiology

Problems with the transatlantic telegraph line played a critical role in the development of electrophysiology. The "string" galvanometer was developed to allow the visualization of the tiny signals emerging from their oceanic passage. The Dutch physiologist Einthoven (1860–1927) then improved the string galvanometer and made the first practical EKG. Helmholtz, physician-physiologist, developed sensitive instruments for measuring cellular membrane potentials.

Secondly, the transatlantic cable signals suffered such distortion that only one or two letters per minute could be transmitted. In a brilliant triumph of applied physics Lord Kelvin used Maxwell's equations (largely derived from Faraday and Ampere) to derive the "telegraph" equation and explain the source of the distortion.

Prior to World War II, Curtis and Cole [8] showed all cells to possess electrical properties that sustain potentials of as much as 0.1 volt across their membranes. Special long cells such as motor nerves also exhibited signal transmission characteristics similar to those known to exist in oceanic coaxial telegraph cables. Before each cell could be investigated by microelectrodes, their operating electrical characteristics were imputed by the use of the telegraph equation to measure transmission velocities, and to divine the cell membrane-specific electronic characteristics (capacitance and resistance) for many cell types. World War II briefly interrupted research in electrophysiology but produced a revolutionary advance in bioinstrumentation. The prewar mathematical models were then validated by Hodgkin and Huxley who impaled squid axon cells with microelectrodes and measured the electrical characteristics directly. These mathematical models are used today to optimize the CEW waveform for safety and effectiveness.

2.2.3 The Electric Fence

The first electrified fences were hastily built during World War I on European battlefields to temporarily contain refugees and prisoners during advances and retreats [6]. Electrification was achieved by the 220 volt 50 hertz AC power source that is the European standard utility power.

After the war a somewhat improved and allegedly safer electric fence was exported to cattle countries such as Australia and the western United States. In 1936, because of the reported deaths of several children who supposedly became immobilized and subsequently electrocuted by the unremitting AC fences, the newly empowered Underwriters Laboratory of Chicago began a 3-year intense study[6]. Relying on prewar safety experiments on animals by Ferris et al. [5], and the Russian Kiselev, Dalziel (UC Berkley) used a human model, namely his captive audience of electrical engineering students [4]. At that time his work was 60 hertz specific although he later expanded it to include pulse waveforms [9].

Near-perfect execution of his classic protocol resulted in abundance of critically important results. The overlap of UL work with Dalziel's experiments was ideal. Both studies were inextricably entwined in nonlethal weapon development. UL's Research Bulletin 14 (1939) titled: "Electric Shock as it Pertains to the Electric Fence" appears to have had the most direct linkage to Jack Cover's research and development of the TASER CEW. Research Bulletin 14 has since become an ANSI/UL Standard now labeled *Electric Fence Standard #69*. It was most recently revised in 2005.

2.3 Cover's Challenges

The results of Jack Cover's safety and efficacy studies are not found in the scientific literature as one might expect. They do, however, have an effective presence, now in the public domain, through his detailed arguments with examiners during the prosecution of his patent applications in the years between 1970 and 1982.

In one challenge, the 1852 "Electric Whaling Apparatus" patent #8843 was repeatedly held out by the patent examiners to be an example of disqualifying prior art despite the fact that electronic technical revolution had occurred in the intervening 100 years. Ultimately, the rejection was overcome as the claims were fine-tuned and countervailing arguments were accepted.

During the pendency of his first patent Cover introduced his fully functional, 7-watt, twin-projectile TASER device to the LAPD. Chief Daryl Gates, Capt. Greg Meyer, and other deeply involved department members facilitated what was then an unfamiliar test and evaluation protocol into their enormous system. Included in the department's assignments for a device's performance evaluation was a comprehensive medical report which later appeared as a scientific research product of the Los Angeles County Hospital [10].

To further the commercialization of his device, Cover sought product validation through some form of industrial or government listing. Initially, he was shuttled from one office to another apparently due to the ballistic nature of his device and its gunpowder-propelled dart electrodes. Units offered for testing exceeded the mass limit for an exemption from the "exploding charge" rule. The Bureau of Alcohol, Tobacco, and Firearms (BATF)—despite an earlier official decision that "tethered projectiles" allowed escape from the "gun" rules—fell back on the gunpowder rule to restrict the TASER device to "firearm" classification. This was a particularly bitter pill especially in view of the parallel decision of the Consumer Public Safety Commission (CPSC) that the electrical output of the device had to be proven citizen-safe. This determination was made after several years of rigorous scientific testing by the Commission's outside expert Theodore Bernstein (professor of electrical engineering, University of Wisconsin, Madison). Although Bernstein had developed an international reputation for scientific analyses of electrical accidents, especially in rural-industrial safety matters, he had never before been confronted with a battery-powered, self-contained electric generator. Selection of a suitable testing protocol was complicated by the pulsatile nature of the output energy since existing safety standards were largely based on continuous utility sine-wave power. Bernstein found it expedient to apply complex waveform analysis to the microsecond pulses Cover had crafted to fulfill his incapacitation objectives. Although the classical Fourier approach employed by Bernstein was by no means new to mathematicians and engineers, this unusual physiological safety application in the early 1970s was revolutionary. Bernstein had written extensively on this dilemma and was confident of his analytic approach [11]. Had this been a medical device, the path would have been clearer; the FDA would definitely have been the primary authority and UL might have had a supporting authority. The CPSC, without argument, had no experience or expertise in this technology, and the BATF was satisfied with the firearm registration.

After months of bureaucratic juggling Professor Bernstein executed a testing protocol capable of satisfying both Jack Cover and the CPSC that the safety evaluation was adequate. All the familiar electrical safety work in the mid-1970s was related to 60 hertz. The Cover TASER device's pulse was closer to 60 kilohertz—1,000 times the frequency. By Bernstein's calculations, the output of Cover's TASER CEW was equivalent to 8.7 milliamperes of utility power, a value less than 1/10 the relevant ventricular fibrillation threshold for alternating current powered consumer devices. A 10:1 safety factor was deemed acceptable by national and international standards authorities including the CPSC. However, the BATF made no official use of this result and simply continued its firearm classification.

Currently, TASER International uses a compressed gas propellant in place of explosive powder, which avoids the BATF restriction. Furthermore, the pulse generator design is vastly different from the Cover device and would only in the most general way be sheltered under the Bernstein safety claims of a quarter century past. A more modern understanding of the biological effects of higher frequencies and short pulses—supported by the newer international standard IEC 479-2—suggests the safety margin of Cover's original TASER was actually over 100!

The Douglas County Sheriff in Omaha had a minor interest in the stun gun being pitched to his department and he also had an implanted pacemaker. His chief training officer had heard about the devices and had several stun guns that the sales person had left behind. They approached the local medical school, which had physicians, biomedical engineers, and students that were eager to learn. This led to the first scientific study of the effects of stun guns. A total of 20 officers were exposed. Each of the officers had a small reaction at each stun gun electrode site. All cleared without a trace of scar in 1–10 hours. Three animals were also tested. None showed any evidence of cardiac rhythm capture effects from ten applications of 10 seconds each. The stun gun current was delivered to the unshaved skin directly over the heart. Stun gun pulses were also applied via pacemaker leads to inside of the heart without problems [12,13].

The remainder of this volume takes the story from there.

References

1. Victor [pseud] Appleton, *Tom Swift and His Electric Rifle*, 1911 Available at www Gutenberg. org/etext/3777
2. Resnick, B. and J. Murray, *A Guide to TASER Technology*: Stun guns, Lies and Videotape, 1997 Whitewater Press, Whitewater, CO 81527
3. Meese, E. III, *The Attorney General's Conference on Less Than Lethal Weapons*, June, 1986 Attorney General, U.S. Department of Justice, NIJ and FBI, Washington, DC
4. Dalziel, C.F. and F.P. Massoglia, Let-go currents and voltages, 1956 *AIEE Trans.*, vol. 75, part II, pp. 49–56
5. Ferris, L.P., B.G. King, P.W. Spence and H.B. Williams, Effects of Electric Shock on the Heart, May, 1936 *Elec. Eng.*, pp. 495–515
6. Electric shock as it Pertains to the Electric Fence, December, 1939 H.B. Whitaker, *Bulletin of Research, Underwriters' Laboratories, Inc.*, Number 14
7. *Electric Whaling Apparatus, United States Patent Office, Patent No.8843*, March 30, 1852 Dr. Albert Sonnenburg and Philipp Rechten of Bremen, Germany
8. Cole, J.S. and H.J. Curtis, Electric Impedance of the Squid Giant Axon During Activity, 1939 *J. Gen. Physiol.*, vol. 22, p. 649
9. Dalziel, C.F., Study of the Hazards of Impulse Currents, 1953 *AIEE Trans.*, vol. 72
10. Kornblum, R.N. and Reddy, S.K., J *"Effects of the Taser in Fatalities Involving Police Confrontation,"* 1991 Forensic Sciences, Mar; vol. 36, No. 2, pp. 434–438
11. Reynolds, T.S. and T. Bernstein, The Damnable Alternating Current, 1976 *Proc. IEEE*, vol. 64, No. 9, Sept
12. Stratbucker, R.A. Assessment of Cardiac Hazards in Handheld Electronic Law Enforcement Devices,1986. *Proc. 8th Conf. of the IEEE Eng. in Med. Biology,* Ft. Worth TX, Nov.
13. Stratbucker, R.A. and M.G. Marsh, Relative Immunity of Skin and Card. Vasc. Systems to Stun Gun Electrical Pulses, 1993. *Proc.13th Conf. of IEEE E.M.B.S.*, San Diego CA, Oct.

Chapter 3
Conducted Electrical Weapons and Resolution of Use-of-Force Encounters

Charlie Mesloh, Mark Henych, and Ross Wolf

The mission of law enforcement agencies can be considered tenuous in American society. While police officers are charged with maintaining the peace and order, this is complicated by a myriad of factors unique to each and every situation. When an officer responds to a citizen call for assistance or reacts to problems observed in the field, the officer is charged with either quelling the disturbance or apprehending a suspect, sometimes through the use of force. Usually, when utilized, the use of force is justifiable and legal, particularly when overcoming resistance during arrests or in course of protecting themselves or others from harm.

Some of the largest leaps in modern law enforcement have been the result of the deployment of new technologies. For example, the widespread use of the law enforcement cruiser and radio communications allowed officers to respond to situations quickly, and to apprehend those who fled. The implementation of mobile data terminals, in-car video, and computer-aided dispatch are all other examples of the positive effects of technology on law enforcement. However, some of the most successful and widespread developments in law enforcement technology have been in the area of nonlethal weapons. Since officers often had to resort to hand-to-hand tactics (which have high incidence of injury) technological advancements have allowed officers to minimize injuries while still maintaining order.

Nonlethal weapons are designed, as their name implies, to cause compliance by discomfort or incapacitation rather than to kill. While there are many different types of nonlethal weapons, arguably the greatest technological advancements have been in the development of the Conducted Electrical

---

The project was supported by Award No. 2005-IJ-CX-K050, by the National Institute of Justice, Office of Justice Programs, United States Department of Justice. The opinions, finding, and conclusions or recommendations expressed in this publication are those of the author and do not necessarily reflect the views of the Department of Justice.

C. Mesloh (✉)
Florida Gulf Coast University
e-mail: cmesloh@fgcu.edu

M.W. Kroll, J.D. Ho (eds.), *TASER® Conducted Electrical Weapons: Physiology, Pathology, and Law*, DOI 10.1007/978-0-387-85475-5_3,

Weapon (CEW). These devices, widely known by the names of their manufacturers such as TASER® and Stinger®, have changed the nonlethal environment and sparked controversy and debate in academia and professional, trade, and medical journals. CEWs were initially heralded as a Star Trek "Phaser-like" device that incapacitated without lasting debilitating effects. However, as reports grew in the media of subjects that died after CEW application, they have also now become tied to very negative unpleasant things, such as "excited delirium," "cocaine psychosis," and "in-custody death."

This chapter examines the use of the CEW at the event level, the complicated circumstances in which officers deployed them, and the response of the subject. Researchers examined 4,303 police reports where force was utilized. These documents were collected from two major central Florida law enforcement agencies that use CEWs, employing a total of over 2,000 officers. Of specific interest for this review was the context in which the CEWs were used and the outcomes on suspect and officer injuries.

3.1 Legal Review

3.1.1 Defining Police Use of Force

Use of force can be defined as the "exertion of power to compel or restrain the behavior of others,"[1] or when used in the context of policing, "acts that threaten or inflict physical harm on suspects."[2] Generally, police force can be classified into several modal categories including: (1) deadly versus nondeadly, (2) physical versus nonphysical, and (3) reasonable versus excessive[1]. "Deadly force" is used to define force that is likely to cause death or some serious bodily injury;[2] conversely, "nondeadly force," "less-than-lethal force," or "nonlethal force" is the application of force that is not likely to result in death or serious bodily injury.[3,4] "Physical force" implies the touching, prodding, redirection, or physical manipulation of a subject to comply with demands,[5] whereas "nonphysical force" implies the use of threats or other verbalization techniques to gain compliance.[6] "Reasonable force" is applied force which is necessary to achieve a legal goal, while "excessive force" is applied force which is disproportionate to what is necessary to achieve a legal goal.[7] Bittner,[8] Garner et al.,[9] Reiss,[10] Scharf and Binder,[11] and Sherman[12] have all discussed the hypothesis that the capacity to use non-negotiable coercive force is at the core of the police role in society. So basic is the element of force to the police that some researchers claim that the main reason citizens call the police is based on the belief that force may be necessary.[13]

The decision of police officers to intervene, or apply force, in a given incident is a subset of discretionary choices facing them everyday. As noted by Davis,[14] "a police officer may be said to exercise discretion whenever effective limits of his or her power leave the officer free to make choices among possible choices of action or inaction." Conclusions to use force, and decisions concerning the extent of force to be used, are within the discretion of police officers. Thus, an individual officer must decide in each situation whether to ignore, or confront and coerce a citizen to follow his direction. Through studies that have examined police use of force reports, citizen complaint reports, and police/citizen surveys, it has become clear that police officers today rarely apply physical force.[15]

Discretionary decisions regarding when, where, and how much force to use is a cumulative process.[16] Once a course of action is decided upon, additional discretionary choices follow that may lead an officer to either increase or decrease the level of force used. Terrill[17] examined the complexity of police–citizen encounters involving force. He reported that when verbal commands are considered as a use of force, force occurs in more than half of all encounters. He also reported that the inclusion of suspect resistance into police force studies offers a "more complete picture within the context of how officers apply varying forms of force." Terrill based his study on previous observational and data-collection studies;[18,19] these studies underscored the importance of understanding force in varying degrees and levels, from verbal commands to the use of deadly force. These studies also included suspect resistance levels as a measure to understand the police use of force.

An officer's decision to use force is not an arbitrary one; rather it is shaped by the US Constitution, the US Supreme Court and local court decisions, state laws, agency policy, and training. As such this chapter has included a section on the most influential court decisions that have the greatest impact on current police use of force practices.

3.1.2 How the US Courts Have Framed Police Use of Force

For a very long time, nearly all federal circuits framed police use of force within the 14th Amendment substantive due process "shocking to the conscience" standard articulated by the Second Circuit in *Johnson v. Glick*[20]. Because this case focused on establishing the intent of the officer rather than the reasonableness of their actions, there remained no clear standard for evaluating claims of excessive force by the police. Later, the Court addressed the use of deadly force in the case of *Tennessee v. Garner,*[21] possibly the Supreme Court case with the largest effect ever on police policy[22]. The Court ruled in this case that the state can legally "seize the life of an individual" only when an officer believes that a suspect's actions place either the life of the officer or the lives of other citizens nearby in imminent jeopardy. The significance of an officer's intent gave way to the reasonableness of the Fourth Amendment in cases where "seizures" of an individual are deemed to have occurred[23].

The Supreme Court also interpreted excessive force with the decision of *Graham v. Connor.*[24] In this case, the Court established the "objective reasonableness standard," mandating that actions of officers involving questions of use of excessive force be "judged from the perspective of a reasonable officer coping with a tense, fast-evolving situation." These US Supreme Court decisions, while providing a general standard for the efficacy of police behavior, continued to fail to provide specific criteria that officers may use when deciding whether and how much force should be applied.

In more recent case law, *Brosseau v. Haugen,*[25] the Court remained ambiguous, allowing that even unwise use of force may be legal, and there is a "sometimes hazy border between excessive and acceptable force"[26]. Even more recently, in *Scott v. Harris*, the court's conclusion regarding an officer's right to use force may continue to provide controversy to this topic:

> A police officer's attempt to terminate a dangerous high-speed car chase that threatens the lives of innocent bystanders does not violate the Fourth Amendment, even when it places the fleeing motorist at risk of serious injury or death[27].

Thus, there remains an indistinct and obscure understanding on the appropriate use of force by case law, leaving much room for interpretation and perception; police agencies therefore create use of force continuums and policy to more clearly define accepted agency practices.

3.1.3 The Use of Force Continuum

To appreciate the complexity of situations where the police utilize CEWs as a nonlethal force alternative, one must conceptualize force not as a static concept but rather as a series of responses, ranging from verbal commands, as a minor exertion of force, to deadly force, the maximum amount of force possible to apply[28–31]. These continuums provide officers with means to measure the escalation or deescalation of force within agency policy that can be legitimately used in a confrontation. While use of force continuum policies provide officers with the ability to jump from one step to another based on the officer's interpretation of the suspect's level of resistance, the underlying philosophy centers on protecting both the officer and suspect[32].

The use of force continuum relies on the concept of multiple categories of increasing officer perceptions of suspect resistance linked to similar groupings of the officer's response to those perceptions. As law enforcement officers are expected to make split second decisions based on rapidly evolving situations, the incorporation of a use of-force continuum into departmental policy provides guidance to officers in making force decisions; albeit the continuum is reflective of public opinion and agency liability concerns rather than a maximum legal level standard. Law enforcement agencies incorporate these force continuums into preservice and on-the-job training programs in order to be able to identify varying levels and severity of resistance[33]. While these continuums within agency policies are not universal, indeed there are almost as many as in American policing as there are agencies, they all rely on legally and publicly acceptable responses by the police[34]. These continuums propose that officers should progressively examine and react to each situation, deescalating once resistance has declined or stopped[35].

Although continuums are useful for training and policy setting, they provide little knowledge for academicians who delve into the subject, quite simply because there is very little information on the actual levels of noncriminal resistance that police officers encounter. Additionally, the fact that there are so many different agency-adopted continuums makes studies difficult.

To further complicate the study of police use of force, Conner[36] found that 95–97% of all police–citizen contacts involve cooperative subjects. Alpert and Dunham[37] reported that 61% of the suspects who were being placed under arrest did not resist the officer at all; 18% offered only slight resistance. Even though the vast majority of citizens that police interact with on a daily basis can be classified as cooperative, many observational studies have found "disrespectful" or "uncooperative" citizens to be arrested more often[38–42]. From both a legal and policy perspective, perceived suspect resistance is a decisive factor in police use of force[43].

While the use of nonlethal force is agency specific, most agencies utilize a use of force matrix[44] system with similar delineations. Progressively, Levels 1–5 indicate more serious levels of force that are not apparently deadly in intent, while Level 6 indicates force that could be interpreted by the officer as deadly. The force continuum example pictured in Fig. 3.1 offers an interval level of measurement of uses of force that are bidirectional. The arrows in the center of

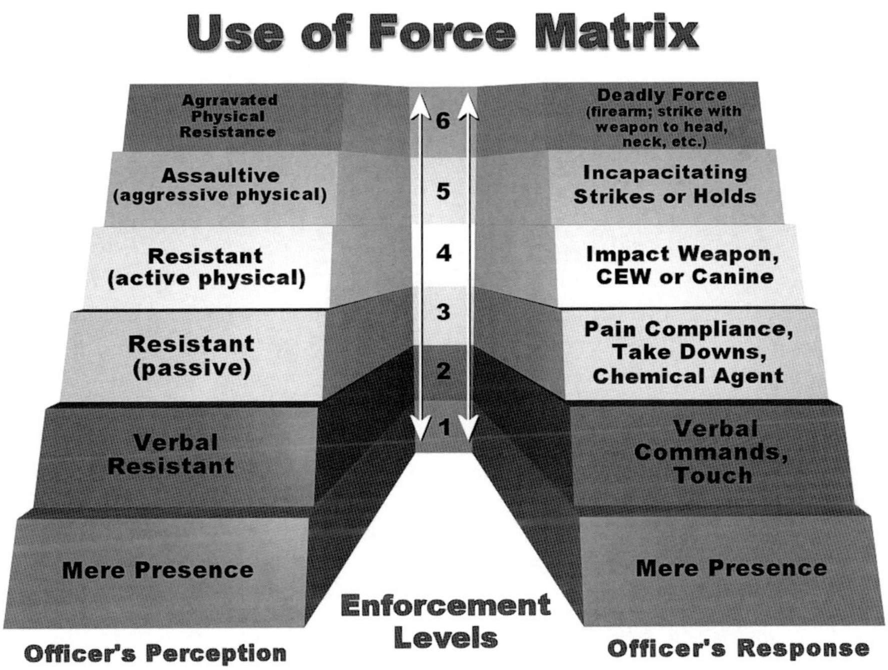

Fig. 3.1 Example of a use of force matrix
Note: Reprinted with permission[70].

the diagram indicate the level of force used by the suspects (as perceived by the officer) and the authorized, or acceptable, levels of force in response. Using a similar bidirectional approach to understanding use of force, this chapter examines variables included from officer use of force reports so as to examine the levels of officer force and suspect resistance at event level where conflict takes place.

3.1.4 Measuring Police Use of Force

Studies that examine police force do not "always specify clearly how force was defined or measured, and the definitions and measures of force tended to be unique to each study"[45]. Additionally, research on police use of force has focused on several theoretical perspectives: situational, organizational, psychological, or neighborhood characteristics. However, none of these theoretical perspectives have appeared in all studies, and are often not even measured or reported[45].

Scholarly efforts have been able to determine that police force, and its intensity, is commonly affected by the context in which the police and citizens meet[45–47]. Thus, to better understand officer definitions of appropriate police force, it is necessary to explore the impact of theoretically relevant individual, situational, and community factors[48].

Garner[49] explained that police use of force research, while expansive, has varying approaches that are fraught with "limited strengths and substantial weaknesses." A review of the literature reveals that there are numerous accepted ways to gather information about police use of force. These include examinations of agency policy,[50,51] observational accounts of police force incidents,[52–55] analysis of official police records and use of force reports,[56,57] citizen complaints about the use of force,[58,59] and surveys of police officers or arrested persons[60,61]. Regardless of the research strategy employed, one constant finding is that police force utilization is uncommon and its improper use is exceedingly rare[62–65].

While each type of data collection has strengths and weaknesses, the review of police records may have certain advantages over other categories. The review of police records may provide more organized data on more use of force incidents than do interpretations of police work through observations[66]. Additionally, review of police report data provides a wider view of police behavior over the studied jurisdictions than can normally be captured through observational accounts. However, a major weakness of police report review in the context of police force is that these reports suffer from bias provided by the officers who wrote the reports. It has been reported that this approach may be most suitable for interjurisdictional comparisons rather than intrajurisdictional comparisons[66].

3.1.5 Conducted Electrical Weapons

Conducted Electrical Weapons (CEWs) are nonlethal weapons or devices designed to deploy electric current through the body of the target to temporarily cause loss of muscle control. Over the past several years, the technology for these devices has become more user-friendlier than the original, more rustic devices, allowing the user to apply the device from greater distances, with more accurate application.

The use of CEWs in the context of more humane policing has allowed law enforcement agencies to use a technology that allows them to stop a confrontation that otherwise would have resulted in a "Level 6," or deadly force, response. Findings from this study tend to corroborate anecdotal evidence of numerous events wherein officers could have justifiably and legally used deadly force and would have had immunity from prosecution, but instead utilized a CEW, thus potentially saving lives[67].

TASER International is the company best known for producing CEWs. Their product has become so well known that the name "TASER" has become synonymous with "CEW," much like BandAid® is to a plastic bandage. However, additional CEW manufacturers, such as Stinger, have entered the law enforcement marketplace and are seeing use in some departments.

TASER CEWs were being tested or used in over 7,200 law enforcement, military and correctional agencies throughout the United States and abroad in 2005,[68] and are reportedly used by over 11,000 agencies in 2007[69]. TASER International continues to advertise their device as among the safest and most effective less-than-lethal force choice available, claiming that TASER use reduces officer shootings and suspect injuries[70]. The darts fired from the TASER can reach from 15 feet (civilian model) to 35 feet (law enforcement model). Despite the length of the wire, recent best practices guides by the Police Executive Research Forum (PERF)[71] suggest restricting targeting to less than 15 feet as the darts travel at an angle to each other, limiting accuracy beyond this distance. This is consistent with other studies, which indicate that beyond 15 feet accurate placement of probes is difficult[72]. Early studies indicated this weapon's effectiveness at incapacitating a subject ranged from 50% to 85%[73] when deployed. Police agencies have reported that since the TASER weapon was deployed to officers in the field, the use of deadly force by officers and the number of officers injured during arrest confrontations has been dramatically reduced[74].

3.2 CEW Effectiveness and Officer/Suspect Injuries

3.2.1 The Use of Force Report

To examine the use of CEWs at the event level and develop an understanding of possible outcomes in the escalation or deescalation of force as a result their application, the authors designed a research methodology that would embrace an understanding of levels of force and resistance, agency policy, and the limitations of CEWs. Resulting injuries to officers or citizens were also an important part of this research.

To capture this data the researchers used the "use of force report," a regular tool that most law enforcement agencies utilize when a citizen/officer encounter requires the intervention of police force. This reporting tool captures a considerable amount of data and allowed this research endeavor to begin at the event level. Examples of data captured by the report include the following: suspect demographics, specific information about the type of force used in an encounter, and the type of resultant injuries. The use of force reports were collected via public information requests at the respective agencies and coded into SPSS for analysis.

Some data were not collected in regards to incidents that resulted in a death. This was due to ongoing litigation regarding the individual incidents, and as a result of sealed settlements. In order to comply with state law and court orders, the researchers did not pursue the agencies for cases involving these incidents. An additional factor not examined within this study is the threatened use of force by officers. Use of force reports from both agencies fail to illustrate this useful variable, and as a result are not captured. In order to maintain coding consistency, if the report did happen to state that a weapon was drawn, displayed, or threatened, it was coded as "No Force" as no actual force was used against the suspect.

3.2.2 Data Collection

Both the Orange County Sheriff's Office (OCSO) and the Orlando Police Department (OPD) provided photocopies of use of force reports that were dated between January 1, 2001 and December 31, 2005. OCSO provided the researchers with 2,450 reports (57.1% of all reports reviewed); and OPD provided 1,843 reports (42.9% of all reports reviewed) for incidents that occurred during the identified time period.

Orlando is America's 27th-largest metropolitan area, but the jurisdictional limits of the City of Orlando "proper" have a population of 217,327. The City of Orlando is the largest municipality within the jurisdictional limits of Orange County, which has a total population of over 1.04 million. In addition to the resident population, the Orlando Metropolitan Statistical Area acts as host to over 47 million tourists a year, creating a need for additional government resources, which include policing resources.

The Orlando Police Department has a mayoral appointed Chief of Police and serves a jurisdiction of approximately 94 square miles. The population living in the city of Orlando is 61% white, 27% African-American, 17.5% Hispanic, 2.7% Asian, and 0.4% other. The median age of the population is 32.9 years, and 40.8% of the population owns their own home. Of the 25 year-or-older population, 82.2% have a high school (or equivalent) education or higher, 19.9% have a bachelor's degree, and 8.3% have a graduate or professional degree. The median annual household income in Orlando is $35,732 with 19.9% living under poverty[75].

The Orange County Sheriff's Office has an elected Sheriff, and serves a total jurisdiction of approximately 907 square miles. Within this jurisdiction, however, there are 13 separate municipalities, each run by their own governments and most with their own police agencies. Home to Disney World, the population served by the Orange County Sheriff's Office (those residents that are not in the city limits of a municipality) is 680,687. Those living in unincorporated Orange County are 68.6% white, 18.2% African-American, 18.8% Hispanic, 3.4% Asian, and 0.4% other. The median age of the population is 33.3 years, and 60.7% of the population owns their own home. Of the 25 or older population, 81.8% have a high school (or equivalent) education or higher, 18.3% have a bachelor's degree, and 7.9% have a graduate or professional degree. The median household income is $41,311 and 12.1% live under the poverty limit[75].

3.2.3 *Analysis of the Data*

The force used by the police in a police-subject encounter does not occur in a vacuum. "Virtually any inquiry concerning how or why officers use force is augmented by the inclusion of citizen resistance. Knowing an officer used force tells us very little without knowing the specific type of force used, how many times it was used, and what the citizen behavior was prior to each use."[76] (p. 157) Klinger[77] noted that prior attempts to study nonlethal force in police encounters failed to examine that multiple levels of force may be used within a single encounter. To overcome this issue, this study decomposed confrontations at the event level into a series of iterations, representing a single suspect action and officer reaction. If the confrontation was not brought to resolution within the first iteration, it then progressed into second and third iterations.

A total of 4,303 uses of force were examined during this time period. Both agencies utilized TASER CEW as their first choice, although during the course of this study some transition was made from the M-26 to the X26 model. Over half of the suspects (55.6%) were subdued at the end of the first iteration, which allows us to examine 2,394 cases to determine which less lethal weapons were most effective in bringing conflict to resolution. Slightly less than 30% (29.3%, n = 1,262) of the confrontations ended at the second iteration, while 15% (n = 647) ended in the third iteration. There were no confrontations that extended beyond three iterations of force, although there were some cases where the suspect escaped and could not be identified (Fig. 3.2).

In cases which ended at the first iteration, it was possible to show linkage between the injury and the choice of the officer's weapon or tactic. Of key interest to this study were the resulting injuries to both officers and suspects in their confrontations. However, it was difficult, if not impossible, to assign responsibility for the injury to a specific officer action in cases that surpassed the initial iteration, as in many cases multiple techniques or less lethal weapons

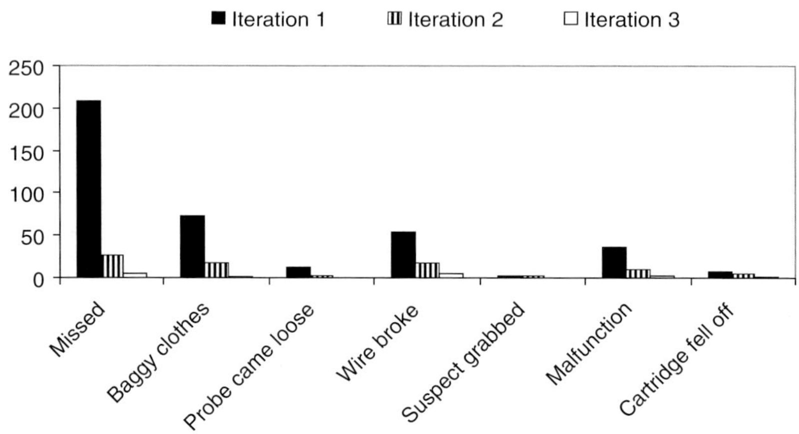

Fig. 3.2 Deployment problems through three iterations

were used. As an example, it was possible for an officer to use a control technique in the first iteration, a chemical agent in the second, and a CEW in the third (see Table 3.1).

3.2.4 Application of the CEW

In examining the effectiveness of CEWs, specific attention was paid to the method for coding effectiveness, as this is highly controversial measure. For the purpose of this study, a CEW deployment was coded as effective if after a 5-second application a suspect became immediately compliant. To ensure clarification, this study defines a CEW deployment as a single application of the

Table 3.1 Effectiveness of CEWs at various iterations

	Iteration 1		Iteration 2		Iteration 3	
	N	%	*N*	%	*N*	%
Missed	209	9.9	26	3.2	5	1.8
Baggy clothes	73	3.5	18	2.2	1	0.4
Probe came loose	13	0.6	2	0.2	0	0
Wire broke	54	2.6	17	2.1	5	1.8
Suspect grabbed	3	0.1	3	0.4	0	0
Malfunction	37	1.8	10	1.2	2	0.7
Cartridge fell off	8	0.4	5	0.6	1	0.4
Ineffective	452	21.4	176	21.9	36	13.3
Effective	1,264	59.8	548	68.1	219	81.5
Total	2,113		802		270	

Note: Due to rounding, percentages may not total to 100%.

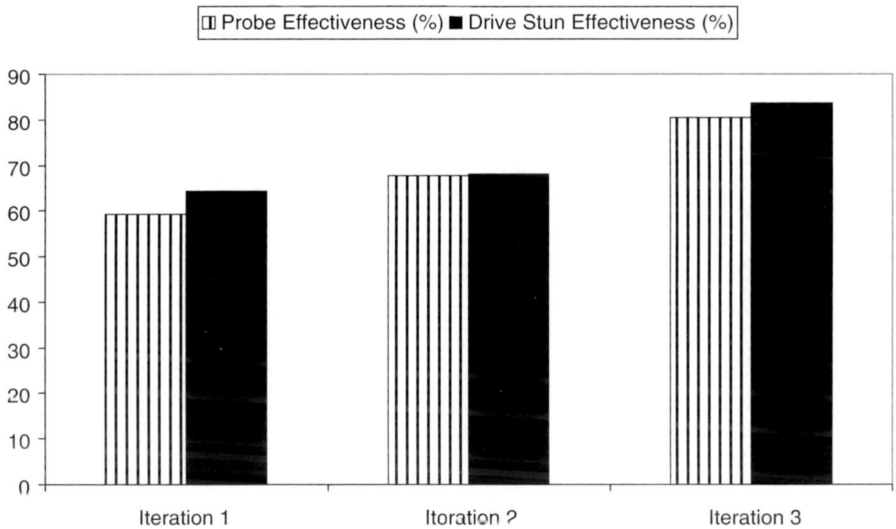

Fig. 3.3 Probe and drive-stun effectiveness by iteration

CEW (which entails pulling the trigger a single time and deploying the probes; a drive-stun, which is described later, was also considered a single application) (Fig. 3.3).

It must be acknowledged that TASER International training suggests the use of multiple applications until compliance of the subject is achieved. For the purpose of this study, the researchers have viewed each application of a CEW as a single unique event and subsequent deployments are coded and captured as "Iteration 2" or if applied again, "Iteration 3." In light of negative media coverage over multiple applications of CEWs, it was prudent to capture the data in this manner. However, it must be clear that the coding of "ineffective" in a single application does not necessarily mean that in the context of the complete encounter the CEW was ineffective, rather only that a single use did not gain immediate suspect compliance.

Viewed in this light CEWs may be disproportionately weighted, as a grapple or compliance hold may be applied for a significant amount of time in order to gain acquiescence, whereas a CEW can only be applied for 5-second intervals. Conversely, in some cases the probes may have missed, the CEW may have malfunctioned, or baggy clothing may have prevented a proper application of the CEW, yet the suspect may still have surrendered for any number of reasons.

3.2.5 CEW Drive-Stun (Touch-Stun)

The data collected in this study captured information from the use of force reports where CEWs were deployed in a drive-stun or touch-stun manner. A drive-stun is achieved by removing the cartridge from the CEW, activating the CEW, and then physically maintaining contact with the subject and the CEW's contact points. As the name implies the CEW is pushed or driven into the subject aiming for a muscle mass or group. This contact can be 5 seconds or less in duration as the operator has the ability to retract the device thereby discontinuing or shortening the duration of the application. Situations included forcing belligerent noncompliant suspects into handcuffs or into patrol cars after they were handcuffed, but remained noncompliant.

This application has advantages and inherent disadvantages; one advantage is that by removing the weapon the CEW can be deployed in environments where firing the probes is inappropriate. In addition, the operator can easily reduce the duration of the current if the subject becomes immediately compliant. As a disadvantage, however, the distance between the probes is comparably small, resulting in less loss of physical control by the subject.

In reviewing the data collected, the number of drive-stuns was analyzed; in the first iteration a total of 176 (4.1% of all uses of force, or 8.3% if all TASER CEW deployments) contacts, or drive-stuns, were made. In officer/suspect confrontations that extended into Iteration 2 and Iteration 3, the drive-stun function was used 159 (3.7% of all uses of force) times and 67 (1.6% of all uses of force) times, respectively.

Table 3.2 Comparison of probe deployment and drive-stun effectiveness

	Probe		Drive-stun	
	N	%	*N*	%
Iteration 1	1,151	59.4	113	64.2
Iteration 2	365	67.7	108	67.9
Iteration 3	131	80.4	56	83.6

Note: Due to rounding, percentages may not total to 100%.

3.2.6 Suspect and Officer Injuries

Substantially more suspects sustained injury than law enforcement officers. While 23% of suspects were injured during force encounters only 3% of officers were. Injuries to both officers and suspects were more likely to occur during traffic stops and disturbances, with the majority of injury types comprised of bruises and abrasions. However, a trend emerged once the data was deconstructed at the event level. Injuries to both suspects and officers increased in proportion to the length and duration of the confrontation. This in itself is not startling; it is common sense to expect that more injuries would occur in longer confrontations. Traditional force, wielded by either officer or suspect, may be cumulative and the more applications substantially increase the possibility of injury to either or both. A prior study[78] indicated that both officer and suspect injuries are closely related to the nonlethal weapon deployed.

In the first iteration, 19% of suspects were injured in relation to only 1% of officers. However, in the second iteration, 25% of suspects and 3% of officers were injured. In the final iteration, 33% of suspects were injured in comparison with 11% of officers. Clearly, as the confrontation continues, the rate of injuries to both officers and suspects continues to climb. The researchers therefore examined CEWs in the context of their effectiveness at reducing conflict by ending a confrontation.

3.2.7 CEW's Potential to Reduce Injuries

When examining conflict at the event level, this research focused on the CEW's ability to end officer and suspect confrontations; this is inherently a measure for their effectiveness. A total of 2,391 use of force reports described the conflict as ended at the first iteration. CEWs were deployed 2,113 times in the first iteration, and out of those deployments, 1,460 conflicts were ended at this level (representing a 69% conflict resolution success rate). In comparison, other nonlethal weapons and tactics were not nearly as successful. Impact weapons represented 45% success rate, compliance holds were successful 16% of the time, takedowns had a 41% success rate, and chemical agents were 65% effective in stopping conflicts before they escalated to a higher level, or before resorting to alternative tactics or weapons.

3.3 Limitations

Because this study used two large departments in a single geographic region, the results may not be able to be generalized to agencies that do not fit this demographic. However, there is no reason to believe that results from these other departments would be significantly different since law enforcement training, tactics and use of force is fairly generalized across America. This work dealt only with the TASER brand of CEWs. It is unclear if these statistics can be generalized across the entire field of CEWs. The data were analyzed retrospectively and the conclusion is only as accurate as the data collected or recorded.

3.4 Conclusion

CEWs play an important role in law enforcement. This research shows that CEWs are deployed more frequently than other nonlethal weapons and tactics, but they also appear to have higher success rates in conflict resolution. This success in bringing officer/suspect confrontations to an end is invaluable, as it has the effect of reducing injuries to all persons in the conflict. When confrontations continue into multiple iterations, the result is a much higher injury rate for both suspects and officers. This immediately begets the conclusion that the law enforcement community has a duty to use sufficient levels of force (equal or greater to that of the subject's level of resistance) quickly and decisively at the onset of a conflict. This may cause concern to some, especially if there is community distrust in the police; however, when properly administered in the hands of a legitimate police organization, they may in fact be reducing injuries to all parties.

The fact that CEWs offer society the best "set phasers on stun" solution currently available makes them extremely appealing to police in use of force situations. In a police use of force confrontation, the most humane weapon or tactic would be one in which the resultant injury would be the least severe. While CEWs are not injury free (puncture wounds from dart probes, or skin burns from drive-stuns), the alternative (broken bones from batons, burning pain from pepper spray, and potential death from firearm) makes them a preferential choice. Clearly this research has shown that CEWs are very effective at ending conflict situations quickly, this in turn leads to less injuries to both suspects and officers.

Notes

1. Garner JH, Schade T, Hepburn J, Buchanan J. Measuring the continuum of force used by and against the police. *Criminal Justice Review*. 1995;20(2):146–168.
2. Fyfe JJ. Police use of deadly force: Research and reform. *Justice Quarterly*. 1988;5(2):166–205.
3. Klinger DA. The micro-structure of nonlethal force: Baseline data from an observational study. *Criminal Justice Review*. 1995;20(2):169–186.
4. Pate AM, Fridell LA, Hamilton EE. *Police use of force: Official reports, citizen complaints, and legal consequences* (Vols. 1 and 2). Washington, DC: United States Department of Justice, National Institute of Justice. 1993.
5. Garner J, Buchanan J, Schade T, Hepburn J. *Understanding the use of force by and against the police*. Washington, DC: United States Department of Justice, National Institute of Justice. 1996.
6. Clede B. Police nonlethal force manual: Your choices this side of deadly. Harrisburg: Stackpole Books. 1987.
7. Petrowski TD. When is force excessive? *FBI Law Enforcement Bulletin*. 2005:74(9):27–32.
8. Bittner E. *The functions of police in modern society*. Washington, DC: Government Printing Office. 1970.
9. Garner JH, Maxwell CD, Heraux CG. Characteristics associated with the prevalence and severity of force used by the police. *Justice Quarterly*. 2002;19(4):705–746.
10. Reiss AJ Jr. *The police and the public*. New Haven, CT: Yale University Press. 1971.
11. Scharf P, Binder A. *The badge and the bullet: Police use of deadly force*. New York, NY: Praeger. 1983.
12. Sherman L. Perspectives on police and violence. *The Annals of the American Academy of Political and Social Science*. 1980;452:1–12.

13. Langworthy RH, Travis LP. *Policing in America: A balance of forces* (2nd ed.). Upper Saddle River, NJ: Prentice Hall. 1999.
14. Davis KC. *Discretionary justice: A preliminary inquiry*. Baton Rouge: Louisiana State University Press. 1969.
15. National Institute of Justice. *Use of force by the police: Overview of national and local data*. Washington, DC: U.S. Department of Justice, Office of Justice Programs. 1999.
16. Goldstein H. *Policing a free society*. Cambridge, MA: Ballinger Publishing. 1977.
17. Terrill W. Police use of force and suspect resistance: The micro process of the police-suspect encounter. *Police Quarterly*. 2003;6(1):51–83.
18. Klinger DA. The micro-structure of nonlethal force: Baseline data from an observational study. *Criminal Justice Review*. 1995;20(2):169–186.
19. Garner JH, Schade T, Hepburn J, Buchanan J. Measuring the continuum of force used by and against the police. *Criminal Justice Review*. 1995;20(2):146–168.
20. *Johnson v. Glick, 481 F.2d 1028 (1973)*.
21. *Tennessee v. Garner, 471 U.S. 1 (1985)*.
22. Hicks WL. Constraints in the police use of force: Implications of the just war tradition. *American Journal of Criminal Justice*. 2004.28(2):255–270.
23. Alpert GP, Smith WC. How reasonable is the reasonable man? Police and excessive force. *Journal of Criminal Law & Criminology*. 1994:85(2):481.
24. *Graham v. Connor, 490 U.S. 386 (1989)*.
25. *Brosseau v. Haugen, 125 S.Ct. 596 (2004)*.
26. Petrowski TD. When is force excessive? *FBI Law Enforcement Bulletin*. 2005;74(9):27–32.
27. *Scott v. Harris, 127 S. Ct. 1769 (2007)*.
28. Garner JH, Schade T, Hepburn J, Buchanan J. Measuring the continuum of force used by and against the police. *Criminal Justice Review*. 1995;20(2):146–168.
29. Garner J, Buchanan J, Schade T, Hepburn J. *Understanding the use of force by and against the police*. Washington, DC: United States Department of Justice, National Institute of Justice. 1996.
30. Klinger DA. The micro-structure of nonlethal force: Baseline data from an observational study. *Criminal Justice Review*. 1995;20(2):169–186.
31. Terrill W. Police use of force and suspect resistance: The micro process of the police-suspect encounter. *Police Quarterly*. 2003;6(1):51–83.
32. Hicks WL. Constraints in the police use of force: Implications of the just war tradition. *American Journal of Criminal Justice*. 2004:28(2):255–270.
33. Terrill W. Police use of force: A transactional approach. *Justice Quarterly*. 2005;22(1): 107–138.
34. Garner JH, Schade T, Hepburn J, Buchanan J. Measuring the continuum of force used by and against the police. *Criminal Justice Review*. 1995;20(2):146–168.
35. Williams GT. Force continuums: A liability to law enforcement? *FBI Law Enforcement Bulletin*. 2002;71(6):14–19.
36. Conner GJ. Use of force continuum: Phase II. *Law and Order*. 1991;39(3):30–32.
37. Alpert GP, Dunham RG. *The force factor: Measuring and assessing police use of force and suspect resistance. Use of force by the police: Overview of national and local data*. Washington, DC: U.S. Department of Justice, Office of Justice Programs. 1999.
38. Sherman L. Perspectives on police and violence. *The Annals of the American Academy of Political and Social Science*. 1980;452:1–12.
39. Peterson E. Conception of criminality illustrated by a stochastic process model for deviant behavior. *Journal of Research in Crime and Delinquency*. 1972;9(1):31–45.
40. Friedrich RJ. *Impact of Organizational, Individual, and Situational Factors on Police Behavior* (Vols. 1 and 2). University of Michigan Doctoral Dissertation; Ann Arbor, MI: UMI Dissertation Services. 1977.
41. Reisig MD, McCluskey JD, Mastrofski SD, Terrill W. Suspect disrespect toward the police. *Justice Quarterly*. 2004;21(2):241–268.

42. Worden RE, Shepard RL, Mastrofski SD. On the meaning and measurement of suspects' demeanor toward the police: A comment on "demeanor and arrest". *Journal of Research in Crime and Delinquency*. 1996;33(3):324–332.
43. Terrill W. Police use of force and suspect resistance: The micro process of the police-suspect encounter. *Police Quarterly*. 2003;6(1):51–83.
44. Wolf R. *Use of force matrix*. "Police and Society" Presentation on Use of Force, University of Central Florida; Orlando, Florida. 2006, October 24.
45. Garner JH, Maxwell CD, Heraux CG. Characteristics associated with the prevalence and severity of force used by the police. *Justice Quarterly*. 2002;19(4):705–746.
46. Reisig MD, McCluskey JD, Mastrofski SD, Terrill W. Suspect disrespect toward the police. *Justice Quarterly*. 2004;21(2): 241–268.
47. Weidner RR, Terrill W. A test of Turk's theory of norm resistance using observational data on police-suspect encounters. *Journal of Research in Crime and Delinquency*. 2005;42(1):84–109.
48. Friedrich RJ. Police use of force: Individuals, situations, and organizations. *The Annals of the American Academy of Political and Social Science*. 1980;452:82–97.
49. Garner JH, Maxwell CD, Heraux CG. Characteristics associated with the prevalence and severity of force used by the police. *Justice Quarterly*. 2002;19(4):705–746.
50. Adang OMJ, Mensink J. Pepper spray: An unreasonable response to suspect verbal resistance. *Policing: An International Journal of Police Strategies & Management*. 2004; 27(2):206–219.
51. U.S. Government Accountability Office. *Taser weapons: Use of Tasers by selected law enforcement agencies*. (Publication No. GAO-05-464). Washington, DC: U.S. Government Printing Office. 2005.
52. Klinger DA. The micro-structure of nonlethal force: Baseline data from an observational study. *Criminal Justice Review*. 1995;20(2):169–186.
53. Terrill W. Police use of force and suspect resistance: The micro process of the police-suspect encounter. *Police Quarterly*. 2003;6(1):51–83.
54. Terrill W. Police use of force: A transactional approach. *Justice Quarterly*. 2005;22(1):107–138.
55. Weidner RR, Terrill W. A test of Turk's theory of norm resistance using observational data on police-suspect encounters. *Journal of Research in Crime and Delinquency*. 2005;42(1):84–109.
56. Ross DL. Assessing the patterns of citizen resistance during arrests. *FBI Law Enforcement Bulletin*. 1999;68(6):5–11.
57. Morabito EV, Doerner WG. Police use of less-than-lethal force: Oleoresin Capsicum (OC) spray. *Policing: An International Journal of Police Science and Management*. 1997;20(4):680.
58. Hickman MJ. *Citizen complaints about police use of force*. Bureau of Justice Statistics Special Report, June 2006. Washington, DC: United States Department of Justice, Office of Justice Programs. 2006.
59. McCluskey JD, Terrill W. Departmental and citizen complaints as predictors of police coercion. *Policing: An International Journal of Police Science and Management*. 2005; 28(3):513–529.
60. Garner JH, Maxwell CD. *Measuring the amount of force used by and against the police in six jurisdictions*. Washington, DC: U.S. Department of Justice, Office of Justice Programs. 1999.
61. Garner J, Buchanan J, Schade T, Hepburn J. *Understanding the use of force by and against the police*. Research in Brief, Washington, DC: United States Department of Justice, National Institute of Justice. 1996.
62. Klinger DA. The micro-structure of nonlethal force: Baseline data from an observational study. *Criminal Justice Review*. 1995;20(2):169–186.
63. National Institute of Justice. *Use of force by the police: Overview of national and local data*. Washington, DC: U.S. Department of Justice, Office of Justice Programs. 1999.

64. Reiss AJ Jr. *The police and the public*. New Haven, CT: Yale University Press. 1971.
65. Worden RE, Shepard RL, Mastrofski SD. On the meaning and measurement of suspects' demeanor toward the police: A comment on "demeanor and arrest". *Journal of Research in Crime and Delinquency*. 1996;33(3):324–332.
66. Garner JH, Maxwell CD, Heraux CG. Characteristics associated with the prevalence and severity of force used by the police. *Justice Quarterly*. 2002;19(4):705–746.
67. Mesloh C, Henych M, Wolf R. *Lethal weapon effectiveness, use of force, and suspect & officer injuries: A five-year analysis*. Washington, DC: United States Department of Justice, National Institute of Justice. 2007.
68. U.S. Government Accountability Office. *Taser weapons: Use of Tasers by selected law enforcement agencies*. (Publication No. GAO-05-464). Washington, DC: U.S. Government Printing Office. 2005.
69. Taser International Inc. *Corporate History*. Retrieved July 3, 2007 from http://www2.taser.com/company/Pages/factsheet.aspx
70. Taser International Inc. *Facts about Taser*. Retrieved June 8, 2005, from www.taser.com/facts.
71. Police Executive Research Forum. *PERF Conducted Energy Device Policy and Training Guidelines for Consideration*. Washington D.C. PERF Center on Force and Accountability. 2004.
72. Mesloh C, Henych M, Hougland S, Thompson F. TASER and less lethal weapons: An exploratory analysis. *Law Enforcement Executive Forum*. 2005;5(5):67–79.
73. Donnelly T. *Less lethal technologies: Initial prioritization and evaluation*. London, United Kingdom: Police Scientific Development Branch, Home Office Policing and Crime Reduction Group. 2001.
74. Hopkins P, Beary K. *TASER use and officer/suspect injuries*. Orlando, Fl: Orange County Sheriff's Office. 2003.
75. Goltz J. *Police Organizational Performance in the State of Florida: Confirmatory Analysis of the Relationship of the Environment and Design Structure to Performance*. Unpublished doctoral dissertation, University of Central Florida. 2006.
76. Terrill W, Alpert GP, Dunham RG, Smith MR. A management tool for evaluating police use of force: An application of the force factor. *Police Quarterly*. 2003;6(2):150–171.
77. Klinger DA. The micro-structure of nonlethal force: Baseline data from an observational study. *Criminal Justice Review*. 1995;20(2):169–186.
78. Hougland S, Mesloh C, Henych M. Use of force, civil litigation, and the Taser: One agency's experience. *FBI Law Enforcement Bulletin*. 2005;74(3):24–30.

Chapter 4
Nonlethal Weapons: The Broader Context

Peter J. Cuenca and John G. McManus

Recent experiences in peacekeeping missions, developments in nonlethal technology, and the current Global War on Terrorism have resulted in increased development and fielding of nonlethal weapons (NLW) within the U S military [1–3]. Such weapons are designed to complement and extend diplomatic and military options beyond the use of more traditional lethal weapons. Non-lethal weapons allow the military to accomplish dangerous objectives within the framework of very restrictive rules of engagement (ROE) typically associated with missions such as peacekeeping, peacemaking, detention operations, and humanitarian assistance. Soldiers executing operations other than war (OOTW) today are armed with a variety of non-lethal weapons and munitions allowing them to control the amount of force necessary to accomplish the mission. Such flexibility strengthens peacekeeping and peacemaking missions, reduces the possibility of innocent casualties, and, most importantly, protects the individual soldier by allowing greater control of tactical situations.

The impetus to develop and employ nonlethal technology within the US military did not arise until the post cold war era and the shift of the United States' national focus from conventional war to peacekeeping operations, humanitarian operations, and regional conflicts (OOTW) [4]. However, nonlethal technology has been available since 1960, as described by a Central Intelligence Agency (CIA) report that discusses weapon systems which could be characterized as employing nonlethal technology [5]. Also, the US National Science Foundation Report on Non-lethal Weapons in 1972 contained a list of 34 different weapon systems that could be characterized as nonlethal [6].

The US Department of Defense (DOD) established the Joint Non-Lethal Weapons Program (JNLWP) in 1996 to research, develop, and acquire nonlethal capabilities [7]. Nonlethal technologies have had successful limited use by US armed forces in Somalia, Haiti, Panama, and the Balkans [8]. In 1996, US

P.J. Cuenca (✉)
Department of Combat Medic Training, Department of Emergency Medicine, Brooke Army Medical Center

M.W. Kroll, J.D. Ho (eds.), *TASER® Conducted Electrical Weapons: Physiology, Pathology, and Law*, DOI 10.1007/978-0-387-85475-5_4,

Marines equipped with NLW safeguarded the withdrawal of over 2,500 United Nations peacekeepers from Somalia without a death among the peacekeepers, the Marines, or the populace [9–11]. In 2000 and 2001, Task Force Falcon in Kosovo was able to deal with several large-scale civilian riots involving women and children without loss of civilian or military life [8]. US forces executing nation building and peacekeeping operations in Iraq and Afghanistan have also implemented nonlethal technologies [12].

The US military defines NLW as "weapons that are explicitly designed and primarily employed so as to incapacitate personnel or materiel while minimizing fatalities, permanent injury to personnel, and undesired damage to property and the environment." Additionally, "unlike conventional lethal weapons that destroy their targets principally through blast, penetration, and fragmentation, nonlethal weapons employ means other than gross physical destruction to prevent the target from functioning. Nonlethal weapons are intended to have relatively reversible effects on personnel and materiel" [13].

In the US military, the term "nonlethal" does not equate to zero mortality or nonpermanent damage; these are the goals and not guarantees of these weapon systems. NLW are systems primarily designed and employed to prevent the target from functioning or taking effective action while minimizing casualties and collateral damage. NLW achieve desired target responses using means other than gross physical destruction. NLW provide target discrimination and mitigate collateral damage through precision and accuracy, as well as in type, magnitude, and duration of effect and degree of reversibility/recovery [14].

NLW enhance the capability of US forces to discourage, delay, or prevent hostile actions, limit escalation, take military action in situations where use of lethal force is not the preferred option, improve force protection, and temporarily disable equipment, facilities, and personnel [13]. Presence of NLW on the battlefield does not mean they must be used. In all cases, the on scene commander retains the option for immediate use of lethal weapons when deemed appropriate, in accordance with the standing rules of engagement and standing rules for the use of force (SROE/SRUF) [15].

The US military employs the continuum of force concept to provide its Servicemembers a guide in determining the appropriate level of force that is appropriate and necessary in intensity, duration, and magnitude based on the tactical situation they are faced with (see Fig. 4.1). The continuum of force concept states that there is a wide range of possible actions, ranging from presence and verbal commands to application of deadly force that may be used to counter resistance or a threat in any situation [16–17]. Military operations vary in nature, and threat levels can rise and fall several times based on the actions of both the Servicemembers and the individuals or group involved. As a guide, the purpose of a force continuum concept is to serve as a graphic training aid to assist Servicemembers in use-of-force decision making consistent with ROE/RUF, commander's intent, and the force options available. The force continuum is not policy. It does not replace ROE/RUF. It is a training tool the

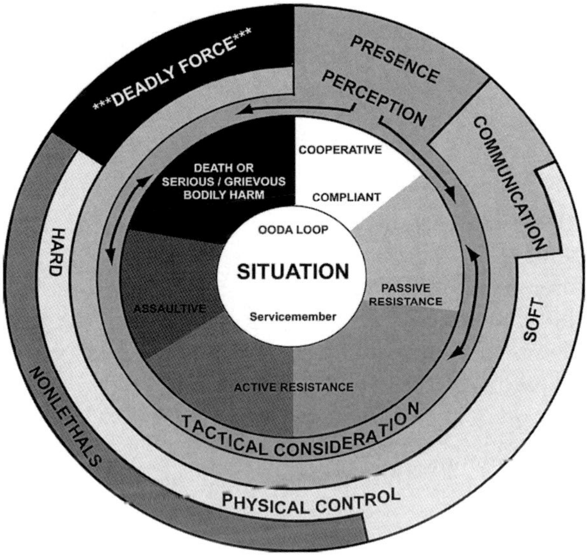

Fig. 4.1 Force continuum graphic
(Source: *Multi-Service Tactics, Techniques, and Procedures for the Tactical Employment of Nonlethal Weapons (FM 3-22.40/MCWP 3-15.8/NTTP 3-07.3.2/AFTTP(I) 3-2.45)*)

military uses for ROE/RUF and, as a training tool, assists Servicemembers in determining when and how they use force [15].

The force continuum centers on the SITUATION which includes the Servicemember and his/her utilization of the observe-orient-decide-act (OODA) loop which drives the function of the force continuum. Outside the SITUATION circle is a circular band of five broad categories of subject behaviors [15,18]:

(a) Cooperative/compliant – follows verbal and nonverbal instructions and direction given.
(b) Passive resistance – behavior that creates no immediate danger of physical harm; often encountered as refusal to comply to verbal or nonverbal instruction or direction, refusing to move or going limp or "dead weight".
(c) Active resistance – nonassaultive physical action to resist; examples include pulling, walking, or running away to avoid control.
(d) Assaultive – attempting to or actually striking, kicking, wrestling, or biting a Servicemember without the use of actual weapon of any kind.
(e) Serious bodily harm or death – actions that the Servicemember reasonably believes are intended to, or likely to cause serious bodily harm or death to any person; guns and knives are the most obvious weapons, but improvised weapons such as pipes, chains, hazardous materials, or any tool that can be used as a bludgeon or cutting instrument may pose a serious or deadly threat.

Between the subject behavior band and force options band in the force continuum graphic lies the tactical considerations and perception band. Factors that Servicemembers bring to a situation are unique to that individual. The Servicemember's perception of a situation may affect his/her orientation (from OODA) and in turn his/her tactical considerations. Situational-specific ROE/RUF are overriding tactical considerations for all military personnel.

The outer band of the force continuum graphic contains five broad categories of force options [15,18]:

(a) Presence – the visible presence of the Servicemember does have an effect that can calm a situation or be a controlling force; military personnel can represent possible consequences on the subject's undesired behavior. An example of this is when a motorist is speeding down the road, comes over a hill and spots a marked police car on the side of the road. The most probable reflex action the motorist will take is to step on the brake and slow down. In this example, the presence of the marked patrol car influenced or "forced" the driver to a desired outcome (slow down) and the only "force" used was presence.
(b) Communication – verbal and nonverbal communication can be utilized to control or resolve the situation. Like presence, communication is present throughout the entire force continuum. Even as other categories of force are employed, communication should be present in the form of instructions, directions, orders, or warnings.
(c) Physical control – use of manual force; any physical technique used to control a subject that does not involve the use of a weapon. The purpose of the force continuum is *control*: maintaining it or gaining it over the *situation*. The concepts of defense and offense are immaterial. Physical control includes two subcategories: soft techniques and hard techniques.

    Soft techniques are control oriented, have lower probability of causing injury, and include restraining techniques, joint locks, and compliant flex cuffing.

    Hard techniques are intended to stop a subject's behavior, have a higher probability of causing injury, and include empty hand strikes (punches and kicks), any type of "submission" hold or technique that risks dislocation of joints or fractured or broken bones, any physical hold technique intended to bring about or cause unconsciousness or is likely to produce unconsciousness. These techniques may be raised to the level of deadly force.

(d) Nonlethal force – all weapons and devices that are under the DOD definition of "nonlethal" (DODD 3000.3) and are nonlethal by design. These include batons and OC spray. Within the military, NLW are purposely not listed individually in order to avoid creating a "nonlethal force continuum." Either NLW are assessed as appropriate to the situation or they are not. When they are assessed appropriate to the situation, the type of nonlethal force to be used is dictated by the tactical situation (distance, time available, number of subjects involved, ROE/RUF). Further, not all Servicemembers and not all units will have all the NLW available, therefore their options within this category may be very limited. The ROE and RUF predetermine what NLW options are available to military units; however, their employment is strictly controlled by the commander.

(e) Deadly force – defined in DOD Directive 5210.56, November 1, 2001, as: "Force that a person uses causing [or that a person knows or should know would create a substantial risk of causing] death or serious bodily harm." The same directive defines "serious bodily harm" as "Does not include minor injuries, such as a black eye or a bloody nose, but it does include fractured or dislocated bones, deep cuts, torn members of the body, serious damage to internal organs, and other life-threatening injuries."

NLW are available in the military for use in a variety of conflict scenarios, from humanitarian and peace operations to major combat operations. Service-unique nonlethal capability sets (NLCS) support platoon company size units and contain a mix of counter-personnel (CP) and counter-materiel (CM) systems, protective equipment, enhancement devices, and training devices. Some of the NLCS CP items include 12 gauge, 40 millimeters, and 66 millimeters nonlethal munitions or grenades, TASER CEWs, Oleoresin Capsicum (OC) spray/gas, and CM devices, including tire spikes (caltrops) and the portable vehicle-arresting barrier (PVAB) (both of which are used to deny vehicles access to critical infrastructure at roadblocks and entry control points) [19]. One of the nonlethal CM "smart weapons" used during the Gulf War was a special warhead adapted for the Tomahawk Cruise missile that dispersed thousands of carbon fibers after exploding over an electrical power station target. After the carbon fibers drifted down and settled, they would cause the power station to short circuit. By using this type of non-lethal weapon, the United States was able to neutralize several of the Iraqi electrical power stations without permanent damage [20,21].

The JNLWP currently is sponsoring several acquisition programs including Acoustic Hailing Devices (AHD), the Improved Flash Bang Grenade (IFBG), Vehicle Lightweight Arresting Device (VLAD), Joint Non-Lethal Warning Munitions (JNLWMs), the Individual Serviceman Non-Lethal System (ISNLS), the Airburst Non-Lethal Munition (ANLM), the MK19 Non-Lethal Munition (MK19 NLM), and the Mission Payload Module – Non Lethal Weapon System (MPM-NLWS) [22].

AHDs are nonlethal, nonkinetic, long-range hailing and warning devices. The devices use advanced directed acoustic energy technology to provide a nonlethal warning capability at a greater range than many other nonlethal systems available to US forces. AHDs are capable of producing highly directional sound beams that allow users to project warning tones and intelligible voice commands beyond small arms engagement range. The capability will enable US forces to more effectively determine the intent of a person or crowd at a safe distance and potentially deter them prior to escalating force [23,24].

The IFBG development program aims to improve the effectiveness and safety of currently fielded nonlethal flash bang munitions by eliminating perchlorates in their formulation. Perchlorates are strong oxidizers used in fireworks, explosives, rocket propellant, flash compositions, and other pyrotechnics. The IFBG will initiate and eject a metal powder payload that reacts with oxygen in the air to create a large, bright, long-duration flash and loud noise. A longer duration of the flash and bang increases temporary incapacitation. The IFBG will support missions such as hostage rescue, room clearing, and other operations in complex urban terrain [22].

JNLWMs are nonlethal small arms cartridges capable of projecting clear, unambiguous warning signals out to distances of 100, 200, and 300 meters. The JNLWMs are designed to be employed as warning signals that will enable bystanders to determine the intent of unidentified vessels, vehicles or personnel. The projectiles are not intended to strike downrange targets. They will minimize fatalities, protect the innocent, limit collateral damage, and temporarily incapacitate. JNLWM cartridges are shoulder fired with standard military 12-gauge shotguns or 40-millimeter launchers. The projectile airbursts at a fixed distance to provide a light flash, loud report (bang), and smoke [25].

ISNLS is an evolving non-lethal weapon concept currently supported by the commercial-off-the-shelf FN303 Less Lethal Launcher. The ISNLS gives the Servicemember the ability to engage targets with nonlethal force at greater distance and accuracy than is currently available. The FN303 Less Lethal Launcher is a compressed-air powered semiautomatic launcher designed to fire nonlethal projectiles at established nonlethal ranges. The launcher is made from a durable, lightweight polymer with flip-up iron sights and an integrated Picatinny 1913 rail for mounting red dot sights (included) or other accessories [26].

ANLM is a US army-led program. The ANLM is designed to enable a precision airburst delivery of nonlethal munitions. It is intended for use in area denial and hostile crowd scenarios and is being developed for use with several different weapons, including the MK19 and the M203 (see Fig. 4.2) [27].

The MK19 NLM is being developed to provide the military area denial, counter-personnel capability. The MK19 NLM is shown in Fig. 4.3 and fires blunt trauma projectiles (Fig. 4.4) in rapid-fire mode. The shell casing of the munition is ejected by the MK19 launcher, and the ring air foil projectile is launched as a nonlethal, blunt-impact round [28].

The MPM-NLWS provides the military a capability that delivers nonlethal counter-personnel effects to control crowds, deny and defend areas, control access, and engage threats while providing increased standoff range for protection of friendly forces. The MPM-NLWS will be mounted onto the High

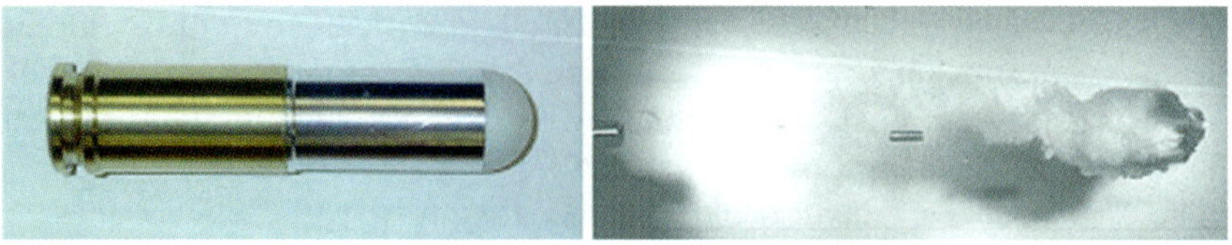

Fig. 4.2 The airburst nonlethal munition

Fig. 4.3 Personnel Mk19 short-range nonlethal munition (SRNLM)

Fig. 4.4 Mk19 nonlethal munition vortex ring

Mobility Multipurpose Wheeled Vehicle (HMMWV) and Navy surface water vessels through an evolutionary acquisition process. Effects such as obscuration and illumination will be added to address future and emerging capability gaps. The weapons platform will be capable of providing greater range, area coverage, precision, and scalability than current NLW systems. The MPM-NLWS is designed to operate in all environments and is particularly well suited for asymmetric warfare, urban environments, maritime security, homeland defense, and decisive combat engagements. Potential nonlethal payloads include the following: warning flash bangs, incapacitating flash-bangs, point-effect blunt impact, illumination, and smoke [29]. Figure 4.5 shows such a system mounted on an unmanned vehicle.

NLW have proven their value during conflict. The potential offered by NLW is increasingly being recognized by those inside and outside the Department of Defense. The Council on Foreign Relations' 2004 report stated, "Nonlethal Weapons and Capabilities could substantially improve the United States' ability to achieve its goals across the full spectrum of modern war," and it called for an increased commitment to realizing this potential. Additionally, the Secretary of Defense initiated the US military transformation process in order to achieve and maintain military advantages over potential adversaries. This transformation includes a greater emphasis on the availability of NLW [30]. NLW fill the gap between shouting and shooting and allow Servicemembers to perform their jobs without loss of life.

Fig. 4.5 Metal storm, a potential MPM-NLWS, mounted on the TALON mobile unmanned ground vehicle (UGV)

References

1. Department of Defense, 1996. *Directive Subject: Policy for Non-Lethal Weapons,* Department of Defense Directive 3000.3, Washington, DC, 9 July 1996, 1–3.
2. Department of Defense, 2001. *Defense Planning Guidance*. Washington, DC.
3. Committee for an Assessment of Non-Lethal Weapons Science and Technology, National Research Council, 2001. *An Assessment of Non-Lethal Weapons Science and Technology*. Washington, DC: National Academic Press, 2001, 101.
4. Duncan JC, Scott RC, McClain RS, Stephens D, 1998. "A Primer on the Employment of Non-Lethal Weapons." *Naval Law Review*, Volume 45, 1998. Accessed at http://handle.dtic.mil/100.2/ADA359487.
5. Aftergood S, 1994. "The Soft-Kill Fallacy." *The Bulletin of the Atomic Scientists*. September /October 1994. 40.
6. Wright J, 1994. "Shoot Not to Kill." *The Guardian*. May 1994.
7. Joint Non-Lethal Weapons Program Website: History Section. Accessed at https://www.jnlwp.com/History.asp.
8. Alexander JB, 2002. "Non-lethal weapons to gain relevancy in future conflicts. (Commentary)." *National Defense*. 1 March 2002.
9. Sheehan JJ., General, USMC. "Non-Lethal Weapons: Let's Make It Happen" Non-Lethal Defense Conference. 7 March 1996.
10. Council on Foreign Relations, 2004. "Lack of Nonlethal Weapons Capabilities Hindering U.S. Efforts in Postwar Iraq; Experts Urge Department of Defense to Increase Spending Seven-Fold." 26 February 2004. Accessed at http://www.cfr.org/publication.html?id = 6794.
11. National Defense University, 1996. Chapter 12: Unconventional Military Instruments. Strategic Assement 1996. Accessed at http://www.ndu.edu/inss/Strategic%20Assessments/sa96/sa96ch12.html.
12. Burgess L. "Troops Train to Use Nonlethal Weapons to Control Crowds, Reduce Civilian Deaths." *Stars and Stripes – Stripes Sunday Magazine*, 22 December 2002.
13. Department of Defense Directive (DODD) 3000.3. *Policy for Nonlethal Weapons*, July 1996. Certified current as of November 2003.
14. Jackson AE. "The Army's Nonlethal Weapons: An Overview." *Infantry Magazine*, May–August 2000.
15. FM 3-22.40/MCWP 3-15.8/NTTP 3-07.3.2/AFTTP(I) 3-2.45. *Multi-Service Tactics, Techniques, and Procedures for the Tactical Employment of Nonlethal Weapons*. 24 October 2007.
16. Willingham S. "Peacekeeping Duties Bolster Demand for Kinder Weapons." *National Defense* May 2000.
17. Simpson S. Power Point Presentation: "The U.S. Department of Defense Joint Non-Lethal Weapons Program – Training." March 2000. Accessed at http://www.dtic.mil/ndia/nld4/simpson.pdf.
18. Shupe PK. "Nonlethal force and rules of engagement." *Military Police* April 2003.
19. Current Non-Lethal Capabilities. Joint Non-Lethal Weapons Program Website. Accessed at https://www.jnlwp.com/capabilities.asp.
20. Lewer N. "Non-Lethal Weapons – A New Dimension." *Bulletin of Arms Control*, September 1996. 1.
21. Fulgham DA. "Secret Carbon-Fiber Warheads Blinded Iraqi Air Defenses." *Aviation Week & Space Technology*, April 1992. 18–20.
22. Joint Non-Lethal Weapons Program. Vehicle Lightweight Arresting Device (VLAD) Fact Sheet. Accessed at https://www.jnlwp.com/Resources/FactSheets/VLAD%20Oct%202006.pdf.
23. Joint Non-Lethal Weapons Program Public Affairs Media Release. "Acoustic Hailing Device Contract Awarded." Posted June 22, 2007. Release #: 2007-109.

24. Joint Non-Lethal Weapons Program. Acoustic Hailing Devices (AHD) Fact Sheet. Accessed at http://www.jnlwp.com/Resources/FactSheets/AHD%20Oct%202006.pdf.
25. Joint Non-Lethal Weapons Program. Joint Non-Lethal Warning Munitions (JNLWM) Fact Sheet. Accessed at https://www.jnlwp.com/Resources/FactSheets/JNLWM%20Oct%2006.pdf.
26. Joint Non-Lethal Weapons Program. Joint Non-Lethal Warning Munition (JNLWM) Fact Sheet. Accessed at https://www.jnlwp.com/Resources/FactSheets/JNLWM%20Oct%2006.pdf.
27. Joint Non-Lethal Weapons Program. The Airburst Non-Lethal Munition (ANLM) Fact Sheet. Accessed at https://www.jnlwp.com/Resources/FactSheets/ANLM%20Oct%2006.pdf.
28. Joint Non-Lethal Weapons Program. Mk19 Short Range Non-Lethal Munition (NLM) Fact Sheet. Accessed at https://www.jnlwp.com/Resources/FactSheets/MK19Fact%20Sheet30APR07FINAL.pdf.
29. Joint Non-Lethal Weapons Program. Mission Payload Module – Non-Lethal Weapon System (MPM-NLWS) Fact Sheet. Accessed at https://www.jnlwp.com/Resources/FactSheets/MPM%20Oct06.pdf.
30. Cohen W. Secretary of Defense. Annual Report to the President and Congress. Chapter 15. 1998.

Chapter 5
Transcutaneous Muscle Stimulation

James D. Sweeney*

The high-voltage, brief pulse width stimulus train applied by the latest generations of conducted electrical weapons (CEWs), such as the TASER® M26 and X26 CEWs, are intended primarily to strongly activate skeletal muscle contraction (thus disabling the target individual through incapacitation of their ability to move and to stand), while secondarily also eliciting strong sensations of pain and/or exhaustion. TASER CEW stimuli applied through transcutaneous darts which have contacted or penetrated the surface of the torso are inherently protective against cardiac events because current needs to penetrate deep within the torso to reach the heart itself, and because stimulus pulse widths needed to activate the heart are longer in duration than those needed to stimulate skeletal muscle or nerve.

This chapter focuses upon the theory of action of electrical waveforms delivered by conducted electrical weapons for effectively incapacitating human subjects via skeletal muscle activation, using the M26 and X26 CEW waveforms as examples.

5.1 The TASER CEW Waveforms

The TASER X26 CEW is a pistol-shaped device that shoots two tethered darts and delivers 19 very short duration pulses per second with a typical peak voltage of 1,200 volt (1,000–1,500 volt) over a 5-second burst. The device generates open-circuit (arcing) voltage of up to 50,000 volt to arc through air or across thick clothing but that voltage is never seen in, or delivered into, the body. The pseudomonophasic waveform delivered by the X26 (Fig. 5.1a) is a specially

---

**James D. Sweeney reports serving as a member of the Scientific and Medical Advisory Board of TASER International, Inc. No other potential conflict of interest relevant to this chapter was reported.*

J.D. Sweeney (✉)
Department of Bioengineering, Florida Gulf Coast University
e-mail: jsweeney@fgcu.edu

M.W. Kroll, J.D. Ho (eds.), *TASER® Conducted Electrical Weapons: Physiology, Pathology, and Law*, DOI 10.1007/978-0-387-85475-5_5,

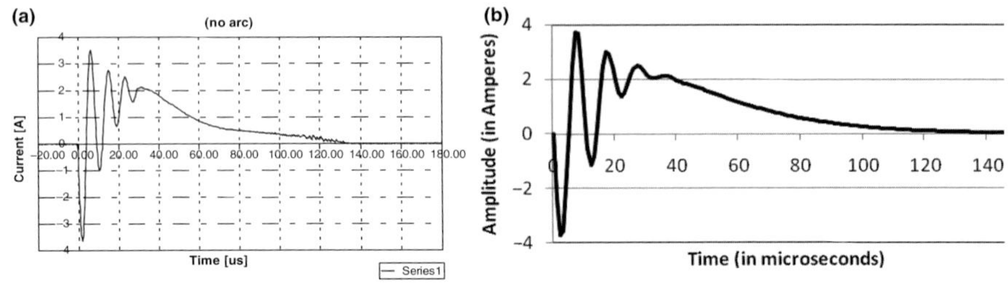

Fig. 5.1 (**a**) Output test waveform of TASER X26 ECD into a typical tissue load. The output circuit attempts to deliver a constant charge per pulse. Thus the output current is largely load independent. (**b**) Simulated X26 waveform consisting of a damped 100 kilohertz leading phase superimposed upon a slower monophasic component

designed and relatively complex pulse with an overall waveform duration of about 100 microseconds. Initial cycles of the arcing phase of the waveform are well approximated as a damped 100-kilohertz sinusoid that rides upon the lower frequency, monophasic component (Fig. 5.1b) of about 70 microseconds duration. As we shall see, it is this latter part of the waveform which is primarily responsible for activation of skeletal muscle force. The peak current of the normal pulse is about 3.3 amperes (typical range of 3.0–3.6 amperes) with a typical interdart body impedance of 400 Ω (ohms), this would correspond to a peak voltage of 1,300 volt.

The TASER M26 CEW delivers damped sine wave current pulses, each having a peak amplitude of about 15 amperes and a fundamental frequency of about 50 kilohertz (kHz) as seen in Fig. 5.2a (actual waveform) and Fig. 5.2b (simulated waveform). The output voltage varies with contact impedance and may exceed 50,000 volt to arc through air and/or clothing when the probes are not directly contacting the body; dropping to a delivered voltage of about 3,000–5,000 volt. The first, main cycle (or primary phase) is about 10 microsecond long and delivers

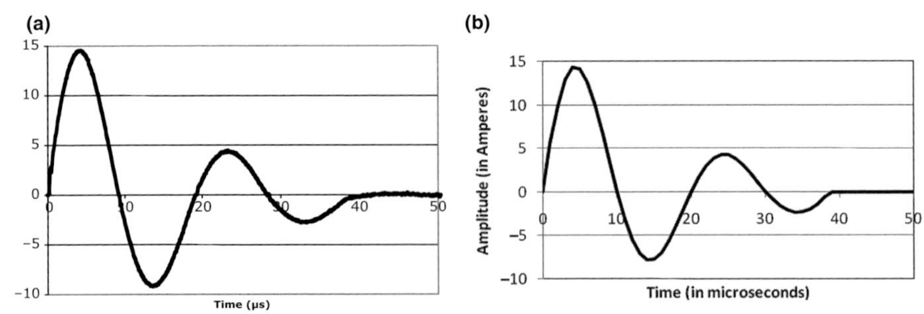

Fig. 5.2 (**a**) Advanced TASER M26 output waveform into a typical test load. (**b**) Simulated M26 waveform as two cycles of a damped 50 kilohertz sine wave

about 85 microcoulombs of charge. As we shall see, the remaining cycles of the M26 waveform contribute only to a lesser extent in muscle electrical activation.

5.2 Theory and Applications of Muscle Stimulation by Electric Fields and Currents

In comparison to medical devices intended to deliver therapeutic currents transcutaneously [1,2], which often utilize stimulus pulse widths on the order of several hundreds of microseconds or more, the TASER M26 and X26 CEW waveforms are relatively brief in duration (by design). In general, skeletal muscle activation by electrical stimulation is elicited by excitation of α-motor neurons that innervate such muscle fibers. This fact often comes as a surprise, in that skeletal muscle cells are themselves excitable. Skeletal muscle excitability, however, is less than that of motor neuron cells in that both rheobase and chronaxie values of skeletal muscle are higher than those of the myelinated nerve axons which innervate them – rheobase being defined as the minimum stimulus strength necessary (generally occurring at long stimulus durations or pulse widths) and chronaxie being defined as the stimulus duration needed at twice the rheobase stimulus strength (as a measure of how steeply the strength–duration relationship rises for brief stimulus durations). Therefore, immediately adjacent to TASER dart locations it is possible that skeletal muscle fibers might be "directly" stimulated but any significant distance away from the darts one would expect skeletal muscle to be "indirectly" activated through its nervous innervation. Sensations of pain and discomfort in response to TASER stimuli would be expected to result from a host of sensory nerve fiber types, to some extent dependent upon the specific locations of TASER dart attachment to the body (as well as the specific tissues located between and near the darts in what might be called the "capture" zone of the darts where excitable cells are activated).

5.2.1 Electrical Stimulation of Motor and Sensory Nerves

Electrical stimulation of motor or sensory nerves can be achieved in general if an imposed electric field near such nerves is sufficient in amplitude, timing and spatial extent to depolarize (i.e., shift the cell's membrane potential in a positive direction) the cell to "threshold" (where an action potential(s) is elicited). Rattay's "activating function" modeling approach [3] to nerve excitation implies that relatively long, straight cell processes (such as the axons of α-motor neurons which may be up to a meter in length as they pass from their cell bodies out to muscles in the extremities) are stimulated according to the shape of the E field (i.e., electric field) imposed along their lengths where *changes* in the E field (which will also correlate to *changes* in the current density J) bring about stimulation. In this chapter, we focus only upon a corollary to Rattay's theory which states that where nerve cells start and stop (i.e., at their cell body or at their connections to other cells such as at skeletal muscle innervation sites or at the end receptors of sensory cells, etc.) or where such cells bend with respect to the imposed E field (e.g., near the TASER darts where current spreads out away from the dart penetration sites) then stimulation thresholds tend to correlate well to the *magnitude* itself of the imposed E field (or current density J). We will refer to this mode of electrical excitation as "ends and bends stimulation" (see also [4]).

The largest diameter, myelinated α-motor neuron axons (which innervate skeletal muscle fibers) tend to have relatively low electrical thresholds. This is because, in general, threshold inversely correlates with cell diameter (so larger diameter cells are easier to stimulate). From the simulation work of J.P. Reilly on uniform field excitation of myelinated nerve using his SENN (spatially extended nonlinear node) model [4], we can estimate that large 20 micron diameter α-motor neuron axons require an electric field minimum of about 6 volt per meter (or 60 millivolt per centimeter) along each cell for end-stimulation threshold, which in theory then scales linearly with diameter (threshold doubles for fibers half the size, etc.). Current thresholds for relatively brief stimulation pulses, such as those of the TASER X26 and M26, then rise up from minimal (rheobase) levels according to the well-known "strength–duration" relationship (Fig. 5.3). The empirical "hyperbolic" form of the strength–duration curve, first derived by Weiss [5], can be stated as:

$$I_{\mathrm{T}} = I_{\mathrm{R}}(1 + \tau/\mathrm{t})$$

where "I_{T}" is a threshold current for a rectangular waveform at a given pulse width "t", and "τ" is the strength–duration time constant (and is also equal to the chronaxie for this formulation). "I_{R}" is the rheobase current value. Similarly, threshold charge Q_{T} can be described as:

$$\mathrm{Q_T} = \mathrm{Q_0}(1 + \mathrm{t}/\tau)$$

where Q_0 is the minimum charge threshold (in theory, found as pulse width t approaches zero).

Consideration of the sizes of the full range of myelinated axons of motor nerves predicts the following approximate rheobase E field threshold levels, corresponding strength–duration time constants, and strength–duration adjustments to rheobase levels in order to predict needed TASER X26 and M26 CEW threshold E field values, assuming for simplicity that stimulation derives mainly for the X26 CEW from the 70 microsecond pseudomonophasic waveform component, and for the M26 CEW from the first 10 microsecond half-cycle (Table 5.1).

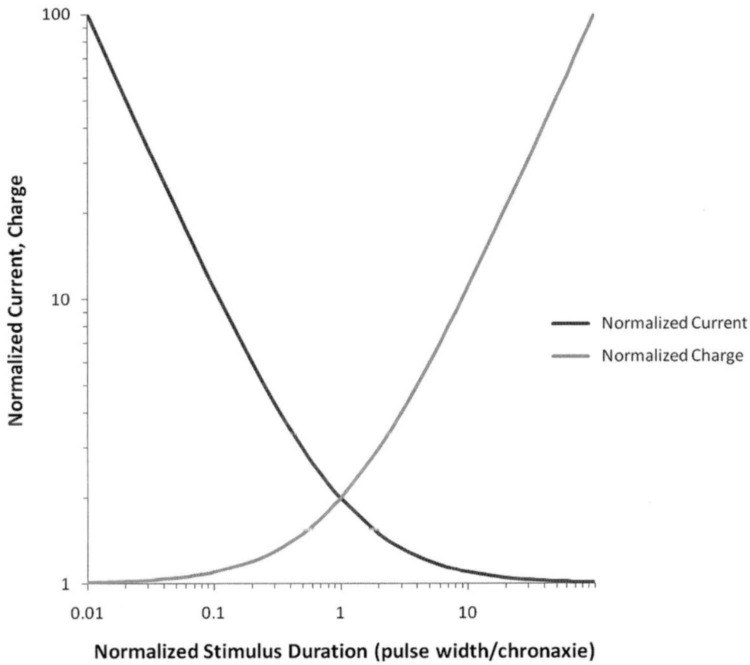

Fig. 5.3 Strength–duration and charge-duration curves for nerve excitation by rectangular pulses. In this normalized form, the rheobase current (current needed for large pulse widths) is normalized to unity, as is the minimum charge threshold (charge needed for short pulse widths). The strength–duration time constant τ is also normalized to 1

Perceptions of vibration and fine touch are carried by myelinated type II Aβ and Aγ nerve fibers, which given their similar size and chronaxie values to the motor Aa fibers (diameters about 2–15 microns; τ about 30 milliseconds, based on [6]), would be expected to be activated by the X26 and M26 CEWs

Table 5.1 Ranges of α-motor nerve fiber parameters and corresponding electric field threshold stimulation estimates for the TASER X26 and M26 CEW waveforms

Efferent motor nerve fiber diameters	About 13 microns (range 8–20 μm)
Rheobase threshold estimates for rectangular stimuli	About 9 V/m rheobase threshold (range 15–6 V/m)
Strength-duration time constant	τ about 140 μs (range 80–150 μs)
E field threshold estimates for X26	Needed X26 threshold E field: about 27 V/m = (9 V/m)*(1 + 140/70)
	X26 E field range of about 18–45 V/m
E field threshold estimates for M26	Needed M26 threshold E field of about 135 V/m = (9 V/m)*(1 + 140/10)
	M26 E field range of about 90–225 V/m

with similar threshold values. "Sharp" pain and discomfort are carried by higher threshold myelinated type III Aδ fibers (diameters about 1–5 microns; τ about 650 milliseconds, based on [6]), while small (about 0.5–2 microns) nonmyelinated type IV C axons are thought to be responsible for conveying dull, aching diffuse pain. From the work of Koslow [7], we can estimate that C fibers have stimulation thresholds about 20 times higher than those of sensory A fibers.

By comparison to motor or sensory myelinated nerves, the heart's excitability (as in pacing of the heart) is relatively low for brief, transcutaneous stimuli. This is because, despite the fact that the minimum (rheobase) level of stimulus needed for long-duration pulses is probably similar between cardiac cells and myelinated nerves, cardiac strength–duration time constants are about 2–3 milliseconds or higher (i.e., at least 10–20 times higher than the α-motor neuron fibers which control skeletal muscle contraction). The heart is also located deep within the torso (as opposed to the skeletal muscle which comprises much of the superficial layers of the torso and into which TASER darts may penetrate if they embed just below the skin). Thus, as discussed further in Chapter 6 relatively little current density will pass through the heart in this situation.

5.2.2 Computer Modeling of the M26 and X26 Waveforms and Expected Stimulation of Peripheral Motor Nerves as a Function of Waveform Shape

We have assumed thus far that the main stimulation effect of the X26 CEW waveform on motor neuron excitation, and therefore activation of skeletal muscle force, derives from the pseudomonophasic component of the waveform and not from the leading higher frequency damped sinusoid cycles. Similarly, we assume that the stimulation effect of the M26 CEW waveform is elicited mainly by the first, largest amplitude half-cycle of the overall damped sinusoid. Just as computer models of peripheral nerve excitation by extracellular electric fields have been used to better understand a wide variety of both medical and accidental modes for electrical stimulation (for a detailed overview and survey see [4]), modeling can be used to investigate the expected theoretical actions of waveforms such as those produced by the X26 and M26 CEWs, as well as various subcomponents of the waveforms.

The mammalian, myelinated nerve model of McIntyre, Richardson, and Grill (MRG) [8] has been used to study threshold excitation by X26 and M26 CEW simulated waveforms via "ends stimulation." Specifically, a 21 node of Ranvier implementation of this multicompartment double-cable model of mammalian axons, coded in NEURON [9], has been used to consider excitation at one termination of the cable. (A basic implementation of the MRG model in

NEURON is archived and available at: http://www.neuron.yale.edu/neuron/.) Threshold currents for extracellular, cathodic X26 and M26 CEW stimulation (and stimulation with waveform subcomponents, as well as with pure sinusoidal currents) have been normalized with respect to rheobase level currents for rectangular, cathodic stimuli – a common point-of-reference for waveforms in medical applications. While the M26 CEW waveform is readily modeled as two complete cycles of a 50 kilohertz sine wave that is damped by two exponential terms (Fig. 5.2b), the X26 CEW simulated waveform (of Fig. 5.1b) can be constructed from an initial 100 kilohertz waveform (damped by a hyperbolic tangent term) superimposed upon a slower, monophasic wave (constructed from two hyperbolic tangent terms). This approach to simulation of the X26 CEW waveform is seen in Fig. 5.4.

We can see from the simulation results depicted in Fig. 5.5 that our assumption that the main excitatory effect of the composite X26 CEW waveform derives from the monophasic component of the waveform, and not the leading damped 100 kilohertz sinusoid is correct. This figure also compares the M26 CEW waveform (a damped 50 kilohertz sine wave) to a simpler nondamped 50 kilohertz sine (two cycles). We see that the damping of the M26 CEW waveform by comparison has relatively little effect. The inherent strength–duration dependence within both the X26 and M26 CEW waveforms is also apparent in this data set. Thresholds for both waveforms sit only modestly above equivalent rectangular stimuli, primarily because rectangular stimuli contain more

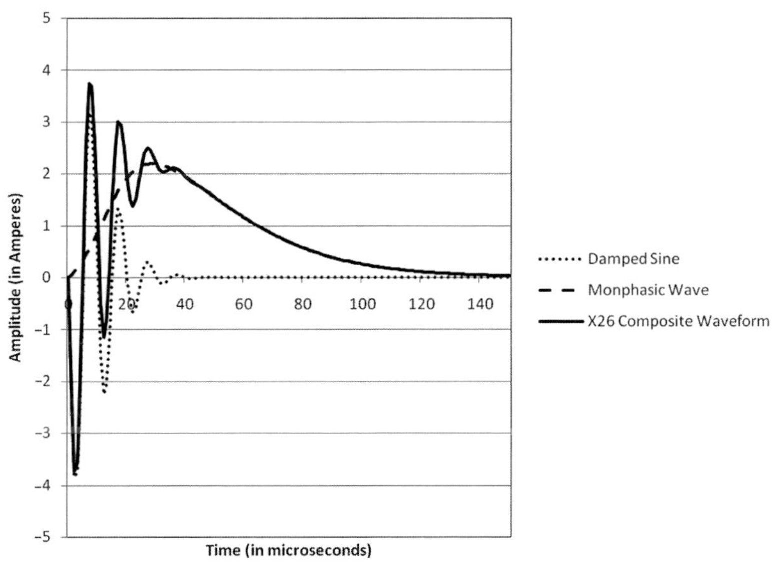

Fig. 5.4 The simulated X26 composite waveform (of Fig. 1b) can be separated into two main subcomponents for computer modeling. A leading 100 kilohertz damped sine wave, combined with a slower monophasic wave, yield the overall composite waveform

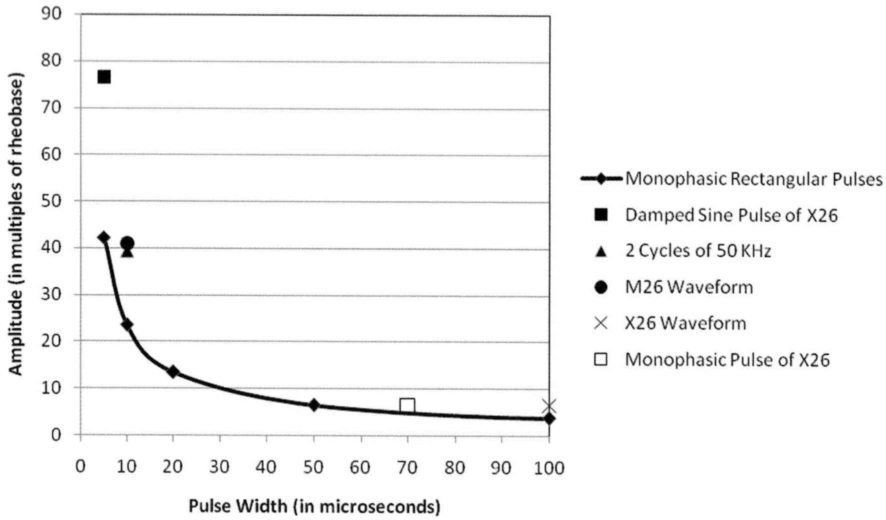

Fig. 5.5 Strength-duration simulation results comparing necessary waveform threshold amplitudes (in multiples of the rheobase current needed for long-duration, cathodic rectangular pulses) for "end stimulation" of a 10 micrometer peripheral nerve model fiber. The complete X26 waveform threshold is arbitrarily plotted at a pulse width of 100 microsecond (the approximate total duration of the X26 waveform), while the pseudomonophasic subcomponent of the X26 waveform is plotted at a pulse width of 70 microsecond. The damped sine components of the X26 waveform are plotted at a pulse width equal to 5 microsecond (the effective pulse width of one half-cycle of the X26's leading sinusoidal phase). M26 waveform threshold results are plotted at a pulse width of 10 microsecond (the effective pulse width of one half-cycle of the M26's sinusoidal component), as are threshold results for two non-damped cycles of a pure 50 kilohertz stimulus

charge within the same pulse width. With this model, in fact, the M26 CEW waveform behavior is dominated by achieving the minimal charge needed to stimulate the axon.

Figure 5.6 allows a closer look at the X26 and M26 CEW waveforms in comparison to 50 and 100 kilohertz sine wave stimuli over varying numbers of cycles. These simulations confirm that much of the effect of the M26 CEW waveform derives from excitation caused within the first (and largest amplitude) half-cycle. For both the 50 and 100 kilohertz sine wave cycle, note that every half-cycle thresholds rise due to anodic current flow (causing hyperpolarization of the axon termination) following the prior half-cycle of cathodic current flow (which causes depolarization). With both sine wave frequencies, there is some benefit in the first few cycles where thresholds decrease about 20–30% toward an eventual asymptote due to nonlinear rectification, i.e., there is some cancellation but also some summation within the axon membrane. As in Fig. 5.5 we see that the damped sine component of the X26 CEW waveform, by itself, is a relatively poor stimulus while the monophasic component of the X26 CEW pulse effectively yields the lower threshold of the composite X26 CEW waveform.

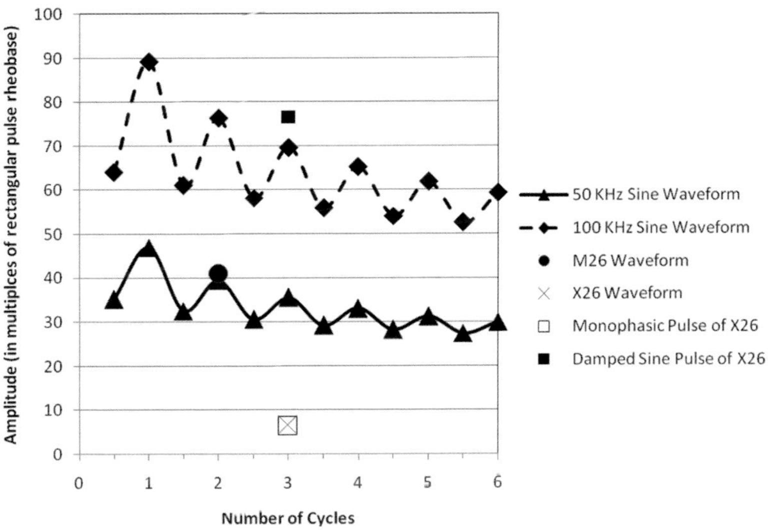

Fig. 5.6 Additional simulation results, detailing model stimulus threshold amplitudes as a function of the number of cycles of sinusoidal current fluctuation for the X26 and M26 waveforms in comparison to subcomponents of the X26 waveform, as well as to simple, undamped 50 kilohertz and 100 kilohertz waveforms. X26 results are arbitrarily plotted at three cycles (given that about three damped cycles are contained within the typical X26 waveform). The M26 result is arbitrarily plotted at two cycles (given that the typical M26 waveform contains about two complete, damped cycles)

5.2.3 Skeletal Muscle Force Recruitment by Electrical Stimulation

An α-motor neuron fiber and the skeletal muscle fibers it innervates are known as a "motor unit" [10]. The X26 TASER CEW delivers 5-second bursts of stimuli delivered at 19 hertz, by comparison twitches generated by slow-twitch motor units are known to fuse at about 5–10 hertz and to reach tetanic fusion at 25–30 hertz, while fast motor units can require 80–100 hertz stimulation to reach complete fusion [11,12]. (The M26 pulse rate is 15–25 pulses per second depending on battery type and charge state. The majority of human skeletal muscles typically comprise a relative balance of slow-twitch and fast-twitch fibers (about 40–70% of one type versus the other) [13]. Muscles with a predominately postural function will normally tend to have higher proportions of slow-twitch fibers, while muscles responsible mainly for phasic activities (such as eye movement) will tend to have higher percentages of fast fibers. Significant interindividual variations in skeletal muscle slow versus fast muscle fiber type distributions also exist and are dependent on gender, age, health, and genetic influences.

In normal physiological usage, motor unit recruitment and firing patterns are effectively optimized for performing the variety of motor tasks encountered in life. Maximal motor neuron firing rates vary physiologically across a wide range, and are dependent upon specific muscles and tasks [14]. In general, motor neurons that innervate more fatigue-resistant, slow-twitch motor units have average physiological firing rates that are lower than those innervating more fatigable, fast-twitch units. Monster and Chan [15], for example, found that almost all motor units in the human extensor digitorum communis muscle, regardless of motor unit type, began firing at about 8 hertz and increased their firing rates to a maximum of 16–24 hertz for maximal voluntary force levels. We expect then that the TASER X26 and M26 CEW stimulation rate of 19 hertz should be well suited for evoking powerful, incapacitating levels of skeletal muscle force production on the order of those found for strong voluntary contractions, but well below force levels that might be evoked with higher frequency electrical stimulation.

In fact, Ding and colleagues have studied in depth the force–frequency relationships seen with electrical stimulation of the human quadriceps musculature, in order to identify the most appropriate stimulation patterns for use in clinical applications of functional electrical stimulation [16]. The quadriceps is a useful example for estimation of TASER CEW activation of other large, mixed fiber type muscles such as the superficial back muscles the trapezius and latissimus dorsi, or various muscles of the extremities. In studies on six, healthy human subjects, Ding found that brief, nonfatiguing constant frequency stimulus bursts at 20 hertz evoked on average about 66% of the isometric, peak quadriceps force brought about by comparable 100 hertz stimulation (as estimated from Fig. 5.4 in [16]). Implementation of a mathematical model also developed by Ding and colleagues [16–18] provides us with a tool for theoretically comparing the 19 hertz TASER stimulation rate with higher and lower frequency patterns. Figure 5.7 details simulated maximal isometric force responses for 5 second bursts at 19 hertz (the duration and rate of the TASER X26 and M26 CEWs), as well as comparable 1, 10, 50, and 100 hertz stimulus trains (model implemented in MATLAB®; assuming typical human subject parameter values and utilizing equations from [17]). As expected, 19 hertz stimulation evokes simulated peak forces on the order of about half (specifically, 46% for this example) of those for the comparable 100 hertz pattern. While 19 hertz stimulation then presumably evokes peak forces on the order of those that a subject could elicit through strong voluntary contractions (see above), we expect that significantly higher frequency bursts (e.g. 50 or 100 hertz) could generate excessive forces in subjects beyond those needed to incapacitate. Lower frequency patterns, such as those seen for 10 hertz and below might fail to generate powerful, well-fused contractions sufficient to immobilize.

5.3 Conclusions and Summary

While the TASER X26 and M26 CEW waveforms exhibit some similarity to stimulation waveforms utilized in medical devices, these systems necessarily incorporate leading high-frequency sinusoidal components with open-circuit

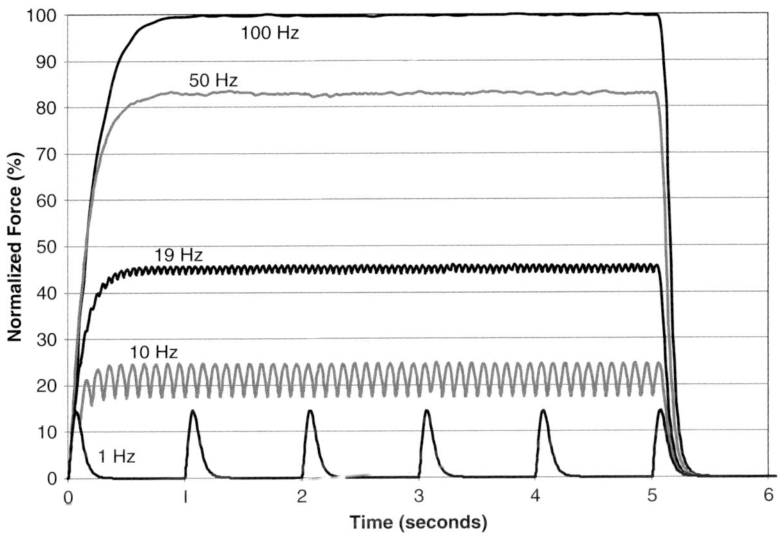

Fig. 5.7 Simulated maximal isometric force responses of human quadriceps muscle to 5 second bursts of electrical stimuli at frequencies of 1, 10, 19, 50 and 100 hertz. The response at 19 hertz is indicative of the anticipated force pattern for TASER activation of human skeletal muscles with mixed fiber types. Underlying model based upon mathematical force and fatigue model system of [17] assuming typical subject parameters. Force responses are normalized to the peak of the 100 hertz results

voltage amplitudes designed to produce arcing between the TASER CEW darts and subjects. TASER CEW waveforms are also necessarily brief in duration so as to insure cardiac safety while still delivering sufficient current so as to "capture" volumes of skeletal muscle that effectively incapacitate subjects.

In this chapter, we have focused analysis upon the predicted electric field strengths that should be needed within the body in order to stimulate the α-motor neurons that innervate skeletal muscle. We have also presented computer modeling of the strength–duration behavior of the TASER X26 and M26 CEW waveforms, contrasting their predicted threshold stimulation levels against each other and in comparison to simpler sinusoidal and rectangular stimuli, as well as the X26 CEW waveform subcomponents. Consideration of pertinent experimental results, as well as implementation of a modeling approach to prediction of mixed, fiber-type skeletal muscle evoked forces due to varying frequencies of electrical stimulation, confirms also that the frequency of TASER CEW stimulation is appropriate for generation of powerful muscle contractions within physiological ranges. Chapter 6 further details analysis of the effectiveness and safety of the TASER X26 and M26 CEW waveforms via modeling and consideration of electric fields and current densities created within the body via typical dart locations.

References

1. Lake, D.A., *Neuromuscular electrical stimulation. An overview and its application in the treatment of sports injuries*. Sports Med, 1992. **13**(5): pp. 320–36.
2. Rushton, D.N., *Electrical stimulation in the treatment of pain*. Disability & Rehabilitation, 2002. **24**(8): pp. 407–15.
3. Rattay, F., *Analysis of models for extracellular fiber stimulation*. IEEE Trans Biomed Eng, 1989. **36**(7): pp. 676–82.
4. Reilly, J.P., *Applied bioelectricity: from electrical stimulation to electrical pathology*. 1998, New York: Springer. 1–563.
5. Weiss, G., *Sur la possibilite' de rendre comparable entre eux les appareils survant a l'excitation electrique*. Arch Ital de Biol, 1901. **35**: pp. 413–46.
6. Li, C.L. and A. Bak, *Excitability characteristics of the A- and C-fibers in a peripheral nerve*. Exp Neurol, 1976. **50**(1): pp. 67–79.
7. Koslow, M., A. Bak, and C.L. Li, *C-fiber excitability in the cat*. Exp Neurol, 1973. **41**(3): pp. 745–53.
8. McIntyre, C.C., A.G. Richardson, and W.M. Grill, *Modeling the excitability of mammalian nerve fibers: influence of afterpotentials on the recovery cycle*. J Neurophysiol, 2002. **87**(2): pp. 995–1006.
9. Carnevale, N.T. and M.L. Hines, *The NEURON book*. 2006, Cambridge; New York: Cambridge University Press. xix, 457p.
10. Monti, R.J., R.R. Roy, and V.R. Edgerton, *Role of motor unit structure in defining function*. Muscle & Nerve, 2001. **24**(7): pp. 848–66.
11. Burke, R.E., *Firing patterns of gastrocnemius motor units in the decerebrate cat*. J Physiol, 1968. **196**(3): pp. 631–54.
12. McPhedran, A.M., R.B. Wuerker, and E. Henneman, *Properties of motor units in a heterogeneous pale muscle (M. Gastrocnemius) of the cat*. J Neurophysiol, 1965. **28**: pp. 85–99.
13. Johnson, M.A., J. Polfar, D. Weightman, and D. Appleton, *Data on the distribution of fibre types in thirty-six human muscles. An autopsy study*. J Neurol Sci, 1973. **18**(1): pp. 111–29.
14. Enoka, R.M., *Morphological features and activation patterns of motor units*. J Clin Neurophysiol, 1995. **12**(6): pp. 538–59.
15. Monster, A.W. and H. Chan, *Isometric force production by motor units of extensor digitorum communis muscle in man*. J Neurophysiol, 1977. **40**(6): pp. 1432–43.
16. Ding, J., A.S. Wexler, and S.A. Binder-Macleod, *A mathematical model that predicts the force-frequency relationship of human skeletal muscle*. Muscle Nerve, 2002. **26**(4): pp. 477–85.
17. Ding, J., A.S. Wexler, and S.A. Binder-Macleod, *A predictive fatigue model-I: Predicting the effect of stimulation frequency and pattern on fatigue*. IEEE Trans Neural Syst Rehabil Eng, 2002. **10**(1): pp. 48–58.
18. Ding, J., A.S. Wexler, and S.A. Binder-Macleod, *Development of a mathematical model that predicts optimal muscle activation patterns by using brief trains*. J Appl Physiol, 2000. **88**(3): pp. 917–25.

Chapter 6
Current Flow in the Human Body

Dorin Panescu and Robert A. Stratbucker*

Nonlethal weapons provide military and law enforcement personnel with a tool to resolve conflict with a proportionate, lawful, appropriate, and necessary use of force [1,2]. The CEW method of incapacitation is through electrical activation of skeletal muscle tissue innervated by peripheral nerves within the electric field created by the CEW [3]. The stimuli from a will override the motor nervous system and block the command and control of the human body. Conventional stun devices stimulate sensory neurons for pain compliance and can be overridden by a focused individual. The CEW directly stimulates preendplate motor nerves, causing incapacitation regardless of subject's mental focus, training, size, or drug-induced dementia [4]. The most popular TASER CEW models supplied to law enforcement agencies are the M26 and X26. Their typical output waveforms are shown in Figs. 6.1 and 6.2, respectively. The M26 delivers a leading biphasic pulse pair of about 10 microsecond duration for each half-cycle. The X26 delivers waveform that to a first approximation appears as a pseudomonophasic (half sinusoid) pulse of about 100 microsecond. Table 6.1 provides a specification summary for these two devices [5,6].

The goals of this chapter are to analyze the distribution of CEW currents in the body for different electrode configurations, to understand the safety and efficacy of TASER CEWs electro-muscular incapacitation, and to assess the TASER CEW risk in triggering ventricular fibrillation.

**Robert A. Stratbucker reports serving as a full-time employee of TASER International, Inc. and as a medical and scientific consultant to TASER International, Inc. Dr. Stratbucker reports as a stockholder of shares of TASER International, Inc. and with patents assigned to TASER International, Inc. No other potential conflict of interest relevant to this chapter was reported.*

D. Panescu (✉)
Chief Technical Officer with NewCardio, Inc., Santa Clara, CA
e-mail: dpanescu@NewCardio.com

M.W. Kroll, J.D. Ho (eds.), *TASER® Conducted Electrical Weapons: Physiology, Pathology, and Law*, DOI 10.1007/978-0-387-85475-5_6,

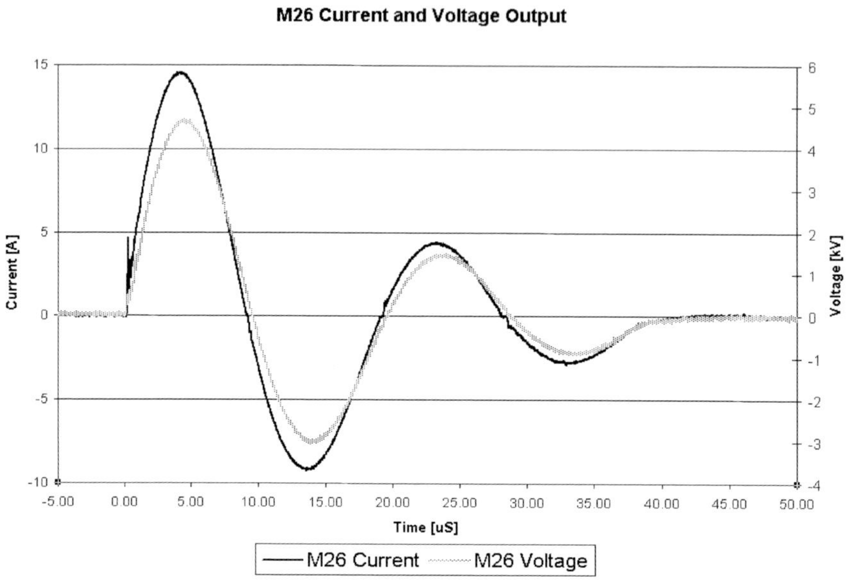

Fig. 6.1 M26 TASER CEW output for a 400 ohm load

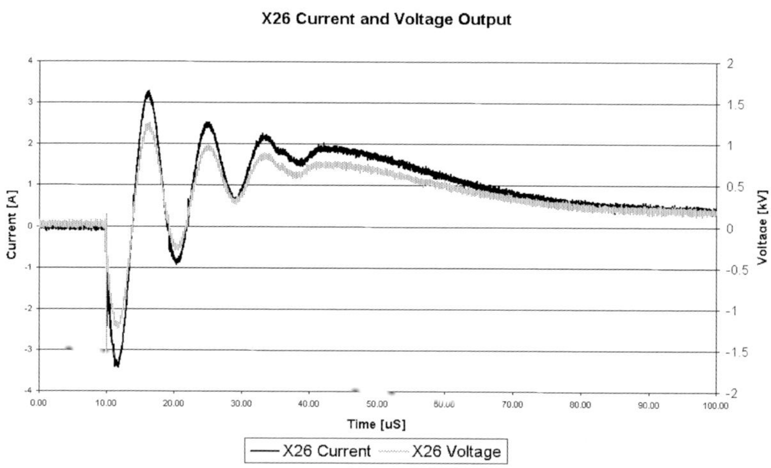

Fig. 6.2 X26 TASER CEW output for a 400 ohm load

6.1 Modeling of CEW Currents in the Human Body

Many people have received a CT or MRI scan but most do not appreciate that there is no direct imaging as there is with an X-ray. Instead, indirect measurements are made and then combined with large-scale mathematical computation

Table 6.1 Specifications of M26 and X26 TASER CEWs

Specification	M26	X26
Average pulse voltage (V)	3,400	400
Open-circuit peak voltage (kV)	50	50
Output peak voltage in typical load (kV)	5	1.2
Overall pulse duration (μs)	40	100
Main phase duration (μs)	10	100
Total per second discharge ("on") time (ms)	0.8	1.9
Energy delivered in typical load (J/pulse)	0.5	0.07
Nominal internal power rating (W)	26	7
Power delivered in typical load (W)	10	1.3
Charge in the main phase (μC)	85	100
Net charge (μC)	33	88
Pulse rate (pulse/s)	20 ± 25%	19
Total delivery duration (s)	5	5
On-demand delivery termination	Yes	Yes
Power source	Battery of 8 AA NiMH rechargeable or alkaline cells	Two 3 volt lithium CR123 cells

to derive the image that is finally presented. Related mathematical techniques are used everyday to calculate where electrical currents flow through the human body. The precise mathematical technique used is called finite element modeling (FEM) and it begins with CT or MRI images which show precisely where the different tissues are located in the human body. To assess the distribution and effects of TASER CEW currents, we developed several FEMs that analyzed skeletal muscle and motor nerve activation, cell electroporation and current density distributions in cardiac tissue.

6.1.1 Skeletal Muscle Activation by Pulsed Electric Fields

As discussed in the previous chapter, skeletal muscle activation by electrical stimulation is elicited by excitation of α-motor neurons which innervate such muscle fibers. Skeletal muscle excitability is less than that of motor neuron cells. Both rheobase and chronaxie values of skeletal muscle are higher than those of the myelinated nerve axons which innervate them. Therefore, immediately adjacent to CEW electrode locations it is possible that skeletal muscle fibers might be "directly" stimulated but at any significant distance away from the probes the skeletal muscle would be "indirectly" activated through its nervous innervations. Sensations of pain in response to CEW stimuli would be expected to result from a host of sensory nerve fiber types, to some extent dependent upon the specific locations of electrode attachment to the body. To elicit muscle activation, each CEW pulse has to inject current such that the generated electric fields capture sufficient volume of skeletal muscle, through indirect stimulation via motor nerves. At the same time, to avoid direct tissue damage, the current densities (J) and electric field strengths (E) have to be lower than thresholds that may produce cellular electroporation. Based on existing modeling and experimental literature, we have assumed the following J and E thresholds for excitation:

- Motor neuron chronaxie: 140 microseconds
- Motor neuron rheobase E field: 0.06–0.15 volt per centimeter for excitation at axon terminations such as motor endplates [7]
- Strength–duration correction of needed E field strength for the M26: $(1 + 140/10) \times (0.06\text{–}0.15$ volt per centimeter$) = 0.9\text{–}2.25$ volt per centimeter;
- Strength–duration correction of needed E field strength for the X26: $(1 + 140/70) \times (0.06\text{–}0.15$ volt per centimeter$) = 0.18\text{–}0.45$ volt per centimeter.

Reilly [8] and Su et al. [9] expressed excitation thresholds for uniform fields with single monophasic stimuli. For short pulses ($t \leq 5$ microseconds), the field-times-duration criteria $(Et)_{min}$ were 2.98×10^{-3} (volt × second per meter). If the E threshold values above 0.18–2.25 volt per centimeter were adjusted for the duration of M26 or X26 CEW pulses then the resulting field-times-duration ranges would be equivalent to those of Reilly and Su et al. In comparison to these expected field strength values needed for neuromuscular activation, Gehl et al. have reported that for irreversible electroporation field strengths of 1,600 volt per centimeter are needed [10].

Based on these values, it is estimated that the E field required for successfully activating motor nerves with the M26 and X26 TASER CEWs has to exceed 0.18–2.25 volt per centimeter, whereas to avoid irreversible electroporation E has to be less than 1,600 volt per centimeter. This yields a worst-case range for the E field strength of 2.25–1,600 volt per centimeter to ensure successful activation with either device while avoiding electroporation.

6.1.2 FE Modeling of J and E Distributions for Neuromuscular Capture

To understand the J and E distributions generated by CEW electrodes, we studied two FE models. These were FEM #1 (typical adult) and FEM #2 (anorexic body habitus) with the following characteristics:

- Regions (dimensions reflect region thickness)
  - Epidermis – 3 millimeters in FEM #1, or 1 millimeter in FEM #2.
  - Dermis – 6 millimeters in FEM #1, or 2 millimeters in FEM #2.

  - Fat – 5 millimeters in FEM #1, or 3 millimeters in FEM #2.
  - Muscle – 6 millimeters in FEM #1, or 10 millimeters in FEM #2.
  - Deeper tissue – 6 millimeters only in FEM #2.
  - CEW electrodes – 9 millimeters long, 2 millimeters diameter.

- Nodes: 45,360 in FEM #1, or 45,900 in FEM #2.
- Elements: 41,080 and 41,272 hexahedral elements in FEM #1 and FEM #2, respectively.
- Model: 25-centimeter long, 5-centimeter wide, 2-centimeter thick. All regions above were 25-centimeter long and 5-centimeter wide. Their respective thickness is listed above. The overall model thickness was 2 centimeters.
- CEW electrodes were modeled 5, 10, 15, or 20 centimeters apart.
- Voltage boundary conditions: 1,000 volt.
- Both models computed steady-state solution.
- Model was conservative as it assumed no rib resistivity.

The regular model, FEM #1, was used to analyze typical skin and fat layer configurations. To estimate worst-case J and E magnitudes, the "anorexic" model, FEM #2, simulated conditions corresponding to subject constitutions with unrealistically thin layers of fat. The electrodes were modeled based on dimensions provided by TASER CEW specifications [5,6] and were considered either fully penetrated into tissue or slightly embedded into the skin surface – simulating drive-stun mode CEW probe deployment. The interelectrode distance was varied between 5 and 20 centimeters [3,5–6]. The analyses focused on current flow within 2 centimeters of the skin. Shown in Table 6.2, the FE region resistivities were based on previous published reports [11,12].

The conditions above represented typical and worst-case scenarios for J and E distributions in the skin, fat, muscle, and deeper tissue layers. Actual-use values would not be expected to exceed the FEA prediction results. Since Cosmos, the FE software used in this study [13], solved for steady state, rather than for transient solutions, the applied voltage was set at 1,000 volt, in the range of output voltages for X26 devices. The actual pulse voltage of the X26 averages 400 volt. The use of 1,000 volt was justified because, given its longer pulse duration and higher net charge, the X26 model should have a higher probability of tissue stimulation than the TASER M26.

Table 6.2 Finite element model material properties

Region	Resistivity ($\Omega \cdot$cm)
Epidermis	1,000,000
Dermis	500
Fat	2,200
Skeletal muscle	$\rho_x = \rho_y = 200$
	$\rho_z = 1{,}000$
Body tissue	200
Probe electrodes	0.001

6.1.3 FE Modeling of J and E Distributions in the Heart

Previous studies analyzed cardiac safety of CEW devices in animals [14,15]. For example, Fig. 6.3 summarizes a very important finding: CEW electrical discharges, while effective, did not affect the systemic blood pressure. After an average of 26 discharges per animal, all of the 9 subject animals remain hemodynamically stable [14]. Each CEW discharge was 5-second long.

While this finding addresses the safety of The TASER CEW from a hemodynamics perspective, it is also important to understand the likelihood of electrically induced arrhythmias. Figure 6.4 shows typical strength–duration curves for current, charge, and energy [16].

Parameter c represents the chronaxie and equals 0.693 τ, where τ is the myocyte membrane time constant. Based on a literature survey, Sun et al. found that the rheobase current density (i.e., for very long durations – or $d/c > 10$ in Fig. 6.4) required to induce ventricular fibrillation (VF) equals 7 milliamperes per square centimeter [9,17]. As presented in Figs. 6.1 and 6.2, the longest duration of the main phase current is about 100 microseconds for the X26 model. The cardiac

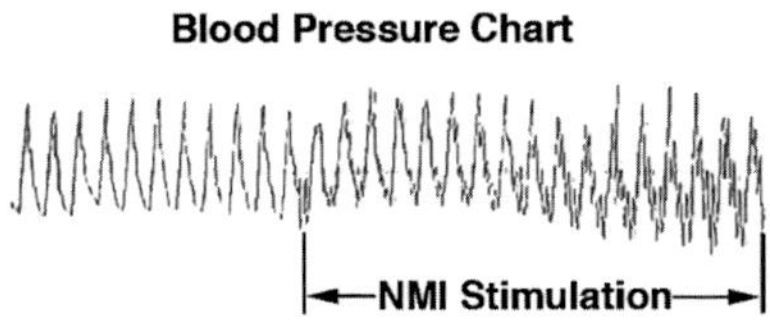

Fig. 6.3 Blood pressure did not change during application of CEW 5 second pulses

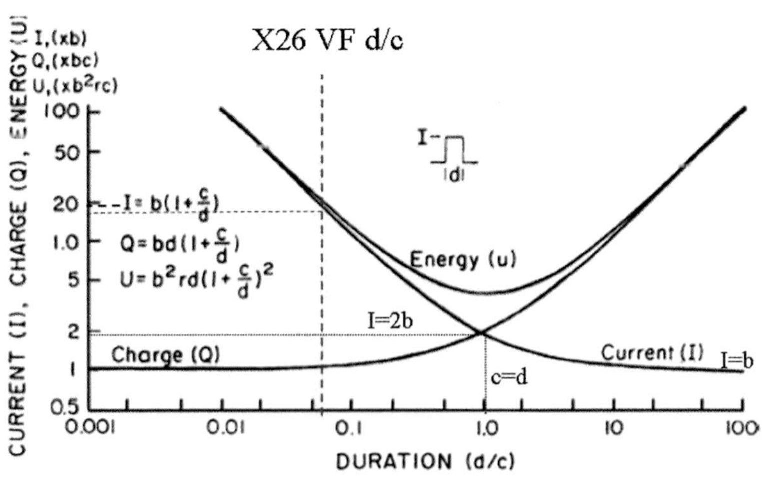

Fig. 6.4 Strength–duration curves for I, Q, and U and the d/c ratio and rheobase multiple required for VF induction using the X26 waveform parameters

myocyte chronaxie used was 1.2 milliseconds for a VF induction model [16,17]. (This is a conservative model as many authorities use a value of 2 milliseconds for far-field stimulation.) Thus, the corresponding d/c value is 0.08 (for a c/d value of 12). Therefore, using Fig. 6.4, the corresponding current density thresholds required to induce VF is 91 milliamperes per square centimeter (i.e., $91 = 7^*(1 + 12)$). As such, for increased cardiac safety, the CEW current density in the heart volume would have to be less than 91 milliamperes per square centimeter.

In order to understand how CEW discharges compared to this VF current density threshold, we used a FEM that approximated whole-body human anatomy. The structure of the FEM, shown in Fig. 6.5, was also in accordance with recent studies of cardiac arrhythmias initiation by CEW currents [18]. The FEM had the following properties:

- Regions:
  - Muscle (neck, shoulder, limbs)
  - Heart
  - Bone (spine, ribcage)
  - Lungs
  - Skin/fat (~20–30 millimeter thick)
  - Abdomen
- Elements: 8,640 hexahedral elements
- Model: human body, about 176-centimeter long
- Electrodes: on body surface at various distances
- Applied voltage: 1,000 volt

The mesh size was nonuniform. Smaller regions of the body, such as the heart, had a finer mesh size. The average size of an element in the heart region was approximately 1.25 centimeter × 1.25 centimeter × 1.25 centimeter. As shown in Table 6.3, tissue resistivities were assigned using values published in previous work [12]. For a worst-case scenario, the probe electrodes are placed at chest locations over the heart, approximately 7 centimeters apart. This approximates the minimal CEW probe separation, as seen in drive-stun mode [5,6,9]. Voltage-type

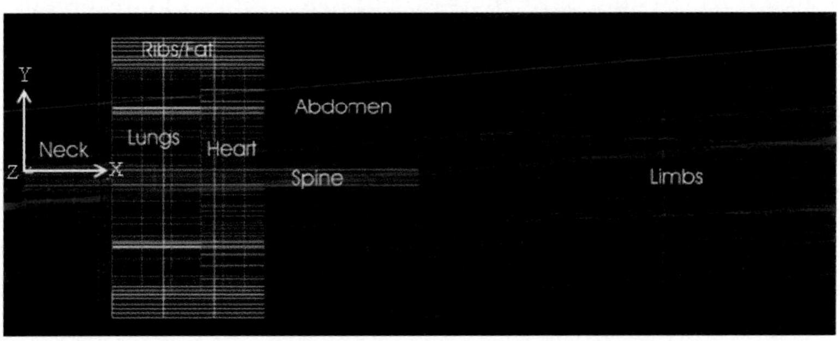

Fig. 6.5 The mesh of the whole-body FEM comprised of seven regions

Table 6.3 Finite element model material properties

Region	Resistivity (Ω·cm)
Skin/fat	2,200
Lungs	1,100
Bone	5,000
Heart	450
Abdomen	200
Skeletal muscle	300
Electrodes	0.001

boundary conditions are set at nodes corresponding to the assumed probe placement. Since Cosmos [13] solved for steady state, rather than for transient solutions, the applied voltage was set at 1,000 volt, even though the average output pulse voltage for X26 CEW is 400 volt. As explained above, this was justified because, given its longer pulse duration, the X26 CEW would have a higher theoretical probability of cardiac stimulation than the TASER M26.

6.1.4 Neuromuscular Stimulation

Figure 6.6 shows the mesh of FEM #1, described in FEM of J and E Distributions for Neuromuscular Capture (the name of the section you are referring to), and the corresponding *J* distribution. The CEW probes were fully penetrated into tissue and located 15 centimeters apart. The magnitude of *J* is listed in ampere per square millimeter.

Figure 6.7 presents a close-up of the current distribution values in layers proximal to the electrode. Note that *J* decreases dramatically 2 millimeters (2 grids) away from the electrode. The transversal *J* seen in the muscle layer was 15.63 milliamperes per square centimeter. In the same FEM #1, if we removed the anisotropy of the skeletal muscle and lowered the resistivity of the fat layer (i.e., thus converting the fat and muscle layers into homogeneous layers) then the transversal *J* increased to 45.49 milliamperes per square centimeter, or about threefold. Table 6.4 summarizes the effects of skeletal muscle anisotropy and of high-resistivity fat. Since these two layers prevent a significant portion of

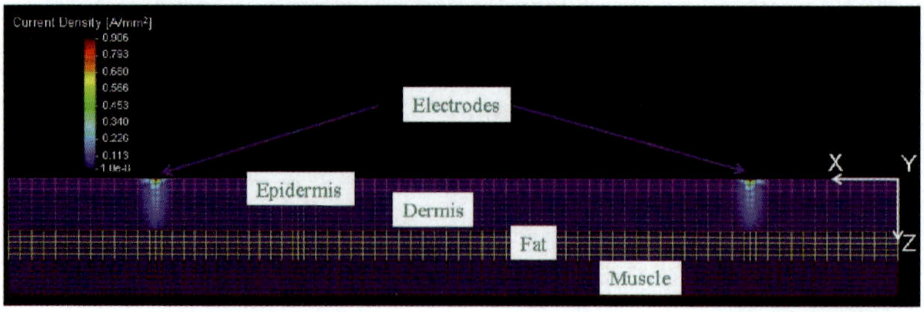

Fig. 6.6 FEM #1 mesh and overall current density distribution

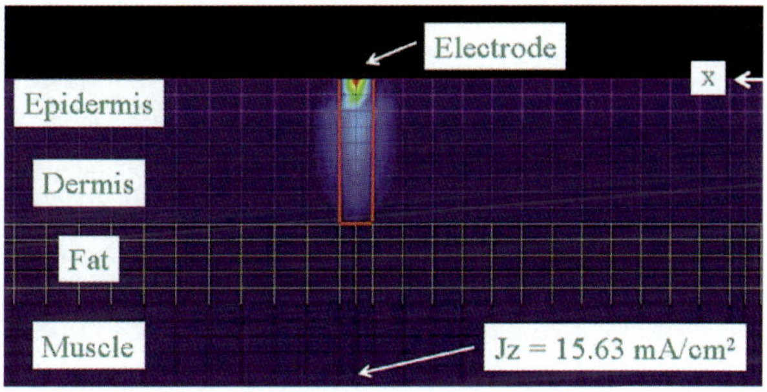

Fig. 6.7 FEM #1 current density distribution around a CEW electrode. Current density color codes are the same as in Fig. 6.7

Table 6.4 Electric shell effect of fat and skeletal muscle

Condition	Transversal J (mA/cm^2)	Ratio of longitudinal J to transversal J	Comments
Thin body with 5 mm fat and anisotropic muscle layers	15.63	8	88% of current is diverted away from deeper tissue layers by fat and longitudinal muscle electrical conduction
Same but with muscle anisotropy removed	20.81	5	Removing muscle anisotropy increases current into deeper tissue layers by 30%
Same but with high-resistivity fat and muscle anisotropy removed	45.49	2.9	Removing fat increases current into deeper tissue layers by 200%

the current from reaching into deeper layers of tissue, they provide the equivalent of an electric shell. Removal of skeletal muscle anisotropy was achieved by assigning $\rho_x = \rho_y = \rho_z = 300$ ohm·centimeter. To remove the high resistivity of fat, $\rho = 500$ ohm·centimeter was assigned to the fat layer. Resistivities of all other FEM regions were assigned values as shown in Table 6.2.

Figure 6.8 shows the distribution of E in a transversal cross-section through the center of a probe, along a direction perpendicular to the electrode–electrode line. The values are expressed in volt per millimeter. Maximum E values of 67.8 volt per millimeter were reached within a volume of approximately 4 cubic millimeter, located about 1 millimeter beneath the electrode. These values were far lower than 1,600 volt per centimeter the E threshold required for irreversible electroporation [10]. In the muscle layer, the E magnitude decreased rapidly with distance toward a minimum of about 30 volt per centimeter. The values of E in the muscle layer were greater by a significant margin than 2.25 volt per centimeter the threshold required to capture the motor neurons responsible for muscle activation.

Given that certain suspects may have a very thin body build, we studied J and E distributions in FEM #2, an "anorexic" model that used an unrealistically

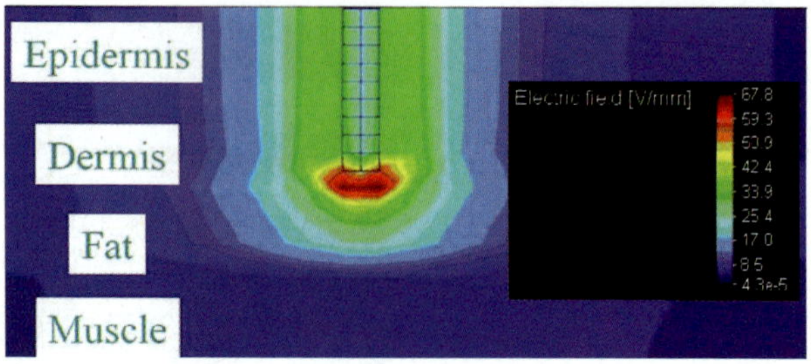

Fig. 6.8 FEM #1 electric field strength distribution in volt per millimeter around a CEW electrode

thin layer of fat. Concerns have been raised about use of TASER CEWs in drive-stun mode in such subjects. Figure 6.9 shows the current density distribution produced by electrodes deployed in drive-stun mode. The current density was significantly attenuated at the point it reached into deep body tissue layers. Table 6.5 summarizes the absolute maximum E and J values reached in various layers of FEM #2. These maxima were not corrected for the artificially high transient increments in the numerical solution that were caused by discontinuity in electrical resistivity at the border between regions.

The results show that the skeletal muscle upper bound E maximum was significantly lower than the threshold for irreversible electroporation. Additionally, even if the deep body tissue included cardiac tissue, the upper bound J maximum in such layer would have been significantly lower than 91 milliamperes per square centimeter, the J threshold required for ventricular fibrillation. As such, if TASER CEW electrodes were deployed in drive-stun mode on a thin-body suspect it would be highly unlikely fibrillation could be triggered.

Figure 6.10 explains how E and J maximum values, as computed based on FEM #2, varied when the distance between CEW electrodes increased. As expected, both E and J decreased with increasing interelectrode distance. In

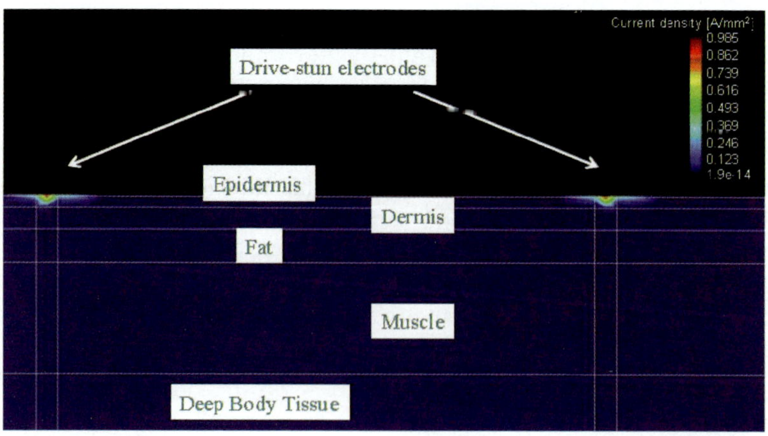

Fig. 6.9 FEM #2 current density distribution with CEW electrodes in drive-stun mode

Table 6.5 Maximum E and J in various layers of FEM #2

Tissue	Max E (V/cm)	Max J (mA/cm^2)
Dermis	1,253	2,506
Fat	1,154	524
Skeletal Muscle	241	255
Deep body tissue	7	37

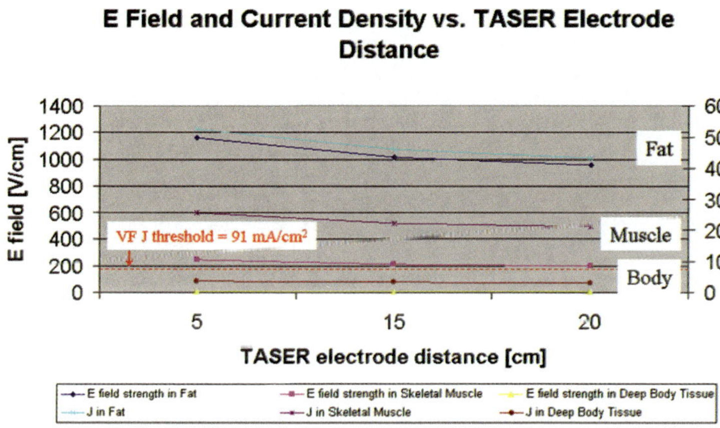

Fig. 6.10 Dependency of E and J with distance between CEW electrodes in FEM #2

the deep body tissue layers, J maximum values were always less than 91 milliamperes per square centimeter, the theoretical threshold required for VF induction. At the same time, regardless of the interelectrode distance, E maximum values were always less than 1,600 volt per centimeter, the threshold required for irreversible electroporation.

FEM #2 was also used to analyze at what distance from the CEW probe tip the estimated tissue current density dropped below the theoretical VF threshold. This distance estimated how far away from the skin the heart would have to be in order to minimize the probability of causing VF during CEW discharges. Considering the probe embedded 5 millimeters into tissue (2/3 through the fat layer), model FEM #2 predicted that J dropped below the threshold for VF at about 3.7 millimeters from the probe tip. This location would be about 8.7 millimeters from the surface of the skin. The model used realistic fat and anisotropic skeletal muscle resistivities, as presented in Table 6.2. Figure 6.11 presents the resulting current density profiles in a cross-section at probe level.

With the CEW probe fully embedded, J dropped below 91 milliamperes per square centimeter, the theoretical VF threshold, at 5.7 millimeters from the tip, or 14.7 millimeters from the skin surface. Figure 6.12 presents the resulting current density profiles, with the probe fully embedded into tissue, in a cross-section at probe level. These distances were shorter than similar distances reported by Sun and Wu et al. [19,20,21]. Their results indicated

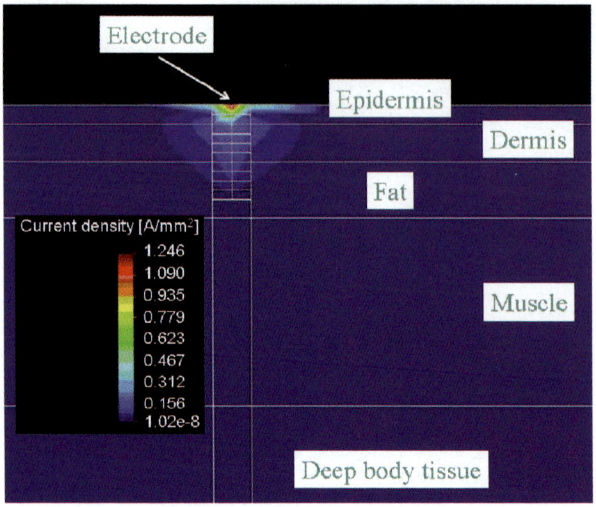

Fig. 6.11 Current density around a CEW probe embedded 5 millimeters into tissue

that tip-to-heart distances of 5.8–17 millimeters (or 14.8–26 millimeters from skin surface, for a 9-millimeter long probe) were required, on average, to induce VF in anesthetized pigs. The removal of fat tissue and the replacement of the anisotropic skeletal muscle layer with an isotropic conductive gel could explain differences with respect to the FE modeling results presented above.

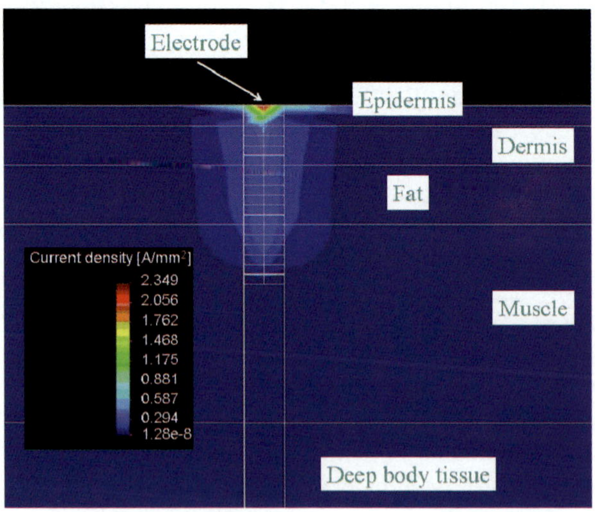

Fig. 6.12 Current density around a CEW probe fully embedded into tissue

For a body mass index (BMI) of 30 kilograms per square meter (the average BMI for in-custody death suspects [22]). Wu et al. [20] reported that the epicardial surface of the heart was located about 30 millimeters under the skin surface, about *twice* the 14.7 millimeter skin-heart distance predicted above. Given that current density drops rapidly with distance from the current source, the twofold distance difference provides additional margin of safety against induction of VF.

6.1.5 Current Density Distributions in the Heart

These simulations used the FEM presented in Fig. 6.5. While the geometry of this FEM was a simplification of the human whole-body anatomy, the model provided useful information about the effects of various layers of tissue on the distribution of CEW currents. The model was also helpful in providing information about the effects of different CEW electrode configurations on current distribution. Figure 6.13 shows the voltage distribution predicted by the FE solver when TASER CEW electrodes were applied in drive-stun mode over the frontal chest area, very close to the heart location. The voltage in the model reached its maximum in the electrode region. Then, it decreased rapidly with distance from electrode. Similarly, Fig. 6.14 displays the current density distribution in a cross-sectional view at the heart level. The current density reached its maximum of about 603 milliamperes per square centimeter in the tissue region beneath electrodes then it decreased rapidly. Underneath skin/fat and muscle layers J was in the range of 20–40 milliamperes per square centimeter.

As such, this model also indicated that the skin, fat and muscle layers significantly attenuated CEW currents before they reached the epicardium. In the heart volume shown in Fig. 6.14 the maximum current density is 2.7 milliamperes per square centimeter, about 34 times lower than the 91 milliamperes per square centimeter VF threshold.

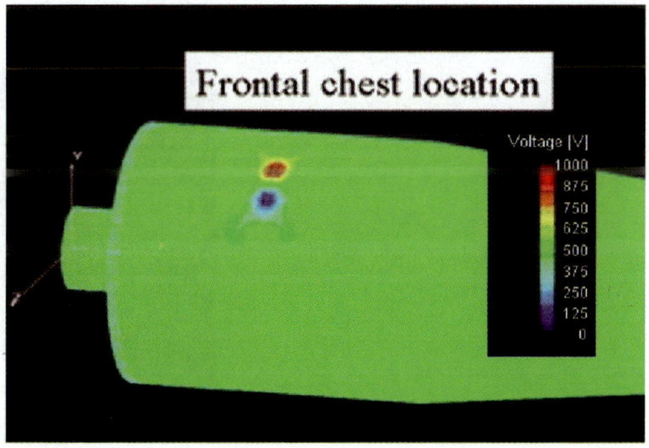

Fig. 6.13 Voltage distribution with CEW electrodes in drive-stun mode over frontal chest, in proximity to the heart volume

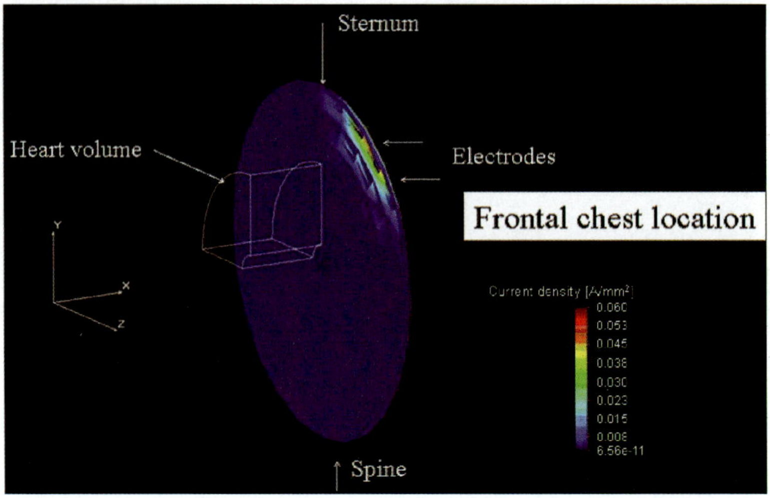

Fig. 6.14 Current density distribution in ampere per square centimeter with CEW electrodes at frontal location

Figure 6.15 shows the simulated voltage distribution when TASER CEW electrodes were applied to the suspect's dorsal area, approximately 20 centimeters apart. As with the previous electrode placement, the voltage in the model reached its maximum in the electrode region. Then, it decreased rapidly with distance from electrode. Figure 6.16 displays the current density distribution in a cross-sectional view at the heart level. The maximum current density in the heart volume was 0.064 milliampere per square centimeter, about 1,421 times lower than the VF threshold.

Table 6.6 summarizes the current density values in the heart volume at three different locations and the respective safety margins with respect to the 91 milliamperes per square centimeter threshold required for induction of ventricular fibrillation.

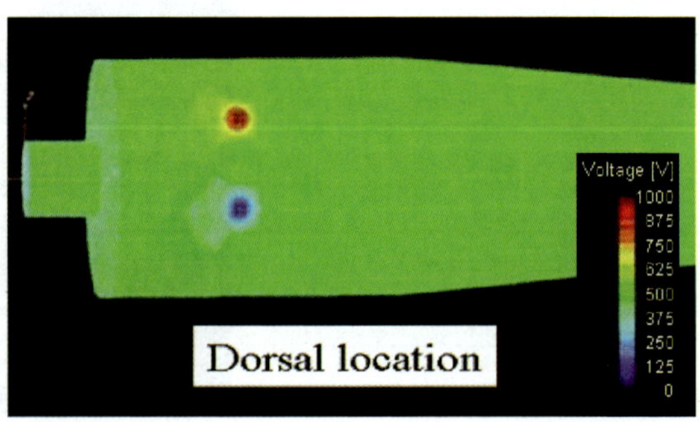

Fig. 6.15 Voltage distribution with CEW electrodes at dorsal location, 20 centimeter apart

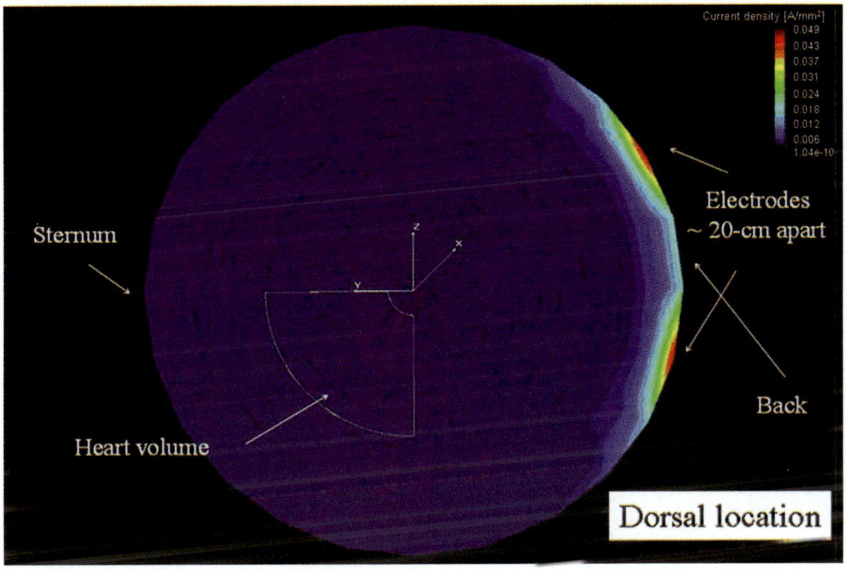

Fig. 6.16 Current density distribution in ampere per square centimeter with CEW electrodes at dorsal location

Table 6.6 Current density in the heart and safety margins, as predicted by FEM

CEW electrode separation and position	Maximum current density in the heart (mA/cm^2)	Safety margin with respect to VF threshold
20 cm – over dorsal area	0.064	1,421 times
20 cm – left nipple to left thigh	0.24	379 times
7 cm – frontal chest, straight over heart	2.7	34 times

6.1.6 Implantable Pacemakers

The question of whether pulses from TASER CEWs can capture or pace the heart is also an interesting one. The model discussed in Fig. 6.5 may be used to address this question. The cardiac myocyte excitation threshold is reported to be between 2 and 5 volt per centimeter [23–28]. Given the assumed heart tissue resistivity of 450 ohm·centimeter, for the worst-case current density shown in Table 6.6, the resulting electric field strength is 1.215 volt per centimeter. This strength constitutes a margin of 2.9 times with respect to the average excitation threshold of 3.5 volt per centimeter cited above. Table 6.7 shows safety margin for cardiac capture of pacing. Consequently, TASER CEW currents are not likely to capture the heart, although the capture margin can vary significantly with the application vector.

Table 6.7 Safety margins for cardiac capture or pacing

CEW electrode separation and position	Maximum current density in the heart (mA/cm^2)	Cardiac capture safety margins
20 cm – over dorsal area	0.064	120 times
20 cm – left nipple to left thigh	0.24	31 times
20–25 cm – left axilla to back of neck	0.89	8.7 times
7 cm – frontal chest, straight over heart	2.7	2.9 times

Some suspects may wear cardiac implantable devices, such as pacemakers or implantable defibrillators. Per the EN 60601-1 [29], an international standard that stipulates general safety requirements for electrical medical devices, implantable devices are required to have a high-voltage protection network. The protection network protects the devices from high-voltage external fields, such as those that may be encountered during defibrillation or radiofrequency ablation. Given its magnitude, it is possible that voltages from TASER CEW probes near the pacemaker may actuate the protection network of such implanted devices. A typical implanted defibrillator uses a protection network that consists of high-voltage switches connected in series with the electrodes [28]. These switches become open circuit in the presence of high voltages. Consequently, the presence of an implantable defibrillator is not expected to alter the distribution of TASER CEWs electric fields. Similarly, TASER CEWs are not expected to have any damaging effects on implanted defibrillators.

A typical implanted pacemaker uses a protection network such as that shown in Fig. 6.17 [28]. When the voltage across the back–back diodes exceeds about 12 volt, the diodes effectively clip the voltage between pacing electrodes right ventricular (RV) tip or ring and the can of the device [28]. Although TASER CEWs are not expected to damage the circuitry inside implanted pacemakers, it is important to understand whether the clipping effect of the protection network has any effects on the distribution of TASER CEWs electric fields. Figure 6.18 shows a FEM where one of the TASER CEW probes was placed close to the location where pacemakers are implanted. The location of a hypothetical pacemaker can is marked in Fig. 6.18. The other TASER probe

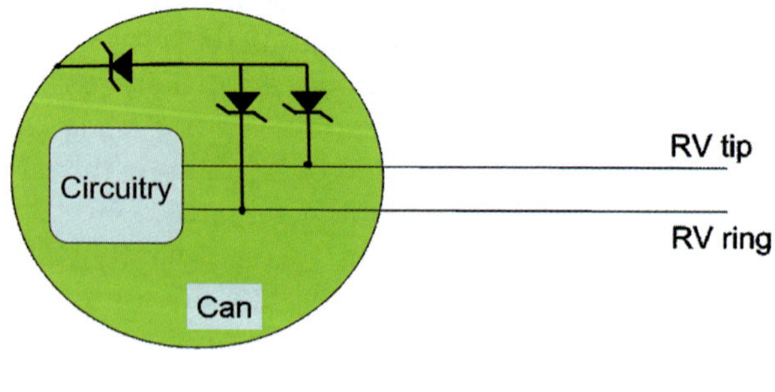

Fig. 6.17 Illustration of a typical implanted pacemaker protection network

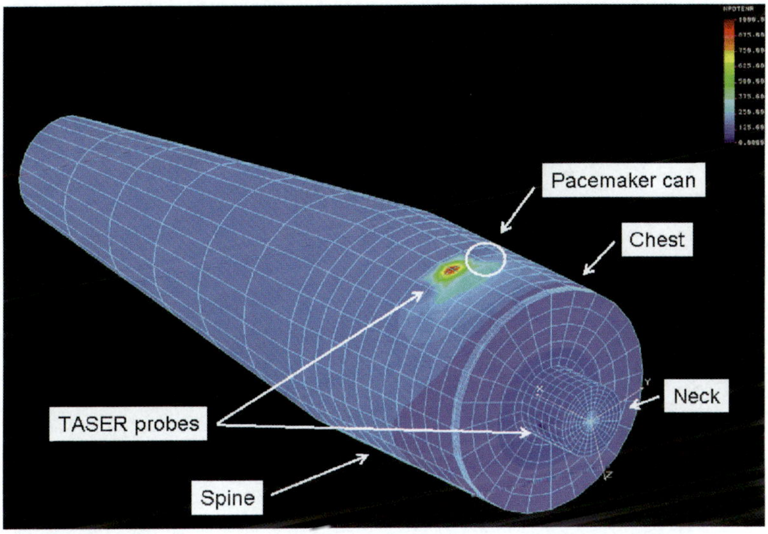

Fig. 6.18 Voltage distribution with TASER CEW probes on the *left* lateral chest and on the neck. The voltage drops rapidly with distance. The location of hypothetical pacemaker can is *marked*

was located on the back of the model's neck. The voltage boundary condition was maintained at 1,000 volt (i.e., approximately the TASER X26 peak output voltage), same as in the other models presented above. With no pacemaker assumed inside the model, the maximum electric field strength reached in the heart volume was 0.4 volt per centimeter, producing a safety margin of 8.7 with respect to the average excitation threshold of 3.5 volt per centimeter. Figure 6.19 shows the E field distribution through a cross-sectional view at the heart level. As with other models, the heart volume is shown in a white outline. Within the heart volume, the maximum electric field strength was reached at the element drawn in dark red. Then, a 12 volt difference was assumed across the regions of the FEM that correspond to the location of an implanted pacemaker. Given that the voltage at the hypothetical location of the can (shown in Fig. 6.18) was about 500 volt, or about half the applied voltage, the boundary voltage condition at the apex of the heart was forced to 488 volt. The heart apex is the typical location of the RV tip electrode. This forced voltage conditions increased the maximum electric field strength in heart to 2.16 volt per centimeter, slightly higher than very low end of reported cardiac excitability thresholds. Figure 6.20 shows the E field in a cross-sectional view. The increase in E field is notable at the apex of the heart volume, as indicated by the lighter blue colors. With respect to the average cardiac myocyte excitation threshold of 3.5 volt per centimeter, the presence of a pacemaker protection network reduces the capture safety margin from 8.7 to 1.6. As presented above, these margins vary with the TASER CEW application vector.

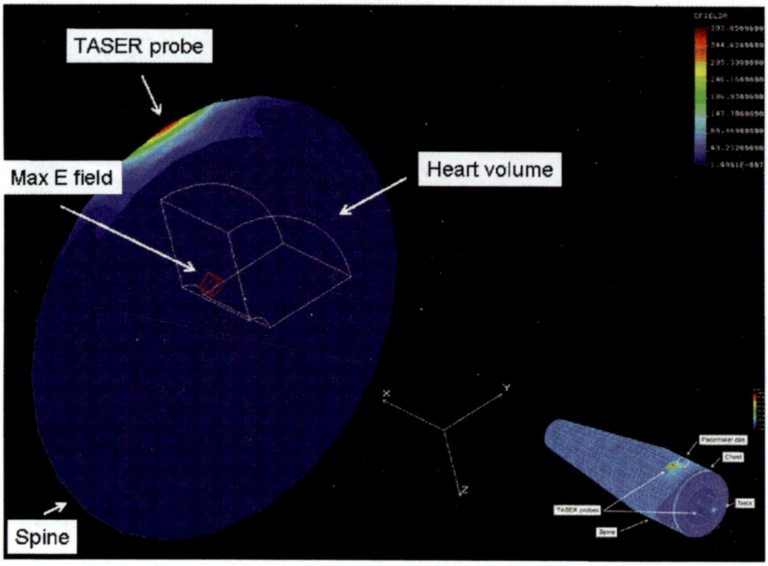

Fig. 6.19 Cross-sectional *E* distribution (V/cm). For orientation purposes, the whole-body model of Fig. 6.18 is shown in the inset. The max *E* in the heart is 0.4 volt per centimeter

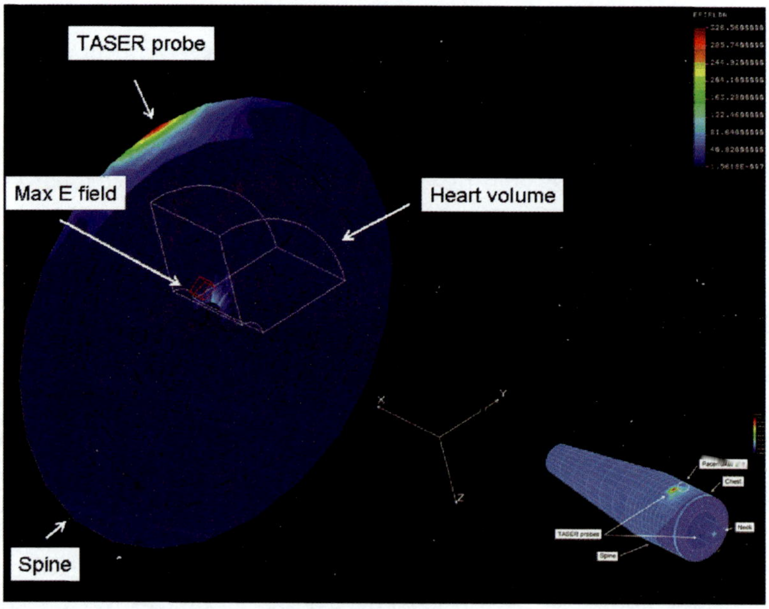

Fig. 6.20 Cross-sectional *E* distribution (V/cm) in the presence of an implanted pacemaker. For orientation purposes, the whole-body model of Fig. 6.18 is shown in the inset. The *light blue color* at the heart apex shows an increased max *E* of 2.16 volt per centimeter

6.2 Discussion

Research conducted on defibrillation and pacing devices [12,30,31] showed that – even under optimal electrode placement configurations – only a low fraction of the current that entered the human thorax reached the heart. For example, studies found that more than 66% of the input voltage dropped across portions of the thorax within 4 centimeters under cardiac electrodes that were optimally placed [12,30]. For same optimal electrode cardiac placement, less than 10% of the input voltage dropped across the left ventricle [31]. The high resistances of skin, fat, and thoracic cage reduced the electric field across the heart. Consequently, the current density at the heart level was significantly reduced with respect to values measured at electrode levels. Similarly, Deale and Lerman studied the ratio of transcardiac to transthoracic threshold currents in dogs [32]. They reported that the thoracic cage shunted 82% of the input current and that the lungs shunted 14%. Only the remaining 4% of the input current passed through the heart.

Consistent with research cited above, our FE analyses – conducted for worst-case scenarios (i.e., fat and skeletal muscle layers with unrealistically thin dimensions) – showed that the skin, fat, and anisotropic skeletal muscle layers attenuated and diverted a large portion of CEW currents into longitudinal directions, allowing just a fraction to penetrate transversally into deeper layers of tissue. However, the FEMs predicted that CEW J and E values for the muscle region were efficacious because they were higher – by a significant margin – than thresholds required for neuromuscular activation. The maximum values for J and E were safe because they were lower, by at least a factor of 7, than levels reported to produce permanent cellular electroporation or tissue damage. The maximum CEW current density in the heart, 2.7 milliamperes per square centimeter, was about 34 times lower than 91 milliamperes per square centimeter – the threshold required to induce VF. Therefore, we conclude that TASER CEWs are efficient and safe in producing neuromuscular activation for temporary suspect incapacitation. Showing a substantial cardiac safety margin, even under worst-case electrode placement, CEW devices are not likely to generate currents in the heart that are high enough to trigger VF.

It may be useful to place the theoretical risk of electrically induced VF associated with TASER CEWs in the context of risk levels accepted by medical device safety standards, such as the EN 60601-1. TASER International Inc. reported that its devices were used in more than 700,000 human volunteers and over 600,000 human suspects during actual law enforcement field deployments for a total of over 1.3 million [33]. In all of these situations, no scientific or medical evidence has been provided that TASER CEWs have ever induced VF. As such, the theoretical overall risk of VF induction with The TASER CEW is estimated at less than 1/(1.3 million) = 0.000 000 7. Sun et al. estimated the probability to induce VF during CEW discharges to be in a higher range of about 0.000 006 1 [9,19]. The EN 60601-1 international standard stipulates accepted regulatory requirements for the safety of electrical medical devices [29]. This standard sets the allowed threshold for patient leakage currents generated by medical devices that have direct contact with patients' heart. Citing from the standard [29]:

> The allowable value of patient leakage current for type of applied parts in normal condition is 10 μA_{rms} which has a probability of 0.002 for causing ventricular fibrillation or pump failure when applied through small areas to an intracardiac site. Even with zero current, it has been observed that mechanical irritation can produce ventricular fibrillation. A limit of 10 μA_{rms} is readily achievable and does not significantly increase the risk of ventricular fibrillation during intracardiac procedures.

This implies that under normal device operation, the allowed maximum patient leakage current is 10 μA_{rms} *to the inside of the heart concentrated at the small electrode tips of a pacemaker lead.* Although this patient leakage current was thought to have a 0.002 probability of causing VF or pump failure in humans, the standard accepts this probability as being within reasonable expectations for safety. Regulatory bodies, such as the US Food and Drug Administration (FDA) or Germany-based TUV, certify electrical medical devices as being safe for use in intracardiac clinical procedures if they comply with the patient leakage current limit above. Therefore, by accepting requirements of EN60601-1 regulatory bodies, including the US FDA, accept that a probability of causing VF of 0.002 represents an extremely low risk. This risk probability of 0.002 is significantly higher than the demonstrated and modeled risk of VF from a TASER CEW.

References

1. Non-lethal Force, Wikipedia. Available at: http:// en.wikipedia.org/wiki/Non-lethal_weapon
2. Council of Foreign Relations, *Less-lethal Weapons and Capabilities*, New York, NY, 2004. Available at: http://www.cfr.org
3. TASER *Technology Summary*. Available at http://www.taser.com/facts/qa.htm.
4. Smith PW, Hand-held stun gun for incapacitating a human target, US Patent 6,636,412, October 21, 2003.
5. TASER International, *M26E Series Electronic Control Device Specification*. 2006.
6. TASER International, *X26E Series Electronic Control Device Specification*. 2006.
7. Reilly JP, Freeman VT, and Larkin WD. Sensory effects of transient electrical stimulation: Evaluation with a neuroelectric model. *IEEE Trans Biomed Eng* 1985; 32(12); 1001–1011.
8. Reilly JP. Applied bioelectricity: from electrical stimulation to electropathology. New York: Springer, 1998.
9. Sun H and Webster JG. Estimating neuromuscular stimulation within the human torso with Taser® stimulus. *Phys Med Biol* 2007; 52; 6401–6411.
10. Gehl J, Sorensen TH, Nielsen K, Raskmark P, Nielsen SL, Skovsgaard T, and Mir LM. In vivo electroporation of skeletal muscle: threshold, efficacy and relation to electric field distribution. *BBA-General Subjects* 1999; 1428(2–3); 233–240.

11. Panescu D, Webster JG, and Stratbucker RA. A nonlinear finite element model of the electrode-electrolyte-skin system. *IEEE Trans Biomed Eng* 1994; 41(7); 681–687.
12. Panescu D, Webster JG, Tompkins WJ, and Stratbucker RA. Optimization of cardiac defibrillation by three-dimensional finite element modeling of the human thorax. *IEEE Trans Biomed Eng* 1995; 42(2); 185–192.
13. Structural Research & Analysis Corporation (SRAC), division of SolidWorks Corporation, COSMOS/M: http:// www.cosmosm.com/pages/products/cosmosm.html.
14. McDaniel W, Stratbucker RA, Nerheim M, and Brewer JE. Cardiac safety of neuromuscular incapacitating defensive devices. *PACE* 2004; 28; S1–S4.
15. McDaniel W, Stratbucker RA, and Smith RW. Surface application of TASER stun guns does not cause ventricular fibrillation in canines. *Proc IEEE-EMBS Ann Intl Conf* 2000.
16. Geddes LA and Baker LE. Principles of applied biomedical instrumentation, 3rd ed. New York: John Wiley & Sons, 1989.
17. Sun H, Wu JY, Abdallah R, and Webster JG. Electromuscular incapacitating device safety. *Proc IFMBE, 3rd EMBE Conference, Prague* 2005; 11(1).
18. Lakkireddy D, Wallick D, Ryschon K, Chung MK, Butany J, Martın D, Salıba W, Kowalewski W, Natale A, and Tchou PJ. Effects of cocaine intoxication on the threshold for stun gun induction of ventricular fibrillation. *J Am Col Cardiol* 2006; 48; 805–811.
19. Sun H. Models of ventricular fibrillation probability and neuromuscular stimulation after Taser® use in humans. PhD thesis: University of Wisconsin, 2007. Available online: http:// ecow.engr.wisc.edu/cgi-bin/get/ece/762/webster/.
20. Wu J-Y, Sun H, O'Rourke A, Huebner S, Rahko PS, Will JA, and Webster JG. TASER dart-to-heart distance that causes ventricular fibrillation in pigs. *IEEE Trans Biomed Eng* 2007; 54; 503–508.
21. Wu J-Y, Sun H, O'Rourke A, Huebner S, Rahko PS, Will JA, and Webster JG. TASER blunt dart-to-heart distance causing ventricular fibrillation in pigs. *IEEE Trans Biomed Eng* 2007; in press.
22. Stratton SJ, Rogers C, Brickett K, and Gruzinski G. Factors associated with sudden death of individuals requiring restraint from excited delirium. *Am J Emerg Med* 2001; 19; 187–191.
23. Walcott GP, Walker RG, Cates AW, et al. Choosing the optimal monophasic and biphasic waveforms for ventricular defibrillation. *J Cardiovasc Electrophysiol* 1995; 6; 737–750.
24. Ideker RE and Dosdall DJ. Can the direct cardiac effects of the electric pulses generated by the TASER X26 cause immediate or delayed sudden cardiac arrest in normal adults? *Am J Forensic Med Pathol* 2007; 28; 195–201.
25. Knisley SB, Smith WM, and Ideker RE. Effect of field stimulation on cellular repolarization in rabbit myocardium. Implications for reentry induction. *Circ Res* 1992; 70(4); 707–715.
26. Knisley SB, Smith WM, and Ideker RE. Prolongation and shortening of action potentials by electrical shocks in frog ventricular muscle. *Am J Physiol* 1994; 266(6 Pt 2); H2348–2358.
27. Bien H, Yin L, and Entcheva E. Calcium instabilities in mammalian cardiomyocyte networks. *Biophys J* Jan 6, 2006.
28. Zarlink Semiconductor, *Medical Surge Protection: Technical Documentation*, http:// www.zarlink.com/cps/rde/xchg/zarlink/hs/products.htm.
29. BSI British Standards. BS EN 60601-1:2006 Medical electrical equipment. General requirements for basic safety and essential performance. 2006.
30. Panescu D, Webster JG, Tompkins WJ, and Stratbucker RA. Optimization of transcutaneous cardiac pacing by three-dimensional finite element modeling of the human thorax. *Med Biol Eng Comput* 1995; 33(6); 769–775.

31. Panescu D, Webster JG, and Stratbucker RA. Modeling current density distribution during transcutaneous cardiac pacing. *IEEE Trans Biomed Eng* 1994; 41(6); 549–555.
32. Deale OC and Lerman BB. Intrathoracic current flow during transthoracic defibrillation in dogs. *Circ Res* 1990; 67(6); 1405–1419.
33. TASER International, "Facts." Available at: http:// www.taser.com/facts/index.htm.

Chapter 7
Animal Studies

John G. Webster

Because of the potential danger of ventricular fibrillation (VF), numerous TASER conducted electrical weapon (CEW) studies have been performed on animals. Holden et al. [1] injected M26 and X26 TASER waveforms to an electrode on the ventricular epicardial surface of guinea pig isolated hearts, but were unable to induce VF. However, it is known that inducing sustained VF in very small animals (such as guinea pigs) is difficult and impossible for some stimulus paradigms, justifying the need for considering data from larger animals, such as the live pig, more similar to humans, despite the obvious increases in experimental complexity. In addition, pigs are inexpensive and do not have the animal rights concerns attached to them that pet animals such as dogs do. However, anesthetized pigs may have different susceptibility to VF than conscious pigs. Pig studies have focused on whether or not TASER CEW darts on the back or near the heart on the front can cause VF. Early human studies were performed with TASER darts far from the heart, such as on the back.

VF is a cardiac rhythm disturbance that is fatal without treatment. VF is a very fast, chaotic heart rate in the lower chambers of the heart. VF can occur spontaneously generally caused by heart disease or after receiving external stimulation such as an electric shock of sufficient intensity. The instant VF begins, effective blood pumping stops, leading to a dramatically decreased blood pressure, lack of oxygen in the brain tissue, rapid collapse of the subject, and—in the absence of defibrillation—death within minutes.

Several reports of TASER CEW use by law enforcement personnel have involved repeated shots to a single individual in a short period of time. Repeated exposure to CEW, such as the TASER, could result in repeated, sustained muscle contraction, with little or no muscle recovery period. Therefore, rhabdomyolysis and other physiological responses, including acidosis and hyperkalemia would be areas of concern.

J.G. Webster (✉)
Department of Biomedical Engineering, University of Wisconsin
e-mail: webster@wisc.edu

M.W. Kroll, J.D. Ho (eds.), *TASER® Conducted Electrical Weapons: Physiology, Pathology, and Law*, DOI 10.1007/978-0-387-85475-5_7,

7.1 Stun Guns Applied to Exposed Pig Hearts

Roy and Podorsky [2] tested five types of stun guns on two anesthetized pigs weighing 40 and 52 kilograms. The stun gun with the highest output yielded a damped sinusoidal output. The first half sine wave had a peak current of 9 amperes and a duration of 9 microseconds to yield a charge of 51 microcoulombs. This study showed that stun guns did not cause VF from the skin surface of normal pigs. They had to expose the heart and apply the stun gun electrodes directly to the heart in order to stimulate VF. They also implanted a pacemaker lead screwed into the outside of the heart connected to a pacemaker. Under these conditions with a pacemaker lead carrying the current directly to the heart, VF could be stimulated from the skin surface.

7.2 High Multiples of TASER CEW Charge Applied to Skin of Pigs

McDaniel et al. [3] asked the question, "Can TASER X26 darts inserted into the skin of pigs cause VF?" They found that they could not cause VF, so built a multiple charge research unit that increased the output to an intensity that could cause VF. Figure 7.1 shows that the research unit could deliver a charge up to 48 times the normal charge. The animals were anesthetized with isoflurane and their arterial blood pressure, oxygen saturation, respiration, and heart rate were continuously monitored until sacrifice.

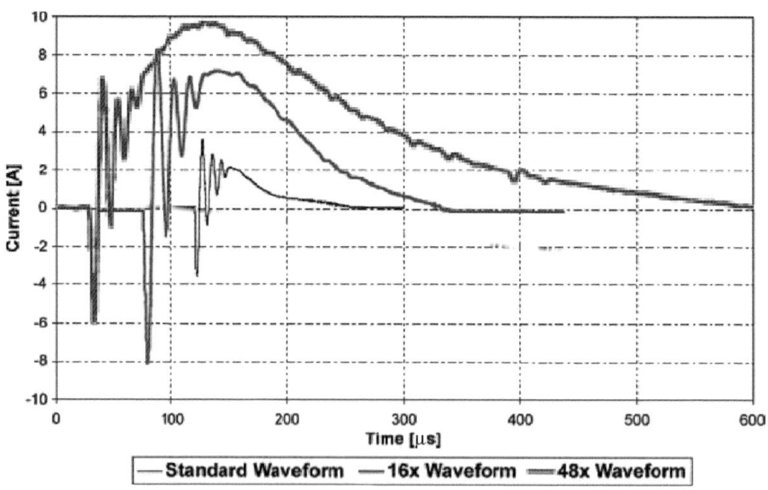

Fig. 7.1 Experimental TASER X26 waveforms. The standard waveform, 16 times standard discharge, and 48 times standard discharge are in the highest, middle, and the lowest one, respectively (McDaniel et al. [3])

The research unit allowed the output capacitance to vary as a multiple of the nominal capacitance (and charge) for a standard TASER X26 (0.008 microfarad). All experimental TASER X26 discharges were delivered with a fixed voltage of 6,000 volts. The waveform, as a short electrical pulse, was delivered at a repetition rate of 19 pulses per second for 5 seconds. The standard TASER X26 stored charge for the experiment control was (0.008 microfarad $\times$ 6,000 volts) $=$ 48 microcoulombs. The standard TASER X26 discharge represented the same amount of charge (coulombs) delivered by fielded TASER X26 devices. The pulses were discharged across the thorax of the animal, using metallic barbs that matched darts deployed in fielded TASER X26s. One pulse delivery probe was placed at the sternal notch and another on the anterolateral thorax at the point of maximum impulse.

TASER X26 discharges were applied in an up-down method to determine a threshold for VF induction, beginning with a standard TASER X26 discharge. Increasing stored charges were applied to the animal until VF was induced. The stored charge was increased in steps by increasing the size of the output capacitors. Each stepped stored charge had a capacitor value equal to a multiple of the standard capacitance unit (0.008 microfarad), using an increasing number of charge multiples (2 and multiples of 4 from 4 to 48). Following the first VF induction, a decreasing series of capacitance-stepped discharges were then applied until VF was no longer induced by five discharges of equal stored charge. The animals were defibrillated with an automatic external defibrillator. A recovery period of at least 90 seconds was allowed after discharges that did not induce VF. If a discharge did induce VF, a recovery period of at least 5 minutes was allowed following defibrillation.

Minimum fibrillating discharge level determined by the VF threshold procedure was defined as the lowest discharge that induced VF at least once; maximum safe level was defined as the highest discharge which could be applied five times without induction of VF; VF threshold was defined as their average. The safety index was defined as the ratio of the VF threshold to the standard TASER X26 discharge (48 microcoulombs).

Nine experiments were completed. The average weight of the pigs was 60 $\pm$ 28 kilograms, ranging from 30 to 117 kilograms. The safety index for stored charge ranged from 15:1 to 42:1 as weight increased from 30 to 117 kilograms ($p < 0.001$, Fig. 7.2). The VF induction threshold level (1,339 $\pm$ 463 microcoulombs stored charge) was significantly higher than the standard level for applied charge (48 microcoulombs stored charge, $p < 0.0001$). The charge multiple at the VF induction threshold was 28 $\pm$ 10 compared to the standard charge multiple of 1 ($p < 0.0001$). The maximum safe charge multiple was 26 $\pm$ 9 with an average stored charge of 1,227 $\pm$ 423 microcoulombs, and the minimum VF inducing charge multiple was 30 $\pm$ 11 with an average stored charge of 1,451 $\pm$ 509 microcoulombs.

The maximum safe levels and minimum ventricular fibrillation induction (VFI) levels of stored charge for experimental data were regressed linearly for significant trends. The relationship between stored charges as a function of

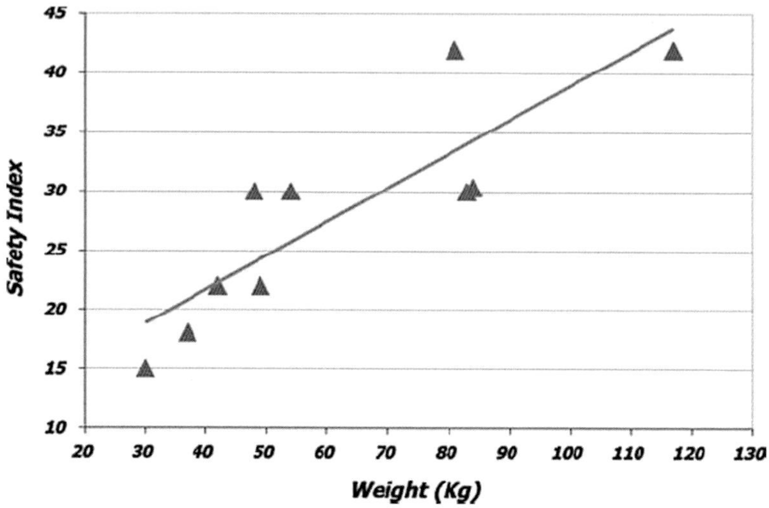

Fig. 7.2 TASER X26 safety index increases with weight (kilogram) (McDaniel et al. [3])

weight was compared to experimental stored charge for minimum VFI discharge. The maximum safe discharge was modeled by 12.5 × (weight (kilograms)) + 473 ($n = 9$, $r^2 = 0.69$) and the minimum VF induction discharge was modeled by 16.5 × (weight (kilograms)) + 460 ($n = 9$, $r^2 = 0.82$). The analysis revealed a linear, increasing relationship of maximum safe and minimum VFI discharge multiples (and therefore safety index) as a function of weight. The relationship further confirmed a significantly greater discharge required to induce VF compared to standard discharge levels for a fielded TASER X26 device.

A TASER X26 discharge that could induce VF required 15–42 times the charge of the standard TASER X26 discharge. Furthermore, this study demonstrated a safety index strongly correlated with increasing weight. In addition, the observation of the hemodynamic stability of the animals suggests that these devices may be safely applied multiple times if needed. Discharge levels output by fielded TASER X26 devices have an extremely low probability of inducing VF.

7.3 Cocaine Effect on VF Threshold

Lakkireddy et al. [4] studied the standard discharge of TASER X26 before and after cocaine infusion. Five adult pigs (4 male, 1 female) with a length of 103 ± 8 centimeters, a weight of 34 ± 8.7 kilograms, and a chest circumference of 65.8 ± 4 centimeters were studied. The animals were sedated with intramuscular ketamine (12 milligrams per kilogram) and

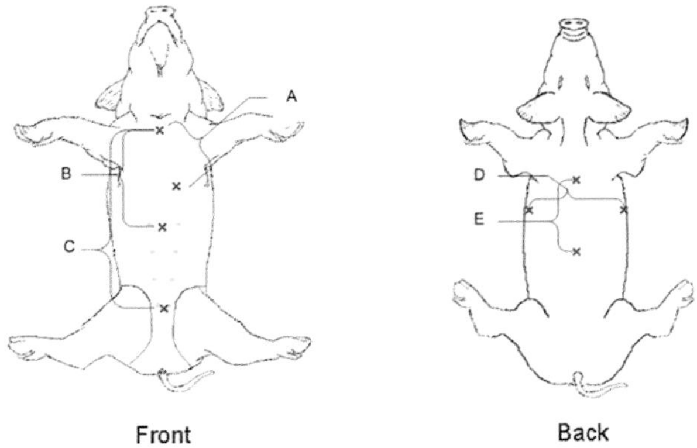

Fig. 7.3 Dart positions on the front and back of the pig. Position A = sternal notch (SN) to point of maximum cardiac impulse (PMI); Position B = SN to supraumbilical region; Position C = SN to infraumbilical region; Position D = side-to-side across the chest; Position E = upper to lower midline posterior torso (Lakkireddy et al. [4])

intubated. Anesthesia was maintained with 1–2% isoflurane mixed with oxygen and nitrous oxide.

Human field experience has shown that the posterior and anterior upper trunk regions were the most common dart attachment sites [5]. Figure 7.3 identifies the five different paired-dart positions tested on the pig body, labeled Positions A through E. Because they hypothesized that current application nearest the heart and along its axis would be the most arrhythmogenic, they tested Position A at the beginnings of the two series. The point of maximum impulse (PMI), typically located slightly left of the xiphoid process, was palpated and confirmed with auscultation and echocardiography. The sequence of testing the remaining four sites was randomized. Two darts were inserted to full depth at the mentioned sites. The mean distance of the PMI dart tip from the epicardial surface measured by echocardiography was 18 ± 4 millimeters.

Standard TASER X26 discharge is a 5-second application. Testing was started with a standard discharge (×1) followed by multiple charge research unit discharges of increasing stored charge in a step-up fashion until VF was induced. The stored charge was increased for each step by multiples of the standard capacitor (×5, ×10, and multiples of ×10 up to ×100). After the first VF induction, the capacitances were decreased in reversed sequence with the addition of ×7 and ×2 when needed until three sequential discharges of equal stored charge did not induce VF.

Minimum VF-inducing multiple (minVFIM) was defined as the lowest TASER X26 discharge multiple that induced VF at least once in three tries. Maximum safe multiple (maxSM) was defined as the highest discharge multiple that could be applied three times without VF induction. Ventricular fibrillation threshold (VFT) was defined as the average of these two values. The TASER X26 discharge multiples at which 2:1 and 3:1 ventricular captures were seen are reported here. These two ratios were chosen because 3:1 capture was the highest capture frequency that did not induce VF, whereas 2:1 capture always induced VF.

All continuous variables were summarized by their means and standard deviations. The effect of cocaine on maxSM, minVFIM, and VFT was tested using the paired *t*-test. A general linear model for repeated measures with a difference contrast was used to compare the TASER and cocaine effects on hemodynamic and metabolic data, cardiac markers, and ECG data. The Bonferroni adjustment was used to correct for between- and within-subjects factors. A level of $p < 0.05$ was considered statistically significant.

Table 7.1 shows the maxSM, minVFIM, and VFT data. The lowest mean maxSM, minVFIM, and VFT were seen at Position A, whereas the highest were seen at Position E. These variables increased 1.5- to 2-fold after cocaine infusion at all positions. The increases were statistically significant in four of the five positions.

Ventricular capture increased with progressive increase in current application strength and ranged from no capture to ≤2:1. The VF was consistently inducible whenever the ventricular capture ratio was ≤2:1. No VF induction was noted when the ventricular capture was ≥3:1. A greater degree of ventricular capture at lower strengths was seen at Position A than at other locations. This correlated with their finding that VF was induced with the lowest minVFIM at this location. Application strength multiples of ×40 and higher were needed on the back (Position E) to accomplish similar ventricular capture ratios. Standard TASER X26 discharge at Position A did not induce VF in any animal despite ventricular capture ratios ranging from 6:1 through 3:1, nor was VF induced with standard ×1 TASER X26 application at any of the other four locations with or without cocaine. Cocaine increased the required strength of TASER X26 discharge that caused a 2:1 or 3:1 ventricular capture ratio at all positions.

There was no significant change in the ECG variables before or after cocaine infusion. No ST-segment or T-wave changes suggestive of myocardial ischemia were seen. There were no significant changes in blood pressure, electrolytes, arterial pH, or blood gases throughout the experiment.

These data suggest that the presence of cocaine decreases the likelihood of TASER X26-induced VF. Cocaine increased the safety margin approximately 1.5–2 times from baseline. The study also showed less myocardial capture after cocaine infusion. This observation is consistent with the hypothesis that rapid capture is the mechanism of VF induction. It is also consistent with the finding that cocaine increases the VF threshold for 4-millisecond stimulation in dogs [6]. The sodium channel blocking effects of cocaine along with its ability to create a hypersympathetic state have been postulated as potential mechanisms behind its arrhythmogenicity [7,8]. However, it is not clear whether these properties in the absence of an appropriate substrate would increase vulnerability to VF. These data also show that TASER darts over the heart have a lower safely margin than darts farther away from the heart.

Table 7.1 Thresholds for ventricular fibrillation induction at different positions before and after cocaine infusion (Lakkireddy et al. [4])

Location	B-maxSM	C-maxSM	p-Value	B-minVFIM	C-minVFIM	p-Value	B-VFT	C-VFT	p-Value
Position A	4.2 ± 1.10	8.6 ± 6.88	0.192	8.0 ± 2.74	15.0 ± 10.00	0.135	6.1 ± 1.92	11.3 ± 8.79	0.260
Position B	12.0 ± 7.58	28.0 ± 4.47	0.030	20.0 ± 10.0	38.0 ± 4.47	0.037	14.5 ± 9.59	33.0 ± 4.47	0.032
Position C	22.0 ± 8.37	50.0 ± 18.71	0.009	32.0 ± 8.37	60.0 ± 18.71	0.009	27.0 ± 8.37	55.0 ± 18.71	0.009
Position D	30.0 ± 7.07	48.0 ± 17.89	0.070	40.0 ± 7.07	58.0 ± 17.89	0.070	35.0 ± 7.07	53.0 ± 17.89	0.070
Position E	38.0 ± 4.47	60.0 ± 14.14	0.011	48.0 ± 4.47	70.0 ± 14.14	0.011	43.0 ± 4.47	65.0 ± 14.14	0.011

7.4 Acute Epinephrine Effects

Nanthakumar et al. [9] studied the effect of another drug, epinephrine, on TASER exposure. Farm pigs weighing 45–55 kilograms were used in this study. They tested two different vectors of discharges (chest: darts placed across the chest/heart, and abdomen: across the abdomen). They also tested TASER X26 and TASER M26 with 5 and 15-second discharges. The guns were initially fired into a cardboard box; the barbs were removed and inserted subcutaneously in two different configurations on the pigs. The skin was pulled up, and 1 centimeter of the barb was inserted parallel to the plane of the skin.

The chest configuration was designed to achieve the largest cardiac potential gradient of stimulation; one of the darts was placed in the right parasternal region, 5 centimeters away from the midline, and another dart was placed in the left lateral border of the thorax. The interdart distance was 26–30 centimeters. This is the average distance between the darts when fired at an object from 2.3 meters away. In the abdomen configuration, the darts were attached below the lowermost ribs on the right and left, such that the vector was oriented across the abdomen.

Because individuals who need to be restrained are usually in a heightened sympathetic state, they also tested TASER X26 discharges during simulated stress by infusing epinephrine. The epinephrine was administered as a continuous intravenous infusion at a dose of 0.1–0.7 microgram per kilogram per minute titrated to increase the animal's heart rate to a 50% increase from the baseline before discharges.

The pigs were sedated with 12-milligram intramuscular ketamine per kilogram of body weight and inhaled isoflurane. The pig was then intubated, and anesthesia was maintained with an inhaled mixture of 1.0–2.0% isoflurane and oxygen. A 6-lead surface electrocardiogram system was attached. Two 7-F venous sheaths were inserted into the right femoral vein, and one 8-F arterial sheath was placed into the right femoral artery. Electrocardiograms, blood pressure, and oxygen saturations were continuously monitored.

Bipolar recording catheters were positioned in the right ventricle and coronary sinus under fluoroscopic guidance, approximately 4.5 centimeters apart. A high-fidelity micromanometer pressure transducer catheter (Millar Instruments, Houston, TX) was introduced into the arterial line and positioned in the descending aorta. All intracardiac signals were also connected to dedicated high-level channels on the recording system to quantify the voltage produced during TASER X26 discharge measured from the catheters (referred to as maximum intracardiac voltage, defined as the maximum voltage difference between the two electrograms at any point during discharge). The signals were sampled at 1 kilohertz, with a low pass filter of 200 hertz and a high pass filter of 0.05 hertz. Voltage was measured by both the mapping system and an oscilloscope with a sampling frequency of 1 million samples per second and a frequency bandwidth of 250 megahertz.

The mean weight of the pigs was 49.9 ± 1.2 kilograms. They studied a total 150 discharges to six pigs; of these, 94 discharges were thoracic and 56 discharges were nonthoracic. In the thoracic vector, 79% resulted in stimulation of the myocardium, compared with 0% in the nonthoracic vector ($z = 22.24, p < 0.0001$) (chi-square = 77.87, $p < 0.0001$).

Table 7.2 shows the effect of the two different TASER X26 models and the effect of duration. There was a significant effect on myocardial stimulation caused by TASER X26 model, with discharges from X26 more likely than those from M26 (98% vs. 54%) to stimulate the myocardium ($z = 3.38, p = 0.0007$).

All discharges during epinephrine infusion were delivered across the chest. There were a total of 16 discharges during epinephrine administration in 4 animals, resulting in 13 episodes of stimulation of the myocardium. The mean dose of epinephrine delivered to achieve the 50% increase in heart rate response was 0.5 microgram per kilogram per minute. One episode resulted in VF. In another animal, an episode of stimulation of the myocardium resulted in nonsustained VT/VF that spontaneously terminated.

When the discharge was vectored across the chest, electrical and mechanical capture of the heart ensued. The cardiac stimulation at high rates persisted during the discharge, and as soon as the discharge ceased there was resumption of normal electrical rhythm. The X26 was more effective in stimulation than the M26. The abdomen discharges did not stimulate the heart or trigger arrhythmias. During simulated stress with epinephrine infusion, presumably because of the shortening of ventricular refraction, some chest discharges resulted in VF and VT.

These findings suggest that there exists the possibility of serious ventricular arrhythmia during TASER X26 discharges in structurally normal hearts during

Table 7.2 Summary of TASER X26 discharges (Nanthakumar et al. [9])

	Capture		
	Total	*N*	%
N discharge	150	74	49.33
Vector effect			
Chest	94	74	78.72
Abdomen	56	0	0.00
Device effect			
X26 Chest	53	52	98.11
M26 Chest	41	22	53.66
M26 Abdomen	29	0	0.00
X26 Abdomen	27	0	0.00
Time effect			
X26 Chest 15 seconds	28	28	100.00
X26 Chest 5 s	25	24	96.00
M26 Chest 15 s	20	12	60.00
M26 Chest 5 s	21	10	47.62

early intense catecholamine stress. Injections of epinephrine in dogs reduce the 2 milliseconds duration VF threshold for about 3 minutes after which the VF threshold increases [10]. In patients with structural heart disease, in which electrophysiological inhomogeneities are present, rapid ventricular stimulation is known to produce catastrophic ventricular arrhythmias. The findings of rapid ventricular stimulation with TASER X26 discharge across the chest suggest a particular risk in individuals with preexisting inhomogeneities caused by structural heart disease.

The greater voltages during discharge across the chest seen in their model support the hypothesis that it is indeed the maximum voltage vector across the heart that results in stimulation of the myocardium. The time constant for membrane depolarization is on the order of 2–5 milliseconds [11]; the strength–duration curve for cardiac stimulation suggests that for very large local voltage gradients effective stimulation may occur.

7.5 Effects of 80-second Exposures

Dennis et al. [12] used 11 standard pigs (six experimental and five sham controls) weighing between 22 and 46 kilograms and anesthetized with ketamine and xylazine. The barbed darts were placed along a line parallel to the cardiac axis. One dart was placed 13 centimeters superior to the xyphoid process and 5 centimeters to the right of the midsternal line. The other dart was placed 7 centimeters to the left of the umbilicus. This dart configuration produced a diagonal separation of approximately 30 centimeters in each pig and is similar to that used by Jauchem et al. [13]. All darts were manually inserted perpendicular to the skin to the maximum depth allowed by the length of the barbed end (9 millimeters) such that the dart tip was located in subcutaneous tissue.

The pigs received a 40-second TASER X26 discharge across the torso with the ventilator shut off. During the next 10 seconds, two ventilated breaths were administered. Then a second 40-second discharge was applied. Heart rate was significantly increased and significant hypotension was noted. Acid–base status was dramatically affected by the CEW discharge at the 5 minutes time point and throughout the 60 minutes monitoring period. Five minutes postdischarge, central venous blood pH (6.86 ± 0.07) decreased from baseline (7.45 ± 0.02; p = 0.0004). PCO_2 (94.5 ± 14.8 millimeter Hg) was significantly increased from baseline (45.3 ± 2.6 millimeter Hg) and bicarbonate levels significantly decreased (15.7 ± 1.04 millimoles per liter) from baseline (30.4 ± 0.7 millimoles per liter). A large, significant increase in lactate occurred postdischarge (22.1 ± 1.5 millimoles per liter) from baseline (1.5 ± 0.3 millimoles per liter). All values returned to normal by 24 hours postdischarge in surviving pigs. A minor, nonsignificant increase in troponin I was seen at 24 hour postdischarge (0.052 ± 0.030 nanograms per milliliter, mean ± SEM). Because the pigs were not able to ventilate during the 80 seconds, they became very acidotic.

One animal in the experimental group (29 kilograms) was found to be in VF after TASER CEW discharge. Cardiac rhythm could not be discerned by EKG during the discharge because of the electrical interference and muscle contractions created by the CEW so it was not possible to verify when the VF first occurred. Since cardiac capture was noted immediately by hemodynamic effects—and induced VF from rapid capture usually occurs in seconds—the VF most likely occurred in the first seconds of the stimulation. Cardiac rhythm was evaluated by echocardiography during the discharge and found to be consistent with ventricular tachycardia. When the discharge ceased, sustained ventricular tachycardia was noted on echocardiography and confirmed by EKG (Fig. 7.4). During the course of the next few minutes, the ventricular tachycardia then degenerated into fatal VF. All surviving experimental animals showed brief atrioventricular (AV) dyssynchrony followed by sinus tachycardia after the discharge. Despite persistent sinus tachycardia, no EKG evidence of acute dysrhythmia was seen in the surviving animals. Echocardiography (echo) showed capture of the ventricular rhythm during CEW discharge but motion artifacts prevented quantitative analysis of cardiac output and ejection fraction. One animal, as described above, went into VF after the discharge as confirmed by EKG and echo. The remaining three animals all showed capture of ventricular rhythm with rapid ventricular contractions seen on echo consistent with ventricular tachycardia (approximate rate of 300 beats per minute). This capture of cardiac rhythm occurred immediately after the start and continued for the duration of the CEW discharge as seen by echo. Sinus rhythm was regained after a brief period of AV dyssynchrony in each of these three animals and sinus tachycardia began within 1 minute after termination of the discharge.

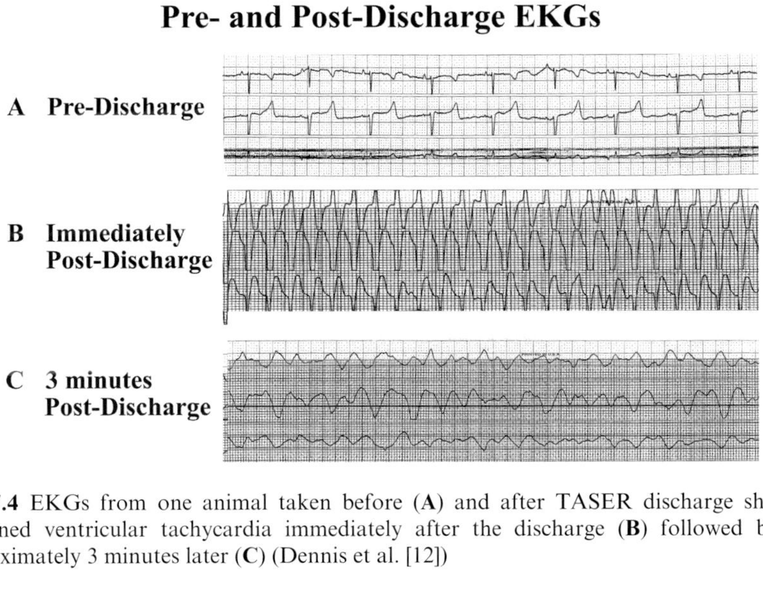

Fig. 7.4 EKGs from one animal taken before (**A**) and after TASER discharge showing sustained ventricular tachycardia immediately after the discharge (**B**) followed by VF approximately 3 minutes later (**C**) (Dennis et al. [12])

In one control pig, before TASER discharge, with the heart exposed via left anterolateral thoracotomy, normal sinus rhythm was directly visualized and confirmed by EKG. When the CEW discharge started, sinus rhythm was immediately (within 1 second) disrupted. In the 31 kilogram animal, the first 40-second discharge resulted in immediate capture of the myocardium producing rapid ventricular contractions consistent with ventricular tachycardia. During this discharge, atrial standstill was also seen. When the first discharge ceased, approximately 15 seconds of dyssynchronous atrial and ventricular contractions were noted, after which sinus rhythm resumed. The second discharge resulted again in immediate disruption of sinus rhythm and ventricular tachycardia. Atrial standstill was again noted during this discharge. However, after 16 seconds, the ventricular tachycardia was replaced by fatal VF.

Thus, immediately after the discharge in 11 pigs, two cases of ventricular fibrillation were seen. In this model of prolonged CEW exposure, clinically significant acid–base and cardiovascular disturbances were clearly seen. The severe metabolic and respiratory acidosis seen here suggests the involvement of a primary cardiovascular mechanism.

7.6 Effect of Dart-to-Heart Distance

In the first study from our group, Wu et al. [14] measured the dart-to-heart distances causing VF for the TASER X26 on anesthetized pigs. The mass of the ten normal healthy pigs ranged from 53.8 to 74.4 kilograms. The pigs were anesthetized and intubated. Blood pressure, oxygen saturation, respiration, heart rate, and ECG of the animal were monitored throughout the experiment. The distances between the heart dart tip and the right ventricle were sequenced from long-to-short distances to determine the threshold distance for VF.

In all experiments, the two stimulation darts setup was used to deliver TASER current to pigs. The stimulation darts were made to match the standard TASER probes which were 9 millimeter long and 0.8 millimeter in diameter.

The heart dart was placed inside a standard French 7 catheter with a conductive lead installed inside to deliver current. One end of the lead was connected to the heart dart and the other end was connected to TASER X26. Figure 7.5 shows that the heart dart in the catheter was attached to a syringe to provide a stable ground for dart-to-heart distance adjustments. The heart dart was placed inside a bluntly dissected path through the intercostal muscle between the third and fourth ribs above the right ventricle. The plunger of the syringe formed an adjustable stop. It was placed against the ribs next to the bluntly dissected path to provide a firm surface to maintain the dart-to-heart distance constant during the vibration caused by the stimulation.

The remote dart, a standard TASER probe, was placed on the abdominal surface ranging from 15 to 54 centimeters caudal from the stimulation dart.

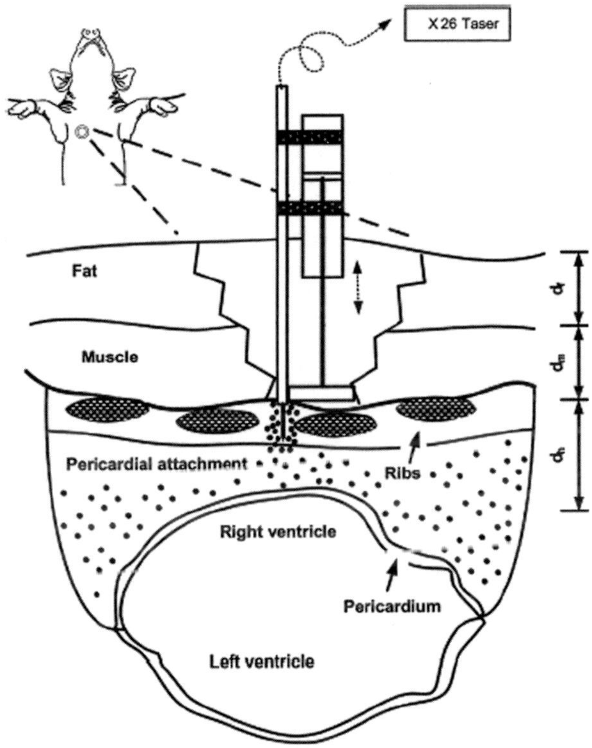

Fig. 7.5 A midsternal dissection exposed the ribs. Above the external intercostal muscles were muscle layers about 25 millimeters (d_m), and fat layers about 15 millimeters (d_f) that were resected to provide the intercostal approach. A bluntly dissected pathway (d_c) contained the heart dart. The heart dart was inserted into a standard F7 catheter with a syringe attached to provide a stable ground to keep the dart-to-heart distance stable during the TASER discharge vibrations (Wu et al. [14])

The 2 millimeter bluntly dissected path was made by bluntly dissecting tissue between the external intercostal muscle wall and the heart. The opening in the muscle created an air gap inside the bluntly dissected path. This air gap between the stimulating electrode and the heart changed the current density distribution due to the higher impedance of the air compared to muscle. In order to maintain a normal electric current distribution, muscle impedance matching gel was injected into the bluntly dissected path to fill the air gap. The gel was fabricated from NaCl, water, and agar. In order to match the 3.0 ohm-meter resistance of intercostal muscle, the concentration 0.2% saline was used as electrolyte and consolidated by agar (10 gram agar with 500 milliliter saline).

Once the anesthetic procedures were complete and the animal was in a homeostatic condition, the sternum was exposed up to the intercostal muscle layer. The most sensitive region of the heart for external stimulation has been found to be close to the midpoint of the right ventricle. This region was under the third and fourth ribs. Thus, they placed the heart dart at this region over the right ventricle. The muscle impedance matching gel was injected into the bluntly dissected path to fill the air gap.

After the dissection, the heart dart setup was inserted into the bluntly dissected path and the remote dart was placed caudally on the abdomen surface separated from the stimulation dart setup with the controlled separation ranging from 15 to 54 centimeters. The bluntly dissected path distance from the external intercostal muscle to the heart was measured to determine the dart-to-heart distance. The dart-to-heart distance was the distance from the tip of the heart dart to the heart. The current from long-to-short distances is applied to determine the distance threshold of VF. In each experiment, the dart-to-heart distance began at 20 millimeters to prevent causing VF immediately. Then TASER X26 current was applied through the heart dart of about 5 seconds with 15–19 pulses per second. The current through a precision resistor was monitored by an oscilloscope.

After the stimulation, the ECG was verified to check if the heart was beating normally. If the heart beating was normal, the dart-to-heart distance was decreased by 2 millimeters for each step closer to the heart. Then the stimulation current was applied again. The same process was applied until the first VF occurred. Immediately, the defibrillator was used to defibrillate the pig with increasing energy from 100, 200, 300, 400 joules. After the defibrillation, a 5-minute recovery period was given to return the pig to a homeostatic condition. After the first VF event, the dart-to-heart distance was increased by 2 millimeters to a less hazardous distance. Then the same process was applied to gather at least 3 VF dart-to-heart distances or the pig was euthanized. During the experiment, the impedance matching gel was refilled whenever there appeared to be an air gap in the bluntly dissected path.

Ten animal tests were performed to study the VF dart-to-heart distance of TASER X26 using the 2-dart configuration. The dart-to-heart distances of all stimulations including VF and no VF events were recorded. Nine VF distances were recorded with average distance 15 millimeters ± 1.22 (SD). Table 7.3 summarizes VF distances of all experiments.

Ten animal experiments, yielded dart-to-heart distances that cause VF in anesthetized pigs for the TASER X26. The dart-to-heart distance that causes VF was 17 millimeters ± 6.48 (SD) for the first VF event and 13.7 millimeters ± 6.79 (SD) for the average of the successive VF distances. Data from only the

Table 7.3 Summary of VF dart-to-heart distances (Wu et al. [14])

Pig	First VF distance (mm)	Average distance (mm)	Average distance without first record (mm)	Number of VF
1	18	15 ± 1.22 (SD)	14.4 ± 2.02 (SD)	9
2	0	4 ± 2.83 (SD)	4 ± 2.19 (SD)	4
3	14	12 ± 2.31 (SD)	11.3 ± 2.31 (SD)	4
4	20	17± 2.58 (SD)	16 ± 2 (SD)	4
5	18	21 ± 2.58 (SD)	22 ± 2 (SD)	4
6	20	19.3 ± 1.15 (SD)	19 ± 1.41 (SD)	3
7	18	18 ± 0 (SD)	18 ± 0 (SD)	4
8	18	18 ± 0 (SD)	*	1
9	20	18 ± 2 (SD)	17 ± 1.41 (SD)	3
10	24	9.3±12.7 (SD)	2 ± 0 (SD)	3
Average	17 ± 6.48 (SD)	15.1 ± 5.28 (SD)	13.7± 6.79 (SD)	

first VF event were used because each successive defibrillation damaged the heart muscle and changed the VF threshold. Standard TASER darts that stimulate a pig may not cause VF because the normal skin-to-heart distance of pigs is about 45 millimeters, which is much farther than the sum of the maximum (most dangerous) VF threshold distance 24 millimeters plus dart length 9 millimeters equal 33 millimeter. Table 7.4 shows that the human skin-to-heart distance is shorter than that of the pig.

Following criticism that the first study allowed a direct path to the heart, the experiment was then repeated by Wu et al. without the use of the conductive gel bluntly dissected path. [15] We used a different specially designed probe with a core conductive rod 0.8 millimeter in diameter to match the TASER X26 dart diameter. The 100-millimeter long skin-to-heart-distance testing probe (Fig. 7.6a) measured the skin-to-heart distance at the stimulation site. The 50-millimeter long blunt probe (Fig. 7.6b) delivered the TASER current. The surface of the blunt probe was insulated except for the first 9 millimeter portion which delivered the stimulation current. The blunt probe was designed to match the structure of standard TASER darts except that the normally sharp tip was made blunt to avoid inadvertently piercing the heart, which would have ended the test.

Two stimulation darts, the blunt probe and the remote dart, delivered stimulation current. The blunt probe was slid through the previously made stimulation site track. The remote dart, a standard TASER probe, was placed on the abdominal

Table 7.4 Echocardiography of erect humans (mixed females and males) shows skin-to-heart distances from 10 to 57 millimeters (Wu et al. [14]) (The low distance subjects were all female except for one male. Also, the dart-to-heart distance assumes a full depth penetration which rarely occurs.)

Dart-to-heart distance (mm)	Skin-to-heart distance (mm)	Number of humans less than skin-to-heart distance (mm)	Probability of human VF for 1 cm^2
17	26	63	0.000087
16	25	56	0.000077
15	24	52	0.000072
14	23	46	0.000063
13	22	37	0.000051
12	21	26	0.000036
11	20	20	0.000028
10	19	14	0.000019
9	18	10	0.000014
8	17	8	0.000011
7	16	6	0.000008
6	15	5	0.000007
5	14	5	0.000007
4	13	5	0.000007
3	12	2	0.000003
2	11	1	0.000001
1	10	0	0

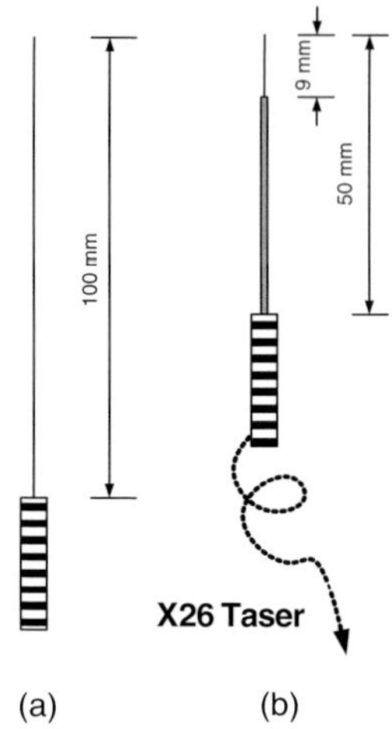

Fig. 7.6 (**a**) The distance testing probe measured the skin-to-heart distance. (**b**) The blunt probe delivered the stimulation current from the exposed 9-millimeter long wire (Wu et al. [15])

surface at a typical TASER X26 separation, 54 centimeters caudal from the blunt probe. Figure 7.7 shows the relation between the blunt probe and the heart. The procedure to measure the distance followed the study shown previously.

Five animals (pig mass = 61.16 kilograms ± 6.23 SD) for 10 VF skin-to-heart distance measurements were made to determine the VF risk. Table 7.5

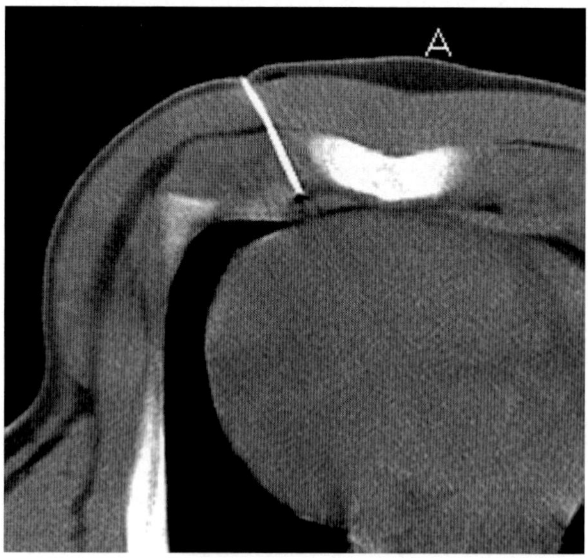

Fig. 7.7 CT image of blunt probe, sternum, and rib (*white*), lung (*dark*) surrounding the heart (*gray*). The blunt probe was inserted through the fat layer, the muscle layer, the intercostal muscle layer between the third and fourth ribs for stimulation site 1 and the fourth and fifth ribs for the stimulation site 2. Typical skin-to-heart distance is 47 millimeters (Wu et al. [15])

Table 7.5 Experimental parameters and results of five animal tests (Wu et al. [15])

Pig	Mass (kg)	Test	Skin-to-heart distance (mm)	VF distance (mm)
1	58	1	51	5
		2	57	7
2	68.8	3	49	6
		4	45	4
3	53	5	45	8
		6	47	8
4	65.6	7	42	4
		8	45	2
5	60.4	9	44	8
		10	45	6
Average	61.16		47	5.8
SD	6.23		4.35	2.04

summarizes the experimental parameters and results of animal tests. The skin-to-heart distance ranged from 42 to 57 millimeters with average 47 millimeters ± 4.35 (SD). The VF dart-to-heart distance ranged from 2 to 8 millimeters with overall average 5.8 millimeters ± 2.04 (SD) and 6.2 millimeters ± 1.79 (SD) for the first VF event in each pig. The first VF distance was longer or equal to the second VF distance except in pig 1. As in Table 7.4, the probability is multiplied by the fraction of humans with skin-to-heart distance less than a chosen distance, as shown in Table 7.6.

Comparing these results with the previous tests, the probability of human VF of the blunt probe is much less than that of the sharp probe, due to the absence of the conducting gel.

Table 7.6 Dart-to-heart distances plus 9 millimeter dart length = skin-to-heart distances; probability estimates include thin females with breasts reflected (Wu et al. [15]) (The low distance subjects were all female except for one male. Also, the dart-to-heart distance assumes a full depth penetration which rarely occurs.)

Dart-to-heart distance (mm)	Skin-to-heart distance (mm)	Number of humans less than skin-to-heart distance (mm)	Probability of human VF for 1 cm^2
8	17	8	0.000011
7	16	6	0.0000083
6	15	5	0.0000069
5	14	5	0.0000069
4	13	5	0.0000069
3	12	2	0.0000028
2	11	1	0.0000014
1	10	0	0

7.7 Effects on pH and SpO_2

Jauchem et al. [13] investigated effects of repeated exposures of the TASER X26 on muscle contraction and resultant changes in blood factors in an anesthetized unventilated pig model.

Effects of the CEW could result either (a) directly from the electrical properties of the applied stimulus or (b) from the resultant muscle contraction. Although such contraction is not synonymous with conventional muscular exercise, there may be some similarities.

Ten domestic pigs with a mean weight of 53.6 kilograms (range 49.5–58.0 kilograms) were used for these studies. For each experiment, a pig was given preanesthetic (atropine 0.05–0.5 milligram per kilogram body weight, subcutaneously) and analgesic (buprenorphine, 0.02 milligram per kilogram body weight, intramuscularly) 10–15 minutes prior to induction of anesthesia. The animals were anesthetized with an intramuscular injection of tiletamine HCl and zolazepam HCl (Telazol1) (6 milligrams per kilogram), followed by oral endotracheal intubation, with the tube secured to the maxilla or mandible. An aural intravenous catheter (18–25 millimeters, 20–22 gauge) was placed and secured with a cyanoacrylate adhesive and tape. Anesthesia was maintained with 100–125 milligrams per kilogram per minute of propofol (Diprivan1) delivered by a Baxter syringe pump. A jugular venous catheter was placed for subsequent blood sampling. Each pig was delivered to the laboratory anesthetized, placed on its dorsal surface in a canvas sling.

The muscle contraction test structure included a metal framing system, a sling (to contain the pig), pulleys, and strain gages. Each anesthetized pig was placed on its dorsal surface in the sling. Twisted polypropylene truck rope was attached to each limb via a neoprene tennis elbow support, while the other end of the rope was attached to a turnbuckle and strain gage. A second set of ropes was attached to each limb with neoprene-blend adjustable wrist/elbow supports. Each of these latter ropes ran through a 400 diameter sheave block and was attached to a 2.27 kilogram mass. The output of the strain gages was quantified, displayed, and stored using instruments and software. Prior to each exposure, the turnbuckles were adjusted to bring the pig's limbs to a standardized anatomical position (stretched maximally), with a baseline force of approximately 44.5 N.

The skin was pierced with standard TASER X26 darts. One dart was placed approximately 5 centimeters to the right of the midline (approximately 13 centimeters cranially from the xiphoid process); the other was approximately 7 centimeters left of the umbilicus (resulting in approximately 30 centimeters separation between darts diagonally).

Six pigs were exposed to the output of the TASER X26 for 5 seconds, followed by a 5-second period of no exposure, repeatedly for 3 minutes. In five of the animals, after a 1 hour delay, a second 3-minute exposure period (again, of 5 seconds on, 5 seconds off) was added.

Venous blood samples (3 milliliters each) were taken from the jugular vein within 1 minute before and 1 minute after each TASER exposure, and at other time points, for measurement of whole blood factors listed above. An additional 9 milliliters of blood was drawn and allowed to clot at room temperature for at least 30 minutes. Within 90 minutes of collection, the samples were centrifuged, and serum was refrigerated until assay. Serum troponin, CPK, and lactate dehydrogenase (LDH) (including isoenzyme forms) were used to provide qualitative estimates of skeletal (or cardiac) muscle damage that might occur as a result of TASER X26 exposure.

In most cases, after several repeated 5-second exposures, the positioning of the straps on the four limbs did not remain consistent. Thus, only the muscle-contraction measurements during the initial portion of the 3-minute exposure periods were considered to be an accurate reflection of limb movement. For the five subjects receiving two 3-minute exposure periods, the mean level of maximal limb flexion was 30.5 kilograms for the first of the two exposures (measured by averaging maximal flexion values during the first second of exposure across all four limbs); the mean level of maximal limb flexion for the second session of TASER X26 exposures was 21.6 kilograms. This decrease in limb flexion was statistically significant as measured by a paired t-test, $t(4) = 5.35$, $p = 0.006$.

Figure 7.8 shows an example of limb contraction tracings obtained from one animal, when exposed to the X26. Each limb exhibited an initial series of clonic (alternating tensing and relaxing of the muscles) contractions, followed by subsequent periodic contractions at slightly lower levels of force. This pattern reflects the discharge of the earlier TASER X26, with a change in pulse rate and power after 2 seconds of discharge.

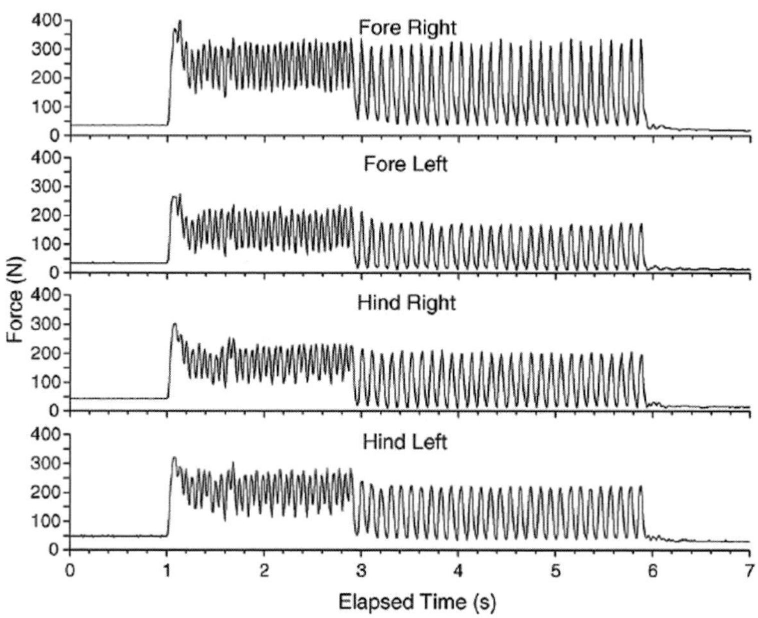

Fig. 7.8 Example of muscle contraction tracings obtained from one animal, when exposed to the TASER X26 (Jauchem et al. [13])

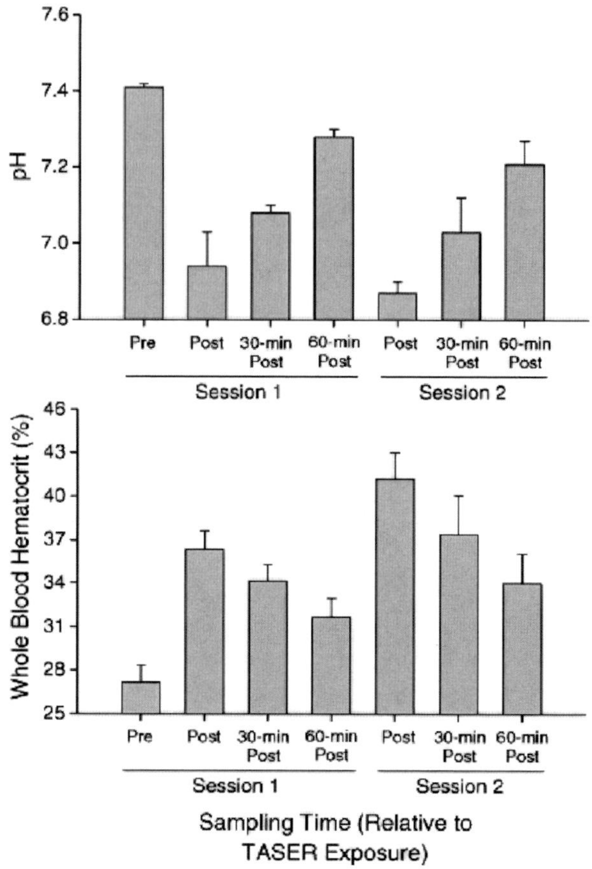

Fig. 7.9 Jugular venous blood pH before, during, and after two sessions of 18 TASER exposures each (Jauchem et al. [13])

From Fig. 7.9, the pH dropped from a normal value of 7.4 down to 7.0 and a pH of <7.2 indicates "severe acidemia." Such a blood pH could be of minor consequence if due to a transient or readily reversible condition. One hour after exposure in the current experiments, pH had returned to above 7.2. After a second TASER X26 exposure session, pH showed a similar pattern.

The degree of muscle contraction generated during Session 2 was significantly lower than in Session 1, shown in Fig. 7.10. This was not surprising, as prolonged activity in skeletal muscle will eventually result in a decline of force production [16]. Both (a) muscle contractions and (b) changes in respiration (see Fig. 7.10) may have contributed to the acidosis. Jauchem described the acidosis as respiratory. Metabolic acidosis (but not respiratory acidosis) can result in a lower threshold for ventricular fibrillation. This is consistent with this study as no ventricular fibrillation was seen.

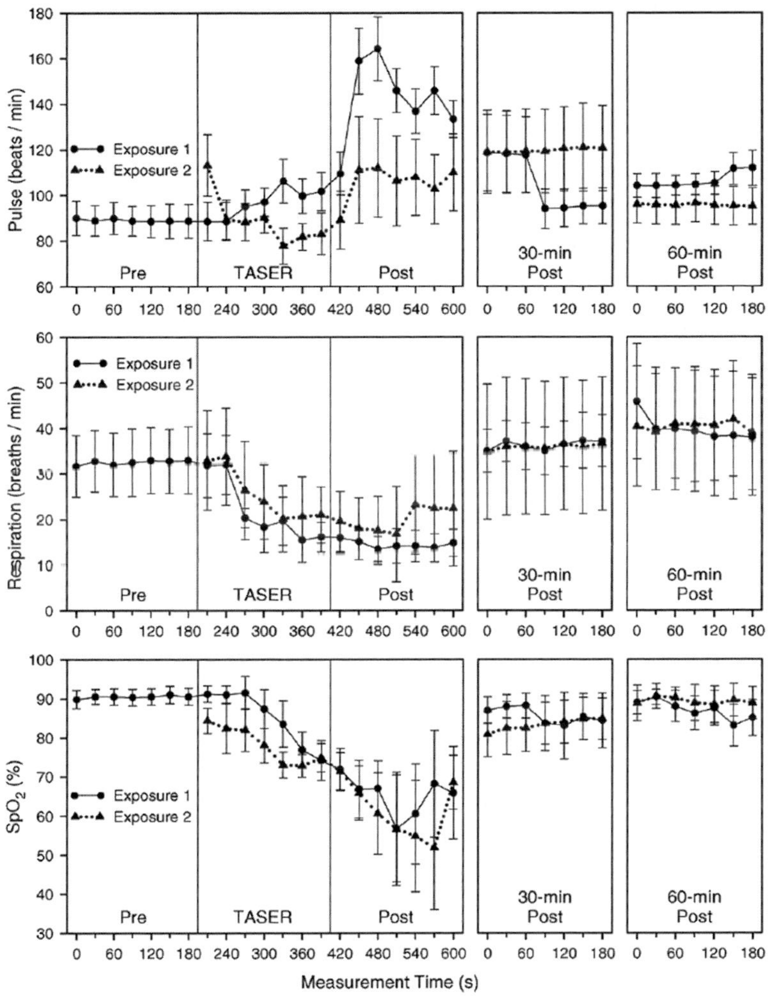

Fig. 7.10 Heart rate, respiration rate, and pulse-oximeter oxygen saturation (SpO_2) before, during, and after two sessions of 18 TASER exposures each (Jauchem et al. [13])

7.8 Effects with No Acidosis

Walter et al. [17] tested Yorkshire pigs (25–71 kilograms) anesthetized, paralyzed with succinylcholine (SCh; 2 milligrams per kilogram), and then exposed to two 40-second discharges from a TASER X26 with a transcardiac current path between the darts. The ventilator was off during each 40-second discharge but on during the 10–15 seconds pause between the discharges. Dart placement was on the ventral body surface of the swine. The superior dart was placed 13 centimeters superior to the xiphoid process and 5 centimeters to the right of the midsternal line. The inferior dart was placed 7 centimeters to the left of the umbilicus. Darts were inserted perpendicular to the skin to their maximum depth (12 millimeters) such that the dart tip was located in subcutaneous tissue. Darts were more than 5 centimeters from the heart. Vital signs, blood chemistry, and electrolyte levels were obtained before exposure and periodically for 48-hour postdischarge. Electrocardiograms and echocardiography (echo) were performed before, during, and after the discharges. p-Values <0.05 were considered significant. Electrocardiograms were unreadable during the discharges due to electrical interference, but echo images showed unmistakably that cardiac rhythm was captured immediately at a rate of 301 ± 18 beats per minute ($n = 8$) in all animals tested. Capture continued for the duration of the discharge and in one animal degenerated into fatal ventricular fibrillation (VF). In the remaining animals, ventricular tachycardia (VT) occurred postdischarge for 1–17 seconds, whereupon sinus rhythm was regained spontaneously. Extreme acid–base disturbances usually seen after lengthy TASER discharges were absent with SCh and blood chemistry values and vital signs were minimally altered postdischarge. Thus, in the absence of systemic acidosis, lengthy transcardiac TASER X26 discharges captured myocardial rhythm, potentially resulting in VT or VF in swine. The acid–base disturbances previously seen after lengthy TASER discharges in swine are largely the result of intense skeletal muscle contractions and are not caused by apnea or depressed cardiac function occurring during or after the discharge.

7.9 Applying the Animal Data

Sun et al. [18] used pig data from Wu et al. [14,15] on dart-to-heart VF distance. They used computer finite element models to determine current density in humans. They used echocardiographic data to determine human dart-to-heart distances. They used police dart-landing reports to estimate the probability of darts landing over the heart. They combined all these data to estimate the probability of TASERs causing human VF. Using data from Wu et al. [14] the estimated probability of human VF was 0.0014. Using data from Wu et al. [15] the estimated probability of human VF was 0.0000061. These are worst case probabilities and may be revised lower by further analysis. They recommended that necessary but not sufficient conditions for human VF are that (1) the dart land in a small region over the heart and (2) the subject has rapid cardiac arrest. Coroners should confirm these conditions before ascribing TASER as a cause of death. Also all training should be done on the back.

7.10 Conclusions

Pigs have a thick band of muscle and fat over the heart to yield a typical skin-to-heart distance of 47 millimeters. In most cases with a TASER X26 dart in the skin, TASER CEW output had to be increased substantially to achieve VF, which suggests the TASER X26 is safe for pigs. For three investigators, VF occurred with a TASER dart in the skin, which suggests that the TASER X26 is not safe for pigs [9,12,17]. Cocaine increases the threshold for VF during the CEW exposure, which may be due to its anesthetic effect. The probability of VF increases with decreasing distance of dart-to-heart. Because only a few humans have skin-to-heart distances of 11 millimeters, the probability of the TASER X26 causing VF is very small.

References

1. Holden, SJ, Sheridan, RD, Coffey, TJ, Scaramuzza, RA, Diamantopoulos, P. Electromagnetic modelling of current flow in the heart from TASER devices and the risk of cardiac dysrhythmias, Phys Med Biol 2007; 52. 7193–7209.
2. Roy OZ, Podorski AS. Tests on a shocking device—the stun gun. Med Biol Eng Comput 1989; 27: 445–448.
3. McDaniel WC, Stratbucker RA, Nerheim M, Brewer JE. Cardiac safety of neuromuscular incapacitating defensive devices. PACE-Pacing Clin Electrophysiol 2005; 28: S284–S287.
4. Lakkireddy D, Wallick D, Ryschon K, Chung MK, Butany J, Martin D, Saliba W, Kowalewski W, Natale A, Tchou PJ. Effects of cocaine intoxication on the threshold for stun gun induction of ventricular fibrillation. J Am Coll Cardiol 2006; 48(4): 805–811.
5. Maier A, Nance P, Price P, Sherry CJ, Reilly JP, Klauenberg BJ, Drummond JT. Human effectiveness and risk characterization of the electromuscular incapacitation device—a limited analysis of the TASER, Part II Non Lethal Weapon—EMI Appendices Public V1 Mar 05 final, Mar 2005, website [Online]. Available: http://www.taser.com/documents/Part%20II%20NLW%20-%20EMI%20Appendices%20Public%20V1%20Mar%2005%20final.pdf
6. Tisdale JE, Shimoyama, H, Sabbah, NH, Webb CR. The effect of cocaine on ventricular fibrillation threshold in the normal canine heart. Pharmacotherapy 1996; 16(3): 429–437.
7. Przywara DA, Dambach GE. Direct actions of cocaine on cardiac cellular electrical activity. Circ Res 1989; 65(1): 185–192.
8. Lange RA, Cigarroa JE, Hillis LD. Theodore e. Woodward award: cardiovascular complications of cocaine abuse. Trans Am Clin Climatol Assoc 2004; 115: 99–114.
9. Nanthakumar K, Billingsley IM, Masse S, Dorian P, Cameron D, Chauhan VS, Downar E, Sevaptsidis E. Cardiac electrophysiological consequences of neuromuscular incapacitating device discharges. J Am Coll Cardiol 2006; 48(4): 798–804.
10. Han J, de Jalon PG, Moe GK. Adrenergic effects on ventricular vulnerability. Circ Res 1964; XIV: 516–524.
11. Mouchawar G, Kroll M, Val-Mejias JE, Schwartzman D, McKenzie J, Fitzgerald D. Prater S, Katcher M, Fain E, Syed Z. ICD waveform optimization: a randomized, prospective, pair-sampled multicenter study. Pacing Clin Electrophysiol 2000. 23(11 Pt 2): 1992–5.
12. Dennis AJ, Valentino DJ, Walter RJ, Nagy KK, Winners J, Bokhari F, Wiley DE, Joseph KT, Roberts RR. Acute effects of TASER X26 discharges in a swine model. J Trauma 2007; 63(3): 581–590.
13. Jauchem JR, Sherry CJ, Fines DA, Cook MC. Acidosis, lactate, electrolytes, muscle enzymes, and other factors in the blood of Sus scrofa following repeated TASER (R) exposures. Forensic Sci Int 2006; 161(1): 20–30.

14. Wu J-Y, Sun H, O'Rourke AP, Huebner S, Rahko PS, Will JA, Webster JG. TASER dart-to-heart distance that causes ventricular fibrillation in pigs. IEEE Trans Biomed Eng 2007; 54(3): 503–508.
15. Wu J-Y, Sun H, O'Rourke AP, Huebner S, Rahko PS, Will JA, Webster JG. TASER blunt dart-to-heart distance that causes ventricular fibrillation in pigs. IEEE Trans Biomed Eng December 2008; 55(12).
16. Adrogue HJ, Madias NE. Management of life-threatening acid-base disorders. First of two parts. N Engl J Med 1998; 338(1): 26–34.
17. Walter RJ, Dennis AJ, Valentino DJ, Margeta B, Nagy KK, Bokhari F, Wiley DE, Joseph KT, Roberts RR. TASER X26 discharges in swine produce potentially fatal ventricular arrhythmias. Acad Emerg Med 2008 Jan; 15(1): 66–73.
18. Sun H. Models of Ventricular Fibrillation Probability and Neuromuscular Stimulation after TASER® Use in Humans, PhD thesis, Dept Elect. Comput. Eng. Madison, WI: University of Wisconsin–Madison, 2007.

Chapter 8
CEW Research Models: Animal and Human Studies

Theodore C. Chan and Gary M. Vilke

Research on CEWs has involved both animals and humans. Multiple investigators have conducted extensive, detailed and complex experimental studies on animal models, as well as human volunteer subjects to measure, monitor, and determine the physiologic effects of CEWs. The findings and results of these animal and human experimental studies have varied and, as a result, the conclusions drawn by investigators as well as other experts have been inconsistent and at times in wide disagreement.

Animal and human subjects each have different strengths and weaknesses as research models. These differences relate to the applicability of each model to actual scenarios and occurrences reported in the field, the different methodologies that must be employed for each, and the relative merits of each model to replicate so-called "real-world" conditions. This chapter will compare experimental research on CEWs conducted in animal models and human subjects. Other sections of this text will review many of these studies in greater detail. In this chapter, we will focus on the different strengths and weaknesses of research conducted in animal models and human subjects, and their implications in terms of assessing the physiologic effects and overall safety of CEWs.

8.1 Animal Versus Human

Experimental research studies on CEWs have been conducted in dogs, pigs, and humans. In terms of these animal models, the basic question remains whether findings from these studies can be extrapolated to humans. Pigs have been favored in this regard because of the animal's similarity to humans in terms of body mass and heart size (Fig. 8.1). On the other hand, while these animals have been used extensively in the study of cardiac arrest and resuscitation, some have

T.C. Chan (✉)
Department of Emergency Medicine, University of California, San Diego Medical Center
e-mail: tcchan@ucsd.edu

M.W. Kroll, J.D. Ho (eds.), *TASER® Conducted Electrical Weapons: Physiology, Pathology, and Law*, DOI 10.1007/978-0-387-85475-5_8,

Fig. 8.1 Relative anatomic differences between dog, pig, and human study subjects. Typical adult dog and pig weights are 30–35 and 60–65 kilograms compared to adult human weight of 70 to 80 kilograms.

argued that pigs make "poor surrogates" for human cardiac physiologic responses to electrical discharges from CEWs [1]. In addition, there are differences in skin, connective soft tissue, muscle mass, and body geometry that may limit the ability to generalize the findings of these studies.

Data are available dating back to the 1930s demonstrating that small swine are sensitive to the electrical induction of ventricular fibrillation (VF) [2]. In dogs and humans the Purkinje fibers are confined to a very thin endocardial layer [3]. In pigs they cross the entire ventricular wall [4]. Recent work has demonstrated that activation in swine proceeds from the epicardium to the endocardium while in dogs and human it proceeds in the reverse direction [5]. Thus, swine are thought to be much more sensitive to the application of external electrical currents.

Pig models also appear to be more sensitive to higher frequency currents. Ventricular ablation with radio frequencies is routinely performed in humans without dysrhythmias, yet ablation in swine will often result in VF [6]. There appear to be differences in the ion channels in swine that may play a role in the development of dysrhythmias [7]. The VF threshold is directly related to the body weight for both utility power and waveforms, so body mass of swine or other animal models need to be considered when evaluating data [2,8–10].

Though swine are very sensitive to electrical currents, it is probably appropriate to use them for relative comparisons. Swine have been used to study the effects of cocaine [11]. Similarly, swine can be used to study the effects of varying body weight on the CEW safety margin [10]. However, as with all swine studies, care must be used when attempting to apply the results to humans.

Undoubtedly, experimental research on human subjects is preferable to animal models in terms of the applicability of any findings to human uses of CEWs (see Table 8.1). However, human investigations, particularly when attempting to simulate "real-world" scenarios such as multiple CEW

Table 8.1 Comparison of experimental research on CEW in animal models and human subjects

Experimental research	Animal model	Human model
Experimental design	Often includes invasive, aggressive experimental intervention	More limited experimental intervention
Observation and data collection	Can include highly invasive monitoring, data collection	Generally non or less-invasive monitoring, data collection
Controlled trial	Occasional	More common
Sample size	Generally smaller	Generally larger
Subject health	Unknown	Generally healthy
Anesthesia or sedation	Yes	No
Applicability to humans in general	Controversial	Yes
Applicability to real-world CEW uses	Limited by applicability of animal models to humans	Limited by difficulty in replicating all field conditions in human experimental research

activations or illicit drug use, may not be practical, feasible, legal or ethically acceptable to perform [12,13]. In fact, research on animal models is longstanding, particularly in medicine, where such investigations often lay the groundwork for further studies in humans. For example, experimental studies in rats, dogs, and pigs were essential in furthering our understanding of cardiac arrest and in developing the nascent science of resuscitation in humans.

From the methodology standpoint, animal models offer distinct advantages in terms of data collection. Because animals are more “expendable,” highly invasive testing and monitoring can be performed, allowing more comprehensive observations to occur and data to be collected. One CEW study conducted in pigs included surgically opening the thoracic cavity to allow invasive cardiac monitoring, as well as the induction of life-threatening VF as part of the study design [4,14]. These types of studies would simply be impossible to conduct in human volunteers and certainly ethically unacceptable to institutional review boards. As a result, data collection in human CEW studies have been limited to noninvasive cardiac and respiratory monitoring and serial blood sampling in subjects.

In addition, because of the more liberal safety standards in animal studies (study subjects are often destroyed after the experiment), more aggressive experimental regimens can be performed. Animal CEW studies have included those involving multiple, serial, and repeated exposures to increasing electrical CEW charges, as well as the close proximity or even direct contact of CEW devices and probes to the heart in animal models [14]. Yet other studies have involved the intravenous administration of epinephrine or cocaine in order to simulate a hyperadrenergic or drug-intoxicated states [11,15,16].

Proponents of these types of experimental designs argue they are necessary because such research cannot ethically be performed on humans. Moreover, research in animal models allows the determination of important physiologic endpoints especially in light of CEW safety concerns [12,13]. For example, the study of 30-minute (minutes not seconds) exposures is simply inconceivable in humans [17]. Alternatively, others have argued that these aggressive regimens in animal models represent the "worst-case" scenario, going well beyond the type and nature of CEW exposures in humans in the field, and are simply not applicable to real-world incidents [18].

The experimental regimens in human studies tend to be just the opposite. Human research on CEWs can be challenging to perform because of the reluctance of human volunteers to participate, as well as the hesitance of research review committees to approve such studies in light of various public perceptions and the media attention regarding the safety of these devices. As a result, the actual experimental design and CEW exposure regimen in these studies is often limited to single, short-duration exposures in otherwise healthy human subjects. Critics argue that these studies, despite being conducted in human subjects, also do not reflect the typical conditions seen in field and, as such, their findings cannot be extrapolated to these situations.

There are a number of other important methodological differences and limitations between animal and human studies. In CEW animal research, subjects are generally sedated or anesthetized, introducing a significant confounding factor to these experimental trials. Alternatively, human study volunteers can undergo physiologic monitoring before, during and after CEW exposure without general anesthesia or sedatives.

Furthermore, for a variety of reasons including the expense of both animals and the laboratory, animal CEW studies have generally involved small numbers of subjects with many protocols having fewer than ten subjects, limiting the overall statistical power of their findings. Human studies have tended to rely on volunteer subjects, who are often forthcoming in greater numbers as a result of CEW training programs. In fact, the number of humans exposed to CEWs as a part of their training has been estimated to be more that 700,000 exposures [19].

Just as important, because of various restrictions, human studies have generally been limited to otherwise healthy subjects. In animal studies, information as to the baseline health and status of the animal subjects is usually not provided. Because animal studies often lack randomization, blinding or adequate control groups, any variability in the selection of animal subjects can have a profound impact on the results [20]. It is likely that this variability, along with the small experimental groups partially explains the often inconsistent findings from animal studies discussed in other chapters of this text.

In the following, we will briefly review some of the seminal animal and human CEW experimental research studies discussing both their strengths and weaknesses. More detailed review of these studies can be found in other chapters of this text.

8.2 Animal Studies

McDaniels and Stratbucker studied the advanced Taser M26 in five anesthetized dogs with an average weight of 25 kilograms [16]. Electrical discharge of the devices placed directly over the chest failed to induce ventricular fibrillation (VF). In 236 discharges, there were no recorded episodes of VF. The authors do note that when both probes were placed directly over the heart they were able to pace the heart similar in action to a pacemaker, but again were not able to induce VF. These data indicate that in dogs with weights and body sizes much smaller than humans, VF does not appear to be induced with Taser darts placed directly over the heart.

A study by Jauchem et al. investigated the metabolic effects of repeated activations of a CEW [21]. Sedated pigs weighing 49.5–58 kilograms received 5-second discharges alternating with 5 seconds of rest for three continuous minutes. Animals demonstrated transient, clinically insignificant increases in potassium and sodium, a significant decrease in blood pH that returned toward normal at 1-hour postexposure, a significant rise in blood lactate that returned to baseline at 2 hours, and a significant rise in whole blood pCO_2 that returned to baseline at 1 hour. Animals then underwent 1 hour of monitoring followed by an additional 3 minutes of the 5 seconds of CEW activation alternating with 5 seconds of rest. Additionally, the authors followed the levels of Troponin I, a cardiac-specific muscle enzyme that is a marker of damage to the heart. While the levels of this enzyme did not approach the predetermined cutoff of 0.35 nanogram per milliliter, there were slight rises after the discharges, though not to the level of statistical significance. The authors conclude that, although a 3-minute exposure as outlined above resulted in significant changes in blood chemistries, most levels returned to preexposure levels within in an hour after exposure [21].

Though humans are often exposed to repeated activations, typically the number of exposures does not approach 18 activations as occurred in this animal study. In this study, the pigs were anesthetized and the ventilator turned off, so the model does not recreate all real-world field conditions, thus the applicability of this study to humans has yet to be determined.

Wu et al performed a study aimed to find the distance from the tip of a Taser dart to the heart that caused VF in repeated attempts [14]. Using ten pigs weighing 53.8–74.4 kilograms, the animals were anesthetized, intubated, and monitored. One dart, they called the "heart dart," was placed over the right ventricle in decreasing distances (from 20 centimeters downward) after skin and muscle were dissected away. This was the distance that the authors measured. The other dart was placed on the abdominal surface, anywhere from 15 to 54 centimeters away from the "heart dart." Five-second Taser discharges were used for the simulation. Starting at 20 centimeters the discharge was delivered. If no VF was noted, then the distance was reduced by 2 centimeters and the discharge repeated. Once they induced VF, the investigators did additional testing to determine the energy for subsequent induction of VF. The dart-to-heart distance that caused VF on the first attempt was 17 millimeters ± 6.48 and for the average of subsequent attempts was 13.7 ± 6.79.

The authors then attempted to extrapolate their findings in pigs to humans. They measured skin to heart distance in 150 healthy human volunteers using echocardiography and found that distance to range from 10 to 57 centimeters. The authors then performed a calculation that shows that the probability of the "heart dart" landing in the 1 square centimeter body area susceptible VF and having a skin-to-heart distance necessary to induce VF, is an extremely low probability event [14].

The authors repeated the study using intact pigs with no gel tunnel directly to the heart and found significantly lower dart-to-heart distance were required to induce VF [22]. The paper then incorporated the findings in pigs with anatomic data obtained in humans in an attempt to probability of inducing VF in humans. As a theoretical model based on mixed subject model data, the results are challenging at best to put into clinical field perspective.

Lakkireddy et al. aimed to study the effects of cocaine on the Taser-induced VF threshold in a pig model, which is the amount of energy required to induce ventricular fibrillation [11]. Using five adult pigs weighing 34 ± 8.7 kilograms, they used a custom device to deliver multiples of the standard Taser discharge from the Taser X-26 in a step-up and step-down fashion in order to determine the VF threshold. They used five different locations of dart placement (three on the ventral surface and two on the dorsum). The results showed that in a single 5-second discharge, VF was not induced in any dart position. When looking for the VF threshold, the most sensitive position for VF induction was the location from the sternal notch to point of maximal intensity (PMI) position. The infusion of cocaine increased the safety margin of the device from 1.5 to 2 times from baseline. Plasma levels of cocaine and benzoylecgonine 30 minutes after infusion were 557 ± 280 U/l and 462 ± 123 U/l, respectively [6].

This study supports the safety profile in terms of VF in pigs, and potentially even greater safety profile in pigs with cocaine in their system, but the true human applicability remains to be determined. These findings appear to contradict clinical experience that cocaine sensitizes the heart and increases the risk for dysrhythmias in humans. However, some studies suggest that the sodium blocking effects of cocaine may increase, rather than decrease, the VF threshold under certain conditions [23].

8.3 Human Studies

Despite the challenges associated with human CEW experimental studies, there is growing body of recent literature and research on the physiologic effects of these devices in human subjects. Most of this work has been conducted by two research teams, Jeffrey Ho's team out of Minnesota (which includes the editors of this text), and a group at the University of California San Diego (UCSD) which includes the authors of this chapter. In large measure, the findings from both these research groups are complimentary and consistent in their findings.

Ho's group conducted a study sponsored by TASER International, Inc. investigating the effect of a standard 5-second Taser X-26 discharge on 66 human volunteers recruited at a training course [13]. Subjects had blood tests and electrocardiographic monitoring for cardiac and metabolic function before and up to 24 hours after CEW discharge. The investigators reported no evidence of significant metabolic or cardiac abnormalities other than in one subject who had a single elevated troponin level at 24 hours. This subject underwent further testing that demonstrated no evidence of cardiac injury. From their results, the investigators concluded they were unable to detect any abnormal cardiac rhythms, direct cardiac damage, or other evidence to suggest a link between CEWs and sudden death [24].

Similarly, the UCSD group conducted a comprehensive prospective trial funded by the US Department of Justice investigating the effects of a standard 5-second TASER X-26 discharge on 32 healthy law enforcement personnel (27 men and 5 women) who were undergoing training on the device and volunteered to receive a TASER activation [25]. Subjects were monitored in terms of their cardiovascular, respiratory, and metabolic physiology to determine the effect of the CEW discharge. In terms of cardiovascular physiology, there was no evidence of ischemia or interval abnormalities noted on the 12-lead ECG before or after TASER discharge. In terms of respiratory function, there was evidence of increased ventilation immediately following the exposure, but no evidence of abnormally low oxygen levels or elevated carbon dioxide levels. In terms of metabolic physiology, there was a transient changes in lactate and bicarbonate levels, but no evidence of acidosis. Otherwise, there were no significant abnormalities or differences in the other electrolyte measurements. The investigators concluded from this study that in healthy individuals, a 5-second Taser X-26 discharge did not result in any clinically significant changes in cardiovascular function, respiratory parameters, or metabolic physiology other than transient changes.

These two studies by Ho and the UCSD group represent the first large experimental human research studies, with a total of nearly 100 subjects, studying the physiologic effects of CEWs. The fact that they are remarkably consistent in their findings on the effects of a single 5-second CEW exposure in humans adds further credibility. However, while these studies have shown that single, short-duration CEW exposures have little risk of detrimental physiologic impact on healthy subjects in controlled clinical investigations, the effect of other factors, such as multiple or prolonged exposures, subject exertion, drug or alcohol intoxication or emotional or psychological stress, commonly encountered in the field setting, have yet to be determined. As such, it remains controversial how applicable the findings of these experimental studies in humans are to real-world CEW uses.

Recent preliminary research has now begun to focus on investigating these other factors experimentally in humans. Work from Ho's group reported on the effects of a continuous prolonged 15-second X-26 discharge (34 subjects) and multiple (three 5-second) exposures (15 subjects) in human volunteers placed in the supine position. The investigators reported that measures of ventilation actually increased during the discharge in both groups with a corresponding increase in oxygenation and decrease carbon dioxide concentrations [26]. The Ho group also studied the effect of CEWs in acidotic subjects following exertion who received either a CEW discharge or sham exposure [27]. The investigators reported no differences between the two groups in terms of respiratory parameters, metabolic lactate levels, or cardiac troponin levels. This same group of investigators also studied the effect of prolonged CEWs (15 seconds) on 25 human volunteers following an exercise regimen of push-ups and treadmill sprint until subjective exhaustion reporting no evidence of dysrhythmias or cardiac injury on 12-lead electrocardiograms [28].

The UCSD group recently conducted a prospective, cross-over controlled trial comparing cardiac, respiratory, and metabolic physiologic parameters in volunteers after exertion alone and exertion followed by a CEW discharge [25]. Initial results indicate no clinically significant or lasting statistically significant changes in blood measures or cardiovascular parameters in human subjects after rigorous exercise and a CEW exposure.

Recently, the Ho group presented data examining the effect of CEWs on intoxicated subjects [29]. In this industry-sponsored study, 26 human volunteers were given mixed drinks in a controlled setting to achieve a blood alcohol level of 0.08 milligram per deciliter, after which 22 of the subjects received a 15-second CEW discharge (4 subjects served as controls). Compared with controls, the CEW group demonstrated a transient increase in lactate and small drop in pH that corrected within 24 hours. There were no changes in markers of cardiac injury and no evidence of elevated troponin levels in either group. Similarly, the UCSD group found no increases in Troponin levels after CEW exposure in 66 subjects [30]. In another study of 31 subjects, Dawes et al. found no increase in core body temperature after a 15-second CEW exposure [31].

Each of these studies has limitations in terms of the experimental interventions performed, small subject numbers and group size, and limited control groups for comparison. None of the intervention factors studied such as exertion or intoxication can completely simulate conditions that occur in the field setting. However, as experimental trials in which human subjects were closely monitored to detect any physiologic changes associated with CEWs, other factors seen in the field, such as exertion or intoxication, or perhaps the combination of the two, these studies provide important data on the safety of these devices and lead the way for more extensive and comprehensive experimental human research on CEWs in the future.

References

1. Pippin JJ. Taser research in pigs not helpful. *J Am Coll Cardiol.* 2007;49(6):731–732; author reply 732–733.
2. Ferris LP, King BG, Spence PW, et al. Effect of electric shock on the heart. *Electr Eng.* 1936;55:498–515.
3. Schnabel PA, Richter J, Schmiedl A, et al. Patterns of structural deterioration due to ischemia in Purkinje fibres and different layers of the working myocardium. *Thorac Cardiovasc Surg.* 1991;39(4):174–182.
4. Howe BB, Fehn PA, Pensinger RR. Comparative anatomical studies of the coronary arteries of canine and porcine hearts. I. Free ventricular walls. *Acta Anat (Basel).* 1968;71(1):13–21.
5. Allison JS, Qin H, Dosdall DJ, et al. The transmural activation sequence in porcine and canine left ventricle is markedly different during long-duration ventricular fibrillation. *J Cardiovasc Electrophysiol.* 2007;18(12):1306–1312.
6. Pak HN, Kim YH, Lim HE, et al. Role of the posterior papillary muscle and purkinje potentials in the mechanism of ventricular fibrillation in open chest dogs and Swine: effects of catheter ablation. *J Cardiovasc Electrophysiol.* 2006;17(7):777–783.
7. Li GR, Du XL, Siow YL, et al. Calcium-activated transient outward chloride current and phase 1 repolarization of swine ventricular action potential. *Cardiovasc Res.* 2003;58(1):89–98.
8. Dalziel CF, Lee WR. Reevaluation of lethal electric currents. *IEEE Trans Ind Gen Appl.* 1968;IGA-4(5):467–476.
9. Geddes LA, Cabler P, Moore AG, et al. Threshold 60-Hz current required for ventricular fibrillation in subjects of various body weights. *IEEE Trans Biomed Eng.* 1973;20(6):465–468.
10. McDaniel WC, Stratbucker RA, Nerheim M, et al. Cardiac safety of neuromuscular incapacitating defensive devices. *Pacing Clin Electrophysiol.* 2005;28 Suppl 1:S284–287.
11. Lakkireddy D, Wallick D, Ryschon K, et al. Effects of cocaine intoxication on the threshold for stun gun induction of ventricular fibrillation. *J Am Coll Cardiol.* 2006;48(4):805–811.
12. Dorian P, Nanthakumar K. Taser research in pigs not helpful (reply letter). *J Am Coll Cardiol.* 2007;49(6):732.
13. Tchou P. Taser research in pigs not helpful (reply letter). *J Am Coll Cardiol.* 2007;49(6):733.
14. Wu JY, Sun H, O'Rourke AP, et al. Taser dart-to-heart distance that causes ventricular fibrillation in pigs. *IEEE Trans Biomed Eng.* 2007;54(3):503–508.
15. Nanthakumar K, Billingsley IM, Masse S, et al. Cardiac electrophysiological consequences of neuromuscular incapacitating device discharges. *J Am Coll Cardiol.* 2006;48(4):798–804.
16. McDaniel W, Stratbucker R, Smith R. Surface application of TASER stun guns does not cause ventricular fibrillation in canines. Paper presented at: *Proc Annu Int Conf IEEE Eng Med Biol Soc.* 2000.
17. Hughes E, Kennett M, Murray W, et al. *Electro-Muscular Disruption (EMD) Bioeffects: A Study on the Effects a Continuous Application of the TASER@X26 Waveform on Swine.* Philadelphia, PA: Penn State University; 2007.
18. Kroll MW, Calkins H, Luceri RM. Electronic control devices and the clinical milieu. *J Am Coll Cardiol.* 2007;49(6):732; author reply 732–733.
19. TASER_International. Field Use and Statistics. Available at: http://www.taser.com/research/statistics/Pages/FieldUseandStatistics.aspx
20. Pound P, Ebrahim S, Sandercock P, et al. Where is the evidence that animal research benefits humans? *Bmj.* 2004;328(7438):514–517.

21. Jauchem JR, Sherry CJ, Fines DA, et al. Acidosis, lactate, electrolytes, muscle enzymes, and other factors in the blood of Sus scrofa following repeated TASER exposures. *Forensic Sci Int*. 2006;161(1):20–30.
22. Wu J, Sun H, O'Rourke A, et al. Taser blunt dart-to-heart distance causing ventricular fibrillation in pigs. *IEEE Trans Biomed Eng*. 2007 Mar;54(3):503–508.
23. Tisdale JE, Shimoyama H, Sabbah HN, et al. The effect of cocaine on Ventricular fibrillation threshold in the normal canine heart. *Pharmacotherapy*. 1996;16(3):429–437.
24. Ho JD, Miner JR, Lakireddy DR, et al. Cardiovascular and physiologic effects of conducted electrical weapon discharge in resting adults. *Acad Emerg Med*. 2006;13(6): 589–595.
25. Vilke GM, Sloane CM, Bouton KD, et al. Physiological effects of a conducted electrical weapon on human subjects. *Ann Emerg Med*. 2007;50(5):569–575.
26. Ho JD, Dawes DM, Bultman LL, et al. Respiratory effect of prolonged electrical weapon application on human volunteers. *Acad Emerg Med*. 2007;14:197–201.
27. Ho J. Physiologic effects of prolonged conducted electrical weapon discharge on acidotic adults. *Acad Emerg Med*. 2007;14(5 Suppl. 1):s63.
28. Ho J. Absence of electrocardiographic change following prolonged application of a conducted electrical weapon in physically exhausted adults. *Acad Emerg Med* 2007;14(5 Suppl. 1):s128–s129.
29. Moscati R, Ho J, Dawes D, et al. Physiologic effects of prolonged conducted electrical weapon discharge on intoxicated adults. *Acad Emerg Med* 2007;14(5 Suppl. 1):s63–s64.
30. Sloane CM, Chan TC, Levine SD, et al. serum troponin i measurement of subjects exposed to the Taser X-26(R). *J Emerg Med*. 2008;35(1):29–32.
31. Dawes DM, Ho JD, Johnson MA, et al. 15-second conducted electrical weapon exposure does not cause core temperature elevation in non-environmentally stressed resting adults. *Forensic Sci Int*. 2008;176(2–3):253–257.

Chapter 9
Cardiac Arrhythmias

Derek J. Dosdall and Raymond E. Ideker

More than a century ago, investigators demonstrated that large electrical stimuli can cause sudden cardiac arrest [1,2]. Direct electrical stimulation of the heart can cause ventricular fibrillation (VF), which is a rapid uncoordinated contraction of the heart muscle that causes a loss of blood flow to the heart, brain, and other tissues. VF rapidly leads to a loss of consciousness which is followed by death unless electrical defibrillation is performed within a few minutes. Many studies have been performed during the past century to investigate the different types and strengths of electrical stimuli that may initiate cardiac arrhythmias [3].

Emergence of the TASER conducted electrical weapon (CEW) as a non-lethal weapon, particularly in the hands of law enforcement officers, has led to renewed interest in this question. Reports of individuals who have collapsed and died after being subdued with a TASER CEW have raised concerns that these devices might be causing cardiac arrhythmias that have lead to sudden cardiac arrest [4]. The few studies that have been published concerning CEWs in animal models have reported contradictory results. A study in pigs by McDaniel et al. using the waveform from the TASER X26 showed that the device did not produce neither cardiac arrhythmias or ectopic beats [5]. Another study in pigs by Nanthakumar et al. demonstrated a high incidence of cardiac capture with both the TASER X26 and TASER M26 [6]. A recent study in pigs showed that while VF was inducible with a TASER X26, VF induction was highly sensitive to dart location, and that VF induction in humans was unlikely [7]. These studies sought to determine the immediate effects of TASER CEW discharges on arrhythmia development, but no studies have been published to determine if TASER CEW use might cause arrhythmias minutes to hours after

R.E. Ideker (✉)
Departments of Medicine, Biomedical Engineering, and Physiology, University of Alabama-Birmingham, Volker Hall B140, 1670 University Blvd, Birmingham, AL 35294-0019, USA
e-mail: rei@crml.uab.edu

M.W. Kroll, J.D. Ho (eds.), *TASER® Conducted Electrical Weapons: Physiology, Pathology, and Law*, DOI 10.1007/978-0-387-85475-5_9,

shock administration, which is a time period during which many of the reported deaths have occurred [8].

This chapter will review modeling and experimental data to determine the likelihood that the use of a TASER CEW may lead either to immediate sudden cardiac arrest or to delayed sudden cardiac arrest minutes or hours later due to effects of the electrical discharge of a TASER CEW with probes in the vicinity of the heart. This analysis will primarily be concerned with the effect of TASER pulses in normal human adults. The effects of TASER pulses in children and in adults with cardiac disease or under the influence of drugs will not be reviewed.

9.1 Causes of Sudden Cardiac Death

Sudden cardiac death is usually caused by an underlying disease such as coronary artery disease, cardiomyopathy, left ventricular hypertrophy, valvular disease, congenital heart disease, electrophysiological abnormalities, or drug use [9]. Electrical stimulation by CEWs has not been shown to contribute to the development of these disease states or to accelerate sudden cardiac death due to any of these underlying mechanisms.

Myocardial infarction is usually the result of the blockage of coronary arteries causing damage or death within ischemic cardiac cells [10]. This coronary occlusion may be caused by atherosclerotic plaque, thrombus formation, or coronary artery constriction. During the acute phase of myocardial infarction, lethal cardiac arrhythmias may arise. Decreased cardiac function may also lead to heart failure or pulseless electrical activity. As the damaged areas heal, unexcitable scar tissue creates an area that may serve as an anchor for reentrant activation fronts or patchy bundles of viable myocardium within the scar may form the substrate for reentry. These reentrant circuits may cause ventricular tachycardia (VT) or ventricular fibrillation (VF).

Heart failure occurs when the heart muscle is weakened and does not pump blood efficiently to the body. Heart failure may contribute to death either directly because of poor cardiac function or indirectly by increasing the probability of sudden cardiac death due to an arrhythmia. Myocardial infarction, coronary artery disease, nonischemic cardiomyopathy, or other conditions that increase the work of the heart, such as high blood pressure, valve disease, congenital cardiac defects, or diabetes may contribute to the development of heart failure [11].

Pulseless electrical activity (PEA) is defined as organized electrical cardiac activity in the absence of a detectable pulse and is often due to severe cardiogenic shock [12]. A drop in blood pressure may lead to poor coronary artery perfusion, which decreases force and velocity of myocardial contraction. This, in turn, leads to decreased aortic pressure and further decreased coronary artery perfusion which continues a downward spiral that may lead to sudden cardiac death. The most common causes of PEA are respiratory insufficiency or arrest, myocardial dysfunction (myocardial ischemia, myocardial infarction, or congestive heart failure), preload reduction (due to blood loss or pulmonary embolism), or an increase in thoracic pressure (tension pneumothorax or cardiac tamponade). PEA often appears during sudden cardiac arrest after successful defibrillation shocks due to prolonged global ischemia which results from the several minutes of VF preceding defibrillation. It is not clear if the shocks actually cause the PEA or merely unearth it as a result of removing the fibrillation electrical activity.

Asystole, the absence of electrical and mechanical activity, and bradycardia, an abnormally slow heart rate, have been associated with defibrillation strength electrical shocks [13]. Again, it is not clear if the defibrillation shock is merely uncovering an underlying asystole. However, these arrhythmias typically occur with shocks much stronger than those associated with TASER CEWs and with transvenous shocking coils that are placed in close proximity to the sinoatrial or atrioventricular nodes [14,15]. Asystole is commonly seen in cardiac arrest deaths.

In the general population, VF, PEA, and asystole are each responsible for about 20–40% of cases of sudden cardiac death [16]. In individuals with excited delirium syndrome and/or who are being taken into custody by the police and experience sudden death, however, VF is uncommon. Approximately 95% of sudden deaths in such individuals are caused by bradycardia, PEA, and asystole [17–21].

9.2 Electrical Stimulation of an Ectopic Heartbeat

Direct electrical stimulation of sufficient strength and duration during portions of the cardiac cycle can initiate an ectopic activation or "cardiac capture" [22]. Implantable cardiac pacemakers make use of this principle by passing an electrical current through an electrode that is in direct contact with cardiac muscle to stimulate activations in patients with slow heart rates and other electrophysiologic problems. If the pacing pulse is of sufficient strength and duration to raise the transmembrane potential above the activation threshold of the cardiac cells immediately adjacent to the pacing electrode, a new action potential is formed [23]. The action potential then propagates away from the electrode site to excite the remainder of the myocardium as an ectopic heartbeat.

When an external current is applied to cardiac cells, the transmembrane potential does not change immediately because the cell membrane has capacitance in addition to resistance [24]. As long as the transmembrane potential is below the activation threshold, the transmembrane potential rises gradually as the membrane capacitance is overcome (Fig. 9.1). During subthreshold stimulation, the time constant is the time required to change the transmembrane potential 63% of the amount that it would change if the stimulation pulse were infinitely long. When a stimulation pulse is turned off before the activation threshold is reached, the transmembrane potential built up by the pulse on the

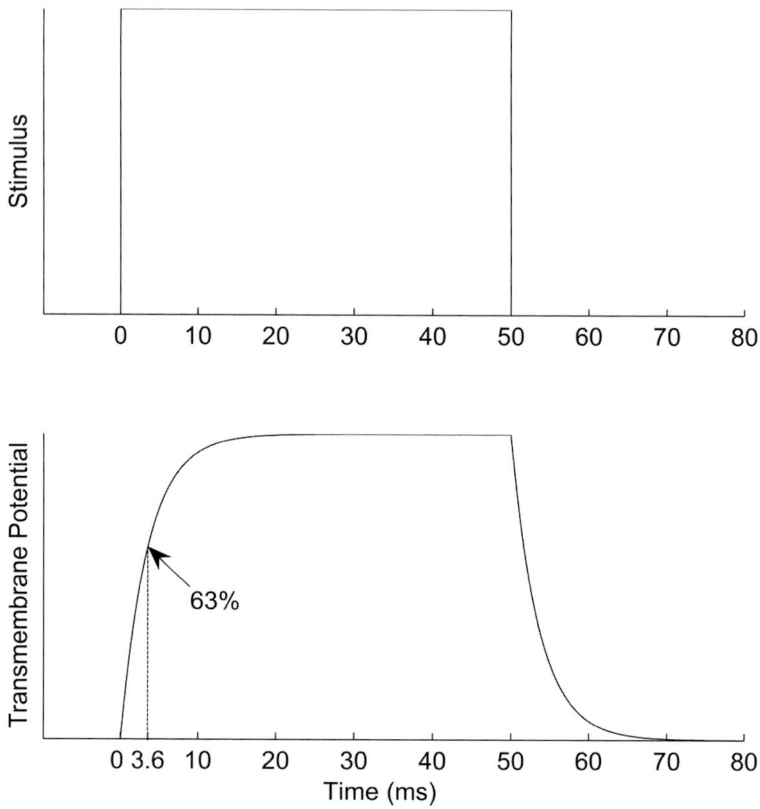

Fig. 9.1 An extracellular electrical stimulus (*top*) and the subthreshold change in transmembrane potential of a cardiac cell modeled as a resistor–capacitor circuit. The stimulus causes a change in transmembrane potential that increases as an exponential function of time. The time constant for stimulating an ectopic beat from the chest wall is taken to be 3.6 milliseconds (adapted from Ideker and Dosdall [60], by permission of Lippincott Williams & Wilkins)

cell membrane decays with a time constant similar to the time constant observed for changing the transmembrane potential during the pulse (Fig. 9.1).

Due to the time constant of the cell membrane, there is a strength–duration relationship to evoke an ectopic cardiac beat with an electrical stimulus. A short duration pulse requires a higher amplitude stimulus to initiate an action potential than does a long duration pulse (Fig. 9.2). This relationship has been shown to extend to very short duration pulses similar in duration to pulses from TASER CEW devices [25]. The time constant for electrodes in direct contact with human cardiac tissue has been determined experimentally [26,27]. However, the time constant for cardiac stimulation from the surface of the chest in humans has not been published, and studies have shown that the effective time constant increases as the distance between the stimulating electrode and cardiac

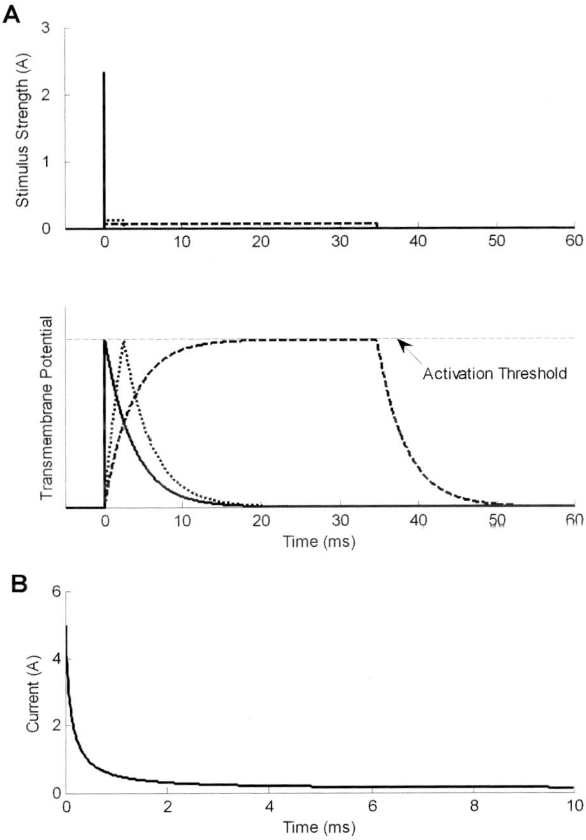

Fig. 9.2 Transmembrane potential is affected by the stimulus strength and duration. (**A**) Pulse durations of 0.1, 2.5, and 35 milliseconds can achieve the same transmembrane potential if the amplitudes of the stimuli are changed appropriately. The 0.1 millisecond pulse must have an amplitude of approximately 37 times the 35 milliseconds pulse to reach the same transmembrane potential. (**B**) The strength–duration curve for a cardiac cell with a time constant of 3.6 milliseconds is shown (adapted from Ideker and Dosdall [60], by permission of Lippincott Williams & Wilkins)

tissue increases [28]. The time constant for external pacing in dogs was determined in a pair of studies to be 2.6 and 4.6 milliseconds, for an average value of 3.6 milliseconds. This time constant will be used below to approximate the time constant for transchest human cardiac stimulation.

9.3 Fundamental Law of Electrostimulation

Early in the twentieth century, investigators developed the Fundamental Law of Electrostimulation. Weiss and Lapique developed strength–duration relationships for excitation [29,30]. Blair developed a simplified resistor–capacitor circuit model that approximated the subthreshold transmembrane response of tissue to a square wave stimulus [31]. Later, applications of the Fundamental Law of Electrostimulation were used to improve pacing and defibrillation waveforms [32–34].

The waveform of the TASER X26 is not a square wave but has a complex shape. An aspect of the Fundamental Law of Electrostimulation is that the stimulatory effect of a waveform, regardless of its shape, may be represented by a square wave with the same duration and with the average amplitude of the original waveform. Therefore, the cellular membrane response to the TASER X26 waveform can be approximated by calculating the cell membrane response of a 1 ampere square wave lasting 0.1 millisecond.

Published data showing the stimulus strength required to stimulate the heart from the chest surface with a pulse duration of 0.1 millisecond could not be found. Historical data describing pacing from the chest wall in human patients with pulses 20–40 milliseconds in duration provide an approximation of the strength required for cardiac stimulation. In ten studies involving a total of 196 individuals, the mean and standard deviation for external pacing was 64 ± 14 milliamperes for square wave pulses with an average duration of 35 milliseconds [35–44]. Based upon the strength–duration relationship with a time constant of 3.6 milliseconds, a square wave 0.1 millisecond in duration would require a stimulus approximately 37 times greater than a stimulus 35 milliseconds in duration (Fig. 9.2). Thus, the Fundamental Law of Electrostimulation predicts that the short pulse from the TASER X26 needs to be approximately 37 times greater than the long pulses required for external pacing, or approximately 2.3 ± 0.5 amperes to stimulate an ectopic beat. Since the TASER X26 delivers a pulse of approximately 1 ampere, which is 2.63 standard deviations less than that required to stimulate an ectopic beat, approximately 0.4% of people could experience an ectopic beat due to a TASER X26 pulse if the electrodes were positioned optimally on the chest as with the ten external pacing protocols mentioned above.

External pacing thresholds are highly dependent on electrode location. An experiment by Geddes et al. demonstrated that as electrodes are placed further away from the optimal pacing sites, the pacing threshold increases significantly [45]. In a protocol in dogs, they reported that moving an electrode on the left anterior wall approximately 4 centimeters from the location with the lowest pacing threshold doubled the pacing threshold, and moving the electrode 8 centimeters from the optimal site tripled the pacing threshold (Fig. 9.3). Therefore, a TASER X26 pulse should not be capable of producing an ectopic beat unless the probes were positioned optimally on the right and left anterior chest.

While the TASER X26 pulses are not typically capable of stimulating ectopic heartbeats, they do create strong contractions of skeletal muscle. The time constant of motor neurons, which in turn activate the skeletal muscle, is much shorter than the time constant for cardiac cells, so that the short duration pulses of the TASER X26 activate motor neurons at much lower stimulus amplitudes than for cardiac tissue [46]. The time constant for stimulation of motor neurons from electrodes on the chest of a dog is approximately 0.24 millisecond, while the time constant for stimulation of cardiac tissue is 3.6 milliseconds [47]. Due to this difference in time constant, a single TASER X26 pulse changes the transmembrane potential in a motor neuron approximately 34% of the amount

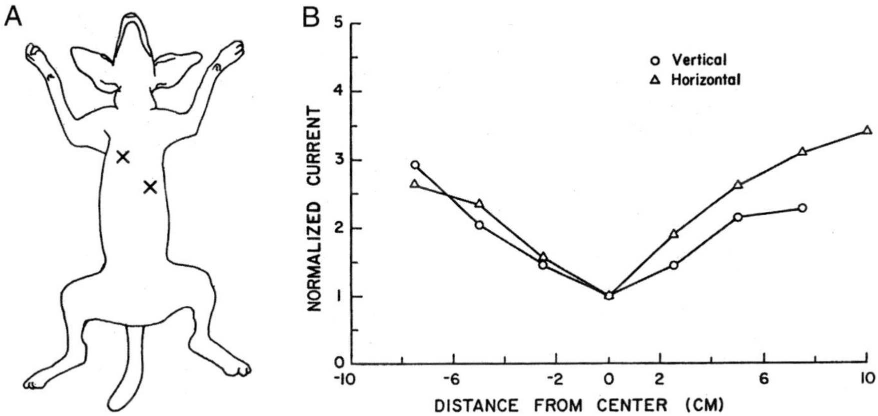

Fig. 9.3 External pacing thresholds are highly dependent upon electrode location. Pacing thresholds rapidly increase (**B**) as the pacing electrodes are moved from the optimal locations for external pacing (shown with Xs in (**A**)) (adapted from Geddes et al. [45], by permission of Blackwell Publishing)

it would change if the pulse were infinitely long, while the transmembrane potential of a cardiac cell changes less than 3% of the amount it would change if the pulse were infinitely long (Fig. 9.4).

Motor neurons are not only excited more easily by TASER X26 pulses, but they are also in closer proximity to the shocking electrodes and so are exposed

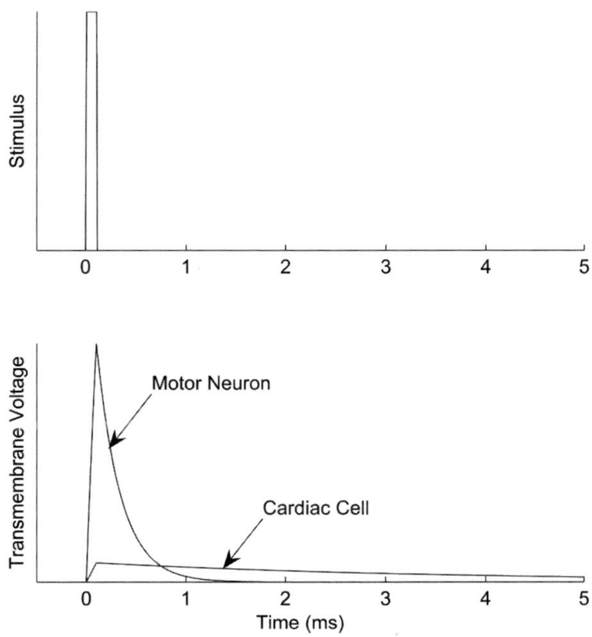

Fig. 9.4 Blair's model predicts that a TASER X26 pulse (**A**) changes the transmembrane potential in a motor neuron more than 12 times as much as in a cardiac cell (**B**) if the pulse strength is the same for the two structures (adapted from Ideker and Dosdall [60], by permission of Lippincott Williams & Wilkins)

to higher current densities than are the cardiac cells. The tissues surrounding the heart, such as the lungs and rib cage, have higher impedances than the more superficial skeletal muscles, which further reduces the current flow to the heart. In fact, studies have shown that even in studies with electrodes designed to deliver current to the heart from the chest surface, only 4–10% of the total current passes through the heart [48,49]. When the electrodes are placed elsewhere on the body, the amount of current that passes through the heart is much less.

A series of electrical pulses delivered in quick succession may potentially have a cumulative effect and increase transmembrane potential more than a single pulse (Fig. 9.4). A TASER X26 delivers pulses at intervals of approximately 53 milliseconds, which is nearly 15 times the time constant of cardiac tissue [5]. Since the transmembrane potential developed by a stimulus decays by approximately 63% for each time constant, the transmembrane potential would return to within 0.0001% of the initial voltage by the time the next TASER X26 pulse is delivered (Fig. 9.5). Therefore, essentially no cumulative effects of TASER pulses on cardiac tissue would be expected.

9.4 Electrical Stimulation of Immediate VF

An electrical shock may initiate VF if the shock causes an ectopic beat that follows the previous beat by a short interval, during a time called the vulnerable period. The stimulation threshold during the vulnerable period is higher than for stimuli delivered after a longer delay [50]. In animal models, the threshold for stimulation during the vulnerable period is approximately 12.6 times the minimum stimulation threshold [51]. Since the Fundamental Law of Electrostimulation predicts that a pulse 2.3 times stronger than a TASER X26 pulse would be required to produce an ectopic beat, a stimulus of approximately 29 times the strength of a TASER X26 pulse would be required to cause an ectopic beat during the vulnerable period, which is required to initiate VF. This estimate is for electrodes optimally placed on the chest for cardiac stimulation, which is presumably not the case in the majority of the TASER CEW applications. Therefore, it is highly unlikely that a TASER X26 pulse could initiate immediate VF by stimulating during the vulnerable period.

9.5 Electrical Stimulation of Delayed VF

VF causes an immediate loss of blood flow to the brain and is consequently accompanied by collapse and loss of consciousness of the subject within 5–10 seconds [52]. There have been reports of subjects who have lost consciousness minutes to hours after the administration of a TASER CEW. One theoretical mechanism for the delayed onset of VF would be if the TASER pulse initiated

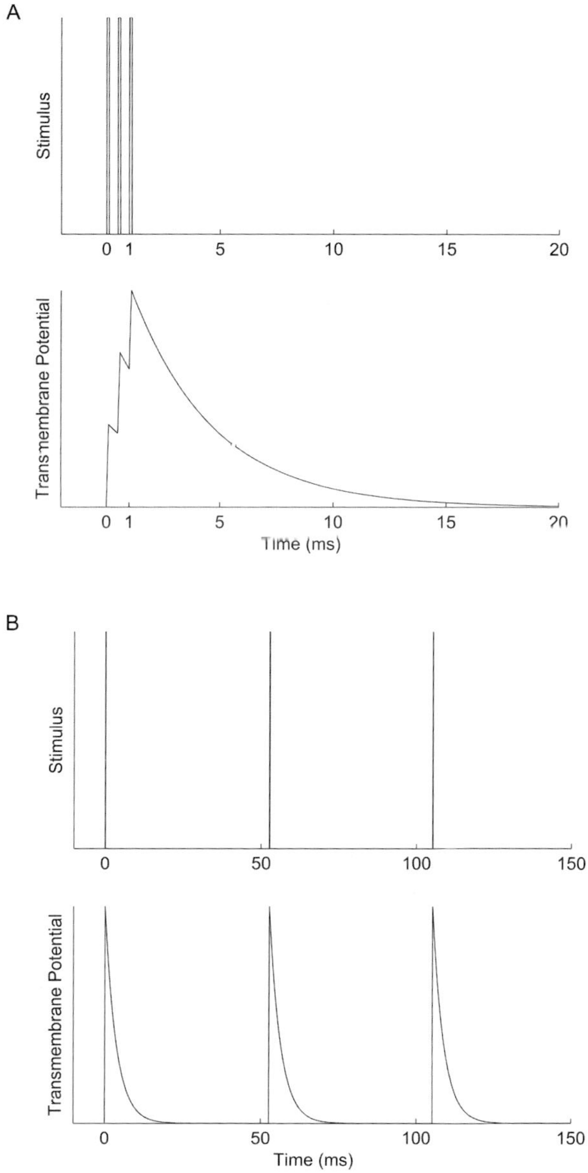

Fig. 9.5 Stimulation pulses that are delivered in rapid succession may have an additive effect on transmembrane potential (**A**), whereas transmembrane potential has time to return to the resting potential when stimuli are delivered with many time constants between pulses (**B**). Pulses with 0.5 millisecond between pulses (**A**) have an additive effect and produce a larger transmembrane potential change than pulses such as the TASER X26 pulses (**B**), which are delivered every 53 milliseconds (adapted from Ideker and Dosdall [60], by permission of Lippincott Williams & Wilkins)

VT, which then degenerated into VF after some period of time. For this to occur, the TASER pulse would still need to cause an ectopic beat to initiate the tachycardia, which is unlikely for the reasons outlined above. Another failing of this mechanism for delayed VF is that in normal human hearts, VT is an unstable arrhythmia that either converts back to sinus rhythm or to VF within a few beats [53,54]. VT is not sustained in hearts without surgical or infarct scar regions. Therefore, if a subject—without a history of cardiac disease—collapses minutes to hours after being subdued with a TASER CEW, it is unlikely that VT was initiated by the TASER pulse that subsequently degenerated into VF.

Another potential mechanism for the delayed onset of VF is if the TASER pulse caused damage to the heart through electroporation or heating and created a substrate that leads to an increased likelihood of arrhythmias. Cardiac arrhythmias and myocardial necrosis have been shown to develop minutes to hours after large electric shocks [55–57]. However, the amplitude of shocks that lead to myocardial necrosis and delayed cardiac arrhythmias is larger than TASER X26 shocks. The transmembrane potential required to cause myocardial necrosis is larger than that required to cause an ectopic beat or immediate VF [58]. Studies have shown that 150 joule defibrillation transthoracic defibrillation shocks, which typically have an amplitude of several amperes and a pulse duration of approximately 10 milliseconds, do not cause myocardial necrosis [59]. If defibrillation shocks of 150 joule, which deliver approximately 50 microcoulombs of charge, do not cause myocardial necrosis, it is unlikely that a TASER X26 pulse (approximately 0.1 microcoulombs) or even a 5-second series of TASER X26 pulses (approximately 5 microcoulombs) would be capable of causing myocardial necrosis and delayed cardiac arrhythmias due to this mechanism.

9.6 Conclusion

A review of relevant research into the mechanisms of shock-induced VF and the application of the Fundamental Law of Electrostimulation indicate that TASER X26 pulses do not reach the threshold for causing immediate or delayed onset VF. This conclusion is based on the accuracy of the Fundamental Law of Stimulation at small pulse widths (0.1 millisecond) and the accuracy of the depiction of the TASER X26 pulse in the paper of McDaniel et al. [5] Unless these assumptions are grossly in error, the safety factors associated with each of the mechanisms for VF discussed make it unlikely that the TASER X26 can cause VF directly due to the electrical effects of the device.

References

1. Hoffa M, Ludwig C. Einige neue versuche über herzebewegung. *Z Rationelle Med.* 1850;9:107–144.
2. Prevost JL, Battelli F. Sur quelques effets des décharges électriques sur le coeur des Mammifères. *Comptes Rendus des Seances, Academi des Sciences.* 1899;129:1267–1268.
3. Reilly JP. Cardiac sensitivity to electrical stimulation. In: Reilly JP, ed. *Applied Bioelectricity: From Electrical Stimulation to Electropathology*. New York: Springer; 1998: 194–239.
4. Strote J, Range Hutson H. Taser use in restraint-related deaths. *Prehosp Emerg Care.* Oct–Dec 2006;10(4):447–450.

5. McDaniel WC, Stratbucker RA, Nerheim M, et al. Cardiac safety of neuromuscular incapacitating defensive devices. *Pacing Clin Electrophysiol.* Jan 2005;28 Suppl 1:S284–287.
6. Nanthakumar K, Billingsley IM, Masse S, et al. Cardiac electrophysiological consequences of neuromuscular incapacitating device discharges. *J Am Coll Cardiol.* Aug 15 2006;48(4):798–804.
7. Wu JY, Sun H, O'Rourke AP, et al. Taser dart-to-heart distance that causes ventricular fibrillation in pigs. *IEEE Trans Biomed Eng.* Mar 2007;54(3):503–508.
8. Ho J, Reardon R, Heegaard W. Deaths in police custody: an 8 month surveillance study. *Annals Emerg Med.* 2005;46 (Suppl) abstract:S94.
9. Zipes DP, Wellens HJJ. Sudden cardiac death. *Circulation.* 1998;98:2334–2351.
10. Bunch TJ, Hohnloser SH, Gersh BJ. Mechanisms of sudden cardiac death in myocardial infarction survivors: insights from the randomized trials of implantable cardioverter-defibrillators. *Circulation.* May 8 2007;115(18):2451–2457.
11. Lane RE, Cowie MR, Chow AW. Prediction and prevention of sudden cardiac death in heart failure. *Heart.* May 2005;91(5):674–680.
12. Aufderheide T. Etiology, electrophysiology, and myocaridal mechanics of pulseless elecrical activity. In: Paradis NA, Halperin HR, Nowak RM, eds. *Cardiac Arrest: The Science and Practice of Resuscitation Medicine.* Baltimore, MD, USA: Williams & Wilkins; 1996:320–337.
13. Waldecker B, Brugada P, Zehender M, et al. Dysrhythmias after direct-current cardioversion. *Am J Cardiol.* 1986;57:120–123.
14. Deakin CD, Ambler JJ. Post-shock myocardial stunning: a prospective randomised double-blind comparison of monophasic and biphasic waveforms. *Resuscitation.* Mar 2006;68(3):329–333.
15. Khan IA. Atrial stunning: basics and clinical considerations. *Int J Cardiol.* Dec 2003;92(2–3):113–128.
16. Nichol G, Thomas E, Callaway CW, et al. Regional variation in out-of-hospital cardiac arrest incidence and outcome. *JAMA.* 2008;300(12):1423–1431.
17. Stratton SJ, Rogers C, Brickett K, et al. Factors associated with sudden death of individuals requiring restraint for excited delirium. *Am J Emerg Med.* May 2001; 19(3):187–191.
18. DiMaio TG, DiMaio VJM. *Excited Delirium Syndrome Cause of Death and Prevention.* Boca Raton: Taylor & Francis; 2006.
19. Paredes VL, Rea TD, Eisenberg MS, et al. Out-of-hospital care of critical drug overdoses involving cardiac arrest. *Acad Emerg Med.* Jan 2004;11(1):71–74.
20. Park KS, Korn CS, Henderson SO. Agitated delirium and sudden death: two case reports. *Prehosp Emerg Care.* Apr–Jun 2001;5(2):214–216.
21. Swerdlow C, Kroll M, Williams H, et al. Presenting rhythm in sudden custodial deaths after use of TASER® electronic control device. *Heart Rhythm* May 2008;5(5):S44.
22. Orias O, Brooks CM, Suckling EE, et al. Excitability of the mammalian ventricle throughout the cardiac cycle. *AJP.* 1950;163:272–282.
23. Wikswo JP, Jr., Wisialowski TA, Altemeier WA, et al. Virtual cathode effects during stimulation of cardiac muscle: Two-dimensional in vivo experiments. *Circ Res.* 1991;68:513–530.
24. Fozzard HA, Arnsdorf MF. Cardiac Electrophysiology. In: Fozzard HA, Haber E, Jennings RB, et al. eds. *The Heart and Cardiovascular Systems: Scientific Foundations.* Second Edition ed. New York, NY, USA: Raven Press; 1992.
25. Pearce JA, Bourland JD, Neilsen W, et al. Myocardial stimulation with ultrashort duration current pulses. *Pacing Clin Electrophysiol.* Jan 1982;5(1):52–58.
26. Fozzard HA, Schoenberg M. Strength-duration curves in cardiac Purkinje fibres: Effects of liminal length and charge distribution. *J Physiol.* 1972;226:593–618.

27. Smyth NP, Tarjan PP, Chernoff E, et al. The significance of electrode surface area and stimulating thresholds in permanent cardiac pacing. *J Thorac Cardiovasc Surg*. Apr 1976;71(4):559–565.
28. Zierhofer CM. Analysis of a linear model for electrical stimulation of axons–critical remarks on the "activating function concept". *IEEE Trans Biomed Eng*. Feb 2001;48(2):173–184.
29. Weiss G. Sur la possibilite de rendre comparables entre eux les apareils servant a l'excitation. *Arch Ital de Biol*. 1901;35:413–446.
30. Lapicque L. Definition experimentale de l'excitation. *Comptes Rendus Acad Sci Paris*. 1909;67(2):280–283.
31. Blair HA. On the intensity-time relations for stimulation by electric currents. II. *J Gen Physiol*. 1932;15:731–755.
32. Irnich W. The fundamental law of electrostimulation and its application to defibrillation. *Pacing Clin Electrophysiol*. 1990;13:1433–1447.
33. Kroll MW. A minimal model of the monophasic defibrillation pulse. *Pacing Clin Electrophysiol*. Apr 1993;16(4 Pt 1):769–777.
34. Walcott GP, Walker RG, Cates AW, et al. Choosing the optimal monophasic and biphasic waveforms for ventricular defibrillation. *J Cardiovasc Electrophysiol*. Sep 1995;6(9):737–750.
35. Luck JC, Grubb BP, Markel ML. Description of the strength-interval relation with external noninvasive pacing. *Pacing Clin Electrophysiol*. 1990;13:2031–2037.
36. Prochaczek F, Galecka J. The effect of suppression of the distortion artifact during transcutaneous pacing on the shape of the QRS complex. *Pacing Clin Electrophysiol*. 1990;13:2022–2025.
37. Altamura G, Toscano S, Bianconi L, et al. Transcutaneous cardiac pacing: Evaluation of cardiac activation. *PACE*. 1990;13:2017–2021.
38. Klein LS, Miles WM, Heger JJ, et al. Transcutaneous pacing: Patient tolerance, strength-interval relations and feasibility for programmed electrical stimulation. *Am J Cardiol*. 1988;62(16):1126–1129.
39. Kelly JS, Royster RL, Angert KC, et al. Efficacy of noninvasive transcutaneous cardiac pacing patients undergoing cardiac surgery. *Anesthesiology*. May 1989;70(5):747–751.
40. Kemnitz J, Winter J, Vester EG, et al. Transcutaneous cardiac pacing in patients with automatic implantable cardioverter defibrillators and epicardial patch electrodes. *Anesthesiology*. Aug 1992;77(2):258–262.
41. Falk RH, Battinelli NJ. External cardiac pacing using low impedance electrodes suitable for defibrillation: A comparative blinded study. *J Am Coll Cardiol*. 1993;22:1354–1358.
42. McEneaney DJ, Cochrane DJ, Anderson JA, et al. Ventricular pacing with a novel gastroesophageal electrode: a comparison with external pacing. *Am Heart J*. Jun 1997;133(6):674–680.
43. Prochaczek FB, Mugica J. Is the new electrode configuration a break point in transcutaneous cardiac pacing tolerance? *European J Biomed Tech*. 1994;16(3/4):98–101.
44. Estes NA, 3rd, Deering TF, Manolis AS, et al. External cardiac programmed stimulation for noninvasive termination of sustained supraventricular and ventricular tachycardia. *Am J Cardiol*. Jan 15 1989;63(3):177–183.
45. Geddes LA, Voorhees WD, 3rd, Babbs CF, et al. Precordial pacing windows. *Pacing Clin Electrophysiol*. Sep 1984;7(5):806–812.
46. Sweeney JD. Skeletal muscle response to electrical stimulation. In: Reilly JP, ed. *Electrical Stimulation and Electropathology*. New York, NY: Cambridge University Press; 1992:285–327.
47. Voorhees CR, Voorhees WD, 3rd, Geddes LA, et al. The chronaxie for myocardium and motor nerve in the dog with chest-surface electrodes. *IEEE Trans Biomed Eng*. Jun 1992;39(6):624–628.

48. Lerman BB, Deale OC. Relation between transcardiac and transthoracic current during defibrillation in humans. *CIRCRES*. 1990;67:1420–1426.
49. Camacho MA, Lehr JL, Eisenberg SR. A three-dimensional finite element model of human transthoracic defibrillation: Paddle placement and size. *IEEE Trans Biomed Eng*. 1995;42(6):572–578.
50. Cranefield PF, Hoffman BF, Siebens AA. Anodal excitation of cardiac muscle. *Am J Physiol*. 1957;190:383–390.
51. Voorhees WD, III, Foster KS, Geddes LA, et al. Safety factor for precordial pacing: minimum current thresholds for pacing and for ventricular fibrillation by vulnerable-period stimulation. *Pacing Clin Electrophysiol*. 1984;7:356–360.
52. McQuillen EN, McQuillen JB. Pain and suffering . . . and unconsciousness. *Am J Forensic Med Pathol*. Jun 1994;15(2):174–179.
53. Morady F, Shapiro W, Shen E, et al. Programmed ventricular stimulation in patients without spontaneous ventricular tachycardia. *Am Heart J*. May 1984;107(5 Pt 1):875–882.
54. Brugada PW, Wellens HJJ. Programmed electrical stimulation of the human heart. In: Josephson MEW, Wellens HJJ, ed. *Tachycardias-Mechanisms, Diagnosis, Treatment*. Philadelphia, PA: Lea & Febiger; 1984:61–89.
55. Wit AL, Janse MJ. Ventricular arrhythmias in the acute phase of myocardial ischemia and infarction. *The Ventricular Arrhythmias of Ischemia and Infarction: Electrophysiological Mechanisms*. Mount Kisco: Futura Publishing Company, Inc.; 1993:161–266.
56. Van Fleet JF, Tacker WA. Cardiac damage from transchest and ICD defibrillator shocks. In: Tacker WA, Jr., ed. *Defibrillation of the Heart: ICDs, AEDs, and Manual*. St. Louis: Mosby-Year Book, Inc; 1994:259–298.
57. Tung L. Electrical injury to heart muscle cells. In: Lee RC, Cravalho EG, Burke JF, eds. *Electrical Trauma: The Pathophysiology, Manifestation, and Clinical Management*. Cambridge: University of Cambridge Press; 1992:361–400.
58. Lee RC, Zhang D, Hannig J. Biophysical injury mechanisms in electrical shock trauma. *Annu Rev Biomed Eng*. 2000;2:477–509.
59. Walcott GP, Killingsworth CR, Ideker RE. Do clinically relevant transthoracic defibrillation energies cause myocardial damage and dysfunction? *Resuscitation*. Oct 2003;59(1):59–70.
60. Ideker RE, Dosdall DJ. Can the direct cardiac effects of the electric pulses generated by the TASER X26 cause immediate or delayed sudden cardiac arrest? *Am J Forensic Med and Path*. 2007;28(4):195–201.

Chapter 10
Electrocardiographic Effects of the CEW

Jeffrey D. Ho*

With the advent of the Advanced TASER® M26 followed soon after by the X26 conducted electrical weapons (CEWs), initial concern for any possible physiologic damage was centered in the area of the cardiac conduction system. While this concern has been theoretical, it makes sense from the standpoint that a large number of people, including educated professionals, do not understand the basics of electrical current except what they were taught by their guardian growing up (e.g., don't stick your finger in the electrical socket or you will kill yourself) or what they have gleaned over the years from literal experience or mass media advertisement (e.g., be careful when working on your roof not to come into contact with the overhead electrical power lines or you will electrocute yourself). What all of these scenarios have in common is that if sudden death from electrical current comes, it arrives in the form of a cardiac arrhythmia. Because of this, the initial focus of research into whether CEWs are causally connected to sudden custodial death was in the cardiac rhythm arena.

It has been theorized that CEWs have been associated with several sudden and unexpected subject deaths while in law enforcement custody. This "in-custody death" (ICD) phenomenon is not new and similar phenomena have been described in psychiatric literature dating back to the mid-1800s [1]. Over the years, there have been attempts to link ICD with single causative factors such as extreme agitation (also known as excited or agitated delirium syndrome), use of chemical irritants (e.g., pepper spray), restraint and positional

---

**Jeffrey D. Ho reports serving as an expert research consultant and medical expert consultant to TASER International, Inc. and reports as a personal shareholder of TASER International stock. No other potential conflict of interest relevant to this chapter was reported.*

J.D. Ho (✉)
Department of Emergency Medicine, Hennepin County Medical Center, University of Minnesota Medical School, Minneapolis, MN, USA
e-mail: hoxxx010@umn.edu

M.W. Kroll, J.D. Ho (eds.), *TASER® Conducted Electrical Weapons: Physiology, Pathology, and Law*, DOI 10.1007/978-0-387-85475-5_10,

asphyxia, structural cardiac abnormalities, or use of illicit stimulant medication [2–15]. Many of these links have been questioned, disproved or found to be absent [16–22]. This has generated more questions than answers in the search for a common cause.

A more recent theory is that of the "CEW-induced" ICD (CEWICD). It is a perception by many that because a CEW incapacitates through the passage of electrical current, it is somehow causing death, presumably from an electrically induced fatal arrhythmia. There have been media sources that have incorrectly compared CEWs to the electric chair used in capital punishment although the electric chair passes current that is magnitudes greater than any CEW on the market [23]. If the CEWICD theory is correct, it would be expected that electrically induced fatal arrhythmias would be inducible in the lab setting.

The data from animal modeling has been confounding. There have been lab studies demonstrating that a CEW applied to an anesthetized and sometimes anatomically manipulated animal model may lead to induction of arrhythmia [24–27]. There have also been animal model studies with opposite conclusions and results demonstrating wide safety margins with regard to ability to induce cardiac arrhythmia [28]. This same study showed that animals with heavier masses required even greater outputs. Additionally, there have been instances when persons of small stature have experienced a CEW deployment without evidence of sudden death [29, 30]. In one case, current had to pass through an 18-month-old baby to save his life when the father had a knife to his throat, thus saving the baby's life [31]. Collectively, these data have generated debate among scientists and investigators. However, it is important to keep in mind that these findings were discovered in nonhuman modeling situations. While animal research is helpful and can often point research in a certain direction, it may not be the best way to search for answers in humans.

10.1 Human Data

There have been several studies that have attempted to look at the CEW effects on human cardiac physiology. Each of these has had a slightly different methodology. However, unlike the animal model data, human research appears to be in agreement with regard to CEW cardiac safety.

Perhaps the first indirect glimpse of human cardiac effect came from a CEW manufacturer (TASER International, Inc., Scottsdale, AZ) itself. This manufacturer has stated that they have data on over 650,000 human volunteers from CEW training exposures with insignificant effects on heart rhythms [32]. A more important indirect observation from this data is that there were no subject collapses immediately following their CEW exposure. This is an important finding since it is understood that if applied electrical current is going to be fatal from a cardiac rhythm standpoint in an otherwise unblemished heart, it will be fatal rather quickly if not immediately since electricity is not somehow stored up in the body to provide a later effect [33]. Therefore, if a CEW application is going to be fatal to a human subject, it would be fatal very soon following the CEW application. The manufacturer data does not support this ever happening. In addition to this, there is custodial death data from Ho et al. presented in 2005 that also speaks against this [34]. In this work, 8 months of custodial death cases were reviewed. In those cases that involved application of a CEW at some point prior to the subject's death, no death occurred immediately following the CEW application. Death occurred minutes, hours, and even days later in all subjects. From an investigative standpoint, this would suggest that a CEW does not appear to have a relationship with causing fatal cardiac rhythms. These two pieces of data were perhaps the first evidence of relative cardiac safety for the CEW although some debate these data since it received partial funding by a CEW manufacturer.

In 2007, Levine et al. presented work demonstrating that the normal heart rhythm seen on an electrical heart tracing remained unchanged after a brief exposure to a CEW [35] and Vilke et al. produced similar work using before and after exposure electrocardiograms [36]. This work was preceded by a comprehensive study conducted by Ho et al. and published in 2006 [37]. All studies used TASER X26 CEW probes fired into human subjects or manually attached by electrodes. Levine examined a cardiac rhythm strip while Vilke and Ho examined 12-lead electrocardiograms (ECGs). None of the studies were able to demonstrate a significant postexposure change in heart rhythm. The criticisms of these studies were that they were only for short-duration exposures (average of 3 seconds in Levine's study and a fixed 5 seconds in Vilke's and Ho's studies) and all studies were performed on resting human subjects. Although no cardiac rhythm changes were demonstrated, there was concern that most CEW exposures in the field are on subjects who are exerting themselves due to fleeing and fighting with law enforcement so that their metabolic status may not be well represented in either of these studies.

Because of this potential metabolic confounder, Ho et al. performed a follow-up ECG study and presented these findings in 2007 [38]. In this study, human volunteers again received baseline ECGs. They were then subjected to a strenuous, anaerobic exercise regimen. This included 30 seconds of timed push-ups followed immediately by a nontimed sprint on a treadmill set at 8 miles per hour and 8° of elevation. For this regimen, the volunteers were instructed to perform as many push-ups as they possibly could in that time period and to run until subjective exhaustion which was measured by their inability to continue to keep pace with the treadmill. Once they completed this regimen, a blood pH sample was taken to gauge their level of exhaustion. This was immediately followed by a 15-second CEW application with a TASER X26 device. This exposure consisted of prepositioned probes placed in various realistic positions involving the thorax and the extremities. Immediately following this exposure, another ECG was obtained to determine if there were any changes from baseline.

This study had two notable points of discussion. First, it used exhausted human subjects which is a model that more closely mimics field situations. And second, the application duration was increased by 3× the durations measured in previous human studies. Intuitively, this greater application time should yield significant findings if any were to be found. The conclusions of this study were that there were no concerning changes in the ECGs found. In fact, one of the ECGs was abnormal at baseline and had normalized after the study protocol had been completed. The conclusions of this study were that theories of CEW induced cardiac rhythm abnormalities are not supported, even in exhausted, acidotic persons receiving a prolonged CEW application.

In looking at the data that exist in the literature regarding CEW association with a dysrhythmia, there has been a single human case report and some studies involving animal models. The single human case report is a letter to the editor of a medical journal of a case of ventricular fibrillation after exposure to a CEW [39]. A violent subject exhibiting signs of excited delirium was briefly subdued with a short ECD discharge. Paramedics were present and found a normal pulse and respiration following the exposure. After a delay of 14 minutes the subject collapsed and had resuscitative measures given. Eventually a recorded rhythm strip showing ventricular fibrillation was obtained. A total of 23 minutes elapsed between the ECD application and the recorded rhythm strip. Since this rhythm did not occur immediately following the CEW application, it appears to be very similar to every other described in the literature in which the CEWICD event occurs proximal to CEW exposure but collapse is not instantaneous. The facts of this case report do not support an electrically induced dysrhythmia.

10.2 Animal Data

One of the animal studies that should be addressed here is the study by Wu et al. from the University of WI, Madison [40]. In this study, a pig model is used. The author's premise is that because pigs have a layer of tissue that is of different density on their chests than what humans have, there is a need to manipulate the model to account for this difference in density. The manipulation of the model includes removal of tissue overlying the animal's chest, the creation of holes into the animal's ribcage, filling of these holes with an electrode conducting gel, and placing the CEW electrodes into the holes. In doing so, upon discharging the CEW they found that they were able to cause the dangerous heart rhythm known as ventricular fibrillation when the electrodes were within 17 millimeters of the heart. The conclusion made from this study is that it is possible to cause ventricular fibrillation with a CEW in humans under very specific and rare circumstances (e.g., very thin, asthenic person with no breast tissue with probe locations in positions on the anterior chest to maximize the proximity to the underlying heart).

The findings of the Wu study should be interpreted with caution. Pigs are not humans and the authors made some significant changes to the anatomy of the animal in order to produce their findings of ventricular fibrillation. It is not certain how to compare a model of a stripped pig with drilled holes and electrode gel to an intact human. In 2006, Lakireddy et al. directly inserted barbs over pig hearts in intact pigs to their full depth over the cardiac apex. They achieved tip-to-heart spacings of 16 $\pm$ 3.3 millimeters which are not significantly different from the Wu spacings [41]. However, they never induced VF. Most illuminating, the same group repeated the experiment with the gel injection and had dramatically different results finding that the barb had to come within 6 millimeters to induce VF [42].

From personal experience, I can also give the following example that points out a flaw in Wu's methodology. At my institution, I teach a critical care class that uses an animal model for numerous procedures. It is used to teach physicians life-saving techniques for emergency patient care. The very last procedure in this class is to open the animal's chest, a technique known as an "emergency thoracotomy." Soon after this procedure is accomplished, the animal is euthanized and the lab ends. One of the techniques we have used to euthanize the animal is to take a 9-volt transistor battery and touch it to the animal's heart. This causes instant ventricular fibrillation each and every time. The point of this is that if I manipulate the animal's anatomy and lessen the distance to the heart, it takes very little electrical current to induce ventricular fibrillation. This appears to be the same principle of Wu's study since it uses significantly altered animal anatomy that lessens the distance to the heart and it, therefore, becomes much easier to create abnormal heart rhythms.

Additional pig studies to be aware of come from the Dennis and Walter et al. group in Chicago [26,27]. These studies again use animal models to demonstrate the ability to cause abnormal rhythms when the animal is subjected to a CEW discharge under artificial lab conditions. The animals used for their studies are much smaller in weight and mass than the classic ICD human subject in the field. Additionally, Ho et al. have reproduced their methodology in humans with opposite results with respect to induction of cardiac dysrhythmias (Fig. 10.1) [43]. It is difficult to consider the pig to be the best model of assessment when human work in this area exists and has an opposite result as has been discussed above.

A final paper that concerns the cardiovascular system and needs to be discussed is the Nanthakumar et al. study from Toronto, Canada [44]. This is an anesthetized animal model study that evaluated multiple CEW discharges across the chest and also the abdomen. It also had the variable of administered epinephrine to simulate the "fight or flight" type of stress seen in persons with agitation. This response was theorized to be similar to that seen in persons who are at highest risk for experiencing a CEWICD. The study found that the CEW was able to "capture" the heart on some of the discharges (meaning that it could cause the heart to beat at a certain non-native rate) and also that it caused two potentially dangerous heart rhythms on two of the discharges after the epinephrine had been administered. The

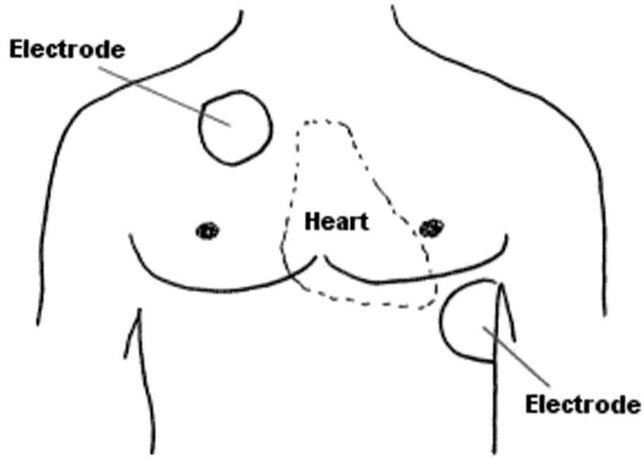

Fig. 10.1 Worst-case scenario positioning of CEW electrodes for Ho et al. study [43]

author's conclusion from this study is that CEWs may pose cardiac risk and further investigation is needed.

While this is a scientific study with a valid methodology, it is important to separate out those items that may lead to a false conclusion. The study was conducted on a pig model which is not necessarily equivalent to a human. Additionally, the animals used in Nanthakumar's study were used over and over again. There were only six pigs subjected to a total of 150 CEW discharges that average out to over 20 discharges per animal. This is a situation that does not reproduce a real-life scenario very well. While there are situations of repetitive CEW discharges on humans in the field, those situations numbering 20 or more would be rare if they have occurred at all. It is difficult to know what the effect of 20+ discharges is on a human since the total number of repetitive discharges on a human undergoing study has been 3, and repetitive human studies have not shown results similar to Nanthakumar [38].

Also, the addition of epinephrine into the animals is a source of confusion. The authors did this to simulate a stress response; however, epinephrine is a potent stimulant. One of the known side effects of epinephrine is induction of dangerous heart rhythms [45]. In Nanthakumar's study, the only two times the animals demonstrated a dangerous heart rhythm were after administration of epinephrine. Because it is difficult to know if the epinephrine dose given to the animals was or was not sufficient to create the simulated condition of stress, it is quite possible that what it did was create a situation of cardiac irritability leading to a dangerous rhythm as a side effect. This study needs to be interpreted with caution when attempting to apply it to the human population. Nanthakumar et al. have also agreed with this approach by stating, "[W]e did not state that NIDs (CEWs) cause ventricular fibrillation in humans, and we agree that we cannot conclude from our study that NID discharges cause arrhythmias in typical use [46]."

Table 10.1 Summary of human CEW studies examining cardiac physiology

Author	Year	Study type	# of subjects	Exposure duration (seconds)	Study method	Results	Citation #
Ho et al.	2006	Serial ECG	32	5.0	RA, FP	NC	[37]
Vilke et al.	2007	Before and after ECG	32	5.0	RA, AE	NC	[36]
Levine et al.	2007	Continuous rhythm strip	105	0.9–5.0	RA, FP or AE,	NC, HRI	[35]
Ho et al.	2007	Before and after ECG	25	15.0	EA, AE	NC	[38]
Ho et al.	2008	Continuous cardiac ultrasound	34	10.0	RA, AE, WC	NRCD, maximal heart rate 158	[43]

RA, Resting adults; EA, Exhausted adults, FP, Fired probes to thorax, AE, Attached electrodes to thorax, WC, Electrodes attached in "worst-case scenario" position on thorax (Fig. 10.1); NC, No ECG rhythm change after exposure reported; HRI, Heart rate increases after exposure reported; NRCD: No rhythm change during exposure reported.

10.3 Conclusion

There is a growing body of research information in the area of electrocardiographic effects of CEWs. While some of the animal data can be confounding, the human data available does not support the theory of CEWs being able to induce clinically significant arrhythmias in humans and is summarized in Table 10.1. This data also does not support CEWICD from an electrically induced arrhythmogenic standpoint. Future research will likely be conducted in this area to validate the current data that can only be seen as helpful as CEWs continue to be investigated.

References

1. Bell L. On a form of disease resembling some advanced stages of mania and fever, but so contradistinguished from any ordinary observed or described combination of symptoms as to render it probable that it may be overlooked and hitherto unrecorded malady. *Am J Insanity*, 1849;6:97–127.
2. Di Maio TG and VJM Di Maio. Excited delirium syndrome cause of death and prevention. 1st ed. Boca Raton, FL: Taylor & Francis Group, 2006.
3. American Civil Liberties Union of Southern California. "Pepper spray update: more fatalities, more questions." *Pepper Spray Update*, June, 1995. (Accessed November 1, 2005 at http://www.aclu-sc.org/attachments/p/Pepper_Spray_New_Questions.pdf).
4. Stratton SJ, Rogers C, Brickett K, and G Gruzinski. Factors associated with sudden death of individuals requiring restraint for excited delirium. *Am J Emerg Med*, 2001;21:187–91.
5. Eckart RE, Scoville SL, Campbell CL, et al. Sudden death in young adults: a 25-year review of autopsies in military recruits. *Ann Int Med*, 2004;141:829–35.

6. Maron BJ, Roberts WC, and SE Epstein. Sudden death in hypertrophic cardiomyopathy: a profile of 78 patients. *Circulation*, 1982;65:1388–94.
7. Ross DL. Factors associated with excited delirium deaths in police custody. *Mod Pathol*, 1998;11:1127–37.
8. Reay DT, Fligner CL, Stilwell AD, and J Arnold. Positional asphyxia during law enforcement transport. *Am J Foren Med Pathol*, 1992;13:90–7.
9. O'Halloran RL and LV Lewman. Restraint asphyxia in excited delirium. *Am J Foren Med Pathol*, 1993;13:289–95.
10. O'Halloran RL and JG Frank. Restraint asphyxia. *Am J Foren Med Pathol*, 2000;21:420–22.
11. O'Halloran RL and JG Frank. Asphyxial death during prone restraint revisited: a report of 21 cases. *Am J Foren Med Pathol*, 2000;21:39–52.
12. Wetli CV, Mash D, and SB Karch. Cocaine-associated agitated delirium and the neuroleptic malignant syndrome. *Am J Emerg Med*, 1996;14:425–8.
13. Karch SB and BG Stephens. Drug abusers who die during arrest or in custody. *J R Soc Med*, 1999;92:110–3.
14. Pestaner JP and PE Southall. Sudden death during arrest and phencyclidine intoxication. *Am J Foren Med Pathol*, 2003;24:119–22.
15. Ruttenber AJ, Lawler-Heavner J, Yin M, Wetli CV, Hearn WL, and DC Mash. Fatal excited delirium following cocaine use: epidemiologic findings provide new evidence for mechanisms of cocaine toxicity. *J Forensic Sci*, 1997;42:25–31.
16. United States Department of Justice, Office of Justice Programs, National Institute of Justice. "The effective ness and safety of pepper spray." *Research for Practice*, April, 2003. (Accessed June 10, 2008 at http://www.ncjrs.gov/pdffiles1/nij/195739.pdf).
17. Chan TC, Vilke GM, Clausen J, et al. The effect of oleoresin capsicum "pepper" spray inhalation on respiratory function. *J Foren Sci*, 2002;47:299–304.
18. Chan TC, Vilke GM, Neuman T, and JL Clausen. Restraint position and positional asphyxia. *Ann Emerg Med*, 1998;32:116–8.
19. Chan TC, Vilke GM, and T Neuman. Reexamination of custody restraint position and positional asphyxia. *Am J Foren Med Pathol*, 1998;19:201–5.
20. Chan TC, Neuman T, Clausen J, Eisele J, and GM Vilke. Weight force during prone restraint and respiratory function. *Am J Foren Med Pathol*, 2004;25:185–9.
21. Kornblum RN and SK Reddy. Effects of the TASER in fatalities involving police confrontation. *J Foren Sci*, 1992;37:956–8
22. Glatter K and SB Karch. Positional asphyxia: inadequate oxygen, or inadequate theory? *Foren Sci Int*, 2004;141:201–2.
23. Anonymous. "Taser packs potent but brief punch of electricity." *USA Today* 3 June 2005, sec A:13.
24. Will JA, Wu JY, Sun H, et al. Can TASERs directly cause ventricular fibrillation? Presented at the meeting of Experimental Biology, April, 2006.
25. Nanthakumar K, Billingsley IM, Masse S, et al. Cardiac electrophysiological consequences of neuromuscular incapacitating device discharges. *J Am Coll Cardiol*, 2006;48:798–804.
26. Dennis AJ, Valentino DJ, Walter RJ, et al. Acute effects of TASER X26 discharges in a swine model. *J Trauma*, 2007;63:581–90.
27. Walter RJ, Dennis AJ, Valentino DJ, et al. TASER X26 discharges in swine produce potentially fatal ventricular arrhythmias. *Acad Emerg Med*, 2008;15: 66–73.
28. McDaniel WC, Stratbucker RA, Nerheim M, and JE Brewer. Cardiac safety of neuromuscular incapacitating defensive devices. *PACE*, 2005;28(supplement): S284–7.
29. United Press International. "Police use stun gun on 6 year old." *The Washington Times*, November 12, 2004. (Accessed November 1, 2005 at http://washingtontimes.com/upi-breaking/20041112-111756-3329r.htm.)

30. Anonymous. Police review policy after tasers used on kids. *CNN*, November 15, 2004. (Accessed November 1, 2005 at http://www.cnn.com/2004/US/11/14/children.tasers/.)
31. Taser International. *Deadly Rhetoric: How the ACLU of Northern California's Fight Against Law Enforcement Control Tools Endangers Communities*. Taser International, 16 February 2006. Available at: http://www.taser.com/savinglives/documents/Deadly%20Rhetoric%20V11.pdf.
32. TASER International website, Physician FAQ section (monitoring). Accessed on June 9, 2008 at: http://taser.com/research/statistics/Pages/FieldUseandStatistics.aspx.
33. Wright RK. Electrical Injures. *eMedicine*. J Balantine, F Talavera, E Legome, J Halamka, and B Brenner (eds.), 2005. (Accessed June 10, 2008 at http://www.emedicine.com/emerg/topic162.htm.)
34. Ho JD, Reardon RF, and WG Heegaard. Deaths in police custody: an 8 month surveillance study. *Annals Emerg Med*, 2005;46(suppl):S94.
35. Levine SD, Sloane CM, Chan TC, et al. Cardiac monitoring of human subjects exposed to the TASER. *J Emerg Med*, 2007;33:113–7.
36. Vilke GM, Sloane S, Levine S, et al. Twelve-lead electrocardiogram monitoring of subjects before and after voluntary exposure to the TASER X26. *Am J Emerg Med*, 2007;26:1–4.
37. Ho JD, Miner JR, Lakireddy DR, et al. Cardiovascular and physiologic effects of conducted electrical weapon discharge in resting adults. *Acad Emerg Med*, 2006;13:589–95.
38. Ho J, Dawes D, Calkins H, and M Johnson. Absence of electrocardiographic change following prolonged application of a conducted electrical weapon in physically exhausted adults. *Acad Emerg Med*, 2007(supplement 1);14:S128–9.
39. Kim PJ and WH Franklin. Ventricular fibrillation after stun-gun discharge. *NEJM*, 2005;353:958–9.
40. Wu J, Sun H, O'Rourke A, Huebner S, Rahko P, Will J, Webster J. Taser dart-to-heart distance that causes ventricular fibrillation in pigs. *IEEE Trans Biomed Eng*, 2007;54:503–8.
41. Lakkireddy D, Kowalewski W, Wallick A, Verma A, Martin K, Ryschon J, Butany J, Natale A, and P Tchou. Cardiovascular safety profile of electrical stun guns (TASER®): Impact of point of delivery on ventricular fibrillation thresholds. *Heart Rhythm* 2006;3:S249.
42. Wu JY, et al. Taser blunt dart-to-heart distance causing ventricular fibrillation in pigs. *IEEE TBME* under review 2007.
43. Ho JD, Dawes DM, Reardon RF, et al. Echocardiographic evaluation of a TASER-X26 application in the ideal human cardiac axis. *Acad Emerg Med*, 2008;15:838–844.
44. Nanthakumar K, Billingsley IM, Masse S, et al. Cardiac electrophysiological consequences of neuromuscular incapacitating device discharges. *J Am Coll Cardiol*, 2006;48:798–804.
45. National Institute of Health. "Epinephrine Injection". *US National Library of Medicine*, April 1, 2003. (Accessed June 10, 2008 at http://www.nlm.nih.gov/medlineplus/druginfo/medmaster/a603002.html).
46. Dorian P and K Nanthakumar. Reply (to the letter: Electronic control devices and the clinical milieu). *J Am Coll Cardiol*, 2007;49:732–3.

Chapter 11
Serum and Skin Effects of CEW Application

Jeffrey D. Ho*

Because there have been some deaths temporally associated with CEW application, there has been intense effort and research to determine if the CEW has been a contributor in these cases or whether it is an item of convenience for blame. Initial theories of contribution centered around direct induction of arrhythmia or electrocution. However, numerous researchers of human subjects have demonstrated a high degree of cardiac safety in various studies [1–5].

Since the demonstration of cardiac insult has not been shown to occur, other theories of CEW contribution to death have emerged. These theories are varied and include hypotheses such as induced hyperkalemia from cellular destruction, induced acidosis from muscular contraction, induced renal failure due to rhabdomyolysis, and insidious late cardiac damage at the microcellular level leading to cardiac dysfunction at a later point in time.

Human research into this area has recently grown and the findings will be discussed below. This research has greatly aided future research directions for CEWs and has allowed the end users to apply this technology in the field with a relative degree of confidence.

11.1 Physiology Review

At the microscopic level, the human body is made up of a network of discrete, organizational structures called cells. Cells make up the basic units of life and are specialized in their function and physiology. Many things can cause damage to cells and CEW exposure has been postulated to be a possible cause. At the

**Jeffrey D. Ho reports serving as an expert research consultant and medical expert consultant to TASER International, Inc. and reports as a personal shareholder of TASER International stock. No other potential conflict of interest relevant to this chapter was reported.*

J.D. Ho (✉)
Department of Emergency Medicine, Hennepin County Medical Center, University of Minnesota Medical School, Minneapolis, MN, USA
e-mail: hoxxx010@umn.edu

M.W. Kroll, J.D. Ho (eds.), *TASER® Conducted Electrical Weapons: Physiology, Pathology, and Law*, DOI 10.1007/978-0-387-85475-5_11,

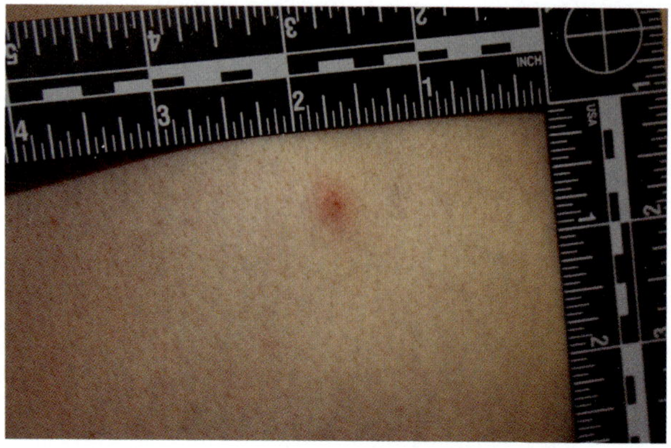

Fig. 11.1 Skin markings from TASER X26 CEW deployed probes at 24 hours after exposure

grossly visual level, the damage to CEW exposed skin areas is obvious. (Figs. 11.1 and 11.2) This makes sense since CEW application involves a delivery of electrical energy to the surrounding tissue. This application is transferred to the surrounding tissue in the form of heat and can result in a clinically insignificant localized burn. However, at the more microscopic level, the significance of cellular damage is not known.

One of the ways that healthcare providers evaluate cellular damage is through an analysis of blood serum biomarkers. These biomarkers typically measure intracellular electrolytes or proteins. When a human is in an undamaged condition, certain levels of these biomarkers are present and acceptable at baseline. As more and more cells become damaged, they break apart and "leak" these biomarkers into the surrounding bloodstream. Therefore, an increase in these biomarkers above certain levels is correlated with cell damage.

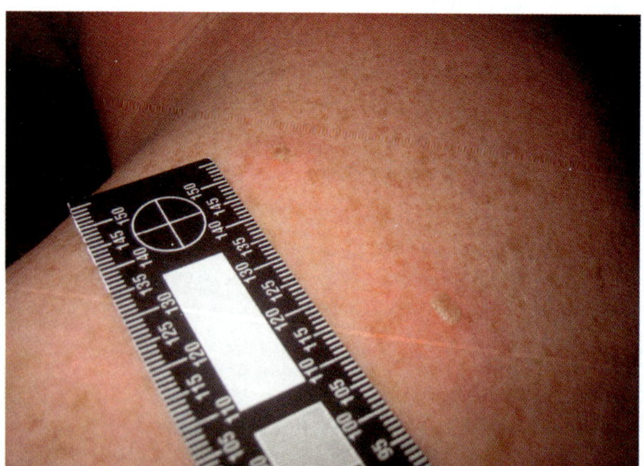

Fig. 11.2 Skin markings from TASER X26 drive-stun at 24 hours after exposure

Table 11.1 Commonly measured biomarkers, normalized values, and interpretation

Biomarker	Explanation	Normal human value
Troponin I (TnI)	Protein whose presence is very specific for cardiac cell damage	0.00–0.09 ng/ml
Potassium (K+)	The major cation in the body found primarily within cells. Levels elevate when cells are damaged or if kidney clearance is impaired	3.5–5.3 mEq/L
Lactate	Acidic by-product of cell metabolism. Levels become high when production exceeds clearance	0.7–2.1 mEq/L
Creatine phosphoKinase (CPK or CK)	Metabolic enzyme released from damaged muscle cells	38–175 units/L
Myoglobin	Protein released from damaged muscle tissue (rhabdomyolysis)	0–85 ng/ml

The relationship is generally thought to be directly proportional to the magnitude of the biomarker measured.

Some of these biomarkers (e.g., K+ or potassium) are not cell specific. This means that if a person's serum potassium level is elevated due to some type of cellular insult, it is impossible to tell from the biomarker alone what type of cell has been damaged. On the other hand, some measurable biomarkers are quite specific and found only in certain types of cells (e.g., Troponin I is found only within cardiac cells). Therefore, when these specific biomarkers are elevated above a certain level, it can be determined exactly what type of cell has been damaged. See Table 11.1 for a matrix of commonly measured biomarkers and what they indicate.

11.2 Biomarker Research

Because CEW applications involve the transfer of electrical energy, initial areas of research into possible contributions of morbidity or mortality were made in the areas of electrocution pathophysiology. Electrocution deaths are fairly well described in the modern day medical literature and it is understood that they are almost always due to induction of an irregular heartbeat that is not compatible with life such as ventricular fibrillation (e.g., exposure to wall sockets) or asystole (e.g., lightening strikes) [6]. Exposure to high power lines can induce both ventricular fibrillation and asystole. In light of this, the initial human CEW research examined cardiac rhythm issues as referenced in the introduction section of this chapter. These studies have failed to show any evidence of significant or lasting cardiac rhythm abnormalities. Because of this, human research has turned to examining various biomarkers in attempts to investigate any correlation with damage at the microcellular level.

11.2.1 Troponin I

One of the first biomarkers to be measured is a specific cardiac protein known as troponin I. Troponin I has an extraordinary specificity for cardiac muscle cells that approaches 100% [7]. It is released into the circulation upon death of the cardiac cell and becomes measurable approximately 6–8 hours after cardiac cellular death. It appears to plateau at approximately 12 hours after cellular death and can remain measurable for several days. It is used by physicians in the clinical setting of evaluating patients for myocardial infarction. Troponin levels have also been noted to be elevated after extreme exercise [8].

There have been human studies evaluating the presence of troponin I following CEW exposure [1,9]. To date, these studies have examined numerous human subjects, both at rest and when physically exerted, who received a CEW exposure for up to 15 continuous seconds. The subjects have been followed for up to 24 hours without a measurable rise in troponin I. In the 2006 study by Ho et al., there was a single subject that demonstrated a rise in their troponin I. This subject underwent extensive evaluation by an independent group of cardiologists without finding further evidence of cardiac damage and it was determined that the cause of this was possibly due to a laboratory error. Based on the current human research available, it is fair to conclude that there is no evidence that CEW exposure leads to any type of measurable cardiac cell damage. This is to be expected based on prior study involving much stronger transthoracic cardioversion current [10].

11.2.2 Potassium

As one of the most abundant ions found in the human body, the majority of potassium is found within the cells themselves. A much smaller amount can be found in the circulating plasma. Potassium has a major role in normal cell function, especially in those cells that have potential for excitability. This role is primarily in determining the resting cell membrane potential. The higher level of potassium found within the cell is maintained through a pump exchange system with sodium ions.

When extracellular potassium levels are out of balance, there can be significant side effects. The primary concern in persons subjected to a CEW exposure is the condition of excess extracellular potassium or *hyperkalemia*. This is typically caused by either cellular destruction (whereby intracellular potassium is released into the serum) or by a condition of acidosis (whereby too much acid is present and the body attempts to buffer this by shifting H^+ ions into the cells at the expense of K^+ ions shifting out of the cells and into the serum). Both of these scenarios have been theorized to be present in persons exposed to CEW applications in the field.

In the first scenario, it has been hypothesized that the actual application of electrical energy and its conversion to thermal energy at the level of the skin and subcutaneous tissue caused cell destruction and subsequent release of potassium. In the second scenario, it is believed that many subjects who receive CEW applications by law enforcement officials in the field do so because of a condition of agitation or resistive behavior. It has been documented in extreme circumstances that this type of behavior is highly correlated to also having an underlying condition of acidosis.[11] Because of this, research has been conducted to evaluate whether or not CEW application is capable of creating a clinically significant condition of hyperkalemia.

The concern for investigating the presence of hyperkalemia with CEW application stems from the dangerous cardiac side effects that are associated with increased potassium. The signs and symptoms seen with this state are dependent upon the quantity of elevation of potassium seen as well as the time frame in which it elevates (short time frames with high elevations are less well tolerated than longer time frames with lower elevations). The clinical conditions that can result range from the subject feeling weak all the way up to a cardiac arrest [12].

Additionally, there is concern that exertion can cause shifts in serum potassium levels in humans. There is literature that demonstrates a rise in the serum potassium level in humans during exercise with a subsequent abrupt fall in the level in the immediate postexercise period.[13] It is theorized that this type of fluctuation could lead to a cardiac arrhythmia [14]. Many CEW applications in the field take place on persons under exertional circumstances (e.g., suspects fighting, fleeing, and otherwise resisting law enforcement). If exerting oneself can create a condition of hyperkalemia, it could be worrisome if a CEW application were to exacerbate this.

Current human research has repeatedly looked at the issue of hyperkalemia. In 2006, Ho et al. examined the effects of a 5-second CEW application on a rested population and found no hyperkalemia [1]. Since then, there have been projects with similar methodology that have validated this initial study [15]. There have also been projects examining this biomarker with prolonged CEW applications under different stressor variables such as exhaustion [9,16]. To date, none have been able to demonstrate an episode of dangerous hyperkalemia following CEW application.

Although critics of this work will call attention to the fact that the longest durations of CEW application have been 15 seconds and there are some CEW field exposures for longer periods of time, none of the human studies demonstrated any trend of rising potassium levels. At this point in time, it is reasonable to conclude that the body of human research does not support a theory of CEW-induced hyperkalemia.

11.2.3 Lactate

In humans, lactate is constantly being produced as a cellular by-product of normal metabolism. Serum lactate generally does not increase until the rate of production exceeds the rate of removal. Lactate production is directly proportional to the level of activity and metabolism that is present. Therefore, higher metabolism states (such as that seen with the use of illicit stimulant medication or agitation) and higher exertional states (as seen when fleeing or fighting with law enforcement) can cause elevated serum lactate levels.

Lactate is an acidic compound. Higher levels within the human body lead to a condition known as *acidosis*. A condition of acidosis can be concerning if left uncorrected because the human body is not designed to operate under conditions of extreme acidosis. There have been sudden deaths in custodial situations associated with a factor of profound acidosis [11].

It had been postulated that because CEW application causes skeletal muscle activation (much like vigorous shivering or exercise), then CEW application might possibly cause a state of profound acidosis in human subjects. Human studies of short duration do demonstrate a slight, clinically insignificant rise in serum lactate [1,15] when CEWs are applied for 5 seconds. However, studies of longer duration comparing CEW exposure to continued exercise (meant to simulate continued fighting or fleeing) show that the CEW application does not cause a worsening condition of acidosis over that seen with continued exercise alone [9].

An interpretation of this for a real-life scenario is that if a subject is engaged in behavior such as fleeing, fighting, or resisting law enforcement, they will develop a similar degree of lactate within their body regardless of whether a CEW is applied to them *or* if they are allowed to continue to flee, fight, or resist for the same period of time in place of a CEW application. If a CEW is not applied to these subjects to gain control of their behavior, it is unlikely that the subject will only continue to flee, fight, or resist for the short period of time equivalent to that of a CEW application. Therefore, there is greater likelihood of a worsening acidosis, the longer that they are allowed to flee, fight, or resist. At this point in time, it is reasonable to conclude that the available body of human research does not demonstrate that CEW application causes worse acidosis than that of continued exertion alone given an equivalent amount of time.

11.2.4 Myoglobin and Creatine Phosphokinase

Myoglobin is a protein found within skeletal and cardiac muscle cells. It is structurally related to hemoglobin which is the component in the blood responsible for binding to oxygen and carbon dioxide that it then transports to various tissues for delivery to sustain life. Myoglobin also binds oxygen and acts as a reserve supply of this within the cell.

Creatine phosphokinase (CPK or CK) is an enzyme used in cellular metabolism, and like myoglobin, is found in certain cells, especially skeletal muscle. It is, therefore, also used as a nonspecific biomarker of cellular damage. In general, it will be elevated when myoglobin is and vice versa with regard to skeletal muscle injury. CK elevations do not necessarily lead to further medical conditions and so the bulk of this section will be spent on discussing myoglobin.

The reason that myoglobin is used as a biomarker is that when skeletal or cardiac cells are damaged or destroyed, their internal supply of myoglobin is leaked out into the surrounding area and eventually finds its way into the blood stream. This process of cellular damage with resultant leakage of myoglobin is called rhabdomyolysis. Under extreme cases of rhabdomyolysis, the kidneys filter out myoglobin from the bloodstream. Myoglobin tends to be too large to pass through the kidney filtration system easily and can cause acute renal failure when too much myoglobin clogs this system.

This is a concern since renal failure can lead to death if severe enough and not treated promptly. Any condition of severe muscle cell destruction can lead to this such as trauma (crush injuries or burns destroying tissue), sustained pressure on a large area of muscle (e.g., an unconscious person lying on their back on a hard surface who is not found for many hours can begin to destroy the muscles in their buttocks, back, and extremities from the pressure of lying on them for so long), or extreme exercise (e.g., constant muscular trauma from extreme weight lifting or marathon running). Acute drug intoxication with cocaine, methamphetamine, heroin, PCP, and some prescription medications has also been implicated as a cause of rhabdomyolysis. It has been postulated that because a CEW might also destroy muscle cells that it could create a situation of dangerous rhabdomyolysis.

Human study does show that CEW exposure does lead to elevations in myoglobin and CK from baseline [1]. This gradual, slow rise is to be expected since CEW exposure activates skeletal muscle. The elevations that have been found are on the order of what has been demonstrated in the sports medicine literature following participation in an athletic event [17]. The elevated CK levels seen were on the order of what would be expected in a human subject after engaging in a strenuous athletic event but much less than what has been reported in the literature for exercise-induced rhabdomyolysis [18]. CEW exposure does not explain the fulminant states of rhabdomyolysis that often accompany subjects in acutely agitated and delirious states.

Severe rhabdomyolysis can lead to death. The resultant renal failure can be fatal all by itself. However, in severe cases, there are usually other complex factors going on such hyperkalemia from muscle cell destruction as well as large shifts in fluids between the intracellular and extracellular spaces that can clinically lead to shock. In general, however, death from severe rhabdomyolysis takes time (hours to days) and does not explain the sudden, unexpected deaths that occur during a confrontation with law enforcement. Based on this, the theory of CEW-induced rhabdomyolysis causing sudden death is not supported.

11.3 Conclusion

Serum biomarkers are used by professionals to evaluate the health status of the human body. The presence of certain biomarkers is very specific in demonstrating damage to very localized organs in the body. An example of this would be that the presence of elevated levels of troponin I indicates damage specifically to cardiac muscle cells. Other biomarkers are much less specific. An example of this would be that the presence of an elevated level of potassium could indicate cellular damage with leakage of potassium into the plasma. However, potassium is found in all cells in the body and it would be difficult to identify exactly which ones the potassium came from on the basis of the biomarker level alone.

In using commonly accepted biomarkers as evaluators of damage to human tissue, an assessment of CEW effect can be made. The current body of research that has examined these markers in a variety of research scenarios does not demonstrate a concerning finding, especially with regard to sudden death. There does appear to be a trend toward elevation of lactate in an exercising population with CEW application but this does not appear to be different in a similar exercising population where no CEW is applied. There is also a trend toward elevation of CK and myoglobin that is consistent with participation in an athletic event and not consistent with leading to sudden death. From a biomarker standpoint, CEW application appears to be a reasonable and safe method of behavioral control.

References

1. Ho JD, Miner JR, Lakireddy DR, et al. Cardiovascular and physiologic effects of conducted electrical weapon discharge in resting adults. *Acad Emerg Med*, 2006;13:589–595.
2. Levine SD, Sloane CM, Chan TC, et al. Cardiac monitoring of human subjects exposed to the TASER. *J Emerg Med*, 2007;33:113–117.
3. Vilke GM, Sloane S, Levine S, et al. Twelve-lead electrocardiogram monitoring of subjects before and after voluntary exposure to the TASER X26. *Am J Emerg Med*, 2007;26:1–4.
4. Ho J, Dawes D, Calkins H and M Johnson. Absence of electrocardiographic change following prolonged application of a conducted electrical weapon in physically exhausted adults. *Acad Emerg Med*, 2007 (supplement 1);14:S128–S129.
5. Ho JD, Dawes DM, Reardon RF, et al. Echocardiographic evaluation of a TASER X26 application in the ideal human cardiac axis. *Acad Emerg Med*, 2008;In Press.
6. Pinto DS and PF Clardy. "Environmental Electrical Injury" in *Up To Date Online, 16.1*. (Accessed June 10, 2008 at http://www.utdol.com/utd/content/topic.do?topicKey = ad_emerg/2283&type = A&selectedTitle = 1~2).
7. Apple FS. Tissue specificity of cardiac troponin I, cardiac troponin T and creatine kinase-MB. *Clin Chim Acta*, 1999;284:151–159.
8. Bakshi TK, Choo MKF, Edwards CC, et al. Causes of elevated troponin I with a normal coronary angiogram. *Int Med J*, 2002;32;520–525.
9. Ho JD, Dawes DM, Bultman LL, et al. Prolonged TASER use on exhausted humans does not worsen markers of acidosis. AJEM, 2008–2009 In Press.
10. Allan JJ, Feld RD, Russell AA, et al. Cardiac troponin I levels are normal or minimally elevated after transthoracic cardioversion. *J Am Coll Cardiol*, 1997;30:1052–1056.
11. Hick JL, Smith SW and MT Lynch. Metabolic acidosis in restraint-associated cardiac arrest: a case series. *Acad Emerg Med*, 1999;6:239–243.

12. US National Library of Medicine and the National Institutes of Health. "Hyperkalemia" in Medline Plus. Accessed June 10, 2008 at http://www.nlm.nih.gov/medlineplus/ency/article/001179.htm).
13. Kilburn KH. Muscular origin of elevated plasma potassium during exercise. *J Appl Physiol*, 1966;21:675–678.
14. Brady HR, Kinirons M, Lynch T, et al. Heart rate and metabolic response to competitive squash in veteran players: identification of risk factors for sudden cardiac death. *Eur Heart J*, 1989;10:1029–1035.
15. Vilke GM, Sloane CM, Bouton KD, et al. Physiological effects of a conducted electrical weapon on human subjects. *Ann Emerg Med*, 2007;50:569–575.
16. Ho JD, Dawes DM, Lapine AL, et al. Prolonged TASER "Drive Stun" exposure in humans does not cause worrisome biomarker changes. *National Association of EMS Physicians Annual Meeting Presentation*, 2008; Accessed on June 9, 2008 at http://www.naemsp.org/documents/2008NAEMSPAbstrats070912revised_2_.pdf.
17. French DN, Kraemer WJ, VanHeest JL, Sharman MJ, Gomez AL, Rubin MR, et al. Physiological[][] damage and stress of a competitive NCAA division 1 football game. *Med Sci Sports Exerc*, 2003 (supplement);35:S320.
18. Sinert R, Kohl L, Rainone T, Scalea T. Exercise-induced rhabdomyolysis. *Ann Emerg Med*, 1994;23:1301–1306.

Chapter 12
Echocardiographic Effects of the CEW

Robert Reardon*

Monitoring of cardiac activity *during* application of conducted electrical weapons (CEWs) is difficult because electrical pulses and muscular activity cause excessive interference and "noise", making electrocardiographic interpretation impossible. Even palpating pulses *during* application of CEWs is difficult or impossible. Invasive monitoring with an arterial line or right heart catheterization is useful in animal models, but is not practical in human subjects. Ultrasound of the heart (echocardiography) *during* application of CEWs is feasible and provides valuable information about cardiac rate and rhythm. Echocardiography has been used to monitor cardiac activity *during* CEW application in both animal and human models with very different results. Echocardiography during CEW exposure in pigs has been reported to demonstrate cardiac "capture" but human studies have found no evidence of cardiac capture, even when probes are placed along the cardiac axis.

12.1 Animal Studies

In a study of prolonged CEW exposure in a pig model, Dennis et al. reported the use of echocardiographic monitoring of four animals each exposed to two 40-second discharges from a TASER X26 device [1]. Study animals weighed between 22 and 46 kilograms. TASER probes were placed into the chest wall along the cardiac axis, at the left lower and right upper thorax. They used a Sonosite 180 ultrasound device with a 2 megahertz probe (Sonosite Inc. Bothell, WA) and made digital video recordings of echocardiograms for further

---

* *Robert Reardon reports serving as a consultant to TASER International, Inc. and receiving consulting fees from TASER International, Inc. No other potential conflict of interest relevant to this chapter was reported.*

R. Reardon (✉)
Department of Emergency Medicine, Hennepin County Medical Center
e-mail: rfreardon@gmail.com

M.W. Kroll, J.D. Ho (eds.), *TASER® Conducted Electrical Weapons: Physiology, Pathology, and Law*, DOI 10.1007/978-0-387-85475-5_12,

analysis. The authors did not report whether they used B-mode, M-mode, or Doppler ultrasound. They did not report which echocardiographic views were obtained and they did not describe how they determined the cardiac rate or rhythm from echocardiographic images. They reported that they observed immediate and ongoing “capture” of the ventricular rhythm, consistent with ventricular tachycardia at a rate of 300 beats per minute (bpm), in all four animals during the entire period of the TASER CEW application. After the CEW discharges, they observed a short period of atrioventricular dyssynchrony then sinus tachycardia in 3/4 animals. In the fourth animal, they reported sustained ventricular tachycardia after the CEW application was complete, which then degenerated into ventricular fibrillation about 3 minutes later.

The same group of researchers recently reported the results of a similar study in which six pigs were each exposed to two 40-second discharges from a TASER X26 device [2]. In this study, the pigs weighed between 22 and 77 kilograms and TASER CEW probes were attached to the chest as in the previous study. They used a LOGIQ 7 ultrasound device (GE Healthcare, United Kingdom). Again, the authors did not report whether they used B-mode, M-mode, or Doppler ultrasound. They did not report which echocardiographic views were obtained and they did not describe how they determined the cardiac rate or rhythm from echocardiographic images. They again reported that they observed immediate and ongoing “capture” of the ventricular rhythm at a rate of 300 bpm in all animals *during* the entire period of the CEW application. They reported that after the CEW exposure was complete three pigs reverted to sinus rhythm, two had ventricular fibrillation and one had ventricular tachycardia. In one case, ventricular fibrillation was fatal and in one case it spontaneously converted to sinus rhythm. The case of ventricular tachycardia also spontaneously converted to sinus rhythm.

12.2 Human Studies

Echocardiographic monitoring of human subjects during CEW exposure was first attempted in 2004 by Ho et al. Twelve volunteers were each given a 5-second exposure from a TASER CEW. Subjects were placed in a supine position and probe leads were secured to the clothing of the subjects legs and lower back with alligator clips. An experienced physician sonographer obtained a parasternal long axis view (Fig. 12.1) of the heart during the CEW application. Echocardiography was performed using a Sonosite Titan with a 2 megahertz sector probe and tissue harmonics. Only B-Mode was used and images were not recorded but interpreted in real time by the physician performing the study. In ten subjects, it was clear that the cardiac rhythm and gross left ventricular function were unchanged before, during and after application of the TASER device. In two cases, the subjects thrashed violently during CEW application and it was physically impossible to obtain echocardiographic images. This initial trial showed that it was possible to obtain echo images during the

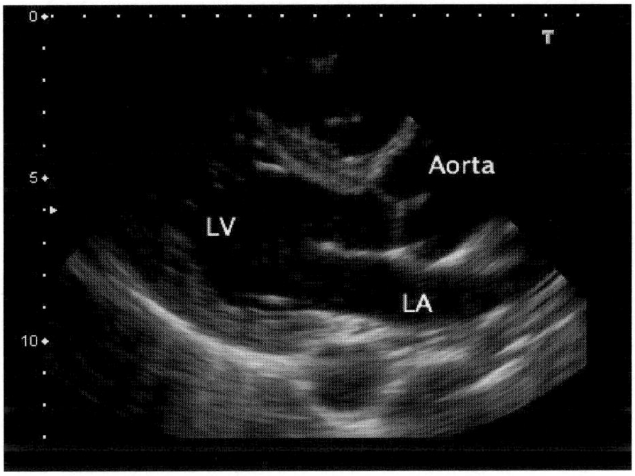

Fig. 12.1 Parasternal long axis view of the heart in diastole. Note that the mitral valve is fully opened and the aortic valve is closed

majority of CEW applications. TASER CEW electrical discharges do not effect ultrasound images or damage the ultrasound equipment. It is very challenging to obtain good echo images during CEW application in awake human subjects. A 5-second exposure period does not allow adequate time to obtain echo images in all patients.

A second CEW echo study by Ho et al. was done in 2005. Sixty volunteers were each subjected to a 10-second exposure from a TASER X26 device. Subjects were placed on the floor in the left lateral decubitus position, leaned against the sonographers thigh, with the sonographer kneeling on the floor adjacent to the subject. An experienced physician sonographer obtained an apical four-chamber view of the heart during the entire duration of the TASER exposure. Echos were obtained using a Toshiba Nemio 20 with a 2–4 megahertz sector probe and tissue harmonics. Only B-mode echo images were obtained. Images were digitally recorded for later evaluation. CEW leads were secured to the subject's upper and lower left back using plastic tape. All images were digitally recorded for later evaluation. The initial intent of the study was to evaluate the effect of CEW exposure on left ventricular function. Several cardiologists blindly read the 10-second video clips that were obtained during the CEW exposure. Although the left ventricle was clearly visualized in all cases the cardiologists reads were very inconsistent, even when the same videos were blindly shown to the same cardiologist twice. It was clear that the quality of the images was not adequate for estimation of left ventricular function in about 10% of subjects. In the remainder of subjects the left ventricular function was judged to be "grossly normal." Despite the inability to determine left ventricular function in some subjects it was possible to determine the heart rate in all subjects during TASER exposure. Heart rates were determined by manually scrolling through the 10-second video clips and counting ventricular contractions. The highest heart rate seen in any subject during TASER exposure was 162 bpm. Although we were unable to determine the cardiac rhythm of subjects in this study there was no evidence of any tachyarrhythmia or cardiac "capture."

Lessons learned in the first two studies of echocardiography during CEW exposure were used to reevaluate the approach and refocus on collecting the most important echocardiographic data. It was apparent from the 2005 study that it was very difficult to obtain adequate B-mode images of the left ventricle during CEW application. Also, it was apparent that the greatest concern about cardiac effects from CEW devices was cardiac "capture" or ventricular tachyarrhythmias. It became clear that the best application of echocardiographic monitoring during CEW exposure was to simply determine the cardiac rate and rhythm.

Echocardiographic monitoring of human subjects *during* CEW application was publically reported for the first time by Ho et al. [3]. In this study, 37 human volunteers were each subjected to an exposure from a TASER X26 device after near maximal anaerobic exertion. Prior to exertion each subject had echocardiographic monitoring to establish heart rate and rhythm. An Ultrasonix CEP ultrasound machine with a 4–2 megahertz phased array probe was used to obtain a parasternal long axis view of the heart (Figs. 12.1 and 12.2). M-mode interrogation through the anterior leaflet of the mitral valve was accomplished and documented by recording digital video clips onto the hard drive of the ultrasound machine (Fig. 12.3). M-mode tracings confirmed that all subjects were in sinus rhythm prior to exertion. Sinus rhythm can be definitively established by seeing distinct E and A peaks on an M-mode tracing of the anterior leaflet of the mitral valve (Fig. 12.4). The E peak corresponds to mitral valve opening with passive filling of the left ventricle and the A peak corresponds to mitral valve opening with atrial contraction (Figs. 12.5 and 12.6). Baseline heart rate was established by measuring the distance between the E peaks and mean preexertion heart rate was 86 (±2.8) bpm. After baseline measurements were

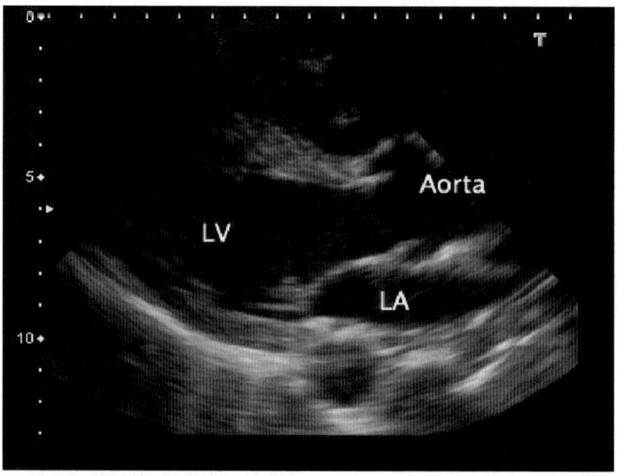

Fig. 12.2 Parasternal long axis view of the heart in systole. Note that aortic valve is fully opened and the mitral valve is closed

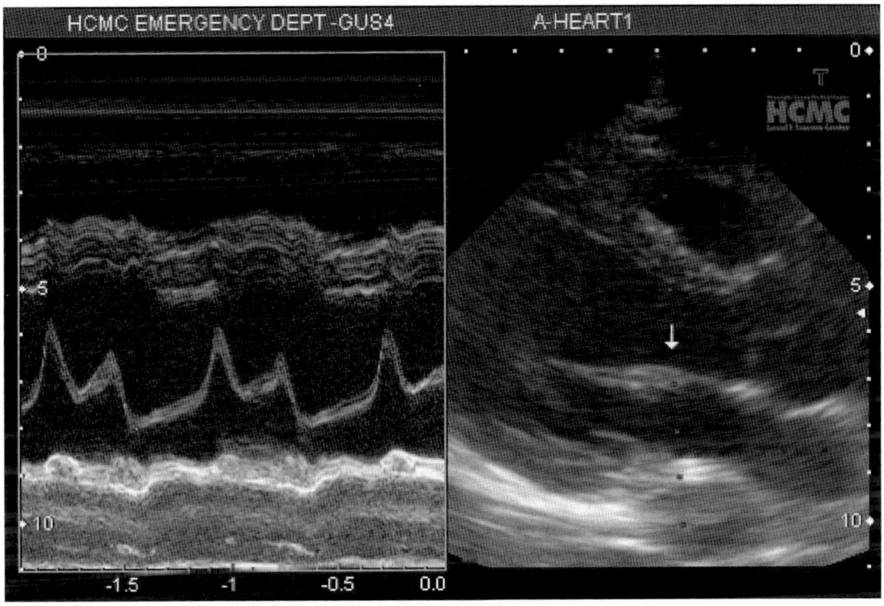

Fig. 12.3 An example of M-mode interrogation through the mitral valve. The *right side* of the screen displays a B-mode (two-dimensional *gray scale*) image and the *left side* of the screen displays an M-mode tracing. Note that on the B-mode image the M-mode cursor is directed through the anterior leaflet of the mitral valve (*arrow*)

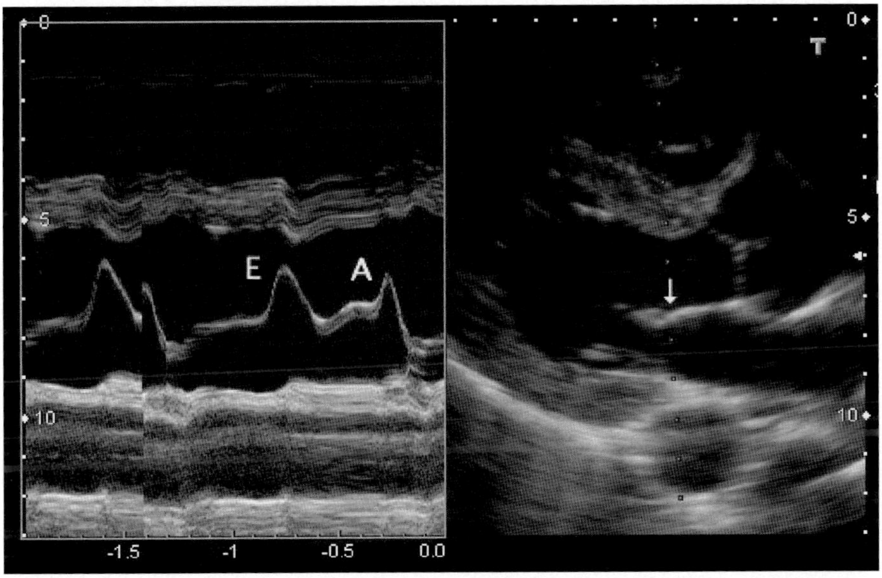

Fig. 12.4 M-mode imaging through the anterior leaflet of the mitral valve. The B-mode image on the *right* shows the M-mode cursor directed through the anterior leaflet of the mitral valve (*arrow*). The M-mode tracing on the *left* shows an E/A pattern consistent with sinus rhythm

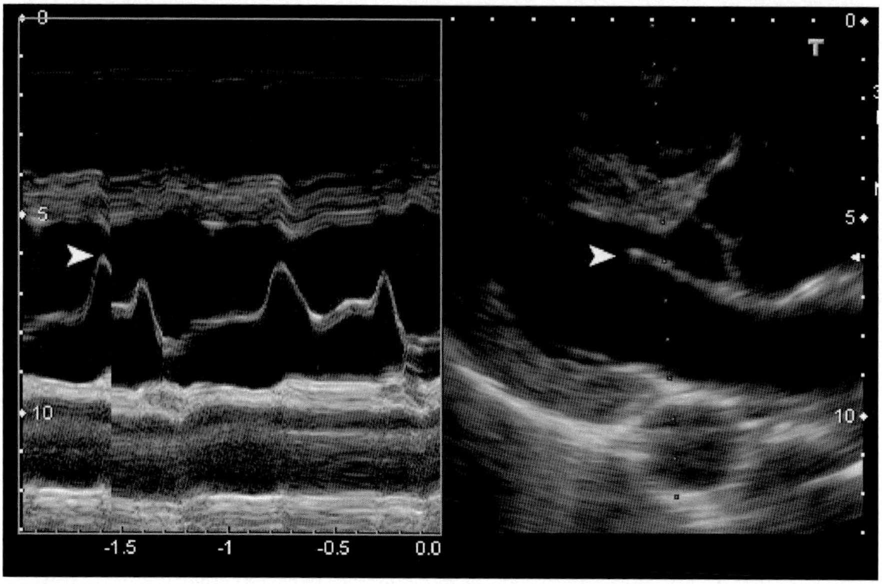

Fig. 12.5 Real-time M-mode tracing showing the E peak (*arrow on left*). The E peak corresponds to opening of the mitral valve during passive filling of the *left* ventricle in early diastole (*arrow on right*)

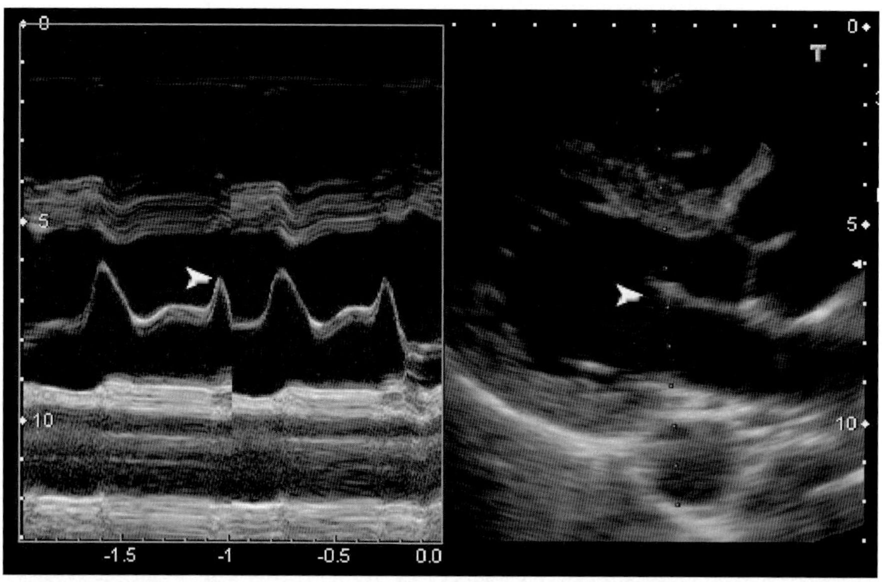

Fig. 12.6 Real-time M-mode tracing showing the A peak (*arrow on left*). The A peak corresponds to opening of the mitral valve during active filling of the *left* ventricle in late diastole (*arrow on right*)

made each subject did a 30-second timed set of push-ups and then ran on an 8% inclined treadmill at 8 miles per hour. Subjects were asked to run until they could no longer maintain pace with the treadmill. Each subject then had TASER leads taped to their upper and lower back and they assumed a supine position on the floor. Echocardiographic monitoring was reestablished and sinus rhythm was again confirmed in all subjects. The mean postexertion heart rate was 153 (±3.0) bpm. Each subject then received a 15-second exposure from a TASER X26 device as echocardiographic monitoring was ongoing. M-mode tracings during CEW exposure were more difficult to obtain and interpret, but digital video clips of the studies were recorded for later review. Muscular activity, gross movement, and wide variations in respiratory effort during CEW exposure made it difficult to obtain echocardiographic images. In many cases, an M-mode tracing of the left ventricular septum and posterior wall was obtained (Fig. 12.7). In these cases, sinus rhythm could not be confirmed but left ventricular rate could be determined by measuring the distance between contractions. In 19 cases, an M-mode tracing showed clear E and A peaks and confirmed that subjects were in sinus rhythm during CEW exposure (Fig. 12.8). In 18 cases, the mitral valve was not visualized and only left ventricular rate could be determined. Determination of heart rate was possible in all cases during TASER CEW exposure. The mean heart rate during the exposure was 140 ±2.6 bpm, a decrease from the preexposure mean rate. In addition, the

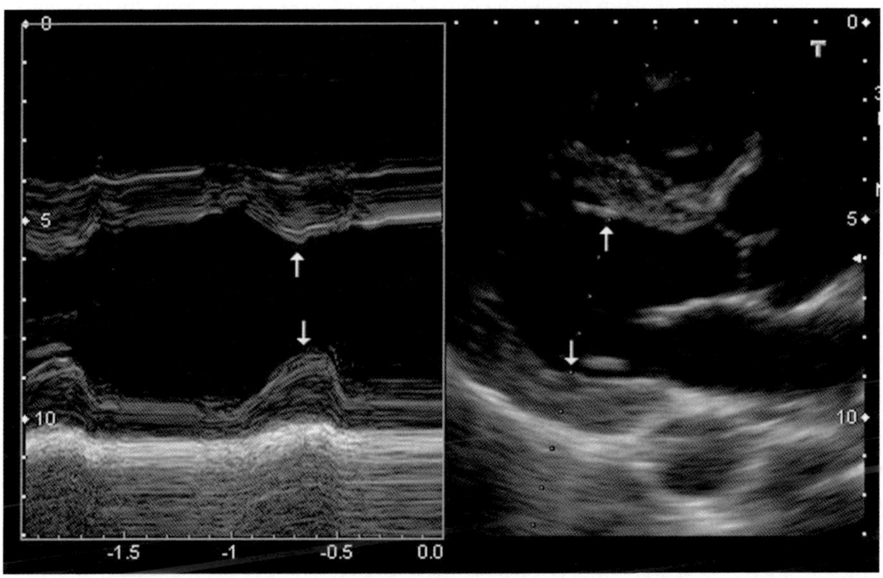

Fig. 12.7 M-mode imaging through the *left* ventricle below the mitral valve. The M-mode image on the *left* shows movement of the septum (*upper arrow*) and the posterior wall (*lower arrow*) during systole. The image on the *right* is a B-mode image with the M-mode cursor directed through the *left* ventricle below the mitral valve (*arrows on right*)

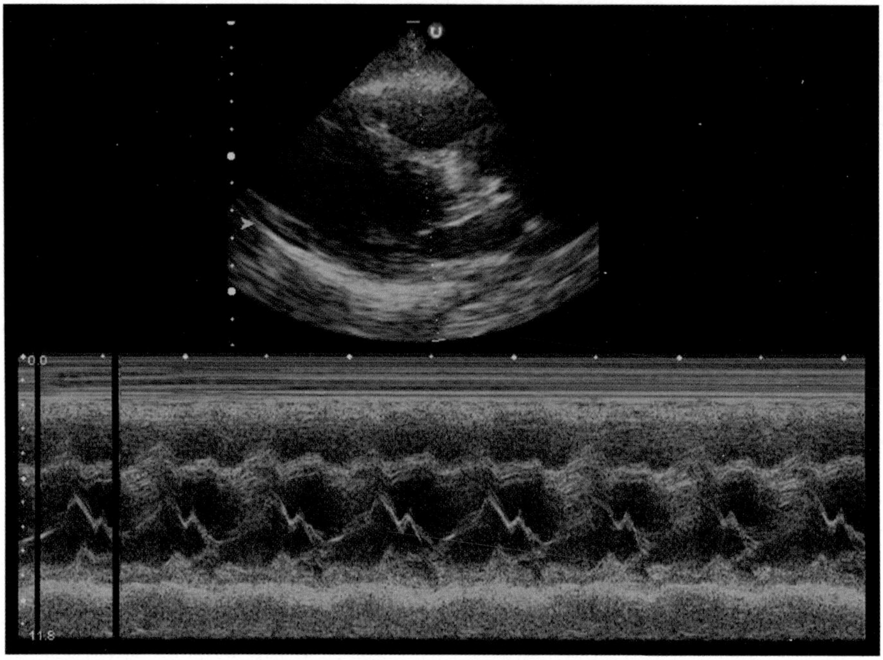

Fig. 12.8 M-mode imaging through the anterior leaflet of the mitral valve during TASER exposure. Note that E/A pattern is present and the E and A peaks are close together because of the elevated heart rate

maximum heart rate recorded in any subject during the exposure was 156 bpm. Echocardiographic monitoring was continued after the exposure. Rate and rhythm were again recorded about 1 minute after the CEW exposure was completed. Postexposure mean heart rate was 111 ±2.2 bpm and all subjects were determined to be in sinus rhythm. The authors concluded that there was no evidence of induced tachyarrhythmia during or after CEW exposure.

These results are much different from the reports of "cardiac capture" and tachyarrhythmias in animal studies. Even though sinus rhythm was only confirmed in about one-half of cases, there was clearly no "capture" at a rate of 300 bpm as was reported in the pig studies. The fact that the mean heart rate *decreased* during TASER CEW exposure is also compelling evidence that no cardiac capture or tachyarrhythmias occurred.

In a more recent study, adult human volunteers ($n = 44$) underwent limited echocardiography before, during and after a 15-second TASER X26 CEW application with preplaced thoracic electrodes [4]. The electrodes were placed in a cardiac axis position at the upper left or right sternal border and the estimated cardiac apex. Ultrasound images were analyzed using M-mode through the anterior leaflet of the mitral valve for evidence of arrhythmia by a trained ultrasonographic emergency physician. Heart rate and the presence of sinus rhythm were determined. Data were analyzed using descriptive statistics. The mean HR prior to starting thc event was 105.6 (range 62–146, 95% CI = 91.2–112.1). During the CEW exposure, the mean HR was 123.3 (range 70–158, 95% CI = 116.8–129.9) and a mean of 93.9 (range 55–121, 95% CI = 68.8–98.9) at 1 minute after CEW exposure. Sinus rhythm was clearly demonstrated in 28 (63.6%) subjects during CEW exposure (mean heart rate 125.2 range 75–158, 95% CI = 117.0–133.4). Sinus rhythm was not clearly demonstrated in 16 subjects due to movement artifact (mean heart rate 119.9, range 70–152, 95% CI = 107.7–132.2), however, the maximal heart rate of 152 bpm was again not consistent with previous animal findings. The authors again concluded that TASER CEW exposure in a vector that would be considered a "worst-case scenario" for an adult human did not show evidence of cardiac capture or induced concerning tachyarrhythmia.

Significant differences between the results of animal studies and practical experience in humans has lead some authors to argue that swine studies of the cardiac effects of CEWs are not helpful [5]. Significant experience using echocardiographic monitoring to evaluate the effects of CEW exposure supports this assertion.

12.3 Conclusions

Echocardiography is an invaluable tool for studying the cardiac effects of CEWs. Echocardiography is currently the only practical means of monitoring cardiac activity *during* CEW exposure in human subjects. Several human studies using echocardiography to evaluate cardiac activity *during* CEW exposure have shown no evidence of cardiac capture or tachyarrhythmia. This technology has allowed researchers to determine that concerning cardiac findings in animals during CEW application could not be reproduced in humans.

References

1. Dennis AJ, Valentino DJ, Walter RJ, et al. Acute Effects of TASER X26 Discharges in a Swine Model. *J Trauma*. 2007;63(3).
2. Valentino DJ, Walter RJ, Dennis AJ, et al. TASER Discharges Capture Cardiac Rhythm in a Swine Model. *Acad Emerg Med*. 2007;14(5 Supp 1):S104.
3. Ho J, Reardon R, Dawes D, et al. Ultrasound Measurement of Cardiac Activity During Conducted Electrical Weapon Application in Exercising Adults. *Ann Emerg Med*. 2007;50(3 Supp 1):S108.
4. Ho JD, Dawes DM, Reardon RF, et al. Echocardiographic Evaluation of a TASER-X26 Application in the Ideal Human Cardiac Axis. *Acad Emerg Med* 2008.
5. Pippin JJ. Taser research in pigs not helpful. *J Am Coll Cardiol*. Feb 13 2007;49(6):731–732; author reply 732–733.

Chapter 13
Rhabdomyolysis

Ronald Moscati* and Samuel Cloud

Conducted electrical weapons (CEW) can conjure images of pulsating currents of electricity coursing their way through the human body in much the way we envision a movie scene of a person in contact with a power transformer. These conscious or subconscious images of CEW and the knowledge that electrocution is associated with rhabdomyolysis, likely form the basis of the fear that CEW could also cause this side effect.

13.1 Physiology of Rhabdomyolysis

Rhabdomyolysis is a clinical condition that develops when skeletal muscle cells are damaged and then release their contents into the interstitium and eventually, the intravascular space. The entity was first described in World War II England as a consequence of the crush injuries that occurred due to the London bombings. Rhabdomyolysis has a myriad of causes, from the mentioned electrical and crush injuries, to extreme physical activity, viruses, alcohol, cocaine use, and numerous less frequent etiologies. The final insult that leads to muscle cell death and rupture is the loss of function of the sodium-potassium-adenosine triphosphatase pump. This leads to a buildup of intracellular calcium from the endoplasmic reticulum with disastrous consequences [1,2].

Rhabdomyolysis is diagnosed when a patient has a serum creatine phosphokinase (CPK) elevation of at least five times the upper limit of normal and an appropriate clinical picture. Symptoms of rhabdomyolysis include myalgias

**Ronald Moscati reports serving as a consultant to TASER International, Inc. and receiving consulting fees from TASER International, Inc. No other potential conflict of interest relevant to this chapter was reported.*

R. Moscati (✉)
Department of Emergency Medicine, SUNY at Buffalo, Erie County Medical Center
e-mail: moscati@buffalo.edu

M.W. Kroll, J.D. Ho (eds.), *TASER® Conducted Electrical Weapons: Physiology, Pathology, and Law*, DOI 10.1007/978-0-387-85475-5_13,

(muscle pain), weakness, and fever. The complications of rhabdomyolysis are what are the most devastating, accounting for its 5% mortality rate. Patients can develop metabolic acidosis, hyperkalemia, acute renal failure, and disseminated intravascular coagulation. Acute renal (kidney) failure is the most noted side effect, occurring in up to one-third of cases [1]. The mechanism of injury is thought to involve obstruction of the renal tubules by myoglobin as well as a direct toxic effect by myoglobin and its breakdown products on the renal endothelium (the thin layer of cells lining the interior surface of blood vessels). This effect seems more pronounced in a state of hypovolemia. Rhabdomyolysis, when recognized early and treated aggressively with IV fluids and perhaps urinary alkalinization, usually has a benign course [2].

13.2 Relationship of Rhabdomyolysis with CEW Exposure

Extreme physical exertion [3] and status epilepticus [4] (continuous seizure activity) can lead to rhabdomyolysis through a mechanism of overuse of muscle fibers leading to muscle cell damage as is described above. Electrical injuries [5], in which there is actual heavy (large power line) current conduction through the body, can cause rhabdomyolysis through direct injury to muscle fibers by conducted electrical currents. Each of these mechanisms is of interest when considering the possibility of CEW-induced rhabdomyolysis. Many "stun-guns" deliver a painful electrical exposure at their point of contact with the skin. This does not generally result in conduction of electrical current through muscle. These types of devices do not cause long term or maximal muscle contractions or overuse. It is unlikely therefore that these devices would result in rhabdomyolysis.

TASER M26 and X26 devices represent different types of CEWs that do result in conduction of current between the points on the body where the separate darts contact skin. They also cause repeated skeletal muscle contractions in the area of exposure for the duration of the discharge. This has led to suggestions that TASER® exposure could lead to muscle damage and subsequent rhabdomyolysis.

An early report on 218 TASER CEW injuries seen in the Emergency Department reported a 1% incidence of associated "mild rhabdomyolysis" with no development of renal complications [6]. The actual laboratory values are not reported in this case series. It is noted, however, that the overwhelming majority of patients included in the series had been using drugs and alcohol and were engaged in bizarre behavior. Each of these factors is recognized as independent causes of rhabdomyolysis. This case series clearly does not establish a cause–effect relationship between CEW exposure and rhabdomyolysis. Given the low incidence and mild nature of rhabdomyolysis reported in this moderately large cohort, it would imply that rhabdomyolysis does not occur routinely after CEW exposure.

13.2.1 Animal Data

In an animal model study conducted by the US Air Force, pigs were given repeated TASER CEW discharges of 5 seconds spaced with 5-second rest intervals over a total of 3 minutes [7]. Baseline blood specimens were drawn prior to exposure and at 1, 30, and 60 minutes after exposure. CPK levels may have risen slightly following exposure but not enough to reach statistical significance. There were significant, although transient, increases in lactate and decreases in pH. In this study of an extended repetitious CEW exposure, there was no evidence of induced rhabdomyolysis. The use of an animal model as well as the lack of later CPK measurements, which may have demonstrated further increases were shortcomings of this study.

13.2.2 Human Data

In a human volunteer study by Ho et al., 66 adults were exposed to a single 5-second discharge from a TASER® X26 [8]. Blood samples were drawn prior to exposure, immediately following exposure and at 16 and 24 hours postexposure. The mean CPK levels reported showed no change immediately following exposure with modest increases to 24% and 32% higher than baseline at 16 and 24 hours, respectively. This rise is consistent with what is seen with mild exertion and well below that seen with extreme exertion [2]. As with the animal study and case series, this study does not support the premise that TASER CEW exposure results in rhabdomyolysis.

Reports of in-custody deaths with or without prior CEW application typically refer to deaths within 24–48 hours following the subject being arrested. The rise in CPK and subsequent secondary effects on renal function can occur within this time frame but would generally take a longer period of time to result in death. The only rhabdomyolysis-mediated mechanism to explain death in this time frame would be a case of massive myonecrosis resulting in hyperkalemia and death from subsequent cardiac arrhythmias. The data in the literature to date does not support such a situation resulting from CEW application.

13.3 Conclusion

Rhabdomyolysis is a consequence of muscle injury from overuse or direct damage as a result of mechanical or electrical trauma. While CEW application can cause exposure to electrical discharges and repeated muscle contraction, the data from case series, animal models, and human studies demonstrate that mild and transient increases in CPK occur without evidence of clinically significant increases leading to rhabdomyolysis.

References

1. Counselman, F. Rhabdomyolysis. In: Tintinalli J, Kelen G, Stapczynski J editors. Emergency Medicine: A Comprehensive Study Guide. 6th edition. New York. McGraw-Hill, 2004: 1749–1752.
2. Sauret J, Marinides G, Wang G. Rhabdomyolysis. Am Fam Physician 2002;65:907–912.
3. Kratz A, Lewandrowski K, Siegel A, Chun K, Flood J, Van Cott E, Lee-Lewandrowski E. Effect of marathon running on hematologic and biochemical laboratory parameters, including cardiac markers. Am J Clin Pathol 2002;118:856–863.
4. Guven M, Oymak O, Utas C, Emeklioglu S. Rhabdomyolysis and acute renal failure due to status epilepticus. Clin Nephrol 1998;50(3):204.
5. Sungur M, Guven M. Rhabdomyolysis caused by electric injury. J Emerg Med 2001;20(2): 195–196.
6. Ordog G, Wasserberger J, Schlater T, Balasubramanium S. Electronic gun (TASER®) injuries. Ann Emerg Med 1987;16:73–78.
7. Jauchem J, Sherry C, Fines D, Cook M. Acidosis, lactate, electrolytes, muscle enzymes and other factors in the blood of *Sus Scrofa* following repeated TASER® exposures. Forensic Sci Int 2006;161(1):20–30.
8. Ho J, Miner J, Lakireddy D, Bultman L, Heegaard W. Cardiovascular and physiologic effects of conducted electrical weapon discharge in resting adults. Acad Emerg Med 2006;13:589–595.

Chapter 14
Effects of CEWs on Respiration

Donald M. Dawes*

Conducted electrical weapons (CEWs) induce neuromuscular incapacitation and pain by the application of low-current electrical discharge with special waveforms. The electrical current is delivered by tethered, gas-propelled probes in the most utilized weapons and stimulates both afferent sensory neurons causing pain and efferent motor neurons causing involuntary regional skeletal muscle contractions. There is controversy in the lay press and the medical literature regarding the use of these weapons and the sudden in-custody death phenomenon. There is speculation that the muscle contractions induced by the electrical current may impair breathing leading to hypoxemia (low oxygen content on the blood) and hypercarbia (high carbon dioxide content in the blood). In this chapter, the current medical literature about the effects of these weapons on respiration will be reviewed.

14.1 Respiratory Physiology

The primary function of the respiratory system is to obtain oxygen from the environment and supply it to the blood, and to extract carbon dioxide, the by-product of aerobic cellular metabolism, from the blood, and release it to the environment. A secondary function of the respiratory system is the maintenance of the acid–base balance; a function accomplished by the control of carbon dioxide release from the blood. Air moves from higher pressure to lower pressure. In order for the respiratory system to obtain air from the environment, an act called inspiration, a pressure gradient must be established

---

**Donald M. Dawes reports serving as an external medical consultant to TASER International, Inc. and reports as a stockholder of shares of TASER International, Inc. No other potential conflict of interest relevant to this chapter was reported.*

D.M. Dawes (✉)
Department of Emergency Medicine, Lompoc District Hospital, Lompoc, CA
e-mail: donalddawes@aol.com

M.W. Kroll, J.D. Ho (eds.), *TASER® Conducted Electrical Weapons: Physiology, Pathology, and Law*, DOI 10.1007/978-0-387-85475-5_14,

between the atmosphere and the alveoli (the terminal gas exchange unit of the respiratory system); specifically, a lower pressure must be established in the alveoli than atmospheric pressure. This is referred to as negative-pressure breathing (to distinguish it from the positive-pressure breathing in mechanically ventilated patients). This lower pressure is established by the contraction of the muscles of respiration with an expansion of the thorax. Expiration, in normal quiet breathing is a passive process in which the muscles of respiration relax [1].

The muscles of respiration include the diaphragm, the intercostal muscles (external and internal), the scalene muscles, and the accessory muscles of respiration. The diaphragm, a large dome-shaped muscle that separates the thoracic and abdominal cavities, is the primary muscle of respiration. It is responsible for about 70% of the air that enters the respiratory system in a supine adult during normal quiet breathing (about 40% in an upright adult). During inspiration, contraction of the diaphragm causes it to descend into the abdominal cavity, increasing the volume of the thoracic cavity. The intercostal muscles and scalene muscles raise and stretch the rib cage during inspiration, also increasing the volume of the thoracic cavity. The accessory muscles, such as the sternocleidomastoid muscles, are not utilized in normal quiet breathing, but can be recruited to additionally increase the volume of the thoracic cavity. The elastic recoil of the lung is sufficient to collapse the thoracic volume, increase the alveolar pressure, and allow for the movement of air into the environment. The abdominal wall musculature, which is compliant during inspiration allowing the abdominal contents to push out in response to the descending diaphragm, can be recruited to enhance expiration [1].

The volume of gas in the respiratory system depends on chest wall and lung factors (potentially altered by disease states) and the activity of the muscles of inspiration and expiration. The lung size is dependent on age, height, weight, and sex. The tidal volume is the volume of air entering (or exiting) the respiratory system per breath. In normal quiet breathing, the tidal volume of a 70-kilogram adult is about 500 milliliters. The resting volume (also called the functional residual capacity) is the volume of air left in the lungs at the end of a normal quiet breathing exhalation. It amounts to about 3.0 liters in a 70-kilogram adult. The residual volume is the volume of air left in the lungs after a maximal forced exhalation (recruiting the abdominal musculature), about 1.5 liters in a 70-kilogram adult. The residual volume is important for keeping the alveoli from completely collapsing at low lung volumes. The expiratory reserve volume is the difference between the resting volume and the residual volume, also about 1.5 liters. The inspiratory reserve volume is the difference between the volume of air in the lungs after a maximal inspiration and the volume of air in the lungs after a normal quiet breathing inspiration, about 2.5 liters in a 70-kilogram adult [1].

The last part of each inspiration and each expiration remains in the large conducting airways of the respiratory system in which no gas exchange occurs. The alveolar ventilation is then the tidal volume less this anatomic dead space, about 150 milliliters in a 70-kilogram adult. In addition to anatomic dead space, there is alveolar dead space representing alveoli that are ventilated, but not perfused with blood. The physiologic dead space is the sum of the anatomic dead space and the alveolar dead space. Exhaled gas includes gas from the ventilated and perfused alveoli as well as gas from this physiologic dead space. The pressure of carbon dioxide in this mixed expired gas can be measured and used to estimate the alveolar pressure of carbon dioxide (and therefore arterial pressure of carbon dioxide). The end-tidal (end of inspiration) gas is used since this usually represents gas from the alveoli and not the anatomic dead space. If there is significant alveolar dead space, such as might occur with low right ventricular output, the estimate of the alveolar pressure of carbon dioxide will be less accurate since the pressure of carbon dioxide in the mixed expired gas will be diluted by gas coming from unperfused alveoli. This is important for understanding the limitations of studies that utilize end-tidal gases to estimate the arterial pressure of carbon dioxide [1].

The pressures of oxygen and carbon dioxide in alveolar air are determined by alveolar ventilation, pulmonary capillary perfusion, oxygen consumption, and carbon dioxide production. Alveolar ventilation is normally adjusted by the medulla to maintain a mean arterial pressure of carbon dioxide (about equal to alveolar pressure of carbon dioxide in healthy adults) of about 40 mmHg. The mean alveolar pressure of oxygen is about 104 mmHg. Since the mixed expired gas contains gas from the dead space, the mean pressure of oxygen is higher and the pressure of carbon dioxide is lower (120 and 27 millimeter hydrargyrum, respectively) [1].

The respiratory system is important in the maintenance of the acid–base balance in the blood. Carbon dioxide is an end product of the oxidation of glucose and fatty acids during aerobic metabolism in the tissues. The hydration of carbon dioxide produces carbonic acid, which can dissociate to hydrogen ion and bicarbonate ion. This process is reversed in the respiratory system with carbon dioxide gas being released into the environment. The equilibrium equation is as follows:

$$CO_2(\text{gas}) \Leftrightarrow CO_2 + H_2O(\text{aqueous}) \Leftrightarrow H_2CO_3 \Leftrightarrow H^+ + HCO_3^-$$

This process is not only important in the removal of carbon dioxide, but is important in maintaining the acid–base balance of the body. Acid–base homeostasis is mostly controlled by the respiratory system and the renal system. The respiratory system, however, is most important for immediate acid–base homeostasis. The equilibrium equation above is useful in conceptualizing this process. Hyperventilation, referring to the increased removal of carbon dioxide gas, will "pull" the equilibrium equation to the left, leading to the removal of hydrogen ions. This will cause a respiratory alkalosis (elevated blood pH). Hypoventilation, referring to the decreased removal of carbon dioxide gas, will "push" the equilibrium equation to the right, leading to the addition of hydrogen ions. This will lead to a respiratory acidosis (lowered blood pH). This dynamic control of the blood pH provides a mechanism to adjust to supranormal conditions and pathological conditions. In the condition of diabetic ketoacidosis, for example, there is an overproduction of the fixed acids (acetoacetic and butyric acids). In order to compensate for the increased hydrogen ions in the blood, hyperventilation occurs, pulling the equilibrium equation to the left, and elevating the pH. Arterial chemoreceptors, located in the carotid and aortic bodies, are utilized to detect changes in the acid–base balance and these inputs are utilized to make the necessary changes in ventilation. The medulla is the primary integrating structure for this function (although higher brain centers can "override" the medullary control of breathing) [1].

14.1.1 Cellular Respiration

The molecular "currency" of energy is ATP (adenosine triphosphate). Cells derive ATP from the metabolism of glucose. The initial step in this process is glycolysis and this occurs in the cytoplasm of all cells. In glycolysis, glucose is metabolized to pyruvate and two molecules of ATP. Under aerobic conditions, this pyruvate enters the mitochondria and is metabolized to carbon dioxide and used to produce an additional 34 (or 36, depending on whether the malate–aspartate or glycerol phosphate shuttle is used) molecules of ATP. Under anaerobic conditions, such as in heavily exerting muscle tissue (and in tissues which do not have mitochondria, such as red blood cells), the mitochondrial metabolism does not occur, and, instead, pyruvate is converted to lactate by the enzyme lactate dehydrogenase. Lactate then moves out of the cells, and combines with extracellular hydrogen ions to form lactic acid. The lactic acid can then be transported to heavily aerobic tissues such as the heart and kidneys where it is used as fuel. It is also taken up by the liver and converted back to pyruvate through the Cori cycle. Under conditions of heavy exertion the lactic acid in the blood can exceed the ability of other aerobic tissues to metabolize it, as well as exceed the buffering capacity of the blood, leading to a metabolic acidosis. Generally, the resting serum lactate level in humans is maintained below 2 millimoles per liter although there is some natural variation to this [2]. With increasing exercise it may stabilize at higher levels up to 10–20 millimoles per liter.

14.1.2 Anatomy

The phrenic nerve innervates the diaphragm, the major muscle of respiration, one branch for each hemidiaphragm. The phrenic nerve derives from the cervical vertebrae of the C3-5 nerve roots. The C5 fibers may not join the phrenic nerve trunk until it is in the chest [3]. In the neck, the phrenic nerve originates at the middle scalene muscle and then crosses to the anterior scalene muscle and descends to the base of the neck. The left phrenic nerve descends into the thorax on the anterior surface of the left subclavian artery, and travels along the pericardium to the diaphragm. The right phrenic nerve descends into the thorax behind the innominate vein, and travels along the inferior vena cava to the diaphragm. Both phrenic nerves branch just proximal to the diaphragm, with the terminal branches innervating the diaphragm [4]. In addition, in a large number of persons there exists an accessory phrenic nerve, derived from C5 and C6 or possibly from the nerve to the subclavius muscle, and this may not join the main phrenic nerve until the root of the neck or the thorax. This means that disruption of the phrenic nerve in the neck may not paralyze the diaphragm. Bigeleisen notes that in a supraclavicular nerve block, only 50% of patients have diaphragmatic paresis, and there is no reduction in forced vital capacity in these patients [5]. In a study by Le et al., the average distance from the skin to the carotid artery was 2 centimeters [6]. Since the phrenic nerve at its most superficial in the neck lies just posterior to the carotid artery, this is a reasonable approximation of the distance to the phrenic nerve from the skin.

14.2 The CEW Interaction

14.2.1 Theory of Injury

Since respiration is dependent on the contraction of the respiratory muscles, it has been speculated that the involuntary regional muscle contraction induced by conducted electrical weapons may impair respiration. This hypothetical situation would lead to a decreased ability to obtain oxygen from the environment, leading to hypoxemia, as well as a decreased ability to extract carbon dioxide from the blood, leading to hypercarbia. The hypoxemia would lead to a metabolic acidosis as aerobic metabolism was shifted to anaerobic metabolism and serum lactic acid increased (this process would also be exacerbated by the severe muscle contractions themselves which would lead to increased lactic acid production as described above). The hypercarbia would lead to a respiratory acidosis as carbon dioxide was shifted to hydrogen ions and bicarbonate ions; the equilibrium equation shifted to the right looks like this:

$$CO_2(\text{gas}) \leftrightarrow CO_2 + H_2O(\text{aqueous}) \leftrightarrow H_2CO_3 \leftrightarrow H^+ + HCO_3^-$$

The combined metabolic and respiratory acidosis could lead to general end-organ dysfunction. Acidosis causes a wide range of end-organ dysfunction including: decreased cardiac contractility, arteriolar dilatation and venous constriction leading to central congestion, increased pulmonary vascular resistance, decreased cardiac output, sensitization to cardiac arrhythmias, decreased respiratory muscle strength and hypoventilation, hyperkalemia, protein degradation, loss of intracellular water regulation, obtundation, coma, and death [7]. This mechanism has been postulated to be causative in some cases of sudden, unexpected custodial death proximate to a CEW exposure.

14.2.2 Animal Data with CEW Exposure

Jauchem et al. conducted a study of swine in which the authors noted that "complete cessation of breathing" occurred during the TASER X26 discharge [8]. The authors attributed this to a direct effect of the current on the muscles of respiration. This study was limited in that the swine were deeply anesthetized with tilatamine zolazepam and maintained on propofol and endotracheally intubated *but not ventilated.* In addition, the swine were placed on their backs. This position alone can compromise their respiratory function. The lead author of this research also warns against the many limitations of his study when attempting to apply its findings to law enforcement field use circumstances. Similar results have been found in two other swine studies [9,10].

14.2.3 Human Data

There are now two published studies that examine the respiratory effects of conducted electrical weapons in humans [11]. Ho et al., subjected 54 volunteers, mostly males, to a 15-second discharge from the TASER X26 (CEW) conducted electrical weapon. The subjects were placed supine with the weapon attached unilaterally to the upper chest wall and the upper abdomen in an effort to create a transdiaphragmatic exposure. The supine positioning and transdiaphragmatic connection was meant to create a worst-case scenario (where the diaphragm is most disadvantaged due to compression of the abdominal contents from both gravity and abdominal wall contraction). Discharges consisted of a 15-second continuously exposed group and a second exposure group that received a 15-second cumulative exposure given in three 5-second bursts (separated by 1 second). A commercial breath-by-breath gas exchange system was utilized to measure various respiratory variables including tidal volume, respiratory rate, and end-tidal oxygen and carbon dioxide.

The results from the interrupted and the uninterrupted exposures were the same so the results are mixed in the figures shown. As seen in Fig. 14.1, the respiratory rate was not increased (with statistical significance) until 1 minute after the exposure. Figure 14.2 shows that the tidal volume was increased significantly during and shortly after the exposure. Minute ventilation actually increased during the exposure but was increased significantly after the exposure as seen in Fig. 14.3. These changes are consistent with the changes expected from exertion and/or pain.

End-tidal CO_2 is shown in Fig. 14.4. This dipped slightly during the exposure barely achieving statistical significance with the pooled data. However, it was elevated at the 1-minute postexposure measurement. As shown in Fig. 14.5, end-tidal O_2 was not statistically significantly different at any stage of the measurements.

In a follow-up study by Dawes et al. of 18 subjects with the same methodology, but with venous blood gas and electrolyte sampling, the results of the respiratory

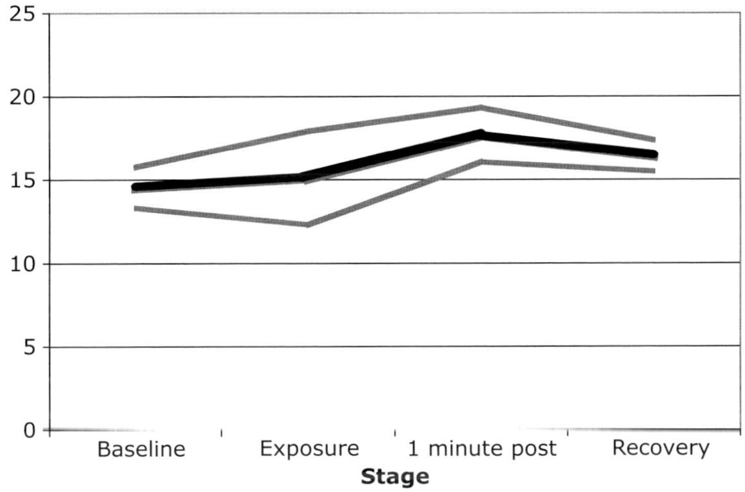

Fig. 14.1 Respiratory rate with 15-second exposures. This plot includes the continuous 15-second exposures with the interrupted 15-second exposures. *Thick middle line* is the mean value while the *thinner lines* are the upper and lower confidence limits with this figure and the others

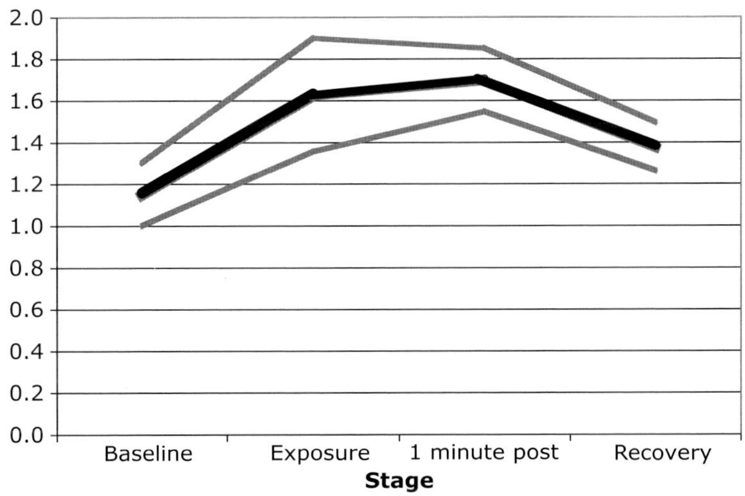

Fig. 14.2 Tidal volume with 15-second exposures

data showed no significant changes before, during, or after the exposure except an increase in respiratory rate during the exposure. The blood chemistries showed a statistically significant decrease in potassium (–0.23), pCO_2 (–5.69), HCO_3 (−3.4), a statistically significant increase in pO_2 (10.8) and lactate (1.8), and no significant change in pH (0.0009) or sodium (0.07) postexposure [12].

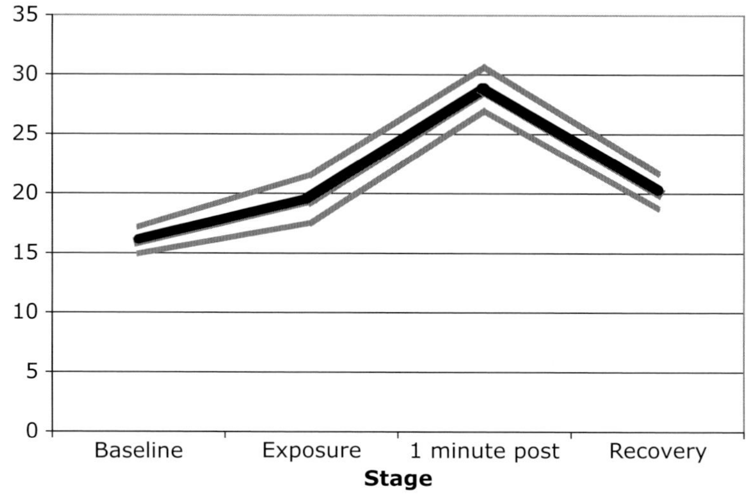

Fig. 14.3 Minute ventilation with 15-second exposures

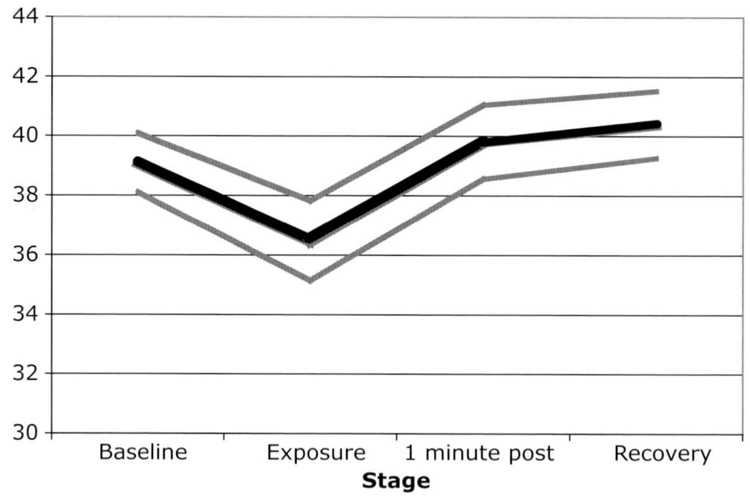

Fig. 14.4 End-tidal CO_2 with 15-second exposure

A study by Vilke et al. reported on a study of 32 subjects who were exposed to a 5-second discharge from a TASER X26. They found that minute ventilation, tidal volume, and respiratory rate were increased at 1 minute, but returned to baseline within 10 minutes. They found that the pH was decreased at 1 minute ((−0.03), not clinically significant), but returned to baseline by 30 minutes. They found no change in oxygen saturation or pO_2, end-tidal CO_2 or pCO_2, and

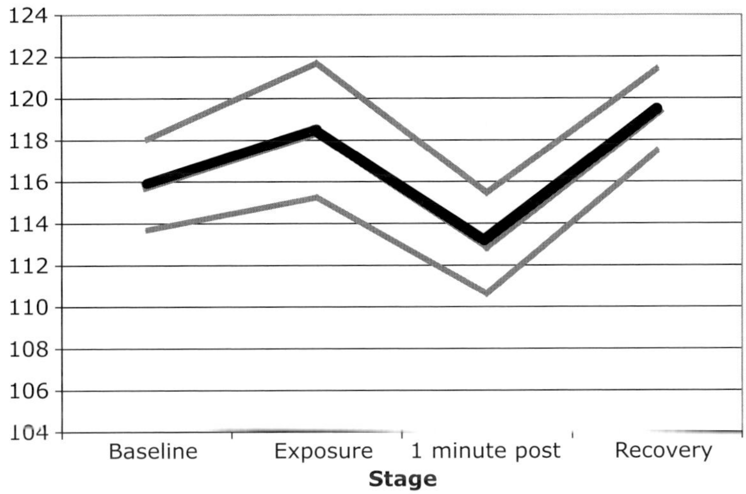

Fig. 14.5 End-tidal O_2 with 15-second exposures

concluded that there was no evidence of hypoxemia or hypoventilation [13]. These findings were consistent with the previous two studies.

While not studies specifically examining respiration, two studies by Ho et al. did examine blood markers of acidosis. In "Cardiovascular and Physiologic Effects of Conducted Electrical Weapon Discharge in Resting Adults," subjects were exposed to 5-second discharges of the TASER X26 [14]. Serum bicarbonate was followed serially. In a "pure" uncompensated respiratory acidosis, the serum bicarbonate should increase. The mean baseline bicarbonate was 22.6. Immediately after the exposure, the mean bicarbonate was 22.0. This does not suggest an uncompensated respiratory acidosis.

In the Ho et al. study "Prolonged TASER® use on exhausted humans does not worsen markers of acidosis," subjects were asked to exercise to physical exhaustion, and were then immediately exposed to a 15-second discharge of the TASER X26 [15]. Venous pH was drawn prior to exertion, immediately after exertion, and immediately after the TASER exposure. In this study, the mean venous pH at baseline was 7.37. After exercise to physical exhaustion, the mean pH dropped to 7.258. After the 15-second continuous TASER X26 discharge, the mean pH remained flat (7.225, not a statistically significant change). The recovery pH returned to baseline (mean pH 7.358). A small control group with the same exertion regimen and a repeat exertion regimen of 15 seconds had similar pH changes. This implies that the 15-second TASER X26 discharge did not exacerbate the acidosis caused by the physical exertion. The control group had similar pH levels after their repeat brief exertion which suggests that the discharge did not delay recovery any more than would a 15-second period of exertion.

These two studies suggest that the exposure to conducted electrical weapons does not affect the acid–base balancing of the respiratory system.

Finally, Ho et al. published a report looking specifically at the ability of a CEW to affect the phrenic nerve if applied in drive-stun fashion to the trapezius muscle [16]. This is a target area that is taught to the CEW user in training and has recently been the subject of a legal claim that theorizes that application to this area can either cause phrenic nerve dysfunction or travel up the spinal cord to cause brainstem (medullary) dysfunction [17]. If either theory were to be true, the expected result would be diaphragm paralysis and respiratory arrest. In this report, ultrasound technology was utilized to visualize the diaphragm in real time during a 10-second drive-stun to the targeted area. No diaphragm paralysis or respiratory cessation was noted. This finding does not support either of the above theories to be true.

The XREP® device is a nontethered, shotgun fired projectile weapon manufactured by TASER International with a similar waveform to the TASER X26 CEW. In a study of this weapon, subjects had venipuncture prior to the application of the CEW and immediately after the exposure, and venous samples were analyzed to obtain venous pH, pCO_2, HCO_3, lactate, as well as Na and K [18]. Breathing data were collected by a breath-by-breath gas-exchange system. All subjects were exposed for a minimum of 15 seconds. Exposure was thoracoabdominal with one lead over the pectoralis major muscle, and the other in the upper abdomen. In 27 subjects, the device was programmed for a 45-second exposure. The subjects could terminate the exposure with a “tap out” button after 15 seconds. In 23 subjects, the exposure was fixed at 20 seconds. In four of these subjects, the device was programmed to deliver two exposures. The first exposure was the standard thoracoabdominal exposure, and the second was between the contralateral abdomen and the thigh.

Fifty subjects completed the study. The analysis was separated into two groups. The first was the self-terminating group (variable time exposure). In this group, respiratory rate and minute ventilation increased significantly during the exposure. End-tidal CO_2 decreased significantly during exposure. Venous pH decreased by 0.023, pO_2 increased by 13.4, HCO_3 decreased by 2.8, lactate increased by 2.4, and potassium decreased by 0.13. The second group was the fixed 20-second exposure (including the 4 with the “double” exposure). In this group, respiratory rate and minute ventilation increased significantly during the exposure. End-tidal CO_2 decreased and end-tidal O_2 increased significantly during exposure. Venous pH did not significantly change. pCO_2 decreased by 4.0, p O_2 increased by 16.3, HCO3 decreased by 3.4, and lactate increased by 2.7. Chemistries had no significant change.

14.3 Conclusion

While the medical literature in this area continues to grow, the current human studies do not suggest respiratory impairment with a continuous discharge of the TASER X26 to the thorax. The available human studies also offer indirect evidence that the acid–base balancing function of the respiratory system is not impaired by a conducted electrical weapon discharge. The likely explanation for these findings is that the skeletal muscle contraction induced by the conducted electrical weapons is regional and not global.

Other experts have demonstrated that current densities fall off precipitously in the tissues. Given the anatomy of the diaphragm (the major muscle of respiration) and its innervation (the phrenic nerve), it is unlikely that this muscle, the most important for respiration, is significantly impaired by these thoracically applied surface currents.

Also, a single, unilateral application of these weapons would not likely affect the contralateral muscles of respiration. As is well known from the medical literature, in unilateral diaphragm paralysis, the patient will usually be asymptomatic at rest, without hypoxemia or hypercarbia (but have dyspnea on exertion) [5–19]. Of course, the effectiveness of the diaphragm does depend on the compliance of the abdominal cavity, which is related to positioning and to the pressure exerted on the abdominal cavity (abdominal wall muscle contraction, weight-force, etc.). The study by Ho et al. did utilize subjects in the supine position (less advantageous for diaphragmatic breathing) and the weapons were placed on the upper abdominal walls, and no hypoxemia or hypercarbia (by indirect measures) was seen.

References

1. Levitsky M. Pulmonary Physiology. 6th Edition, McGraw-Hill, 2003.
2. Cellular Respiration. (http://faculty.clintoncc.suny.edu/faculty/Michael.Gregory/files/Bio%20101/Bio%20101%20Lectures/Cellular%20Respiration/cellular.htm, accessed June 7, 2007).
3. Elefteriades JA, Quin JA, Hogan JF, et al., Long-term follow-up of pacing of the conditioned diaphragm in quadriplegia, PACE: 25 (6), June 2002.
4. Meyers B. and Kozower B. Paralyzed diaphragm in ACS Surgery: Principles and Practice, 2005 (http://www.acssurgery.com/acs/chapters/ch0403.htm, accessed 4/15/07).
5. Bigeleisen P. Anatomic variations of the phrenic nerve and its clinical implication for supraclavicualr block. Brit J Anaesth, 2003; 91(6): 916–917.
6. Le J, Gwak M. and Yang M. A new method of internal jugular vein catheterization using the cricoid cartilage and the external jugular vein as a landmark. Am J Emerg Med, 2006; 24(6).
7. Adrogue H. and Madias N. Management of life-threatening acid-base disorders. NEJM, 1998; 338: 26–34.
8. Jauchem J, Sherry C, Fines D, and M Cook. Acidosis, lactate, electrolytes, muscle enzymes, and other factors in the blood of sus scrofa following repeated TASER exposures. Forensic Sci Int, 2006; 161: 20–30.
9. Dennis AJ, Valentino DJ, Walter RJ, et al. Acute effects of TASER X26 discharges in a swine model. J Trauma, Sep 2007; 63(3): 581–590.
10. Walter RJ, Dennis AJ, Valentino DJ, et al. TASER X26 discharges in swine produce potentially fatal ventricular arrhythmias. Acad Emerg Med, Jan 2008; 15(1): 66–73.
11. Ho JD, Dawes DM, et al. Respiratory effect of prolonged electrical weapon application on human volunteers. Acad Emerg Med, 2007; 14(3): 197–201.

12. Dawes D, Ho J, et al. 15-second conducted electrical weapon application does not impair basic respiratory parameters, venous blood gases, or blood chemistries and does not increase core body temperature. Ann Emerg Med (Supplement), 2007; 50(3): S132.
13. Vilke GM, Sloane CM, Bouton KD, et al. Physiological effects of a conducted electrical weapon on human subjects. Ann Emerg Med. Nov 2007; 50(5): 569–575.
14. Ho JD, Miner JR, Lakireddy DR, et al. Cardiovascular and physiologic effects of conducted electrical weapon discharge in resting adults. Acad Emerg Med, 2006; 13: 589–595.
15. Ho J, Dawes D, et al. Prolonged TASER® use on exhausted humans does not worsen markers of acidosis. Am J Emerg Med 2009 in press.
16. Ho, J., Lapine, A., Joing, S., et al. Confirmation of respiration during trapezial conducted electrical weapon application. Acad Emerg Med, 2009; 15: 398.
17. Neal-Lomax, et al. v. Las Vegas Metropolitan Police Department, et al., United States District Court for the District of Nevada, Case Number 2-05-cv-1464.
18. Dawes D, Ho J, et al. Breathing parameters, venous blood gases, and serum chemistries with exposure to a new wireless projectile conducted electrical weapon in human volunteers. Ann Emerg Med (Supplement), 2007; 50(3): S133.
19. Danill Z., et al. An unusual cause of dyspnea in a 77-year-old man. Chest, 2004; 125(2).

Chapter 15
Neuroendocrine Effects of CEWs

Donald M. Dawes* and Mark W. Kroll**

If these powerful motions prevail, and the bodily forces are fully mobilized for action, and if this state of extreme perturbation continues in uncontrolled possession of the organism for a considerable period, without the occurrence of action, dire results may ensue.

Walter Cannon, Voodoo Death, 1957 [1]

Conducted electrical weapons (CEW) induce neuromuscular incapacitation and pain by the application of a small electrical current. The electrical current stimulates both afferent sensory neurons causing pain and efferent motor neurons causing involuntary regional skeletal muscle contraction. There has been controversy in the lay press with regard to the use of these weapons and sudden in-custody death. Previous research and field data have supported the assertion that these weapons do not cause instantaneous malignant cardiac arrhythmias from the electrical discharge [2–6].

However, there have been claims that this discharge leads to other effects that can lead to delayed cardiovascular collapse. A study by Ho et al. examined serial troponins, electrolytes, and electrocardiograms after a 5-second TASER X26 discharge, and found no clinically important effects [7]. A study by Sloane et al. also found no elevation of troponin at 6 hours [8]. A study by Vilke et al. found no significant or lasting changes in pH, lactate, or electrolytes [9]. Both the Sloane and Vilke studies used a 5-second or shorter discharge. A study by Chan et al. showed no respiratory impairment with a 5-second discharge from a TASER X26 [10]. A

**Donald M. Dawes reports serving as an external medical consultant to TASER International, Inc. and reports as a stockholder of shares of TASER International, Inc. No other potential conflict of interest relevant to this chapter was reported.*

***Mark W. Kroll reports serving on the Corporate and Scientific/Medical Advisory Board of TASER International, Inc. and receives compensation in both roles. No other potential conflict of interest relevant to this chapter was reported.*

D.M. Dawes (✉)
Department of Emergency Medicine, Lompoc District Hospital, Lompoc, CA
e-mail: donalddawes@aol.com

M.W. Kroll, J.D. Ho (eds.), *TASER® Conducted Electrical Weapons: Physiology, Pathology, and Law*, DOI 10.1007/978-0-387-85475-5_15,

study by Ho et al. showed no respiratory impairment with a 15-second discharge from the TASER X26 [11]. There has been speculation that exposure to the discharge of a conducted electrical weapon may induce neuroendocrine effects which might predispose subjects to sudden death. This chapter reviews the current literature of the stress response to conducted electrical weapons.

15.1 Human Stress Response

The human stress response has two parts. The first is the SAM (sympathetic-adrenal-medulla) axis shown in Fig. 15.1. The SAM axis acts when a stressor causes the brain to send a signal down the spinal cord to the adrenal glands

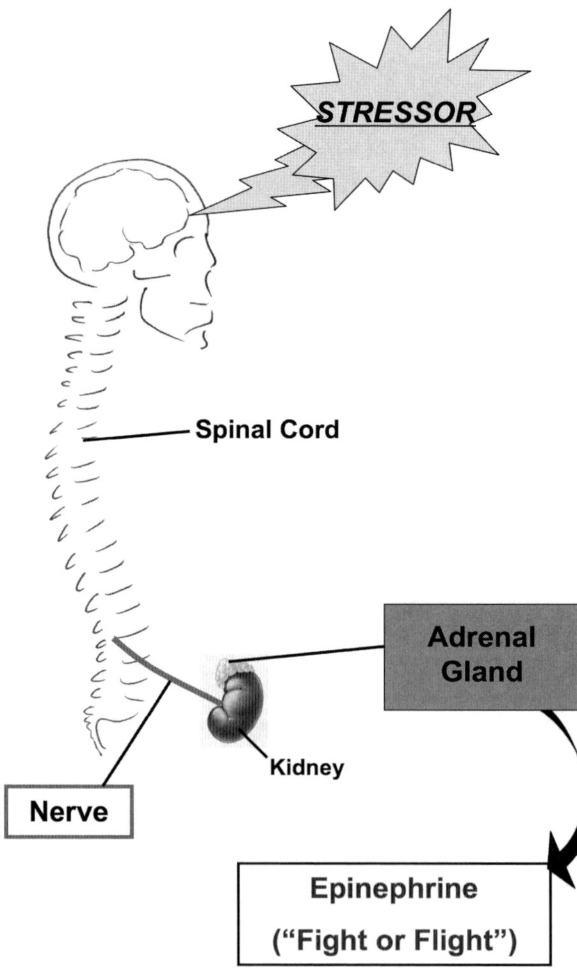

Fig. 15.1 The sympathetic-adrenal-medulla axis. (Lori & Ryan Kroll)

that are located on top of the kidneys. The adrenal glands then release stored epinephrine to initiate the "fight or flight" response. Epinephrine has a number of effects which are very useful for a short-term struggle. It increases both the rate and the strength of contraction of the heart and metabolism. However, there are trade-offs. Stress on the system can cause myocardial ischemia, arrhythmias, hyperthermia, and lactic acidosis.

The second part is the HPA (hypothalamus-pituitary-adrenal) axis is shown in Fig. 15.2. A perceived stressor causes the hypothalamus of the brain to signal the anterior pituitary glands to release two compounds. These are the β-endorphins and ACTH (adrenocorticotropic hormone). β-Endorphins are our body's natural painkillers to prepare one for a fight and reduce the distractions from the pains of a fight. ACTH is a chemical messenger put into the blood stream to trigger the adrenal glands to producing cortisol and mineralocorticoids. Cortisol limits the rate of epinephrine production so that we do not "flame out" by producing too much epinephrine in the expected struggle. The mineralocorticoids increase blood volume so that we can better withstand a injury.

The stress response may contribute to some in-custody deaths. A subject that is paranoid from stimulant drug intoxication or from psychiatric

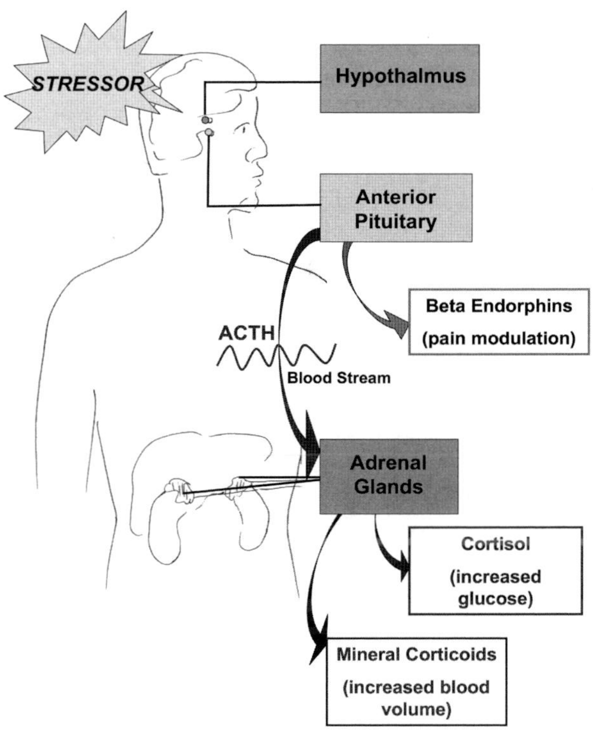

Fig. 15.2 The hypothalamic-pituitary-adrenal axis (Lori & Ryan Kroll)

illness might tend to overreact to law enforcement and have a more pronounced stress response. Lactic acidosis, hyperthermia, and myocardial contraction bands (which can be related to epinephrine surges) are often found in in-custody deaths. This is also seen in medical settings where there are no law enforcement personnel involved [12].

15.2 The Possible Role of the CEW

There is little literature on the stress response induced by the application of a TASER device. A study by Dawes, et al used salivary stress markers to study this response. Salivary markers have been developed to avoid having to use a needle stick (a stressor by itself). A validated marker for the activity in the SAM axis is alpha-amylase. Alpha-amylase levels peak about 10 minutes after the stressor. Whether the alpha-amylase specifically tracks the activity of the SAM axis or just general sympathetic tone is an open issue. Cortisol is a salivary marker for the activity in the HPA axis and its level peaks about 20 minutes after the stressor.

Dawes, et al performed a study to examine the differential stress response between a cold-pressor test (45 s), defensive tactics (1 minute weapon retention drill), TASER X26 CEW and a 5-second spray of OC (pepper spray) directly to the eyes. Salivary samples were taken 10 minutes before the exposure and again at 10–15 and 40–60 minutes afterwards. Results are shown in Table 15.1.

OC spray had a statistically significant increase in the alpha-amylase marker while the CEW did not. All control options had increases in the cortisol marker with defensive tactics having the greatest increase at 40–60 minutes [13].

While this is the only study looking directly at the stress response and CEWs, we can make some inferences from other studies. Sudden emotional distress can cause cardiac dysfunction. Wittstein et al. evaluated 19 patients who had myocardial stunning after sudden emotional stress. All the patients were symptomatic with chest pain, dyspnea, or both [14]. Diffuse T-wave inversion and a prolonged QT interval were seen in most of the subjects. In 17 or the 19, there

Table 15.1 Stress marker responses to cold-pressor, OC spray, defensive tactics, and CEW

Time (minutes)	Marker	Cold-Pressor ($n = 16$)	OC ($n = 10$)	Defensive Tactics ($n = 10$)	CEW ($n = 16$)
Baseline	AA (U/ml)	158	59	264	132
10–15	AA	−35 (NS)	37 ($p = .01$)	64 (NS)	−5 NS
40–60	AA	3 (NS)	8 ($p = .03$)	−85 (NS)	−20 (NS)
Baseline	Cortisol (ng/dl)	170	500	280	470
10–15	Cortisol	70 (NS)	500 ($p = .01$)	250 ($p < .01$)	380 ($p < .01$)
40–60	Cortisol	40 (NS)	10 (NS)	470 ($p < .01$)	320 ($p < .01$)

AA refers to alpha-amylase. Values given after baseline are the median *changes* from baseline.

was a mild elevation in troponin. All the subjects had severe left ventricular dysfunction. In the study by Ho et al., none of the subjects had subjective complaints of chest pain or shortness of breath. None of the subjects had electrocardiogram changes. In the Ho et al. study, one subject had a mild elevation in troponin, but had a complete in-hospital cardiac evaluation and no abnormalities were found [6]. This seems to suggest that conducted electrical weapons are not producing the changes seen in myocardial stunning patients.

There is evidence that exercise is more activating of the stress cascade than pain alone. One study compared high-intensity hand grip at painful durations, cold pressor tests (where the hand is kept in cold water to painful levels), and bicycle exercise. The catecholamine (epinephrine and norepinephrine) levels in the bicycle exercise group were 3–6 times greater than in the two pain stressed groups [15].

In a study by Han et al. in rats, it was found that cocaine combined with exercise increased epinephrine, norepinephrine, and lactate 2–5 times greater than either exercise or cocaine alone, and 11–35 times greater than rest with no cocaine [18]. This study also points to the significant activation of the stress cascade with exertion—but more significantly—demonstrates the much greater effect when exertion is compounded with acute drug use. Other animal studies have shown a relationship between restraint stress and sensitization to drugs of abuse. In a study by Pacchioni et al., it was found that a single restraint exposure was sufficient to cause a significantly increased release of dopamine and locomotor activity with amphetamine in rats [19]. A study by Pudiak et al. demonstrated a significantly higher mortality with cocaine and restraint in rats compared to cocaine alone [20]. These authors concluded that minimizing the stress response may be important in cocaine toxicity. In a study by Pacak et al., rats were subjected to one of five stressors: cold, hypoglycemia, hemorrhage, pain (formulin injection into a limb), and immobilization. The authors found that immobilization increased ACTH the most, followed by hypoglycemia. Cold stress increased norepinephrine the most, followed by immobilization. Hypoglycemia increased epinephrine the most, followed by immobilization. The immobilization stress was consistently one of the highest stressors by these measures [21]. These studies may provide insight into causation in many of these sudden in-custody death cases.

It has been proposed that a stressor, such as the discharge from a CEW, can induce a vasovagal reaction and lead to a bradycardic–asystolic cardiac arrest from a vasovagal response. In a study by Baron-Esquivias et al., no mortality was found in long-term follow-up of their subjects with vasovagal syncope. Their findings were in agreement with the large Framingham study [16]. In addition, in the literature on the cardiac response to conducted electrical weapon exposure, subjects have responded with a tachycardic response [3–6, 10]. Field data also refutes this theory. With this proposed mechanism, it would be expected that the death would occur during or immediately after the conducted electrical weapon discharge. In a review by Ho et al., conducted electrical weapons were never associated with immediate death [17].

The mechanisms in the sudden, unexpected in-custody death are not fully understood. It is likely that many pathological processes are involved. The excited delirium syndrome is an observed phenomenon of behavioral characteristics, signs, symptoms, and clinical sequelae that are commonly seen with custodial deaths. It is not yet clear if this is a specific disease entity related to brain receptor changes associated with chronic drug abuse or chronic mental illness, or is a syndrome with many etiologies. Illegal stimulants particularly cocaine and methamphetamines, are highly associated with in-custody deaths. Physical arrest and restraint itself is highly associated these deaths, as is agitated mental illness. Each of these can contribute significantly to an exaggerated stress response. It is likely that the stress response contributes, in some way, to the in-custody death phenomenon.

15.3 Conclusion

The in-custody death phenomenon is likely a "perfect storm" of extremely physiologic conditions including an exaggerated stress response, potentiated by drugs of abuse and exertion and restraint, with hyperthermia, acidosis, electrolyte changes, and the unmasking of prior health problems. Whether conducted electrical weapons can significantly contribute to the stress response in this milieu is not clear, but its contribution appears to be smaller than or equal to other commonly employed uses of force. Given that the CEW is generally considered more efficacious in the control of subjects with impaired pain perception secondary to drug intoxication or an excited delirium, and that it induces a smaller or equal stress response to other commonly used control options (such as hands-on control or pepper spray), it may be the best option in certain settings to achieve rapid restraint and immediate medical intervention.

References

1. Cannon WB. Voodoo death. Psychosom Med, 1957; 19: 182–190.
2. McDaniel WC, Stratbucker RA, Nerheim M, et al. Cardiac safety of neuromuscular incapacitating defensive devices. PACE, 2005;28: S284–S287.
3. McDaniel W, and R Stratbucker. Testing the cardiac rhythm safety of the thoracic application of Tasers. Europace, 2006; 8: 58P23.
4. Levine SD, Sloane C, Chan T, Vilke G, and J Dunford. Cardiac Monitoring of subjects exposed to the TASER. Acad Emerg Med, 2005; 12 (supplement 1): S71.
5. Vilke G, Sloane C, et al. Does the TASER cause electrical changes in twelve lead ECG monitoring of human subjects? Acad Emerg Med, 2007 (Supplement 1); 14: S104.
6. Barnes Jr. D, Winslow J, et al. Cardiac effects of the TASER X26 conducted energy weapon. Ann Emerg Med, 2006; 48(Supplement): 102.
7. Ho JD, Miner JR, Lakireddy DR, et al. Cardiovascular and physiologic effects of conducted electrical weapon discharge in resting adults. Acad Emerg Med, 2006; 13: 589–595.

8. Sloane C, Vilke G, et al. Serum troponin I measurement of subjects exposed to the TASER X26. Acad Emerg Med, 2007 (Supplement 1); 14: S103–S104.
9. Vilke G, Sloane C, et al. Cardiovascular and metabolic effects of the TASER on human subjects. Acad Emerg Med, 2007 (Supplement 1); 14: S104–S105.
10. Chan T, Sloane C, et al. The impact of the TASER weapon on respiratory and ventilatory function in human subjects. Acad Emerg Med, 2007 (Supplement 1); 14: S191–S192.
11. Ho JD, Dawes DM, et al. Respiratory effect of prolonged electrical weapon application on human volunteers. Acad Emerg Med, 2007; 14: 197–201.
12. Hick JL, Smith SW, and MT Lynch. Metabolic acidosis in restraint-associated cardiac arrest: a case series. Acad Emerg Med, 1999; 6: 239–243.
13. Dawes D, Ho J, Miner J. The neuroendocrine effects of the TASER X26: A brief report. Forensic Sci Int, 2009: in press.
14. Wittstein IS, Thiemann DR, Lima JA, Baughman KL, Schulman SP, Gerstenblith G, Wu KC, Rade JJ, Bivalacqua TJ, and HC Champion. Neurohumoral features of myocardial stunning due to sudden emotional stress. N Engl J Med, 2005 Feb 10; 352(6): 539–548.
15. Stratton J, Halter J, et al. Comparative plasma catecholamine and hemodynamic responses to handgrip, cold pressor and supine bicycle exercise testing in normal subjects. J Am Coll Cardiol, 1983; 2(1): 93–104.
16. Baron-Esquivias G, Errazquin F, et al. Long-term outcome of patients with vasovagal syncope. Am Heart J, 2004; 147(5): 883–889.
17. Ho JD, Reardon RF, and WG Heegaard. Deaths in police custody: an 8 month surveillance study. Ann Emerg Med, 2005; 46 (suppl): S94.
18. Han D, Kelly K, et al. Cocaine and exercise: temporal changes in plasm levels of catecholamines, lactate, glucose, and cocaine. Am J Physiol Endocrinol Metab, 1996; 270: E438–E444.
19. Pacchioni A. et al. A single exposure to restraint stress induces behavioral and neurochemical sensitization to stimulating effects of amphetamine: involvement of NMDA receptors. Ann NY Acd Sci, 2002; 965: 233–246.
20. Pudiak C, and Bozarth M. Cocaine fatalities increased by restraint stress. Life Sci, 1994; 55: 379–382.
21. Pacak K, and Palkovits M. Stressor specificity of central neuroendocrine responses: implications for stress-related disorders. Endocrin Rev, 2001; 22(4): 502–548.

Chapter 16
Electroporation of Cardiac and Nerve Cells

Vadim V. Fedorov, Leonid Livshitz, Geran Kostecki and Igor R. Efimov

It has been recently speculated that CEW pulses might cause a direct injury to cardiac or nerve cells [1,2]. This injury is referred to as electroporation and is the subject of this chapter in which we will explore this phenomenon and investigate the possibility of it occurring with CEW pulses.

Heart and nerve cells use transmembrane pumps and ion channels to allow the traffic of certain ions through their otherwise impermeable membranes. In so doing, the cell sets up an imbalance in both chemical concentration and electric charge across the membrane. The electrical imbalance creates a resting potential of around –85 millivolt in most excitable cells, although there is much heterogeneity between tissues. Small deviations in intracellular voltage are corrected using the mechanisms described above; however, if enough positive charge enters the cell and depolarizes it to a certain threshold, about −65 millivolt, it sets off the opening of sodium ion channels which allows a very rapid influx of sodium ions propelled by both electrical and chemical gradients that cause further depolarization. This depolarization propagates to neighboring cells differently in different tissues. In nerve cells, depolarization propagates down an elongated cell structure called the axon, triggering the release of chemical messengers at the axon terminals that diffuse through the synapse between cells and cause ion channels on other cells to be opened, depolarizing them.

In cardiac cells, structures called gap junctions allow ions and molecules as large as 1,000 Dalton to pass between neighboring cells. Gap junctions are composed of two hexameric structures made up of a group of proteins called connexins. Electrically, gap junctions represent paths of low resistance through which current can easily flow. Because ions can freely commute between cells through the gap junctions, an intracellular ionic concentration change that occurs during an action potential can quickly propagate to neighboring cells,

V.V. Fedorov (✉)
Department of Biomedical Engineering, Washington University
e-mail: igor@wustl.edu

M.W. Kroll, J.D. Ho (eds.), *TASER® Conducted Electrical Weapons: Physiology, Pathology, and Law*, DOI 10.1007/978-0-387-85475-5_16,

propelled by both a voltage and concentration gradient. These propagating electrical impulses are the basis for neuronal and cardiac signal transmission and can be described spatially as a wave with the most depolarized cells located at the wave front. After being quickly depolarized, the cells are repolarized by closing sodium ion channels and using ion pumps and different ion channels to restore the cell to resting potential. During this time—called the refractory period—sodium channels cannot be activated and an action potential cannot occur, ensuring that action potential propagation occurs unidirectionally.

Under normal physiological conditions, a bilayer lipid membrane of cardiac and nerve cells made of lipid extracts of cells is a good barrier for ions and hydrophilic molecules. When an intense transmembrane electric field—exceeding the dielectric strength of a cell membrane—is applied, the membrane conductance will increase dramatically. The principal mechanism of this effect has been found to be electroporation of the lipid bilayer.

Electroporation is cell membrane trauma resulting from high-voltage shocks that, according to recent research, forces highly polar water molecules into the cell membrane, which interrupts the hydrophobic and van der Waals attractions holding it together, creating microscopically visible pores in the membrane. These pores allow all types of ions and even macromolecules through into and from the extracellular space, severely decreasing the electrochemical gradient set up by the cell.

Electroporation of cardiac and nerve cells is a commonly used approach for gene transfer [1]. It can be used as treatment for the creation of a "biological" pacemaker and as treatment for different diseases [2,3]. However, at the present time the use of electroporation for potential therapies is limited to in vitro approaches. Detectable electroporation of the heart cell membrane can occur during clinically relevant intensities of electrical shocks during defibrillation [4–7]. While high-intensity shocks are used routinely, the tissue and cellular responses to large currents are not fully understood. In particular, it remains a subject of debate whether shock-induced electroporation is pro- or antiarrhythmic in clinical settings [8]. It must be stressed that the energy of external defibrillation (up to 360 joule) is many orders of magnitude above that of the most popular CEW, the TASER X26 with 0.7 joule pulses.

16.1 Cellular Responses to Strong Electric Fields

The phenomenon of modifying the cell membrane conductivity electrically has been known since the 1940s. Goldman [9] measured the voltage–current (V-I) characteristics of the membrane of *Chara australia* and found a phenomenon similar to the dielectric breakdown of cell membrane, that is an abrupt increase in the membrane conductance (electroporation) when the membrane was hyperpolarized beyond a certain potential. Rapid-freezing electron microscopy of electropermeabilized cells provided direct evidence of the formation of

Volcano-shaped pores in cell membranes [10]. The pore structures rapidly expand to 20–120 nanometers in diameter during the first 20 milliseconds of electroporation, and begin to shrink, resealing after several seconds.

In whole cell patch clamp experiments, a breakdown in membrane conductance at transmembrane potential thresholds of 0.6–1.1 volt was shown in response to 0.1–1.0 kilovolt per second voltage ramps [11,12], which is unaffected by Na, K, and Ca channel blockers [13]. This result is consistent with the formation of ion-nonspecific membrane pores. Application of this technique for detection of electroporation in the tissue is difficult because the increase in cell membrane conductance translates to only a small decrease in total tissue resistance. Additionally, previous modeling work has shown that electroporation occurs only in a very small region of the tissue, perhaps only in a one-cell layer adjacent to the electrode [14,15]. Yet direct real-time recording or visualization of electroporation of in vivo or in vitro cardiac tissue remains to be developed.

Information about electroporation can be indirectly inferred from staining of the tissue with fluorescent dyes such as propidium iodide (PI) [1] or ethidium bromide (EB) [16], which penetrate cells only through the pores, and subsequent histological imaging of intracellular space. Electroporation-induced electrophysiological changes include the following: depolarization of the cellular membrane during diastolic interval resulting in depression of excitability [8,17–19], reduction of amplitude of action potentials and of the rate of rise of upstroke (dV/dt), and elevation of intracellular calcium concentration [17]. Figure 16.1 illustrates these changes induced by electroporation.

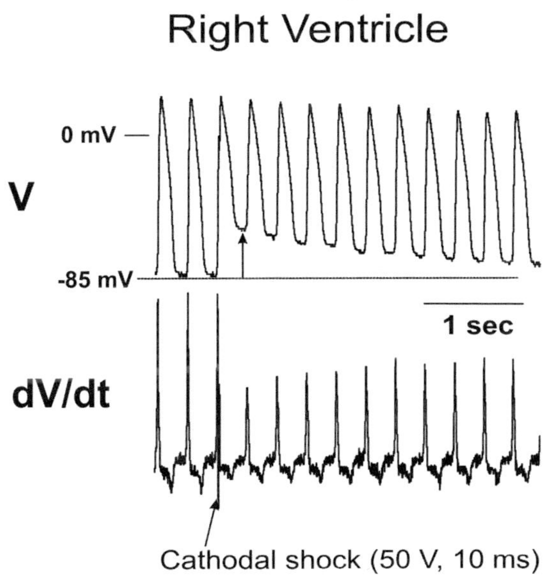

Fig. 16.1 Evidence of shock-induced electroporation in the right ventricular epicardium. Optical recording of transmembrane potentials, V (*upper trace*), shows time-dependent postshock reduction of resting potential and action potential amplitude. Maximal upstroke rate of rise (dV/dt) is also reduced and slowly recovers after shocks (*lower traces*)

16.2 Electroporation Assessment via Shock-Induced Changes of Transmembrane Action Potential Morphology

Electroporation is a frequent consequence of traumatic nerve injury. In order to measure the early secondary effects of different levels of electroporation on axonal structure and function the squid giant axon was studied by Gallant and Galbraith [20] after application of electric fields of strengths of 0.5, 1.0, 1.6, or 3.3 kilovolt per centimeter. Immediately after mild electroporation at 0.5 kilovolt per centimeter, 40% of the axons had no action potentials, but by 1 hour all of the axons had recovered their action potentials. Significant damage and swelling was demonstrated in many cellular organelles such as the mitochondria 1 hour after this mild electroporation. With a 1-hour delay postmoderate electroporation at 1.0 kilovolt per centimeter, most of the axons had no action potentials and most large organelles were swollen and completely nonfunctional. Finally, at severe electroporation levels of 1.65–3.0 kilovolt per centimeter all conduction was lost and even most small organelles had stopped functioning. The structural damage and transport block seen after severe and moderate electroporation were shown to be early secondary injuries because they could be prevented by placing the electroporated axons in an intracellular-type medium (low in Ca2+, Na+, and Cl–) immediately after shock [20].

Recently, we investigated the contribution of electroporation to the nonlinearity of transmembrane potential response (ΔV_m) induced by a strong external electrical field in cardiac cells using optical mapping [7]. We also found that electroporation was voltage dependent and polarity dependent and was significantly more pronounced in the atria versus ventricles ($p < 0.01$), with a summary 50% of effective dose (ED50) for main measured parameters of 9.2 +/– 3.6 volt per centimeter and 13.6 +/– 3.2 volt per centimeter in the atria versus 37.4 +/– 1.5 volt per centimeter and 48.4 +/– 2.8 volt per centimeter in the ventricles, for anodal and cathodal stimuli, respectively. In atria ($n = 5$), shocks of both polarities (27.2 +/– 1.1 volt per centimeter) transiently induced conduction block and reentry around the inexcitable area. Electroporation-induced ectopic activity was a possible trigger for reentry [21].

Double-barrel microelectrode recordings and optical mapping techniques have shown that weak stimuli produce monotonic transmembrane potential changes (ΔV_m) in single cell [22,23], cell culture strands [24], and heart tissue [25,26], as predicted by the cable theory and generalized activating function theory. However, reports on strong shocks of defibrillation strength sharply disagree on the morphology and amplitude of shock-induced responses in ΔV_m.

When a stimulus is applied to a single cell during the early plateau phase of the action potential, the optical recordings show depolarization of the cathodal end and hyperpolarization of the anodal end of the cell [27,28]. We found that at the strong shock-induced hyperpolarization (or more accurately negative polarization) ΔV_m first gradually increases in amplitude but

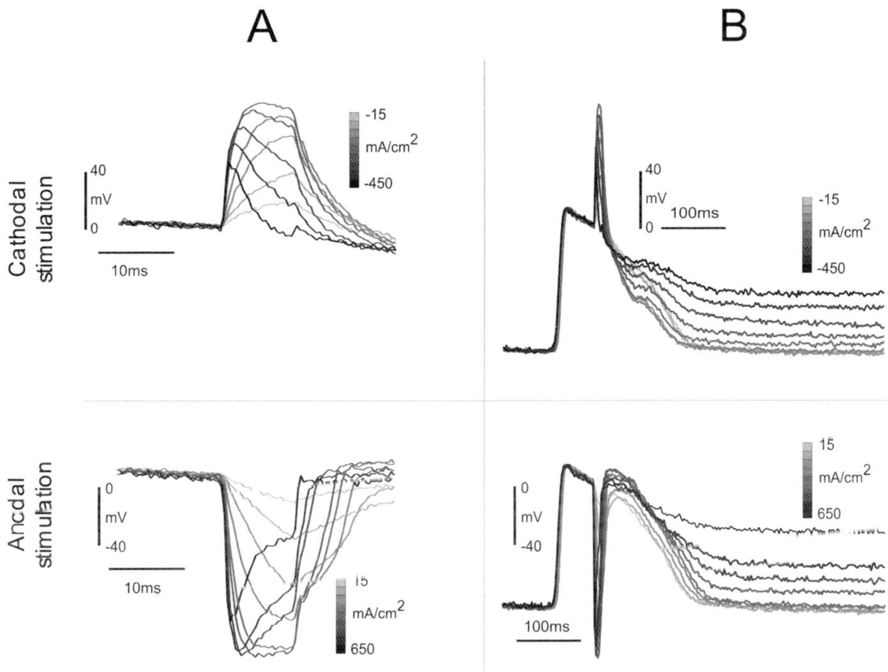

Fig. 16.2 Optical recording of transmembrane potential transients under the electrode during stimulation with different current densities. (**A**) small and (**B**) large time scale. *Arrows* mark stimulus onset and withdrawal. Current strength is *grayscale* coded. Electroporation is evident from saturation of $DV_m$and elevation of the diastolic potential (Reproduced with permission from [7])

soon starts to decay, causing elevation of the average potential of the cell (Fig. 16.2). Previously, similar effects were observed in a single guinea pig ventricular cell [23] and narrow strands of cultured rat myocytes [24]. It was concluded that during application of shocks to cell strands within the AP plateau, passive changes of V_m were followed by two voltage- and time-dependent shifts of V_m, possibly reflecting membrane electroporation [24,29]. We also concluded that the appearance of the second phase in hyperpolarizing transients in our whole heart experiments (see Fig. 16.2A) is a signature of membrane electroporation [7].

Neunlist & Tung [19] presented recordings of frog epicardial ventricular cellular responses recorded from the 150-millimeter diameter area of stimulus application, showing that electroporation reduction of AP amplitude is more pronounced for anodal polarities. Moreover, the maximum ΔV_m during diastolic shock was −200% and +125% of AP amplitude for anodal and cathodal pulses, respectively. Electroporation-induced decrease of AP amplitude was shifted toward lower current densities for anodal compared with cathodal stimulus (half-maximal values 185 and 238 milliamperes per square centimeter, respectively).

Whole heart studies revealed different types of asymmetry for the positive and negative polarizations during strong shocks (see Fig. 16.2) [7]. We also found that these effects are accompanied by epicardial postshock elevation of diastolic potential (DP) (see Fig. 16.2B). In our study, we determined epicardial ΔV_m responses during high-density electrical current stimuli of both polarities applied at a 6-millimeter diameter area of left ventricle using optical recordings (see Fig. 16.1). We detected saturation and subsequent decay of epicardial polarizations during strong cathodal and anodal shocks applied at the area with a size of several space constants (0.8–1.5 millimeters at the epicardium [30]). We did not observe a plateau or an increase in depolarization transients during cathodal stimuli of the same stimulus strengths that caused a decayed hyperpolarization response.

Fast et al. [26] showed that the initial positive polarization in virtual cathode areas of a wedge preparation changes in response to hyperpolarizing responses as the stimuli strength increases to 30 volt per centimeter and above, similar to the behavior of the middle of a single myocyte in studies by Sharma & Tung [23]. Such observations were reported previously by Cheng et al. [31] and Zhou et al. [32], who detected hyperpolarization transients near the cathodal shock electrode. We observed the same phenomena in optical recordings from the epicardium of the rabbit heart [7]. This can explain why depolarization saturation is observed at lower shock current densities than hyperpolarization saturation. Neunlist & Tung [19] stimulated a small area near the electrode that could affect their measurements due to a virtual electrode effect [33] leading to the development of positive and negative polarizations at nearby locations. In contrast, our experiments were designed to overcome this limitation by stimulating a large area relative to the field of view. Also, we did not use transmural sections of tissue, which interrupts fibers and thus could affect the results in a slab preparation [33]. Our study showed that strong cathodal stimuli resulted in hyperpolarizing responses, which were partially reversible. Thus, we concluded that electroporation was the most plausible explanation of these effects. However, it would require voltage-dependent resealing of the pores after the shock application to explain the restoration of membrane resistance.

The correlation between negative ΔV_m and diastolic V_m elevation was recently reported by Fast & Cheek [34] in myocyte cultures. Similar results were shown earlier by Neunlist & Tung [19] and Cheng et al. [31]. Interestingly, Fast & Cheek [34] did not observe nonmonotonic ΔV_m at the cathodal end of the cell strand even at the highest shock strengths. They later demonstrated that the absence of positive ΔV_m decay in the optical recordings of transmembrane potential at the edge of the preparation facing the cathodal electrode during strong shocks resulted from the spatial averaging of polarizations in the neighboring areas [35,36].

In agreement with earlier reports [11,19], we observed that a rather small optical hyperpolarization response could be sufficient for electroporation. Among possible reasons for this were (a) a "dog-bone" virtual polarization near the pacing electrode [19] that could attenuate the response in optical recordings due to optical averaging over areas of opposite polarizations, or (b) insufficient temporal resolution of the optical mapping system [19] that could underestimate the true instantaneous transmembrane voltage produced by a square pulse. To test these hypotheses, we recorded polarization transients during 10-millisecond ramp waveform stimulation in order to negate temporal resolution limitations and positioned the opposite polarization 3 millimeters away from the center recording point in order to prevent virtual polarization effects. We found no significant change in results and concluded that these did not contribute to the low electroporation threshold.

It was shown that Ca-channel blockers did not influence the electroporation threshold in whole cell patch studies [13] and only increased positive ΔV_m in cell culture studies [37]. Application of the Ca-channel blocker nifedipine resulted in an increase in the saturation level of depolarizing responses and did not affect hyperpolarizing responses in cell culture [37]. We also observed that nifedipine increased the saturation levels for positive but not negative ΔV_m during epicardial stimulation. Yet, there is a clear effect of nifedipine on the saturation level for the depolarization signal in our study, which means that other factors (i.e., spatial averaging in optical recordings) could also be responsible for saturation and reversal of the depolarizing responses with the increase in stimulus strength.

16.3 Imaging Electroporation Damage Patterns via Membrane Impermeable Fluorescent Dyes

Another way to characterize electroporation is to analyze the permeation of markers (fluorescent dyes) through cell membrane pores. Uptake depends on dye concentration differences inside and outside the cells and of the net electrical charge of the dye. To investigate electroporation as a delivery method in cardiac tissue, Harrison et al. [1] used three different indicators of electroporation, propidium iodide (PI, 668 Dalton) expression vectors for green fluorescent protein (GFP) and luciferase. Song and Ochi [16] used ethidium bromide (EB, ethidium+, 314 Dalton) as a fluorescent marker that could pass through small pores produced by mild electroporation in isolated rabbit ventricular myocytes. The internal Ca2+ concentration ([Ca2+]i) is also a useful indicator of electroporation, though Ca2+ influx through voltage-gated channels would be expected to contaminate influx through passive pores [36].

Despite the fact that the method of cell permeabilization via electroporation is already a routine technique in vitro, the complete understanding of its mechanisms remains to be formulated. The most fundamental questions remain unknown: what is the size and density of pores created by the shock, do pores grow after the shock, and what is the time of their resealing? Recently, Cheek and Fast detected nonspecific uptake of extracellular molecules using the fluorescence dye PI [29]. They have shown that application of a series of shocks with strength of about 23 volt per centimeter resulted in uptake of membrane-impermeable dye PI. Dye uptake was restricted to the anodal side of strands with the largest negative ΔV_m, indicating the occurrence of membrane electroporation at these locations. PI has been found to be an effective dye for estimation of electroporation with a 20–30 fold increase in fluorescence after binding to nucleic acids.

In our experiments with PI we did not detect an immediate dye fluorescence increase during the shock [7]. It suggests that the amount of PI molecules that penetrated through the electroporation holes during the 20-millisecond stimulus was undetectable in our protocol. This also explains why we did not observe a difference in PI uptake for shocks of different polarities despite the positive charge of the PI molecule. Slow diffusion of PI into the cells takes place when the external electrical field is already turned off, thus fluorescence is continuously rising after the shock during dye perfusion in our experiments [7] as it did in cell culture studies [38]. These data suggest that in our experiments, electroporated cells were repaired within minutes rather than seconds. In our study, we observed the PI dye uptake at shock strength of 700 milliamperes per square centimeter (35 volt per centimeter) when we detected nonmonotonic ΔV_m. No PI uptake was observed at 300 milliamperes per square centimeter (15 volt per centimeter) when ΔV_m was monotonic (Fig. 16.3). We suggest that DP elevation might be a more sensitive indicator of electroporation than PI uptake because maximum of DP elevation can be detected within 1 second after shock application (see Fig. 16.1). However, this method cannot be used to analyze three-dimensional (3D) tissues.

While PI is used widely in electroporation research, these studies are usually conducted on cell suspensions. There were concerns that this molecule may not be well suited for studies in tissues with interconnected cells due to its relatively small molecular weight (668 Dalton) with a radius about 0.6 nanometer, which is smaller than the pore of gap junction channel (about 0.8 nanometer). Thus, it was suggested that PI might diffuse to neighboring cells, creating an appearance of electroporation in intact cells.

Our data show significant differences in the depth of staining with PI, including staining of some interior regions of myocardium, which are isolated from other stained regions. The area of electroporation, identified by PI staining, which occurred in the middle of the papillary muscle confirms that in our experiments diffusion extends less than 0.1 millimeter. Therefore, although diffusion of PI molecules is theoretically possible through gap junctions, perhaps it does not occur over large distances due to rapid binding of PI to the nuclei, which prevents its diffusion in the intracellular space (Fig. 16.4).

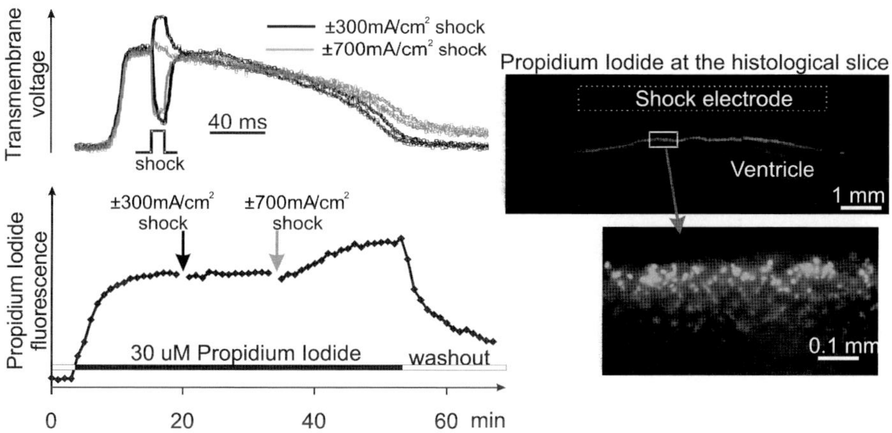

Fig. 16.3 Manifestation of electroporation changes in optical potential recordings is associated with an increase of propidium iodide fluorescence under the stimulation electrode. No increase was observed at sites not under the electrode. Histological images showed typical pattern of nuclear stain in the thin layer of epicardium at the areas where optical potentials had signs of electroporation (Reproduced with permission from [7])

A single cell study showed that during 2 kilovolt per centimeter, 20-millisecond shocks, the cells with irreversible membrane electroporation accumulate a five times larger amount of PI than cells that restored their membrane within 10 minutes after field exposure [38]. It was also shown that 1.8 amperes per

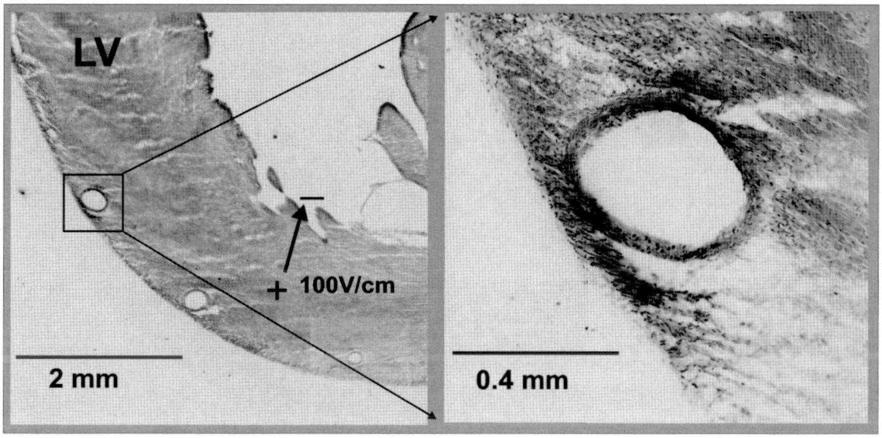

Fig. 16.4 Shock-induced electroporation of coronary blood vessels of the intact heart. To detect sarcolemma electroporation during the shocks rabbit heart was perfused with 30 μM of PI. Then we washed out PI for 30 minutes, cryosectioned the heart from base to apex with 1-millimeter step in 32-micrometer thick slices and analyzed the sections with a confocal microscope

square centimeter stimuli cause irreversible cell damage [5]. This indicates that PI accumulation during the strong shock could be related to other factors (barotrauma, hyperthermia) leading to cell death. If such factors are less dependant on proximity to the tissue boundaries than electroporation [14,15], it can explain the much larger depth of affected tissue after the 1.6 amperes per square centimeter shock in comparison to the 0.7 ampere per square centimeter shock (see Figs. 16.3 and 16.4).

Cardiac structure is a major determinant of electroporation distribution in the ventricles. Recently, we conducted series experiments to employ a simulation/experimental approach [39]. Simulations used a bidomain model of the rabbit ventricles. Distribution of electroporation in the model and experiment was assessed by examining resting potential elevation, an indicator of electroporation, at 200 milliseconds after the shock. Experimental assessment of electroporation distribution was based on shock-induced PI uptake in the intact heart. In simulations and experiments, shocks of varying strength were given via large external electrodes generating uniform field. The combined approach allowed acquisition of the 3D distribution of electroporation early after the shock as well as the distribution of large long-lasting pores that allow passage of PI molecules for many minutes after shock. Simulations and PI fluorescence analysis demonstrated that electroporation occurred not only in the surface layers, but also inside the ventricular walls near blood vessels (Fig. 16.4). The 100 volt per centimeter shock caused transmural electroporation that was particularly pronounced close to the apex, where fiber orientation rapidly changes.

Experimental and clinical studies have shown that electroporation can lead to skeletal muscle tissue necrosis in vivo. The reasons that the accumulation dynamics of electroporation damage at the tissue level are different than for isolated cells include the reduction of membrane lipid mobility caused by adhesion to high-molecular weight biopolymers in the extracellular matrix of tissues. In addition, the distributions of electric fields and, in turn, the V_m in tissue are influenced by the packing density of the cells. Collectively, these results suggest that electroporation is a likely contributor to neurological injury in electrical-shock victims.

16.4 CEW Currents

Some investigations of conducted energy weapons (CEW) in biological models have speculated on a potential for electroporation during TASER CEW application [40,41].

In order to evaluate the field strength which will be capable of reaching neural fibers below the skin, fatty, and muscle tissues, we have conducted a computational investigation using the Livshitz–Einziger–Mizrahi model [42,43], which was previously applied for quantitative analysis of functional electric stimulation of biological tissues [44]. This methodology is described in detail in publication by Livshitz et al. [42–44]. Since the chest has been well studied for CEW purposes we decided to focus on the groin.

Anatomical composition of human body surface in the groin area includes layers of tissues with different electrical properties, which are well known. In our model we considered layers of skin, fat, muscle and connective tissues. In the simulations we considered conservative parameters for thicknesses of various tissues: skin layers = 5 millimeters, fat = 5 millimeters, muscle = 3 centimeters. The following conductivities were considered following Reilly's "Applied Bioelectricity: from electrical stimulation to electropathology" textbook: skin $\sigma_1 = 0.4$ siemens per meter, fat $\sigma_2 = 0.04$ siemens per meter, muscle $\sigma_3 = 0.7$ siemens per meter, and muscle fascia $\sigma_4 = 0.04$ siemens per meter. We considered the distance between the electrodes to be 30 centimeters. We simulated the worst-case scenario: steady-state distribution of the field during application of 5,000 volt, which was based on the older, higher current M26 CEW. Electrodes were simulated as 1-millimeter square rectangular surfaces. Moreover, a more realistic thickness of fatty tissue layer in the supragroin area exceeds 5 millimeters, which would further significantly reduce field penetration in the lower layers of muscle and muscle fascia.

The results of mathematical modeling of a worst-case scenario indicate that a maximum field penetration into the muscle layer and below is 6 volt per centimeter. This field strength is well below the field strength of 35–50 volt per centimeter, which is required for reversible electroporation to occur. More realistic parameters of the model indicate that the field strength reaching the deeper layers of tissue in which nerve fibers of interest are residing would be significantly lower than worst-case scenario of 6 volt per centimeter. Thus, we conclude that reversible or irreversible electroporation in the nerve fibers and tissue layers below skin and fat is highly unlikely to occur.

Most of the energy during TASER CEW application is discharged near the electrode locations. Injury of deeper layer would necessitate accompanied injury to the superficial skin layers at the points of contact and/or penetration of electrodes. Moreover, our simulations indicate that to cause cardiac arrhythmia electrodes should penetrate both fat and skin layers. This evidence also indicates that electroporation of nerve fibers is highly unlikely during "standard" TASER discharge [45].

16.5 Conclusion

Application of electrical shocks is a routine technique to treat cardiac arrhythmias. High intensity fields (>10–50 volt per centimeter) generated inside the cardiac tissue cause transient tissue damage due to electroporation. It does not appear possible for CEW strength shocks to cause electroporation of either cardiac or nerve cells.

References

1. Harrison, R. L., Byrne, B. J., and Tung, L. (1998). Electroporation-mediated gene transfer in cardiac tissue. *FEBS Lett.* **435,** 1–5.
2. Rosen, M. R., Brink, P. R., Cohen, I. S., and Robinson, R. B. (2004). Genes, stem cells and biological pacemakers. *Cardiovasc. Res.* **64,** 12–23.
3. Kim, J. M., Lim, B. K., Ho, S. H., Yun, S. H., Shin, J. O., Park, E. M., Kim, D. K., Kim, S., and Jeon, E. S. (2006). TNFR-Fc fusion protein expressed by in vivo electroporation improves survival rates and myocardial injury in coxsackievirus induced murine myocarditis. *Biochem. Biophys. Res. Commun.* **344,** 765–771.
4. Babbs, C. F., Tacker, W. A., VanVleet, J. F., Bourland, J. D., and Geddes, L. A. (1980). Therapeutic indices for transchest defibrillator shocks: effective, damaging, and lethal electrical doses. *Am. Heart J.* **99,** 734–738.
5. Koning, G., Veefkind, A. H., and Schneider, H. (1980). Cardiac damage caused by direct application of defibrillator shocks to isolated Langendorff-perfused rabbit heart. *Am. Heart J.* **100,** 473–482.
6. Yabe, S., Smith, W. M., Daubert, J. P., Wolf, P. D., Rollins, D. L., and Ideker, R. E. (1990). Conduction disturbances caused by high current density electric fields. *Circ. Res.* **66,** 1190–1203.
7. Nikolski, V. P., Sambelashvili, A. T., Krinsky, V. I., and Efimov, I. R. (2004). Effects of electroporation on optically recorded transmembrane potential responses to high-intensity electrical shocks. *Am. J. Physiol Heart Circ. Physiol.* **286,** H412–H418.
8. Al-Khadra, A. S., Nikolski, V., and Efimov, I. R. (2000). The role of electroporation in defibrillation. *Circ. Res.* **87,** 797–804.
9. Goldman, D. E. (1943). Potential, impedance, and rectification in membranes. *J. Gen. Physiol.* **27,** 37–50.
10. Chang, D. C. and Reese, T. S. (1990). Changes in membrane structure induced by electroporation as revealed by rapid-freezing electron microscopy. *Biophys. J.* **58,** 1–12.
11. Tung, L., Tovar, O., Neunlist, M., Jain, S. K., and O'Neill, R. J. (1994). Effects of strong electrical shock on cardiac muscle tissue. *Ann. NY. Acad. Sci.* **720,** 160–75.
12. O'Neill, R. J. and Tung, L. (1991). Cell-attached patch clamp study of the electropermeabilization of amphibian cardiac cells. *Biophys. J.* **59,** 1028–1039.
13. Tovar, O. and Tung, L. (1992). Electroporation and recovery of cardiac cell membrane with rectangular voltage pulses. *Am. J. Physiol.* **263,** (Pt 2):H1128–H1136.
14. DeBruin, K. A. and Krassowska, W. (1998). Electroporation and shock-induced transmembrane potential in a cardiac fiber during defibrillation strength shocks. *Ann. Biomed. Eng.* **26,** 584–596.
15. Aguel, F., DeBruin, K. A., Krassowska, W., and Trayanova, N. A. (1999). Effects of electroporation on the transmembrane potential distribution in a two-dimensional bidomain model of cardiac tissue. *J. Cardiovasc. Electrophysiol.* **10,** 701–714.
16. Song, Y. M. and Ochi, R. (2002). Hyperpolarization and lysophosphatidylcholine induce inward currents and ethidium fluorescence in rabbit ventricular myocytes. *J. Physiol.* **545,** 463–473.
17. Krauthamer, V. and Jones, J. L. (1997). Calcium dynamics in cultured heart cells exposed to defibrillator-type electric shocks. *Life Sci.* **60,** 1977–1985.
18. Kodama, I., Shibata, N., Sakuma, I., Mitsui, K., Iida, M., Suzuki, R., Fukui, Y., Hosoda, S., and Toyama, J. (1994). Aftereffects of high-intensity DC stimulation on the electromechanical performance of ventricular muscle. *Am. J. Physiol.* **267,** H248–H258.
19. Neunlist, M. and Tung, L. (1997). Dose-dependent reduction of cardiac transmembrane potential by high- intensity electrical shocks. *Am. J. Physiol.* **273,** H2817–H2825.
20. Gallant, P. E. and Galbraith, J. A. (1997). Axonal structure and function after axolemmal leakage in the squid giant axon. *J. Neurotrauma.* **14,** 811–822.

21. Fedorov, V. V., Hepmpill, M., Kostecki, G., and Efimov, I. R. (2008). Low electroporation threshold, conduction block, focal activity and reentrant arrhythmia in the rabbit atria: possible mechanisms of stunning and defibrillation failure. *Heart Rhythm* **5,** 593–604.
22. Gray, R. A., Huelsing, D. J., Aguel, F., and Trayanova, N. A. (2001). Effect of strength and timing of transmembrane current pulses on isolated ventricular myocytes. *J. Cardiovasc. Electrophysiol.* **12,** 1129–1137.
23. Sharma, V. and Tung, L. (2002). Spatial heterogeneity of transmembrane potential responses of single guinea-pig cardiac cells during electric field stimulation. *J. Physiol.* **542,** 477–492.
24. Fast, V. G., Rohr, S., and Ideker, R. E. (2000). Nonlinear changes of transmembrane potential caused by defibrillation shocks in strands of cultured myocytes. *Am. J. Physiol. Heart Circ. Physiol.* **278,** H688–H697.
25. Efimov, I. R., Cheng, Y. N., Biermann, M., Van Wagoner, D. R., Mazgalev, T., and Tchou, P. J. (1997). Transmembrane voltage changes produced by real and virtual electrodes during monophasic defibrillation shock delivered by an implantable electrode. *J. Cardiovasc. Electrophysiol.* **8,** 1031–1045.
26. Fast, V. G., Sharifov, O. F., Cheek, E. R., Newton, J. C., and Ideker, R. E. (2002). Intramural virtual electrodes during defibrillation shocks in left ventricular wall assessed by optical mapping of membrane potential. *Circulation* **106,** 1007–1014.
27. Knisley, S. B., Blitchington, T. F., Hill, B. C., Grant, A. O., Smith, W. M., Pilkington, T. C., and Ideker, R. E. (1993). Optical measurements of transmembrane potential changes during electric field stimulation of ventricular cells. *Circ. Res.* **72,** 255–270.
28. Windisch, H., Ahammer, H., Schaffer, P., Muller, W., and Platzer, D. (1995). Optical multisite monitoring of cell excitation phenomena in isolated cardiomyocytes. *Pflugers Arch* **430,** 508–518.
29. Cheek, E. R. and Fast, V. G. (2004). Nonlinear changes of transmembrane potential during electrical shocks: role of membrane electroporation. *Circ. Res.* **94,** 208–214.
30. Akar, F. G., Roth, B. J., and Rosenbaum, D. S. (2001). Optical measurement of cell-to-cell coupling in intact heart using subthreshold electrical stimulation. *Am. J. Physiol. Heart Circ. Physiol.* **281,** H533–H542.
31. Cheng, Y., Tchou, P. J., and Efimov, I. R. (1999). Spatio-temporal characterization of electroporation during defibrillation. *Biophys. J.* **76**(1), A85.
32. Zhou, X., Ideker, R. E., Blitchington, T. F., Smith, W. M., and Knisley, S. B. (1995). Optical transmembrane potential measurements during defibrillation- strength shocks in perfused rabbit hearts. *Circ. Res.* **77,** 593–602.
33. Neunlist, M. and Tung, L. (1995). Spatial distribution of cardiac transmembrane potentials around an extracellular electrode: dependence on fiber orientation. *Biophys. J.* **68,** 2310–2322.
34. Fast, V. G. and Cheek, E. R. (2002). Optical mapping of arrhythmias induced by strong electrical shocks in myocyte cultures. *Circ. Res.* **90,** 664–670.
35. Sharifov, O. F., Ideker, R. E., and Fast, V. G. (2004). High-resolution optical mapping of intramural virtual electrodes in porcine left ventricular wall. *Cardiovasc. Res.* **64,** 448–456.
36. Fast, V. G., Cheek, E. R., Pollard, A. E., and Ideker, R. E. (2004). Effects of electrical shocks on Cai2+ and Vm in myocyte cultures. *Circ. Res.* **94,** 1589–1597.
37. Cheek, E. R., Ideker, R. E., and Fast, V. G. (2000). Nonlinear changes of transmembrane potential during defibrillation shocks : role of Ca(2+) current. *Circ. Res.* **87,** 453–459.
38. Shirakashi, R., Kostner, C. M., Muller, K. J., Kurschner, M., Zimmermann, U., and Sukhorukov, V. L. (2002). Intracellular delivery of trehalose into Mammalian cells by electropermeabilization. *J. Membr. Biol.* **189,** 45–54.

39. Fedorov, V. V., Constantino, J. L., Nikolski, V. P., Trayanova, N. A., and Efimov, I. R. (2007). Structural determinants of shock-induced electroporation in the ventricles. *Biophys. J.* [Supplement], 285A.
40. Dennis, A. J., Valentino, D. J., Walter, R. J., Nagy, K. K., Winners, J., Bokhari, F., Wiley, D. E., Joseph, K. T., and Roberts, R. R. (2007). Acute effects of TASER X26 discharges in a swine model. *J. Trauma* **63,** 581–590.
41. Walter, R. J., Dennis, A. J., Valentino, D. J., Margeta, B., Nagy, K. K., Bokhari, F., Wiley, D. E., Joseph, K. T., and Roberts, R. R. (2008). TASER X26 discharges in swine produce potentially fatal ventricular arrhythmias. *Acad. Emerg. Med.* **15,** 66–73.
42. Livshitz, L. M., Mizrahi, J., and Einziger, P. D. (2001). Interaction of array of finite electrodes with layered biological tissue: effect of electrode size and configuration. *IEEE Trans. Neural Syst. Rehabil. Eng.* **9,** 355–361.
43. Livshitz, L. M., Mizrahi, J., and Einziger, P. D. (2001). A model of finite electrodes in layered media: an hybrid image series and moment method scheme. *J Appl. Comput. Electromag. Society* **16,** 145–154.
44. Livshitz, L. M., Einziger, P. D., and Mizrahi, J. (2000). Current distribution in skeletal muscle activated by functional electrical stimulation: image-series formulation and isometric recruitment curve. *Ann. Biomed. Eng.* **28,** 1218–1228.
45. Lakkireddy, D., Wallick, D., Verma, A., Ryschon, K., Kowalewski, W., Wazni, O., Butany, J., Martin, D., and Tchou, P. J. (2008). Cardiac effects of electrical stun guns: does position of barbs contact make a difference? *Pacing Clin. Electrophysiol.* **31,** 398–408.

Chapter 17
Eye and Head Injuries

S. Robert Witherspoon, Andreas K. Lauer and Jonathan L. Marinaro

Ocular injuries caused by CEWs are rare occurrences. Some 30 years after its introduction, two case reports have documented the sight threatening potential of these weapons [1,2]. In discussing the impact of ocular injuries by CEWs, understanding pertinent ocular and adnexal anatomy as well as the terminology of ocular injuries is useful in effectively managing individuals requiring extrication of an embedded probe.

In addition to ocular injuries, there have been two case reports of penetrating skull trauma. The case reports, evaluation, and management patients with intracranial penetration are presented in the final portion of this chapter.

17.1 Eye and Adnexal Anatomy

The eye is composed of the globe and surrounding periorbital tissues, either of which may be involved in CEW injuries. The periorbital tissue consists of the skin, underlying orbicularis muscle, fat, glands, cilia, aponeuroses and the nasolacrimal system (Fig. 17.1). The eye or globe is protected by this adnexal soft tissue anteriorly and is housed within the bony orbit. The orbit is a pear-shaped cavity within which the eye is suspended and cushioned by orbital fat, ligaments, and aponeuroses. On average, the eye is approximately 1 inch (25 millimeters) in diameter and is largely fluid filled. The eye wall is mainly comprised of collagen and defined as the cornea and sclera. The cornea is a 5-millimeter thick avascular, clear, and convex disc of approximately 12-millimeter diameter with a highly specialized arrangement of collagen fibers that allows tissue clarity and provides for transmission and refraction of incident light into the eye. The white sclera is also comprised of collagen and its less regular arrangement

A.K. Lauer (✉)
Casey Eye Institute, Department of Ophthalmology, Oregon Health & Science University, Legacy Hospitals
e-mail: lauera@ohsu.edu

M.W. Kroll, J.D. Ho (eds.), *TASER® Conducted Electrical Weapons: Physiology, Pathology, and Law*, DOI 10.1007/978-0-387-85475-5_17,

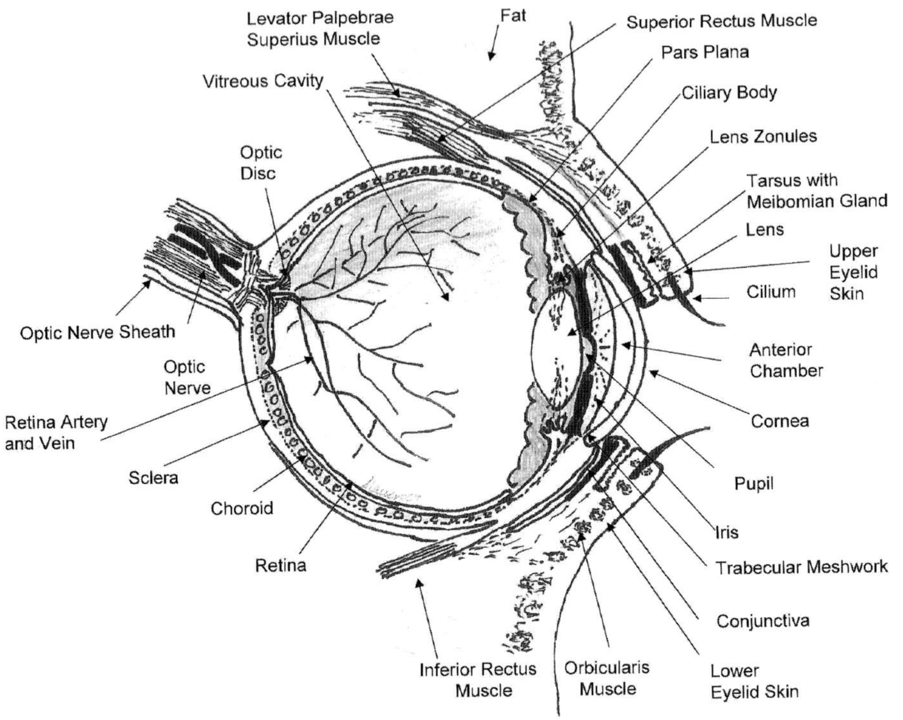

Fig. 17.1 Diagram of the eye and ocular adnexa

confers its opaque appearance. When corneal injury occurs, the regular collagenous arrangement can be compromised during wound healing and can result in opacification. The anterior surface of the eye is protected by conjunctiva that overlies the sclera and separates the orbit from the soft tissues of the eyelid.

Behind the cornea resides a 3-millimeter deep dome-shaped fluid-filled space called the anterior chamber. This space is filled with plasma-like aqueous humor and is limited posteriorly by the pupil and iris. The iris and pupil reside anterior to the natural or crystalline lens and serve to divide the eye into the anterior and posterior segments. The anterior segment represents one-quarter of the eye's volume. A very small fluid-filled space, called the posterior chamber, resides behind the iris and in front of the lens. The ciliary body secretes aqueous humor into the posterior chamber, which then passes through the pupil to fill the anterior chamber. At the junction of the sclera with the peripheral cornea and iris, the aqueous passes through the trabecular meshwork, a drain which communicates with the circulatory system.

The natural or crystalline lens is a convex structure with an annular arrangement of lens cells that is bounded by the lens capsule and suspended by ligaments called lens zonules. Analogous to the cornea, the specialized arrangement of the lens crystalline proteins confers the lens its clarity and capacity to transmit and refract incident light for sharp focus on the retina. All structures posterior to the lens are considered the posterior segment. The vitreous cavity is a 4.5-milliliter space filled with vitreous, a clear gelatinous substance comprised of water, glycosaminoglycans, and scant collagen fibers. The vitreous has contact with retina, optic disc, and lens. The retina and choroid line much or the inner eye wall. The choroid is sandwiched by the retina and sclera and is a highly vascular structure that, amongst other functions, provides the blood supply for the outer two thirds of the retina. The retina is an approximately 0.25-millimeter thick, transparent sheet of neurosensory tissue that serves to transduce light into neurotransmission. Retinal arteries and veins arborize from the center of the optic disc to supply the inner third of the retina. Ganglion cells of the second cranial nerve by located at the inner most layer of the retina funnel at the optic nerve head and project their axons and neurotransmission to the central nervous system (Fig. 17.1). At the optic nerve head, the axons acquire myelin and saltatory nerve conduction projects the information through the optic chiasm to the lateral geniculate ganglia with subsequent radiation to the occipital cortex.

17.2 Eye Trauma

Eye injury is a significant public health problem and is the leading cause for eye-related hospital admissions. Although eye injury can affect any age group, the majority of ocular injuries involve young working age males typically less than 30 years of age. The United States Eye Injury Registry documented that among 8,952 patients with severe eye injuries, 58% were less than 30 years old and the male to female ratio was 4.6:1. More recently 27% of 11,320 patients had vision worse than 20/200 after ocular injury [3,4]. Gun-related eye injury is an important component of ocular trauma and is reported at a rate of 7.5 per 1,000,000 people in 2002. The rates for ocular morbidity due to guns were the highest among individuals 10–19 years of age and males.[3–5] Ocular injury by CEWs such as the TASER® is a newcomer to this form of eye injury.

The Ocular Trauma Classification Group has organized eye injuries using standard terminology to describe various forms of eye injury. The Birmingham Eye Trauma Terminology System (BETTS) serves as a standardized language for ocular trauma that has both diagnostic and prognostic value and has been endorsed by national and international eye care organizations (Table 17.1).[6–8]

17.3 Eye Injuries by Conductive Electrical Weapons

Based on recently reported cases, the eye injuries caused by the TASER CEW probes are related to a mechanical mode of ocular injury rather that an electric one. In recognizing and managing eye injury by the CEW it is important

Table 17.1 Birmingham Eye Trauma Terminology System (BETTS): glossary of terms [6,7,8]

Term	Definition and explanation
Eye wall	Sclera and cornea: Though technically the eye wall has three coats posterior to the limbus, for clinical and practical purposes violation of only the most external structure is taken into consideration
Closed globe injury	No full-thickness wound of eye wall
Open globe injury	Full-thickness wound of the eye wall
Contusion	There is no (full-thickness) wound: The injury is either due to direct energy delivery by the object (e.g., choroidal rupture) or to the changes in the shape of the globe (e.g., angle recession)
Lamellar laceration	Partial-thickness wound of the eye wall
Rupture	Full-thickness wound of the eye wall, caused by a blunt object: Since the eye is filled with incompressible liquid, the impact results in momentary increase of the IOP. The eye wall yields at its weakest point (at the impact site or elsewhere; example: an old cataract wound dehisces even though the impact occurred elsewhere); the actual wound is produced by an inside-out mechanism
Laceration	Full-thickness wound of the eye wall, caused by a sharp object: The wound occurs at the impact site by an outside-in mechanism
Penetrating injury	Entrance wound: If more than one wound is present, each must have been caused by a different agent
Retained or intraocular foreign body	Technically a penetrating injury, but grouped separately because of different clinical implications
Perforating injury	Entrance and exit wounds: Both wounds caused by the same agent

that health professionals understand the structure of the probe as well as the highly specialized anatomy of the eye to limit ocular morbidity and maximize visual preservation. The harpoon-like probe is comprised of a 5.5 millimeter × 1 millimeter shaft typically attached to a 9-millimeter barbed tip. The short barbed tip allows the probe to penetrate and lodge in clothing or skin. The body of the probe creates sufficient resistance preventing deeper penetration and reduces the risk of internal penetrating injury such as a pneumothorax. A longer XP (extended penetration) 13-millimeter probe is sold for northern climates and winter clothing penetration.

Recently, two cases of penetrating eye injury have been reported. Ng & Chehade described the case of a 50-year-old male who suffered a perforation of the right lower eyelid and penetration of the probe into the globe.[1] The probe passed through that sclera, choroid, and retina and the tip resided in the vitreous cavity (Fig. 17.2A and 17.2B). When the probe was extricated the scleral wound was 2 millimeter in size and sufficient to cause the vitreous to prolapse. The wound was sutured closed with repair of an involved extraocular muscle followed by application of transcleral cryotherapy to create a

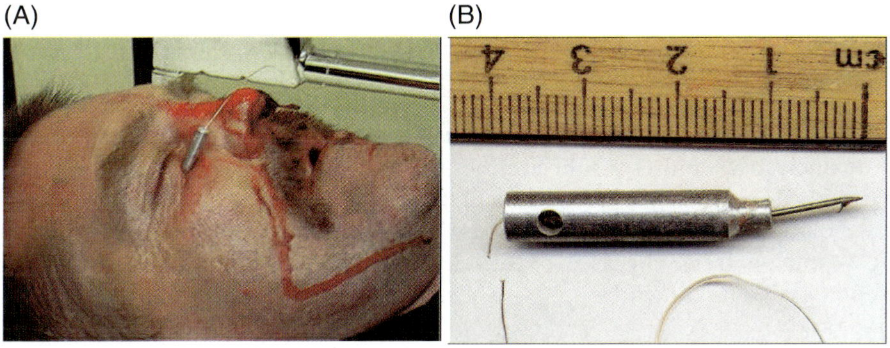

Fig. 17.2 (**A**) Preoperative photograph of TASER probe embedment. The probe perforated the lower eyelid and penetrated the globe. (**B**) Close-up photograph of the extricated probe showing the 4-millimeter barbed tip (reproduced with permission from [1])

chorioretinal adhesion around the wound and to prevent a retinal detachment. The patient was followed for 1 week and had a visual acuity of 20/30.

Chen et al. described a second case in a 21-year-old man who presented with hand motions vision following a TASER probe injury.[2] The retained probe caused a perforating globe injury. The probe entered the eye at the temporal cornea and traversed the iris, lens, and exited the inferior pars plana (Fig. 17.3). A vitreous hemorrhage was present. The probe was removed in the operating room with closure of the perforated globe. The patient achieved a visual acuity

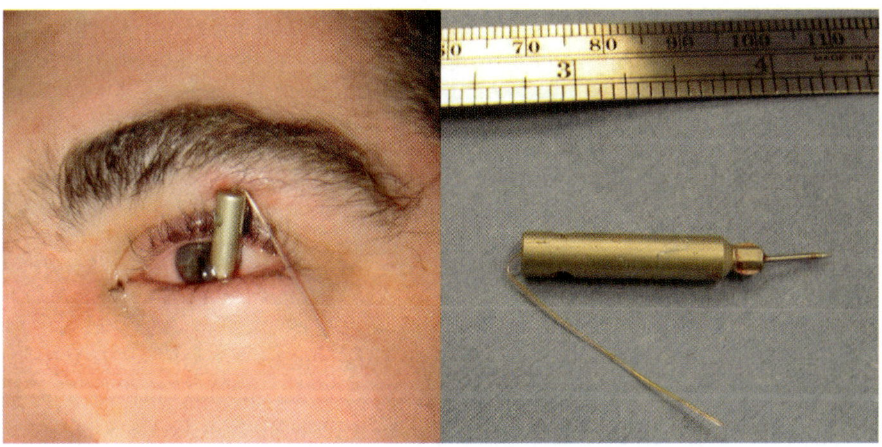

Fig. 17.3 Preoperative photograph of a retained TASER probe that perforated the globe by entering the entering the temporal cornea, traversing the anterior chamber, iris and peripheral lens, with the tip exiting the eye at the pars plana. A vitreous hemorrhage was present. TASER probe electrode as seen upon removal. The needle is 9.5 millimeter in length. The barbed end is not visible in this photograph as can be seen in Ng and Chehade's report (reproduced with permission from [2])

of 20/60 until he developed a retinal detachment 9 months following his injury. The retinal detachment was successfully repaired and the patient was transferred out of the region to prison.

The effect of short pulse electrical currents on the eye has not been reported and is solely speculative at present. It may be presumed that the delivery of high rate of electrical pulses would contract the orbicularis muscles of the eyelid and possibly do the same for the extraocular muscles. Since the 19 pulses per second of the popular X26 CEW is much less than the 70 pps required for tetanizing fast-twitch muscle. The effect of electric current on the neural conduction of the retina is also unknown.

17.4 Management of CEW Ocular Injuries

The management of CEW eye injuries requires special consideration. Any probe injuries involving the periocular region should undergo computed tomography (CT) evaluation of the orbits without contrast. This allows the assessment on ocular involvement and can localize the probe to assist in surgical removal. Any retained probes involving the globe should be removed in the operating room under general anesthesia by an ophthalmologist experienced in ocular trauma. Both reported cases of globe penetration involved combined anterior and posterior segment injury. The first priority of the surgeon should be removal of the probe with watertight closure of the globe. The use of general anesthesia serves to stabilize the globe and immobilize a generally combative patient. The removal of the barbed probe requires surgical principals similar to removing a barbed fishhook. Extraction is performed by gently backing the barbed tip out from the original incision. The patient should be monitored for endophthalmitis, a severe intraocular infection, and for retinal detachment. Secondary reconstruction can be formed at a later time if needed. The patient needs continued follow-up to monitor for retinal detachment that can occur months following the initial injury as evidenced in the second case.

17.5 Head Injuries

A 16-year-old male was struck on the right forehead by a TASER CEW probe while he was resisting arrest.[10] He was reported to be unconscious for almost 5 minutes after the firing of the weapon. Upon presentation to the emergency department, he was found to have the probe penetrating his forehead on the right side. He complained of a mild headache and being shaken but on examination, he was neurologically intact. Because the dart was immobile and denting the skin of the forehead, the patient underwent a computed tomographic scan of the head, which revealed intracranial penetration of the dart and possible dural perforation (Fig. 17.4). The patient was taken to the

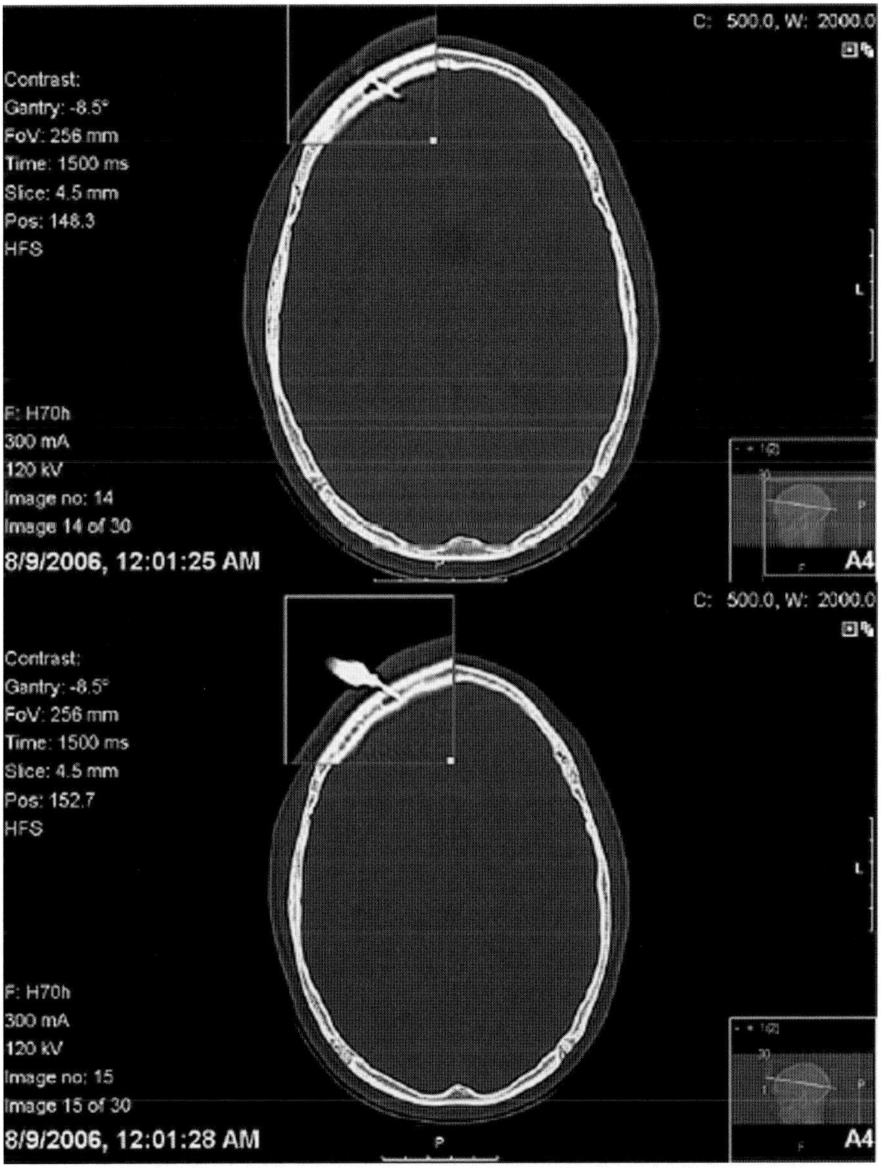

Fig. 17.4 A computed tomographic scan of the head revealing intracranial penetration of the dart and possible dural perforation (reproduced with permission from [10])

operating room. The tip of the probe penetrated the dura and into the brain. There was no active hemorrhage seen at that time. Discoloration of the dura was noticed, possibly due to burning from the electric current. The patient had an uneventful postoperative course and was discharged with no focal neurologic deficit. This case demonstrates that a TASER probe can penetrate the skull and injure the meninges and underlying brain. In a second intracranial injury case reported by Mangus et al., a 24-year-old male was struck in the head by the Taser while he was being subdued and one of the Taser probes penetrated the scalp. Attempt at removing the probe with gentle traction resulted in the fracture of the distal barbed segment from the heavier base. Exploration of the wound failed to disclose the retained barbed hook. A computed tomography (CT) scan of the patient's head revealed a linear metal fragment penetrating the outer cortex and marrow space of the skull and extending into the inner table of the calvarium. Neurosurgical consultation suggested observation, the patient was discharged from the hospital and given outpatient follow-up but there is no subsequent information in this case [11]. Although police personnel, paramedics, and emergency department staff typically remove the barbed probes from the skin, intracranial dart penetration requires neurosurgical consultation. Had an attempt to remove the probe from in the standard fashion occurred, there would have been the risk of breaking the dart and leaving a foreign body in the cranial vault. Anticonvulsant prophylaxis or antibiotics should be considered in consultation with a neurosurgeon.

17.6 Conclusion

The TASER CEW causes ocular injury by mechanical, not electrical damage. Any individual working in the emergency department setting should be aware of the potential for serious eye injury. A CT scan should be performed to assess ocular involvement followed by ophthalmic consultation. If a penetrating or perforating injury is encountered, the probe should be removed in the operating room under general anesthesia. The patient should be followed for late complications such as retinal detachment. As the use of CEWs becomes more prevalent, it is likely that other reports of unintended injury may arise.

The delicate and highly specialized structure of the eye requires health professionals to take special consideration in managing patients with CEWs-induced eye injuries.

Similarly the TASER probe has the kinetic energy to occasionally penetrate the skull. This needs to be considered in the evaluation of these patients if their presentation suggests a probe embedded in the cranium. Appropriate imaging and neurosurgical consultation should be obtained.

References

1. Ng W, Chehade M. Taser penetrating ocular injury. *Am J Ophthalmol* 2005; 139: 713–15.
2. Chen SL, Richard CK, Murthy RC, Lauer AK. Perforatingocular injury by Taser. *Clin Experiment Ophthalmol* 2006; 34(4): 4–80.

3. May DR, Kuhn FP, Morris RE *et al.* The epidemiology of serious eye injuries from the United States Eye Injury Registry. *Graefes Arch Clin Exp Ophthalmol* 2000; 238: 153–7.
4. Kuhn FP, Morris RE, Witherspoon CD, *et al.* The epidemiology of blinding trauma in the United States Eye Injury Registry. *Ophthalmic Epidemiol* 2006; 13: 209–16.
5. McGwin Jr G, Hall TA, Xie A *et al.* Gun related eye injury in the United States, 1993–2002. *Ophthalmic Epidemiol* 2006; 13: 15–21.
6. Pieramici DJ, Sternberg P Jr, Aaberg TM Sr, Bridges WZ Jr, Capone A Jr, Cardillo JA, de Juan E Jr, Kuhn F, Meredith TA, Mieler WF, Olsen TW, Rubsamen P, Stout T. A system for classifying mechanical injuries of the eye (globe). The Ocular Trauma Classification Group. *Am J Ophthalmol* 1997 Jun; 123(6): 820–31.
7. Kuhn F, Morris R, Witherspoon CD, Heimann K, Jeffers JB, Treister G A standardized classification of ocular trauma. Ophthalmology 1996 Feb; 103(2): 240–3.
8. Pieramici DJ, Au Eong KG, Sternberg P Jr, Marsh MJ. The prognostic significance of a system for classifying mechanical injuries of the eye (globe) in open-globe injuries. *J Trauma* 2003 Apr; 54(4): 750–4.
9. Kornblum RN, Reddy SK. Effects of the Taser in fatalities involving police confrontation. *J Forensic Sci* 1992; 36: 434–8.
10. Rehman TU, Yonas H, Marinaro J. Intracranial penetration of a TASER dart. *Am J Emerg Med* 2007; 25: 733 e3-4.
11. Mangus BE, Shen LY, Helmer SD, Maher J. Taser and Taser Associated Injuries: A Case Series. *Am Surg* 2008 Sep; 74(9): 862–5.

Chapter 18
CEW Effects with Illegal Stimulant Intoxication

Patrick Tchou

Conducted electrical weapons (CEWs) such as the TASER® brand devices are increasingly used by law enforcement officers when confronted with combative and uncooperative subjects [1]. These devices minimize direct physical interaction between the police and the subject thus minimizing potential injury to the officer and decrease the need for using more lethal and injury-prone weapons such as the baton or the hand gun. In-custody deaths, however, have raised the speculation that TASER CEWs may contribute to such deaths. Disrupting the cardiac rhythm has been a postulate mechanism of such deaths.

It is beyond the scope of this chapter to describe the electrical activation of the heart, but a brief explanation is helpful to those who are unfamiliar with this subject in order to understand the main substance of this chapter. The muscle of the heart has the capability of transmitting an electric wave along its cell membranes. This wave typically initiates in the right upper chamber of the heart at the natural pacemaker site called the sinus node. This wave then propagates to the lower chamber, the main pumping chambers of the heart. Each wave causes a mechanical contraction of the muscle creating the squeeze of the heart that pumps the blood out and through the body. This electrical wave can also be initiated by an electric pulse applied to the heart. Whether an electric pulse initiates such a wave (captures the heart) depends on the current density that the pulse delivers to the cardiac tissue surface and its duration. This is how an artificial cardiac pacemaker works. Rapid electrical pulses of sufficient current strengths can capture the heart repetitively at a high rate. Should the applied electrical pulses cause very rapid sequential activation of waves, a disorganized rhythm called ventricular fibrillation can be initiated where multiple waves are simultaneously propagating through the heart in a continuous and chaotic pattern. Ventricular fibrillation results in an unsynchronized contraction pattern that mechanically paralyzes the heart's pumping

P. Tchou (✉)
Cleveland Clinic, Section of Cardiac Electrophysiology and Pacing, Department of Cardiovascular Medicine, The Heart and Vascular Institute
e-mail: tchoup@ccf.org

M.W. Kroll, J.D. Ho (eds.), *TASER® Conducted Electrical Weapons: Physiology, Pathology, and Law*, DOI 10.1007/978-0-387-85475-5_18,

function causing collapse into unconsciousness in a matter of 10–20 seconds and irreversible death within a few minutes if not interrupted.

While testing in volunteer humans have generally shown no indication of lethality [2], field reports of deaths surrounding the use of CEWs raise speculation that this electrical stimulation may have cardiac arrhythmogenic effects that may contribute to such deaths. Thus, one possible mechanism by which CEW impulses are postulated to cause death is by initiating such a lethal cardiac rhythm. Several differences between volunteer testing and field conditions could contribute to these differences in observations. Typically, field conditions are under high emotional stresses with high sympathetic tone. Frequently, these conditions are occurring during physical stress as well where the subject has been exerting physically for some time prior to the CEW application. Physical stresses can cause metabolic changes that may predispose a person to arrhythmias as well. Lastly, the subject may be suffering from a delirious condition [3–6], which can be due to the influences of street drugs, alcohol, or psychotic conditions from mental illness or withdrawal from alcohol or other types of sedatives. Testing in human subjects have also involved application of the TASER darts to the back, a site that is more remote from the heart and less likely to stimulate the heart while field use of the TASER tend to have a variety of application sites besides the back. These conditions, which differ from testing of TASER CEW exposures in human volunteers, have been postulated to make human field exposures more vulnerable to develop arrhythmias. Typically, drugs such as cocaine and methamphetamines increase the epinephrine (adrenaline) levels of a person under their influence. There is scant information available on the potential interaction of such stimulant drugs on cardiac vulnerability to the CEW.

We chose to investigate the potential for a TASER CEW to induce ventricular fibrillation (VF). A recent study [7] performed at our institution on anesthetized pigs investigated the effect of cocaine on the threshold for induction of ventricular fibrillation by a CEW capable of varying output charge. Fully anesthetized and mechanically ventilated pigs averaging 34 kilograms in weight had TASER CEW probes inserted into the skin at five different torso positions that mimic commonly seen field application of these probes. These included four anterior torso positions with varying distances from the heart and one posterior position as shown in Fig. 18.1. Cardiac rhythm was continuously monitored with body surface ECGs as well as through bipolar intracardiac electrograms recorded from an electrophysiologic catheter placed transvenously into the right ventricle. The intracardiac bipolar recordings were necessary to detect CEW stimulation-induced cardiac depolarization that could not be appreciated on the surface ECG due to electrical noise generated by the TASER CEW impulses. Standard CEW applications of 5-second durations were delivered. The TASER CEW applications were applied at increasing charge outputs by increasing the size of the capacitor within the device at increments of 5×, 10x, 20x, etc., until ventricular fibrillation (VF) was induced or until a maximum output of 100x. The output was then decreased by the same

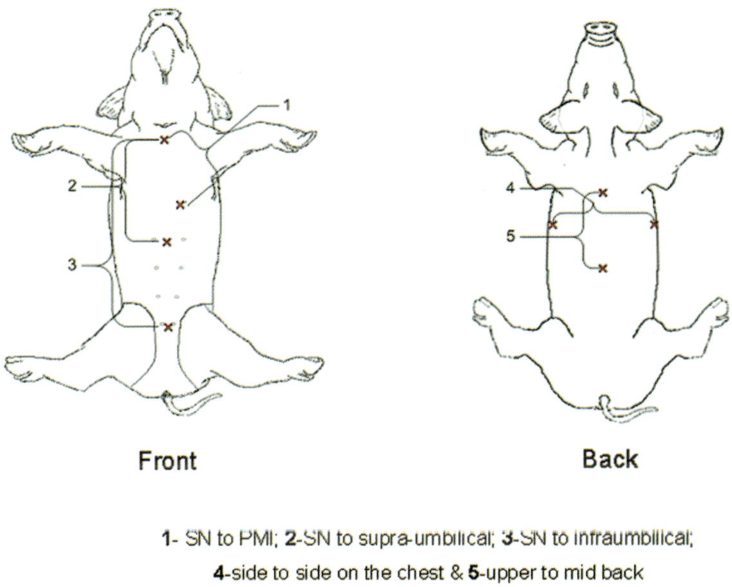

Fig. 18.1 The five positions of paired electrode insertion into the pig's skin. These positions represent common areas of probe contact in human applications. Position 1, the sternal notch (SN) to point of maximal cardiac impulse (PMI) represents to closest probe position to the heart. The PMI probe which is near the apex of the heart, averaged around 1.6 centimeter from the probe tip to the heart. This position essentially bracketed the heart with one probe being close to the ventricular myocardium

increments with the addition of 3x and 2x if necessary until 3 consecutive application of a particular output resulted in no induction of arrhythmias.

The highest output that did not induce arrhythmias was defined as the maximum safe multiple (MaxSM). The lowest output that initiated VF was defined as the minimum VF-inducing multiple (MinVFIM). Since these TASER CEW outputs were changed in a step-wise manner, it is likely that the true minimum VF-inducing output would be somewhat lower than the MinVFIM but somewhat higher than the MaxSM. Thus, another variable, the VF induction threshold (VFT) was defined as the average between the MaxSM and the MinVFIM.

At baseline, prior to infusion of cocaine, each position of paired probes on the torso was tested to determine the above described variables. The results are shown in Fig. 18.2. As hypothesized, the proximity of the paired probes to the heart was related to the strength of the delivered output needed to induce VF. The most sensitive position (Position 1) where the probes were the closest to the heart had the lowest MaxSM, MinVFIM, and VFT. The highest values of these variables were seen when the probes were applied to the back (Position 5), where the probes were farthest from the heart. The other positions had intermediate values proportional to the distances of the probes from the heart. From

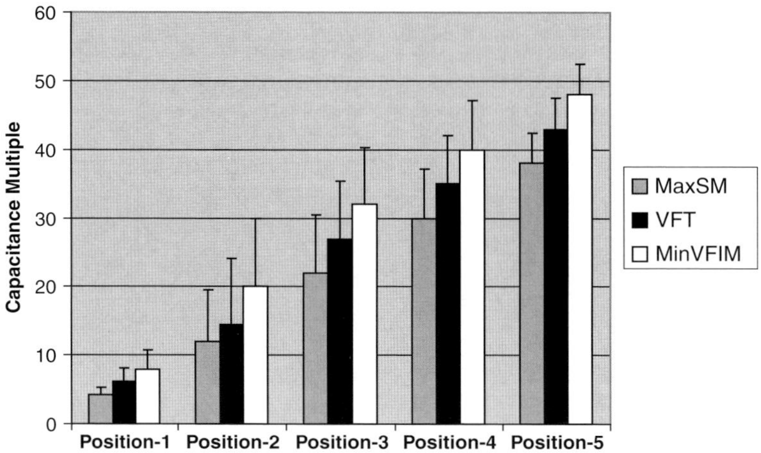

Fig. 18.2 Relationship of probe positions to output thresholds for induction of VF at baseline. The five positions of paired probes are those shown in Fig. 18.1. The capacitance multiple of the standard TASER X26 CEW capacitor is shown on the Y-axis. MaxSM, VFT, and MinVFIM are maximum safe multiple, VF threshold, and minimum VF induction multiple, respectively. See text for full definitions. Position 1 had the lowest values due to its proximity to the heart while Position 5, over the back, had the highest values. All values in different positions were significantly different from each other

our intracardiac recordings, we noted that induction of VF was closely related to capture of the heart at a fast rate by the TASER CEW pulses. TASER X26 impulses are delivered at a rate of 19 pulses per second. Rapid capture of the heart by every 3rd pulse from the CEW output was related to the output strength of the pulses and the distance of the probes from the heart (Fig. 18.3) in a similar manner as VF thresholds. That is, this rapid capture occurred at lower outputs strengths when the probes were closer to the heart. This relationship parallels that of VF induction and is the most likely mechanism of VF induction at increasing outputs. Of note, standard output from a TASER CEW did not induce VF at any of the positions even though capture of the heart could be elicited, especially at Position 1.

We then infused cocaine intravenously into the pigs (8 milligrams per kilogram over 30 minutes). This infusion generated levels that would be considered to have a highly intoxicating effect (557±280 units per /liter cocaine and 462±123 units per /liter benzoylecognine). The TASER CEW testing at the various positions was then repeated in the same fashion as at baseline. Cocaine is well known to have arrhythmogenic effects in the heart [8–11]. Thus, we originally hypothesized that infusion of cocaine will lower the VF induction thresholds. To our initial surprise, our finding revealed that cocaine imparted a protective effect to the heart in resisting the induction of VF by the TASER CEW pulses. Essentially, the MaxSM, MinVFIM, and VFT increased by a factor of 1.5–2 after the infusion. These findings are shown in Fig. 18.4. The

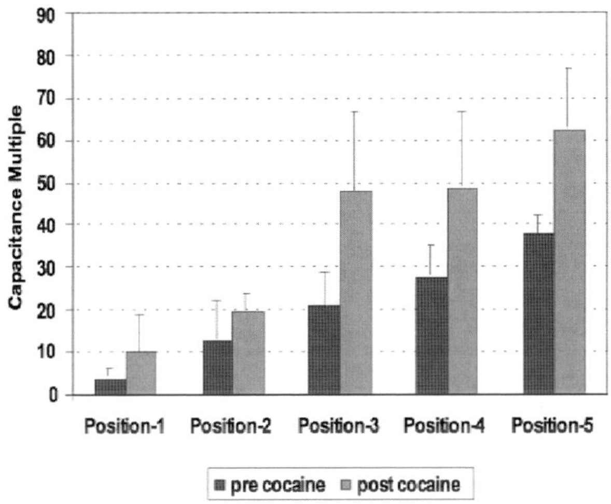

Fig. 18.3 TASER CEW outputs to obtain 3:1 ventricular capture before and after cocaine infusion. The increasing values seen at baseline (precocaine) from Position 1 to Position 5 reflect the changes in distances from the heart. These changes occur in parallel to the changes for ventricular fibrillation thresholds shown in Fig. 18.3 suggesting that rapid ventricular capture is the mechanism for VF induction during TASER CEW application of increasing output. After infusion of cocaine, the capacitance multiples needed to obtain 3:1 ventricular capture increased substantially. These increases were all statistically significant at $p < 0.05$

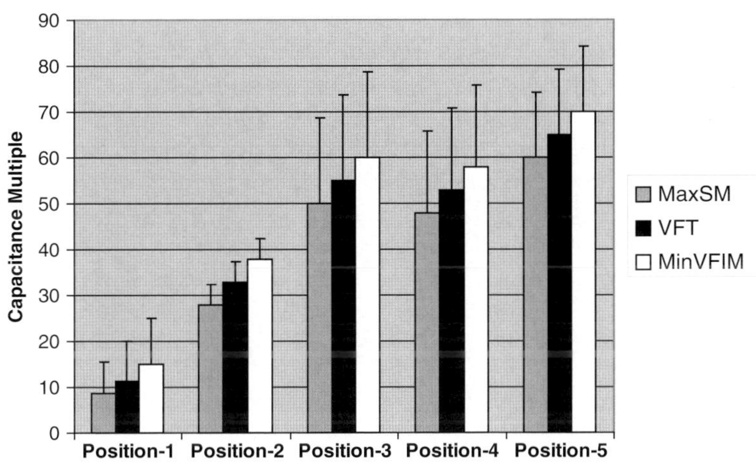

Fig. 18.4 Relationship of probe positions to output thresholds for induction of VF after cocaine infusion. The display format and variables are the same as in Fig. 18.3. However, the values have increased significantly when compared to baseline precocaine values shown in Fig. 18.3. Y-axis adjusted to accommodate higher values

increase parallels a similar increase in the output strength needed to obtain 3:1 capture of the heart after cocaine infusion as shown in Fig. 18.3.

Our observation of the effects of cocaine can be attributed to its "anesthetic" properties. Cocaine has a blocking property on the sodium channel protein within the cardiac cellular membrane [12]. This channel is involved in propagation of the electrical wave through the heart. Thus, blocking this channel makes the heart more resistant to external electrical stimulation. For example, sodium channel blocking drugs are well known to elevate the current threshold needed to stimulate the heart by a pacemaker. It should not be surprising, then, that cocaine provided a protective effect in response to electrical stimulation by the TASER CEW output despite its potential arrhythmogenic properties that have been otherwise observed.

While no studies have yet been reported with drugs such as methamphetamines, one can speculate that those may not have similar protective effects as they do not have sodium channel blocking properties. In addition, other cardiac stimulants that simulate adrenaline effects, or high adrenaline levels per se, may promote the initiation of arrhythmias. Adrenaline, for example, is known to lower *temporarily* current thresholds for cardiac capture by an electrical stimulus [13]. However, after 3–7 minutes the VFT actually rises from the baseline. Nanthakumar et al. [14], for example, found that an adrenaline infusion may acutely facilitate the induction of ventricular arrhythmias by a TASER X26 application in a pig model. Thus, states of high adrenaline that are likely common in delirious conditions, may predispose the heart to higher capture rates and to VF induction. Lastly, one should inject a note of caution in extending animal studies done in an anesthetized state to the field scenarios where law enforcement officers are applying the TASER CEW. The physiologic effects of delirium, high adrenaline states, physical stress, and drug effects may change the propensity for induction of cardiac arrhythmias. Such conditions, by themselves, may produce arrhythmias. Furthermore, cardiac diseases such as advanced coronary disease, prior myocardial infarction, or other afflictions may lower the threshold for electrically inducing arrhythmias. Thus, it has been hypothesized that this might possibly allow a CEW to induce arrhythmias when the probes land close to the heart. Animal studies and observations in normal healthy adults may not be directly applicable to the diseased cardiac state.

References

1. http://www.taser.com/research/statistics/Pages/FieldUseandStatistics.asp
2. Ho JD, Miner JR, Lakireddy DR, Bultman LL, Heegaard WG. Cardiovascular and physiologic effects of conducted electrical weapon discharge in resting adults. Academic Emergency Medicine. 13(6):589–95, 2006 Jun.
3. Stratton SJ, Rogers C, Brickett K, Gruzinski G. Factors associated with sudden death of individuals requiring restraint for excited delirium. American Journal of Emergency Medicine. 19(3):187–91, 2001 May.

4. Ross DL. Factors associated with excited delirium deaths in police custody. Modern Pathology. 11(11):1127–37, 1998 Nov.
5. Pollanen MS, Chiasson DA, Cairns JT, Young JG. Unexpected death related to restraint for excited delirium: a retrospective study of deaths in police custody and in the community. Canadian Medical Association Journal. 158(12):1603–7, 1998 Jun 16.
6. Strote J, Range HH. Taser use in restraint-related deaths. Prehospital Emergency Care. 10(4):447–50, 2006 Oct–Dec.
7. Lakkireddy D, Wallick D, Ryschon K, Chung MK, Butany J, Martin D, Saliba W, Kowalewski W, Natale A, Tchou PJ. Effects of cocaine intoxication on the threshold for stun gun induction of ventricular fibrillation. Journal of the American College of Cardiology. 48(4):805–11, 2006 Aug 15.
8. Chakko S, Myerburg RF. Cardiac complications of cocaine abuse. Clinical Cardiology. 18:67–72, 1995.
9. Schwartz AB, Boyle W, Janzen D, Jones RT. Acute effects of cocaine on catecholamines and cardiac electrophysiology in the conscious dog. Canadian Journal of Cardiology. 4:188–92, 1988.
10. Schwartz AB, Janzen D, Jones RT. Electrophysiologic effects of cocaine on the canine ventricle. Journal of Cardiovascular Pharmacology. 13:253–7, 1989.
11. Inoue H, Zipes DP. Cocaine-induced supersensitivity and arrhythmogenesis. Journal of the American College of Cardiology. 11:867–74, 1988.
12. Przywara DA, Dambach GE. Direct actions of cocaine on cardiac cellular activity. Circulation Research. 65:185–92, 1989.
13. Han J, de Jalon PG, Moe GK. Adrenergic effects on ventricular vulnerability. Circulation Research 1964;14: 516–24.
14. Nanthakumar K, Billingsley IM, Masse S, Dorian P, Cameron D, Chauhan VS, Downar E, Sevaptsidis E. Cardiac electrophysiological consequences of neuromuscular incapacitating device discharges. Journal of the American College of Cardiology. 48(4):798–804, 2006 Aug. 15.

Chapter 19
Alcohol and the CEW

Ronald Moscati* and Jeffrey D. Ho**

Ethanol is the most widely used and abused drug in the United States [1,2]. It is widely available in various forms, legal for the adult population and socially acceptable. This is true despite its recognized role in both acute and chronic physical and behavioral health problems. It is frequently used alone but may be used with other illegal drugs or prescribed medications.

19.1 Alcohol Physiology

Acutely, consumption of ethanol can lead to intoxication with impairment of higher level cognitive functions such as judgment, reasoning, and memory. This, in turn, leads to altered behavior which can include poor interpersonal interactions, aggression, and other activities that can place the intoxicated individual and others in danger [3]. Physically, acute ethanol intoxication leads to increased heart rate, increased blood pressure, GI irritation and, at higher levels, CNS and respiratory depression [4]. Any or all of the above effects can lead to serious illness, injury, or death.

Chronic excessive consumption of ethanol, in addition to repeatedly exposing the individual to the acute risks mentioned above, has additional negative impacts on both behavioral and physical health. Ethanol is physically addictive and induces tolerance with repeated use. As with other addictions, the

---

**Ronald Moscati reports serving as a consultant to TASER International, Inc. and receiving consulting fees from TASER International, Inc. No other potential conflict of interest relevant to this chapter was reported.*

***Jeffrey D. Ho reports serving as an expert research consultant and medical expert consultant to TASER International, Inc. and reports as a personal shareholder of TASER International stock. No other potential conflict of interest relevant to this chapter was reported.*

R. Moscati (✉)
Department of Emergency Medicine, SUNY at Buffalo, Erie County Medical Center
e-mail: moscati@buffalo.edu

M.W. Kroll, J.D. Ho (eds.), *TASER® Conducted Electrical Weapons: Physiology, Pathology, and Law*, DOI 10.1007/978-0-387-85475-5_19,

individual is likely to engage in denial and self-destructive behavior in order to continue drinking. Long-term physical effects impact almost every organ system in the body. Liver failure, cardiomyopathy, seizures, dementia, and brain atrophy are common in long-term alcoholics [4].

Ethanol addiction can induce a potentially life-threatening withdrawal syndrome upon cessation of long-term use [2]. In milder episodes, individuals experience tachycardia, hypertension, tremulousness, and GI upset, with onset typically 12–72 hours following cessation of drinking. As the withdrawal state progresses individuals many also experience agitation, hyper-vigilance, seizures, hallucinations, and delirium. At the extreme, withdrawal can lead to cardiovascular collapse and death. The withdrawal syndrome can last anywhere from 2 days to greater than 1 week. Alcohol withdrawal syndrome is similar to benzodiazepine withdrawal and is typically treated with benzodiazepines in the acute phase followed by gradual tapering of the benzodiazepine.

19.2 Alcohol and CEW

Since alcohol can cause so many physical effects, both long term and short term, it would be difficult to associate specific symptoms to the CEW exposure in an intoxicated individual who has been exposed to a CEW discharge. While not substantiated, concerns have been expressed about CEW exposure leading to arrhythmias, seizures, and clinically significant rhabdomyolysis among other things. Long-term alcohol use can cause cardiomyopathies, which also can lead to arrhythmias. Alcohol use and specifically alcohol withdrawal is associated with seizures. Alcohol is a known cause of rhabdomyolysis. However, none of these entities have ever been reported in a study or even isolated case reports linking their occurrence in the alcohol population with CEW exposure.

The use of a CEW by law enforcement officers to subdue individuals is usually in response to the individual's behavior and failure to follow the directions of the law enforcement personnel. As is noted above, acute alcohol intoxication, chronic alcohol use and the alcohol withdrawal syndrome can all lead to behavioral issues that in turn can lead individuals to have confrontations with law enforcement personnel. In addition, alcohol can be, and frequently is, a cointoxicant with almost any other drug that contributes to situations leading to CEW use.

The altered perception and judgment that accompanies alcohol use can lead to the failure of many "pain-based" less-lethal means of controlling behavior, such as batons, pepper sprays, impact rounds, and traditional stun-guns. Impaired individuals can fight through pain and injuries in this state and in turn inflict injuries on others or themselves. The electrical discharges of the TASER M26 and X26 devices cause moderate skeletal muscle contraction that overwhelms the impaired individual's ability to physically resist. The impaired individual's dangerous behavior can then successfully be stopped and further injury prevented.

Individuals with such behavior as a result of intoxication with alcohol or other substances should be medically evaluated once the situation is controlled. The need for medical evaluation stems from the behavior and intoxication, not from the fact that a CEW was applied to control the situation. As is noted above, seemingly intoxication-related behavior can also result from chronic alcohol use and alcohol withdrawal. Each of these conditions has the potential for serious further medical complications.

19.3 Studies of Alcohol and CEW Use

The literature lacks any alcohol-specific studies of CEW application in acutely intoxicated animal models. The effects of a CEW on pigs infused with cocaine has been studied [5]. Although the study was small, it indicated a protective effect of cocaine in raising the fibrillation threshold. There exist no good animal models for chronic alcohol intoxication or withdrawal.

A study of 218 patients seen in the ED who had received a CEW application during the early 1980s reported three deaths and no long-term morbidity [6]. All three deaths were attributed to acute PCP intoxication. While 26% of the patients seen had laboratory proven alcohol intoxication, none of the deaths occurred in this group.

A single controlled study on the effect of CEW discharge on intoxicated human volunteers has been conducted [7]. Subjects had baseline blood testing including pH, pCO_2, lactate, and troponin prior to alcohol ingestion or CEW exposure. They then ingested alcohol to an average level of 0.11 milligrams/deciliter. pH and pCO_2 decreased and lactate increases after the alcohol ingestion. Subjects were then exposed to a 15-second CEW discharge followed by repeat laboratory testing. The results showed trends of a slight decrease in pH and increase in lactate, neither of which were clinically significant. Follow-up values at 24 hours showed return to baseline for all values. There was no rise in troponin values. No subjects experienced any arrhythmias or seizures. The study, while small, concludes that even with prolonged CEW exposure intoxicated subjects demonstrate clinically insignificant increases in measures of acidosis (that self-corrected) and no change in serum markers of cardiac injury.

19.4 Conclusion

The effects of alcohol use intersect with CEW use primarily as a result of the behavioral alterations in individuals associated with acute and chronic alcohol use. In this respect, CEW application is useful in preventing further injurious behavior to self or others in individuals who are impaired, unreasonable, and resistant to pain deterrence. While alcohol use is also associated with a wide variety of physical maladies, there is no documentation in animal models, individual cases, case series, or controlled studies of CEW exposure inducing or being synergistic with the effects of alcohol. CEW use does not appear to be physically harmful in intoxicated individuals and from the behavioral standpoint may be beneficial.

References

1. Health, United States. With Chartbook on Trends in the Health of Americans. In: Statistics NCfH, ed. Hyattsville, MD: 2006.
2. Al-Sanouri I, Dikin M, Soubani AO. Critical care aspects of alcohol abuse. *Southern Medical Journal.* Mar 2005;98(3):372–381.
3. Giancola PR. Influence of subjective intoxication, breath alcohol concentration, and expectancies on the alcohol-aggression relation. *Alcoholism: Clinical & Experimental Research.* May 2006;30(5):844–850.
4. Rehm J, Gmel G, Sempos CT, et al. Alcohol-related morbidity and mortality. *Alcohol Research & Health: the Journal of the National Institute on Alcohol Abuse & Alcoholism.* 2003;27(1):39–51.
5. Lakkireddy D, Wallick D, Ryschon K, et al. Effects of cocaine intoxication on the threshold for stun gun induction of ventricular fibrillation. *Journal of the American College of Cardiology.* Aug 15 2006;48(4):805–811.
6. Ordog GJ, Wasserberger J, Schlater T, et al. Electronic gun (Taser) injuries. *Annals of Emergency Medicine.* Jan 1987;16(1):73–78.
7. Moscati R, Ho J, Dawes D, et al. Physiologic effects of prolonged conducted electrical weapon discharge on intoxicated adults. *Acad Emerg Med.* May 1 2007;14(5_Supplement_1):S63-b-64.

Chapter 20
Conducted Electrical Weapons and Implantable Cardiac Devices

Subba Reddy Vanga, James L. Vacek, Loren Berenbom, and Dhanunjaya R. Lakkireddy

The cardiac effects of conducted electrical weapons (CEWs) have been a subject of debate for several years with concerns regarding their safety, particularly in patients with active disease processes (such as coronary artery disease or cardiomyopathy) or altered physiologic states such as acidosis or hyperadrenergic conditions [1]. In the past 3 5 years, a large body of animal and human data has clearly established the overall safety of these devices. A recent study by McDaniel et al demonstrated that CEWs did not cause cardiac arrhythmias within the limits of the standard electrical discharge in a porcine model [2]. In the same study, multiple applications of the TASER CEW electrical discharge were reported as safe. Another animal study has shown a high safety margin for ventricular fibrillation (VF) induction by CEWs [3]. A recent publication suggested that despite a low probability of VF induction, TASER X26 current application, close to the heart, had a higher rate of myocardial capture and induced VF on one occasion with epinephrine administration [4]. (This possible effect of epinephrine induction facilitation was not statistically significant.) Another recent study involving healthy human volunteers at rest has demonstrated the general cardiac safety of a TASER X26 application and indicated that there was no significant effect on the electrocardiogram over a 24-hour period [5].

It is well known that implantable cardiac devices such as pacemakers (PM) and implantable cardioverter defibrillators (ICDs) are susceptible to malfunction as a result of electromagnetic interference (EMI). Electromagnetic interference can result in many undesirable consequences, including damage to internal circuitry, oversensing, undersensing, failure to pace, failure to capture, power on reset (POR), triggering of ERI (elective replacement indicator which would suggest some battery depletion), and inappropriate defibrillation

D.R. Lakkireddy (✉)
Clinical Assistant Professor and Staff Electrophysiologist, Mid America Cardiology, University of Kansas Hospitals, Suite G600, 3901 Rainbow Blvd, Kansas City, KS 66160, USA
e-mail: dlakkireddy@mac.md

M.W. Kroll, J.D. Ho (eds.), *TASER® Conducted Electrical Weapons: Physiology, Pathology, and Law*, DOI 10.1007/978-0-387-85475-5_20,

therapy [6–13]. Theoretically it is possible for CEWs to cause EMI in these implantable cardiac devices. This potential interaction between CEWs and implantable cardiac devices, such as permanent pacemakers (PMs) and implantable cardioverter-defibrillators (ICDs) is poorly understood and is subject to intense speculation supported by a few case reports [14–16].

20.1 Case Reports

An ICD shock was reported in a porcine model when exposed to a TASER X26 continuously for 15 seconds. The device detected this as ventricular fibrillation and delivered appropriate shock twice during a 15-second period (Fig. 20.1) [14].

Haegli et al reported similar findings in a patient with an ICD receiving a TASER CEW application but was not shocked by the ICD (Fig. 20.2) [15]. The exposure to the CEW pulse was 5.36 seconds and it terminated before the reconfirmation of VF resulting in no delivered shock. Recently, Cao et al. reported ventricular capture on interrogation of a dual chamber pacemaker in a person exposed to TASER CEW (Fig. 20.3) [16]. This report raised the issue whether CEWs can cause primary myocardial capture or capture only in association with cardiac devices and pacing wires providing a preferential pathway of conduction to the myocardium. This report did not show the induction of sustained ventricular arrhythmia and the recorded tracing was probably the pacing wires sensing the nonmyocardial electrical activity from CEW application. Our group has confirmed the detection of the CEW discharge as rapid noise lasting for the duration of application with appropriate device response similar to findings in some of the prior case reports [17].

We have had three cases of device patients receiving CEW applications [17]. The first case was of a 25-year-old schizophrenic male with a pacemaker for sick sinus syndrome. Aggressive behavior during recovery prompted three CEW applications to his chest with no evidence of EMI behavior or rapid myocardial capture on the electrograms. The second case was of a 45-year-old man with dilated cardiomyopathy who has a DDD-ICD. He had CEW exposures twice to his back without any evidence of ICD shocks, rapid myocardial capture, or EMI behavior. The final case was of a 56-year-old man with a VVI-ICD for ischemic cardiomyopathy. He had four CEW applications to the front with no EGM evidence of ICD shocks, EMI behavior or rapid myocardial capture. Device interrogation at 3 and 6 months did not show any significant change in the lead or generator function.

20.2 Animal Studies

In our study, an anesthetized pig model was prepared with insertion of two probes at the sternal notch and with maximum cardiac impulse separated by 1.5 centimeter from the epicardial surface. A prepectoral pocket positioned between the probes was created to place a device generator that was connected

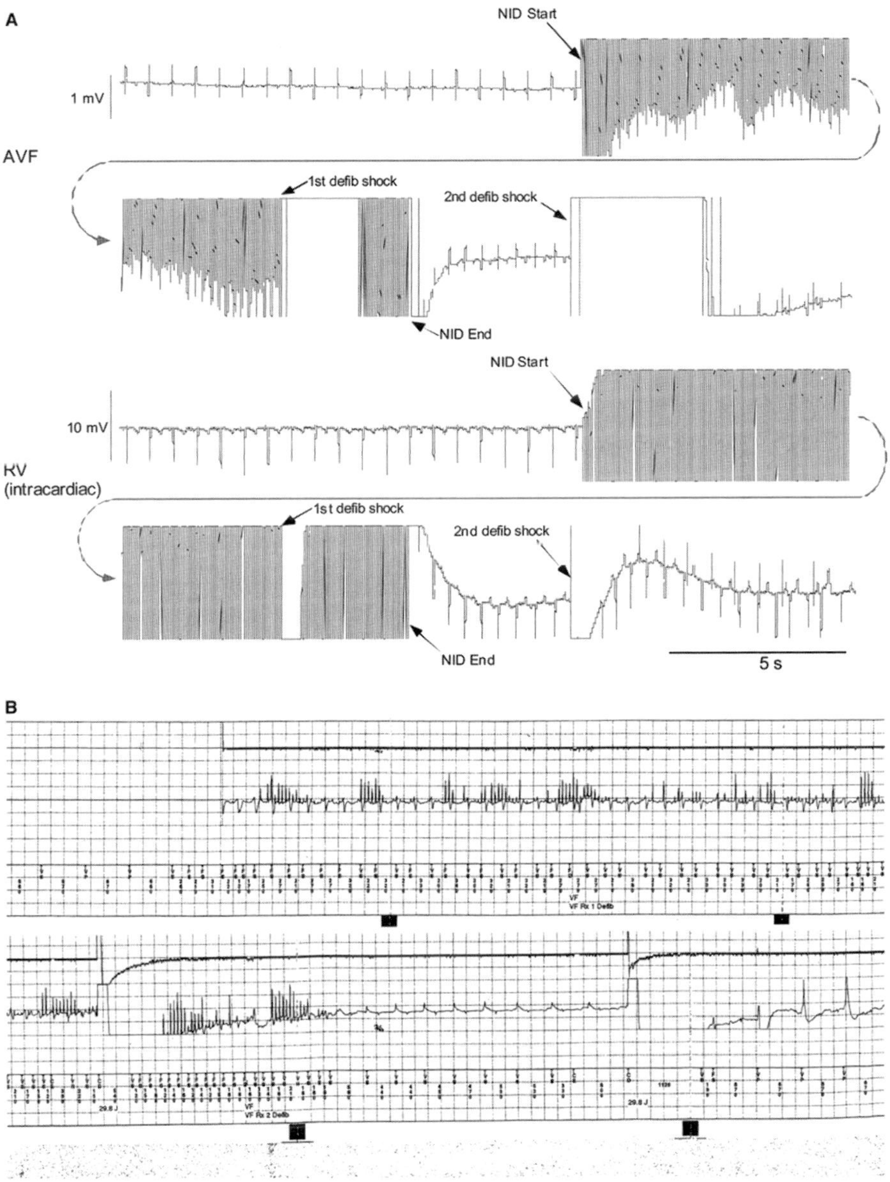

Fig. 20.1 (A) Surface ECG (aVF) and intracardiac (right ventricle) recording 15 seconds of energy delivery by the CEW. **(B)** The near-field electrogram and the marker-channel recording during the episode. The strip started with the marker channel showing the ventricular sense and ventricular pace rhythms at cycle lengths of 660–670 milliseconds. Application of CEW energy resulted in high-frequency electrical signals, with intervals of 120–330 milliseconds representing noise from the delivery of energy by the CEW. This energy is sensed by the device as ventricular fibrillation and resulted in delivery of the first VF therapy (VF R 1 Defib), and a shock is delivered (29.8 joules). As the CEW energy application continued, the second VF therapy was delivered (VF R 2 Defib), and a second shock also was delivered (29.8 joules; lower strip). The upper and lower strips are continuous tracings (from Calton et al. [14] with permission)

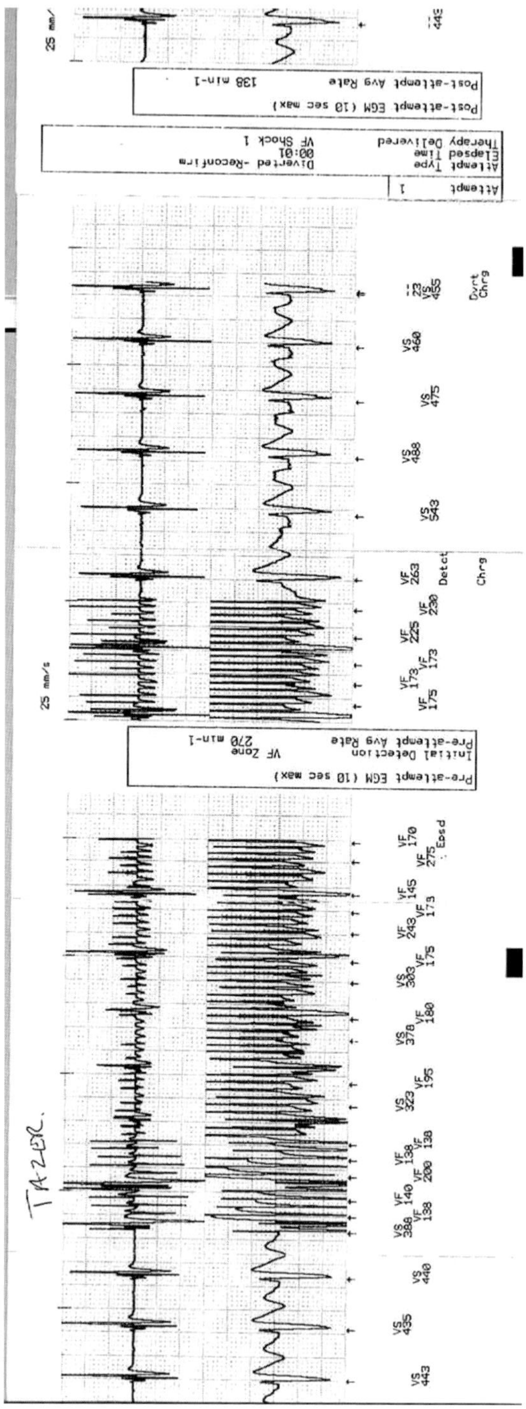

Fig. 20.2 Electrogram tracing from an ICD in a person exposed to TASER CEW. The 'noise' from CEW pulse was detected as VF (detect) and capacitors started to charge. Reconfirmation of VF was failed (Chrg) and it did not result in the delivery shock but in the diversion of the charged energy (from Haegeli et al. [15] with permission)

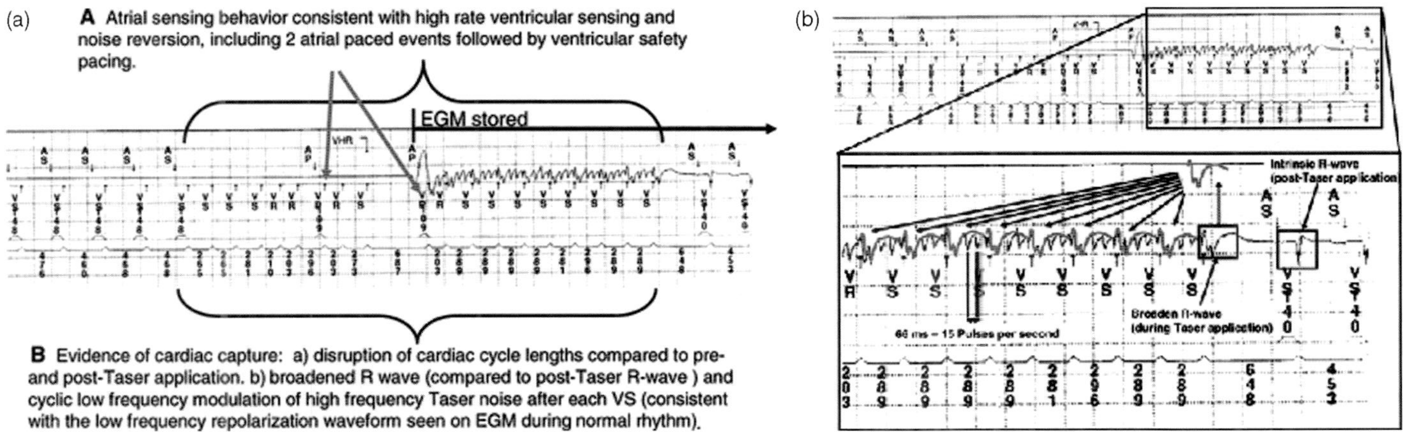

Fig. 20.3 (**A**) EGM recording from a patient with DDR PPM exposed TASER CEW application. The EGM recorded was the summed EGM, which is a combination of atrial and ventricular EGMs into a single tracing. (**B**) Magnified summed EGM tracing stored during CEW application. The high-frequency CEW pulses (15 pps, 66 milliseconds) are labeled on the tracing. The EGM from the last VS during CEW application is superimposed on each prior VS event, showing that the disruption of the high-frequency CEW signal is consistent with a modulation of the signal by a repeating R wave with morphology different than the intrinsic R wave (right side of figure) (from Cao et al. [16] with permission)

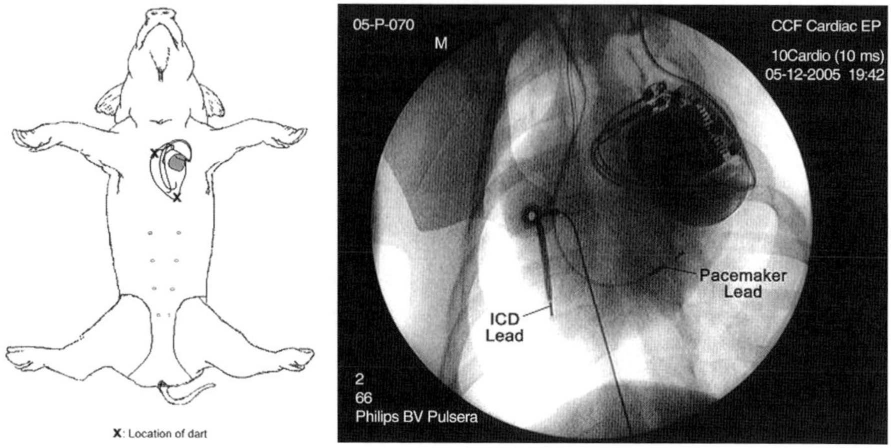

Fig. 20.4 Location of the probes and device with leads in the study conducted by Lakkireddy et al. [17] (from Lakkireddy et al. [17] with permission)

to either a defibrillator lead or to a pace-sense lead placed at the right ventricle apex (Fig. 20.4). Pacemakers were programmed to VVI mode at 60 beats per minute and ICDs for tachycardia detection at 180 beats per minute with provision for up to four shocks at maximum energy levels. The animal was exposed three times with a standard CEW discharge of 5-second duration with each device being tested. A total of seven ICDs and nine pacemakers were checked in the same animal. Before and after each discharge, lead and generator functions were assessed with a device interrogator specific to the manufacturer. Pacing and sensing thresholds as well as pacing and shocking coil impedances were determined before and after each of the three CEW discharges and results were analyzed using the average of three postdischarge values. Results are shown in Tables 20.1 and 20.2. The generators were monitored for abnormal behavior, including oversensing, undersensing, failure to pace, failure to capture, POR, ERI, and inappropriate defibrillation therapy.

Analysis of the experiment results showed that there is no significant difference between the pre and postshock device values. There was no evidence of device malfunction. Telemetry monitoring of the devices showed a consistent electrical artifact during the 5-second period of CEW shock. All ICDs sensed the electrical activity and started charging for a possible shock delivery. Mean cycle length of the artifact detected by ICDs was 176 ± 20 milliseconds corresponding to the rates of artifact stimulus on pace-sensing leads (Fig. 20.5).

However, there was no incidence of VT or VF after the exposure to the CEW current and no ICD delivered a shock in response to a standard 5-second exposure. The minimum charge time to shock delivery for all ICDs used in this study is >5 seconds and this probably can explain the detection, charge, and aborted therapy sequence seen in all models of ICDs in this study.

Table 20.1 Pre and postshock evaluations of ICD systems (from Lakkireddy et al. [17] with permission)

Manufacturer	Model	Bat V preshock	Bat V postshock	R preshock	R postshock	PT preshock	PT postshock	LI preshock	LI postshock	DFCI preshock	DFCI postshock	DCL	CT
Guidant	Vitality DS	3.2	3.2	9.3	7.1	1.4–0.5	0.4–0.5	394	369	54	52	178	7.6
Guidant	Ventak MS	2.58	2.58	8	7.8	0.2–0.5	0.2–0.5	389	397	50	49	160	5.4
Guidant	Vitality DS	3.19	3.19	8	6.9	0.2–0.5	0.2–0.5	352	354	50	52	154	4.9
Guidant	Ventak DR	2.93	2.93	8	7	0.2–0.5	0.2–0.5	367	348	54	51	169	8.4
Medtronic	7273	5.16	5.16	7.5	7.5	2.0–0.4	2.0–0.4	500	474	59	59	210	5.6
St Jude	Atlas DR	3.1	3.1	4.2	4.9	0.2–0.5	2.2–0.5	395	380	44	44	196	5.2
St Jude	Photon VR	3	3	4.3	4.4	0.2–0.5	1.0–0.5	355	375	46	46	165	4.3
Mean		3.31	3.31	7	6.5	0.6	0.9	393	385	51	50	176	5.9
SD		0.84	0.84	2	1.3	0.8	0.9	50	42	5	5	20	1.5

Bat V, battery voltage in V; R, R waves sensing threshold in mV; PT, pacing threshold in volts milliseconds; LI, lead impedance in Ohms; DFCI, defibrillation coil impedance; DCL, detected cycle length in milliseconds; CT, charge time in seconds.

Table 20.2 Pre and postshock evaluation of pacemaker systems (from Lakkireddy et al. [17] with permission)

Manufacturer	Model	Bat V preshock	Bat V postshock	R preshock	R postshock	PT preshock	PT postshock	LI preshock	LI postshock
Medtronic	Insync	2.95	2.95	8	8	0.5–0.5	0.5–0.5	422	409
St Jude	Enpulse	2.75	2.75	5.6	5.6	0.25–0.52	0.75–0.52	417	423
St Jude	Identity DR	2.73	2.71	5	5.3	0.25–0.5	0.25–0.5	334	356
St Jude	Affinity DR	2.75	2.75	7	7	0.25–0.8	0.25–0.8	374	374
St Jude	Integrity AF	2.75	2.76	6.2	6.4	0.25–0.4	0.25–0.4	401	383
St Jude	Affinity DR	2.76	2.76	7	7	0.25–0.5	0.25–0.5	373	403
Medtronic	Insync	2.77	2.77	8	8	0.5–0.5	0.5–0.5	426	422
Guidant	Meridian	2.78	2.76	5.7	5.3	0.3–0.5	0.3–0.5	410	400
Guidant	Pulsar max	2.86	2.85	5.1	5.8	0.2–0.4	0.3–0.4	380	380
Mean		2.79	2.78	6.4	6.48	0.3	0.4	393	394.44
SD		0.07	0.07	1.15	1.07	0.1	0.2	30.15	22.71

Bat V, battery voltage in V; R, R waves sensing threshold in mV; PT, pacing threshold in volts milliseconds; LI, lead charge time in seconds.

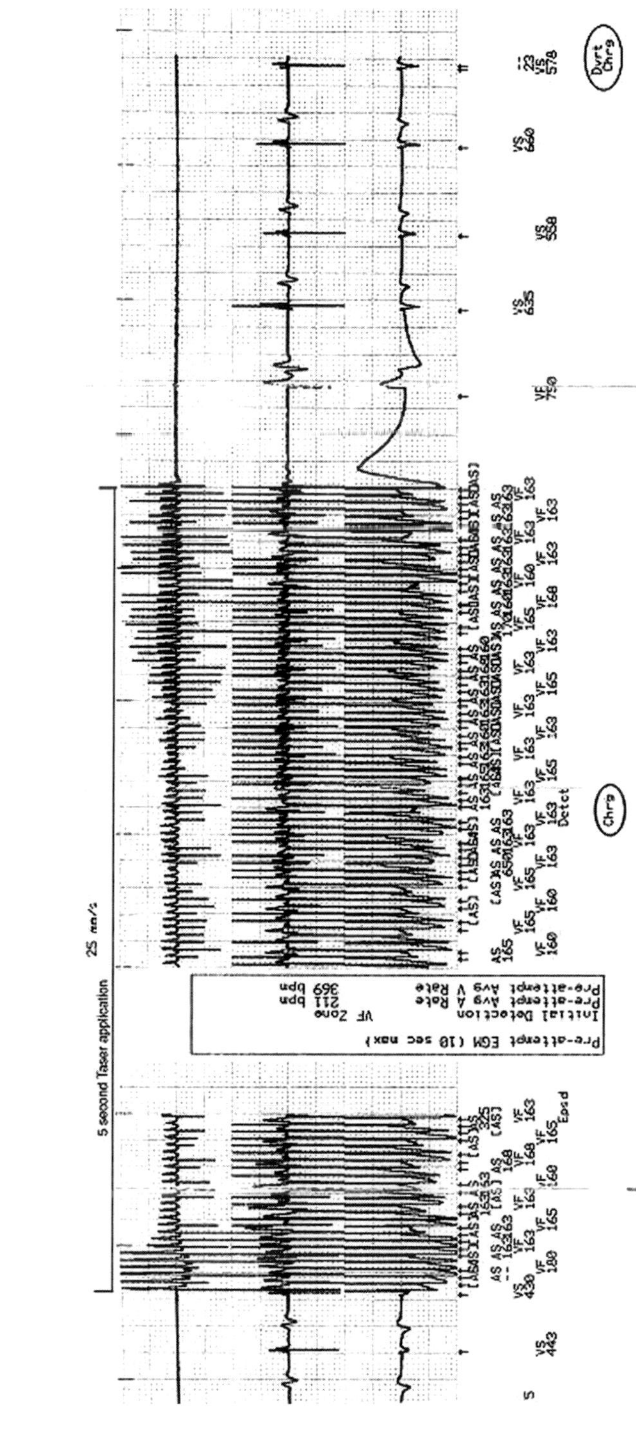

Fig. 20.5 ICD memory record of CEW discharge. The interrogated electrogram strip from the ICD memory after the CEW application showing onset of rapid rate detection with initiation of the application with a detected cycle length corresponds best to the detected CEW pulses rather than the ventricular electrograms even though accelerated ventricular capture can be appreciated visually at cycle lengths around 240 ms (from Lakkireddy et al. [17] with permission)

This study was done using the TASER-X26 CEW and is probably generalized to other CEW models since the X26 has the highest net delivered charge, hence representing a conservative result. An anaesthetized single animal with an acutely placed bipolar ventricular lead was used to test a finite array of devices. Many factors including the effects of anesthesia, variation in duration of lead insertion, utilization of unipolar leads and generators, biological and device variations and duration of applied discharge may impact the reproducibility of the results. Though discouraged by the manufacturer's training guidelines, longer deliveries are possible with this CEW which were not tested in the study. There is the possibility that a law enforcement officer may use a longer application when faced with a difficult subject for whom a standard application was ineffective. The results of these applications in regard to arrhythmia induction, device response, and device function are unknown.

Nevertheless this study has shown that the short-term functional integrity of the implantable devices is not affected by standard CEW application even when exposure is such that the generator is directly between the CEW probes. Field experience suggests that such a close application of probes to the pectoral area is not common. However, it is certainly possible that appropriately placed ICDs may initiate charging of their capacitors for current delivery, as the frequency of CEW pulses may be interpreted as tachyarrhythmias.

Another important potential interaction of devices with electrical energy is switching into reset mode or other alteration of device parameters. It is very well known that external or internal defibrillation sometimes results in activation of reset mode or elective replacement indicator, or transient elevation of capture and sensing thresholds, primarily in older unipolar systems [10–13]. We did not discern this type of interference in our study. An external defibrillator delivers energy of up to 360 joules whereas a TASER X26 pulse delivers 70 millijoule. At 19 pulses per second, the energy delivered per second is only 1.33 joule and the total energy delivered in 5 seconds of standard CEW discharge is about 7 joules. This amount of energy generated during a standard CEW discharge is significantly lower than that of external defibrillation and does not appear to affect device electronic settings. With the delivery of this small amount of energy over 5 seconds and rapid cooling effects of blood with quick dissipation, significant cardiac tissue alteration which would change device threshold parameters is unlikely.

The inappropriate ICD shocks observed during the 15-second exposure in the pig model by Calton et al. [14] is probably due more to the artifact created by the TASER CEW current than true myocardial capture. We have clearly shown in our animal study that the electrical artifact created by TASER CEW current enables ICD detection and if it persists beyond charge and redetect timings may result in an ICD shock despite no true myocardial capture. We believe that the same explanation may explain the ICD shocks in the case reported by Haegeli et al. [15]. Development of better noise reduction algorithms may help in eliminating the inappropriate response to extracardiac electrical activity and can help in differentiating it from true myocardial capture.

20.3 Conclusions

Pacemakers and ICDs are unlikely to be damaged by CEW application. CEW pulses can be recognized incorrectly as cardiac arrhythmias and the potential for "appropriate" inhibitory and shock delivery therapies is present. These events have been recorded in human beings after exposure to TASER CEW pulses. None of the reports so far indicated any physical damage to the device and its electrodes in the short term, but the data on long term and after repeated exposure are lacking. Due to its low energy delivery and the nature of its waveform, TASER CEWs can be regarded as generally safe in most subjects with implanted devices when used in standard applications and as recommended by the manufacturer. Although rare, prolonged application of CEWs might potentially result in inhibition of pacing in patients who are pacemaker dependent and may result in symptomatic bradycardia, however, the noise reversion function should prevent that.

Patients with ICDs have the potential to be inappropriately shocked during prolonged applications as the detected electrical activity from the CEW can be mislabeled as persistent rapid ventricular activity falling in the VF zones beyond the charge and redetection times. It is not clear that the discomfort of the unnecessary ICD shock would add significantly to that of the CEW exposure, especially in the typical subject that is already somewhat anesthetized with illegal drugs or excited delirium behavior. This overall risk must be placed in the context of the exceedingly small overlap in the demographics of the typical pacemaker/ICD patient and the typical violent resister of arrest. This is consistent with our current epidemiological data which finds no reports of complications in pacemaker or ICD patients despite 1.4 million human applications of CEWs. If a cardiologist has a device patient that is suspected of violent behavior or illegal drug use they might be counseled to avoid resisting arrest. That counsel alone might be enough to prevent the occurrence of the theoretically harmful device interaction.

References

1. Strote J, Hutson HR. Taser Use in Restraint-Related Deaths. Prehospital Emergency Care. 2006;10(4):447–450.
2. McDaniel WC, Stratbucker RA, Nerheim M, Brewer JE. Cardiac safety of neuromuscular incapacitating defensive devices. Pacing Clin Electrophysiol. 2005 Jan;28 Suppl 1:S284–S287.
3. Lakkireddy D, Wallick D, Ryschon K, Chung MK, Butany J, Martin D, et al. Effects of cocaine intoxication on the threshold for stun gun induction of ventricular fibrillation. J Am College Cardiol. 2006 Aug 15;48(4):805–811.
4. Nanthakumar K, Billingsley IM, Masse S, Dorian P, Cameron D, Chauhan VS, et al. Cardiac electrophysiological consequences of neuromuscular incapacitating device discharges. J Am College Cardiol. 2006 Aug 15;48(4):798–804.

5. Ho JD, Miner JR, Lakireddy DR, Bultman LL, Heegaard WG. Cardiovascular and physiologic effects of conducted electrical weapon discharge in resting adults. Acad Emerg Med. 2006 Jun;13(6):589–595.
6. Hayes DL, Christiansen J. Electromagnetic interference with implantable devices. In: Ellenbogen KA, Wilkoff BL, ed. *Clinical Cardiac Pacing and Defibrillation*. 2nd ed. Philadelphia: WB Saunders Company 2000:939–952.
7. Hayes DL, Vlietstra RE. Pacemaker malfunction. Annals of Internal Medicine. 1993 Oct 15;119(8):828–835.
8. Barold SS FM, Ong LS, Heinle RA. Interference in cardiac pacemakers: Exogenous sources. In: El-Sherif N, ed. *Cardiac Pacing and Electrophysiology*. 3rd ed. Philadelphia: WB Saunders Company 1991:608–633.
9. Belott PH, Sands S, Warren J. Resetting of DDD pacemakers due to EMI. Pacing Clin Electrophysiol. 1984 Mar;7(2):169–172.
10. Irnich W. Interference in pacemakers. Pacing Clin Electrophysiol. 1984 Nov;7(6 Pt 1): 1021–1048.
11. Levine PA, Barold SS, Fletcher RD, Talbot P. Adverse acute and chronic effects of electrical defibrillation and cardioversion on implanted unipolar cardiac pacing systems. J Am College Cardiol. 1983 Jun;1(6):1413–1422.
12. Toivonen L, Valjus J, Hongisto M, Metso R. The influence of elevated 50 Hz electric and magnetic fields on implanted cardiac pacemakers: the role of the lead configuration and programming of the sensitivity. Pacing Clin Electrophysiol. 1991 Dec;14(12):2114–2122.
13. Yee R, Jones DL, Klein GJ. Pacing threshold changes after transvenous catheter countershock. Am J Cardiol. 1984 Feb 1;53(4):503–507.
14. Calton R, Cameron D, Masse S, Nanthakumar K. Images in cardiovascular medicine. Duration of discharge of neuromuscular incapacitating device and inappropriate implantable cardioverter-defibrillator detections. Circulation. 2007 May 22;115(20):e472–e474.
15. Haegeli LM, Sterns LD, Adam DC, Leather RA. Effect of a Taser shot to the chest of a patient with an implantable defibrillator. Heart Rhythm. 2006 Mar;3(3):339–341.
16. Cao M, Shinbane JS, Gillberg JM, Saxon LA. Taser-induced rapid ventricular myocardial capture demonstrated by pacemaker intracardiac electrograms. J Cardiovascular Electrophysiology. 2007 Aug;18(8):876–879.
17. Lakkireddy D, Biria M, Baryun E, Berenbom L, Pimentel R, Emert M, Kreighbaum K, Kroll M, Verma A. Can Electrical-Conductive Weapons (TASER®) alter the functional integrity of pacemakers and defibrillators and cause rapid myocardial capture? Heart Rhythm. 2008;5:S97.

Chapter 21
Risk Management and the CEW

Greg Bingham

Depending on one's perspective, "risk management" may mean different things. To law enforcement officers in the field, managing risk means using good officer safety tactics to ensure their best chance of survival and safety while bringing their subjects under control (with minimal injury as well). The financial impact of injuries on their agencies are probably the least of their concerns.

Law enforcement administrators, however, spend most of their time and energy managing budgets, manpower, and conduct. Law enforcement agencies throughout the United States and Canada, often struggle with the lack of funding, lack of training, and lack of manpower. If they are able to find ways to minimize risk and reduce the number of injuries to their officers and the public, it will most certainly have a positive impact in all three areas.

Risk management is the term often associated with insurance, which in the end equates to money. While other chapters deal with morbidity and mortality effects of CEWs this one will explore the financial impact.

It is important to recognize that risk can never be eliminated in law enforcement. As a young recruit in our police academy, our training officer told us all that we needed to understand the three rules of war:

Rule #1 – Young men and women die in war.
Rule #2 – There will always be war.
Rule #3 – You can't change Rule #1 or #2.

Make no mistake about it. There is a war going on in our society. Violence and lawlessness are all around us. We ask our peace officers to protect us and our way of life from this violence, knowing they must risk their own lives and well-being to do so. It is our duty as administrators, trainers, and citizens, to see to it that we equip our officers with the most effective and humane tools possible.

There is no perfect tool, weapon, or tactic that will prevent injury or death to everyone all of the time. Some officers will find themselves in situations where

G. Bingham (✉)
Oakland, CA Police Department (Ret.)
e-mail: gkcglb@aol.com

M.W. Kroll, J.D. Ho (eds.), *TASER® Conducted Electrical Weapons: Physiology, Pathology, and Law*, DOI 10.1007/978-0-387-85475-5_21,

they have little choice but to injure or kill in order to protect their own life or the lives of others. Some of our officers will suffer life-changing and career-ending injuries, regardless of what tools they have. Unfortunately, casualties in war are inevitable.

21.1 On-Duty Injuries

On-duty injuries (ODIs) to officers have a major financial impact on an agency and their jurisdiction, whether it is a city, county, or state, costs can be staggering and may go undetected if not correctly assessed.

In California, law enforcement officers of large cities and counties probably have some of the highest salaries and benefit packages in the law enforcement community. Officer's salaries can range from \$50,000–\$95,000 annually, not counting overtime and paid benefits. Depending on their benefit packages, a jurisdiction may be paying an additional 35–90% above the salary.

In the Oakland Police Department on-duty injury costs were conservatively exceeding \$2 million per year. Categories that contributed to the overall costs were; work days lost, overtime needed to fill open positions, the number of officers retired due to on-duty injuries, training new officers, workers' compensation, and types of injuries. After reviewing the above information, an annual average cost was determined. Please note that these costs do not include the city's payments to persons injured while resisting arrest or attacking officers.

21.2 Number of Work Days Lost Due to ODI

The Oakland Police Department, was constantly affected by the lack of manpower, particularly in the Patrol Division. In 2004 and 2005, there were times when as many as 40–50 officers were off work due to injuries. During this same time period, average annual sick leave used by officers was only 2–3 days per year. During the 2004 calendar year, 6,090 days were missed due to on-duty injuries and illness. Since on-duty illnesses were minimal, it was apparent that there was a significant injury problem.

Thus 6,090 work days – approximately 16.5 man-years – of labor were lost. For an agency whose actual number of sworn personnel averaged around 750 officers, this was very significant. In addition, officers were on "light duty" for another 1,648 days. "Light duty" status meant that the officer could have no risk of conflict and their work was usually restricted to clerical work as they were unable to fulfill all their duties as a peace officer.

During this time period (2004), the typical hourly wage for 3 years experience was \$34.53. Using this wage, the chart below is an estimate of what the city paid in wages to injured officers unable to perform their duties. However, the cost of lost wages must also factor in the additional cost of benefit packages. The approximate cost of benefit packages in the Oakland Police Department was

Table 21.1 Annual cost of officer injuries in Oakland California

Days per year	Hours per year	Annual cost
6,090 off work	48,720	$1,682,301
1,648 light duty	13,184	$ 455,243
Total salary cost		$2,137,544
Benefit package 89%		$1,902,414
Total		$4,275,008

89%. As a result, the estimated total cost of lost wages due to on-duty injuries needed to be almost doubled to $4,275,008. See Table 21.1 for calculations. This is an example of the cost that effective risk management attempts to mitigate.

21.3 Additional Costs

When a field officer is injured, it often creates an opening that must be filled by paying another officer overtime. During a 3-month period in the Patrol Division in 2005, the Department spent $711,000 to fill open beats alone. This did *not* include overtime paid to hold officers over from one shift to another to cover manpower shortages, or other reasons. This was specifically for filling open patrol beat positions. As stated before, sick leave was not a significant factor. Since overtime costs were consistently high, I used this figure to determine an average annual cost of approximately $2.8 million, for the one specific type of overtime.

During a 3-year period, approximately 30 officers were retired as the result of injuries they sustained while on duty. The estimated cost to train one recruit officer was $48,000. In addition, a new officer was paid approximately $20,000 in wages during their field training program. These officers were not allowed to work alone and were still in a critical learning period. The field training officers assigned to these new officers were paid a bonus of 5% (approximately $1,110) above their normal salary during this time.

Therefore, the cost to replace an officer who was retired due to an ODI was approximately $69,000. With an average of 10 officers per year retiring due to their on-duty injuries, the overall average annual cost was approximately $690,000 in 2005.

During the fiscal year of 2003–2004, the city's initial workers' compensation cost for sworn officers was $3,383,319. The estimated incurred total cost was expected to rise to $5.2 million dollars.

21.4 Sources of Injuries

A review of over 400 individual on-duty injury reports over a 3 year time period, was performed. Results are shown in Table 21.2.

Table 21.2 Sources of officer injuries

Source	Percentage
Climbing/jumping fences, walls, etc.	6
Training	6
Vehicle collisions/accidents	10
Miscellaneous (tripping, lifting, etc.)	33
Physical conflict (fighting with suspects during arrests and from suspect assaults on officers)	45

21.5 Overall Annual ODI Cost

To determine the full financial impact of on duty injuries, a "high" and "low" estimate were prepared. See Table 21.3. The "high" cost directly reflected all costs in all categories. The "low" cost estimate either eliminated or reduced costs in some categories. The "low" cost estimate shows even to the most skeptical, that on-duty injuries cost the city millions of dollars annually.

Table 21.3 Overall cost estimates

	High estimate	Low estimate
Wages and benefits	$4,275,088	0
Overtime	$2,847,164	$2,000,000
Worker's compensation	$5,222,032	$3,383,319
New officer training	$ 690,000	$ 690,000
Total	$13,034,284	$6,073,319
Cost of conflict (45% of total)	$5,865,427	$2,732,883

21.6 Potential Savings of Full CEW Deployment

I studied several agencies that had fully deployed CEWs and their subsequent reduction in injuries to officers and suspects. Although the percentages did vary, they all reported very significant injury reduction rates for both. In fact, most agencies who had gone to full CEW deployment also shared other common reductions: reduction in citizen excessive force complaints, reduction in other use of other force options, and reduction in lethal force uses, specifically police involved shootings.

It is important to note, that nearly all of these other commonalities do have a financial impact. Just one example would be police involved shootings. Even if an officer is 100% justified in the use of deadly force, it is a near certainty that a lawsuit will follow. Due to the cost of civil litigation, each police involved shooting, that results an injury or a death, will most likely have a significant cost factor involved.

After averaging several agencies statistics, it was determined that there was an overall 50% reduction in injuries to officers and suspects. My personal observation was that those agencies in urban environments, with higher crime

Table 21.4 Estimated savings from conflict injury reduction by adoption of CEWs

Estimated conflict injury reduction rate (%)	Assuming high injury cost estimate	Assuming low injury cost estimate
30	$1,759,628	$ 819,897
40	$2,346,170	$1,093,197
50	$2,932,713	$1,366,496

rates, had higher reductions in injury rates than those of other agencies. Based on this information, and the policing challenges within the City of Oakland, it is very reasonable to conclude that full CEW deployment there would result in a minimum reduction of on duty injuries by 30–50%. Table 21.4 shows the estimated potential savings for the city based on 30, 40, and 50% conflict injury reduction.

The average cost to equip each officer with a TASER X26 CEW with an extended warranty, including air cartridges and a holster was approximately $1,150. To equip approximately 554 field officers would cost the City of Oakland approximately $637,100. Even with only a 30% reduction in the on duty injury rate, based on the low estimate ($819,897) minus the cost of the equipment, in the first year the city would save $182,797.

In the following fours years, after the equipment cost had been paid, even with only a 30% injury reduction rate at the "low estimate", the city could save an additional $819,897 per year, which would equate to approximately $3.2 million dollars.

21.7 Implementation

It is critically important for agencies that choose to equip their officers with CEWs to have an excellent training program and a sound policy in place prior to field deployment. Experience has shown that well-trained officers follow their training and make better decisions. Agencies that invest in good training will reap the rewards of it. Their officers will follow policy and prevent injuries, and in some instances even deaths, all of which will equate to huge savings in medical claims, litigation costs, and civil judgments.

If an agency fails to do either of these, they will be exposing themselves, their officers, and their community, to potential problems and risks that are avoidable.

CEW programs should have a Coordinator, who is extremely well trained and responsible for overseeing and managing the entire program, including training and the selection of instructors. They should also be responsible for keeping all records involving: training and re-certification, tracking equipment, deployments, downloads, and equipment maintenance. In larger agencies, this should be the Coordinator's only job.

This person should also have additional training and expertise in excited delirium and force-options training. It would also be extremely beneficial for the Coordinator to be available on a 24/7 call-out basis to field units, supervisors, press information officers, and the Chief of Police.

Programs need go beyond simply training their officers. It is important to involve, or provide information or training, for city attorney personnel, police commanders, citizens groups, media, medical personnel (both emergency department hospitalstaff and EMS personnel) and medical examiners.

21.7.1 Why Include All These Specialists in the Community?

- An emergency medicine physician upon seeing a TASER CEW probe sticking out of a subject's skin for the first time and is told there is a barb on the other end, may pull out a scalpel and start cutting away. If they have never seen how small one is and how easy it is to safely remove, it's hard to blame them for their lack of experience or understanding in proper probe removal.
- Nurses who work in an emergency department, who have yet to receive any training about excited delirium, its symptoms and its potentially deadly consequences, are at a disadvantage to properly evaluate a subject's need for immediate care.
- A citizen, who's only education of the TASER CEWs have come from the media, or groups with a political agenda, will most likely not be supportive. Someone needs to tell them the truth and it should come from someone knowledgeable.
- City attorney personnel will benefit from technical and practical training with CEW devices and excited delirium, so they can confidently defend false allegations of police misconduct rather than settle prematurely.
- Even today, there are many medical examiners with little training and knowledge of CEWs. Fortunately, this seems to have improved greatly in the past few years. Additionally, timely autopsies (within 12 hours) and specific procedures for brain examination can greatly assist a medical examiner in identifying excited delirium deaths. It is also very important to obtain body core temperature readings as close to the time of death as possible. This all takes cooperation with the coroner's office whose hours of operation may only be Monday to Friday, 9 to 5.

All of these things can add up to save officers, commanders, agencies, and jurisdictions a lot of time, stress, and the unnecessary loss of funds.

Chapter 22
The New York City Experience

Michael D. White and Justin Ready

This chapter seeks to examine all 820 incidents involving the use of TASER CEWs by police officers in the New York City Police Department (NYPD) over 6.5-year period, January 1, 2002 through May 15, 2007. Drawing from police reports of incidents in which an officer used the CEW, the chapter seeks to accomplish several objectives: (1) to track the prevalence of CEW use over time in the study site; (2) to describe the typical situations in which the weapon is used; (3) to examine the demographic and behavioral characteristics of police officers and suspects involved; (4) to assess the effectiveness of the device; and (5) to investigate whether it is used in compliance with departmental guidelines.

Please note that the New York City policy is unique in that CEWs are generally not used for arrest but primarily for emotionally disturbed persons (EDPs). Further discussion will follow.

22.1 National Trends and Experience

Several considerations are important for determining how the TASER CEW is currently applied on the force continuum: (1) when is it appropriate (i.e., reasonably necessary) to use the TASER CEW according to police agencies and individual officers and (2) to what extent is the TASER CEW used as an alternative to lethal force (in situations where lethal force is justified) and as an alternative to other less lethal weapons? It is not yet possible to answer the second question because many police departments have just begun to accumulate detailed records on the circumstances relating to TASER CEW deployments. Additionally, police administrators are still modifying existing policies to incorporate this new technology. There appears to be no clear consensus on where police agencies place the TASER CEW on the force continuum. Many

M.D. White (✉)
School of Criminology and Criminal Justice, Arizona State University
e-mail: mdwhite1@asu.edu

M.W. Kroll, J.D. Ho (eds.), *TASER® Conducted Electrical Weapons: Physiology, Pathology, and Law*, DOI 10.1007/978-0-387-85475-5_22,

police officers have reported that being subjected to pepper spray is a more unpleasant experience because the average recovery time is almost an hour (or weeks if one rubs his/her eyes), whereas the recovery time for the TASER CEW is immediate with typically no after effects [1]. In contrast, a county Sheriff quoted in *The Kalamazoo Gazette* [2] described the shock of the TASER CEW as follows, "They call it the longest 5 seconds of their life. It's extreme pain, there's no question about it. No one would want to get hit by it a second time."

Just as individual officers have different assessments of the TASER CEW, police agencies are also inconsistent in relation to where they place CEWs on the force continuum. A US Government Accountability Office report [3] found that placement of the CEW on the force continuum varied considerably across agencies. For example, the Sacramento Police Department allowed for use of the TASER CEW during harmful situations, such as when a suspect is combative. The Phoenix and San Jose Police Departments permitted use of the TASER CEW at a lower level of force, such as when a suspect is actively resisting arrest. The Orange County (Florida) Sheriff's Department reported allowing use of the TASER CEW when a suspect is passively resisting the verbal commands of an officer. The IACP (International Association of Chiefs of Police) reported that a majority of police agencies place CEWs at roughly the same level of the force continuum as pepper spray.

In response to public concern about the inappropriate use of CEWs, the IACP issued training guidelines for the deployment of these weapons. The Police Executive Research Forum (PERF) also issued a series of policy recommendations. Among other guidelines, it is recommended that CEWs should only be used against those who are actively resisting or exhibiting aggression, and they should generally not be used against pregnant women, children, and visibly frail persons. (PERF has never given a scientific reasoning for the hypothesized increased danger with these individuals. In fact, one could argue that these same individuals might fare far better with a CEW than with baton strikes or pepper spray.)

22.1.1 Effectiveness

In 1991, four Los Angeles police officers were videotaped striking Rodney King more than 50 times with their batons. While not captured on videotape, officers had also used an early version TASER CEW on King, stunning him twice but failing to subdue him. More recently, in September 2004 a police officer in San Jose was forced to shoot and kill a combative suspect after the officer had already used the TASER on the suspect twice. In September 2003, a police officer in Adams County, Colorado involved in a CEW incident wrote in his report, "By the 12th cycle, he appeared physically exhausted and wasn't a problem (*Denver Post,* 9/19/04)." Although these cases suggest that TASER CEWs are not always effective, there are currently few independent empirical studies examining the effectiveness of police officer use of the TASER in the field. Some exceptions are field report analyses produced by *TASER International* and in-house evaluations conducted by a number of police agencies, which are discussed below.

TASER International compared the effectiveness of their CEWs and pepper spray by examining incidents in which officers used both weapons during a single police citizen encounter [5]. The CEW was effective in subduing the suspect during 82.7% of the incidents compared to an effectiveness rate of 33.1 % for pepper spray. Similarly, the Seattle Police Department found that deployment of the CEW effectively resolved 85% of incidents involving the weapon [6]. The Los Angeles Sheriff's Department reported a 94% effectiveness rating for its CEWs, as compared to other less lethal devices such as batons and pepper spray which generally have an 85% effectiveness rating. The LAPD also found that the TASER CEW was associated with fewer injuries during police citizen encounters as reported in the chapter authored by Capt. Greg Meyer.

22.1.2 Safety

Although beyond the scope of the current chapter, research on the health effects of the CEW should be examined in light of findings relating to prevalence and effectiveness. It is important to take into consideration the number of incidents in which the technology is used and the safety and effectiveness of alternative less lethal weapons. For example, research suggests that a blow from a kinetic impact weapon can result in traumatic apnea and cardiac dysrhythmia, possibly resulting in death [7]. Considerations relating to the adoption and deployment of CEWs such as the TASER CEW may benefit from examining the relative effects of other low lethal weapons available for use in comparable settings.

In sum, fundamental questions about the TASER CEW remain unanswered because of limited research studying police use of the device in the field, and selective sampling and reporting data on the circumstances surrounding TASER CEW deployments. While the growing body of medical research on CEWs offers a critical first step in assessing its impact, it is also important to recognize that the technology does not work in a vacuum. Human judgment, stress-provoking conditions, training, and situational cues all play a role in whether or not a CEW deployment will result in a favorable outcome. We initiated this study to begin addressing these considerations by examining all incidents involving the use of the TASER by police officers in the NYPD over a 6.5-year period, focusing specifically on how often the weapon is deployed, the context in which it is used, suspect characteristics, and outcomes (favorable or unfavorable).

22.2 The NYPD Data

The NYPD has NOT issued the TASER CEW to all rank-and-file officers. Rather, the TASER has been issued exclusively to the Emergency Service Unit (ESU) officers (the equivalent of SWAT). Also, all officers who are promoted to the rank of Sergeant or above are trained in its use and are authorized to carry the weapon. Each precinct is equipped with at least one TASER CEW that can be signed out by a supervisor. The Patrol Guide details the specific circumstances in which it is appropriate for a supervisor or ESU officer to use a TASER CEW [8]:

> Patrol supervisors or uniformed members of the service assigned to the Emergency Services Unit may utilize a TASER electronic stun gun or stun device to assist in restraining emotionally disturbed persons if necessary. The TASER/stun device may be used:
>
> a. To restrain an EDP who is evincing behavior that might result in physical injury to himself or others, *OR*
>
> b. To restrain person(s) who, through the use of drugs, alcohol or other mind-altering substances, are evincing behavior that might result in physical injury to himself or others.
>
> Emergency Service Unit personnel will obtain the permission of the Emergency Service Unit Supervisor prior to utilizing a TASER/stun device, except in emergencies. (NYPD Patrol Guide)

As a result, the TASER CEW can only be used in situations involving an EDP or person under the influence of drugs or alcohol who is posing a threat of physical injury, where either Emergency Service officers are deployed or a supervisor is present and has a CEW in his or her possession. The Patrol Guide also offers a definition of an "emotionally disturbed person:"

> A person who appears to be mentally ill or temporarily deranged and is conducting himself in a manner, which a police officer reasonably believes, is likely to result in serious injury to himself or others.

In situations involving an EDP, officers are instructed to create and maintain a "zone of safety" of approximately 20 feet, and call for ESU and a patrol supervisor as well as an ambulance. Officers are NOT to attempt to take an EDP into custody unless

- The EDP is unarmed, not violent, and is willing to leave voluntarily; or
- The EDP's actions constitute an *immediate* threat of serious physical injury or death to himself or others.

22.2.1 Research Design and Data

We examined all reported cases of CEW use by NYPD personnel during a 6.5-year period from January 1, 2002 through May 15, 2007 ($n = 820$). The data have been obtained from the "TASER/Stun Device" report, which must be completed each time an officer uses the weapon. (These reports were provided to the authors by the supervisor of the department's Training Division. Although the form is used primarily for the TASER CEW, there were 33 forms involving use of another type of nonlethal weapon – either a stun device or other similar alternative. Since the focus of this paper is the TASER CEW, these cases were excluded from the analysis.) The report consists of a one-page form that documents information relating to the suspect, the officer, and the incident (i.e., circumstances). Officers who have deployed the weapon are required to complete the form by checking boxes from a range of options, with an additional narrative section where the officer is required to describe the incident in detail. Suspect variables relevant to this study included demographic characteristics, emotional and physical state, behavior during the incident, and presence of weapons. Other variables were officer assignment and attributes of the CEW deployment (e.g., distance, resistance, etc.).

Based on the field reports completed by the responding officers, the authors constructed a data set in SPSS that records 38 variables relating to each deployment. While the unit of analysis is the TASER CEW incident, the data are officer-based. Therefore, if two officers use the TASER CEW on the same suspect, then two separate incidents are recorded. However, if one officer uses the TASER CEW multiple times on the same suspect the case is counted only once in the data set. Variables have been created to indicate when the weapon is used repeatedly by the same officer and the number of times it was used (i.e., number of cycles).

22.2.2 *Analysis*

The research findings involve a descriptive analysis of 820 incidents in which police officers in New York City used the TASER CEW. The first section of the analysis charts annual TASER CEW use during the study period. The second part follows the structure of the department's reporting form, categorizing the research findings into suspect, officer, and incident-related characteristics. The third section considers questions relating to outcomes. Specifically, (1) how often are suspects incapacitated or subdued; (2) to what extent are officers satisfied with the TASER CEW; and (3) is the device being used in accordance with departmental policy?

22.2.3 *Limitations*

A number of limitations warrant discussion. First, the study involves one police department that deploys the TASER CEW in a restrictive and controlled manner. According to *TASER International*, over 12,000 law enforcement agencies have adopted the TASER CEW as a nonlethal alternative, and the ability to generalize the findings here to other departments remains unknown. Second, this study uses official police records as the primary data source, and the findings reflect the responding officer's account of the incident (although there is a supervisory review of each report). Third, the paper consists of a descriptive analysis only. Yet, given the limited research to date, more sophisticated analysis may be premature. Finally, the findings presented here must be placed within the broader context of the other chapters that examine health-related questions. Nonetheless, the authors believe this research offers an important step toward understanding use of the TASER CEW by police officers in the field, by characterizing its use and effectiveness in one police department, and by offering a launching point for asking more in-depth questions about this nonlethal weapon.

22.3 Results

22.3.1 Prevalence

Figure 22.1 displays TASER CEW use by the NYPD from 2002 to 2007 illustrating a steady increase in deployments over time. During 2002 and 2003, fewer than 100 incidents occurred per year; in the following two years, the number of deployments increased to just over 100 per year ($n = 121$ and 130). The number of CEW deployments almost tripled in 2006 ($n = 310$) and incidents in 2007 were projected to peak at more than 350 (approximately 30 per month). The reasons for this increase near the end of the study period are not clear, although we detect a substantial spike in the number of deployments by supervisors not assigned to ESU. (While this increase may reflect a change in

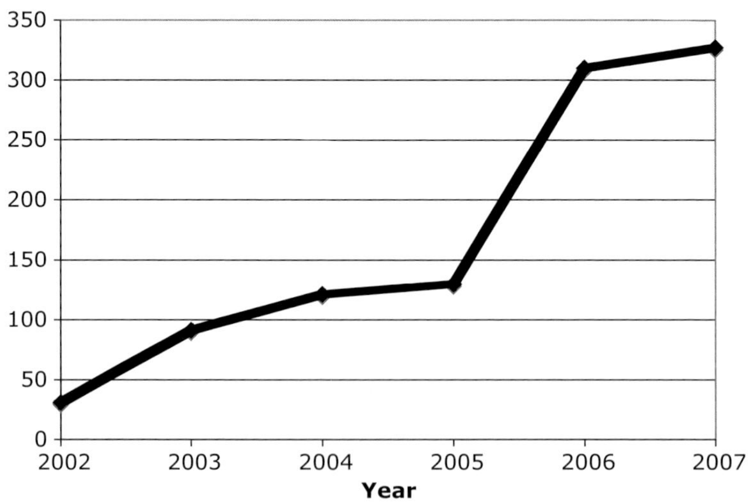

Fig. 22.1 CEW use by the NYPD 2002–2007, actual and projected

practice, there was no policy change that we can document. Supervisors were trained in use of the TASER CEW throughout the study period, with weapons available in each precinct.) Still, given the large number of police citizen contacts in New York City generally, and ESU calls (estimates suggest that ESU receives approximately 30,000 service calls per year), the NYPD's use of the TASER CEW appears to be relatively infrequent.

22.3.2 Suspects

Table 22.1 illustrates suspect-related characteristics from the 820 TASER CEW incidents, including demographic information, the suspect's emotional and physical state (e.g., exhibiting signs of mental illness or intoxication), the suspect's behavior, and whether he or she was armed with a weapon (and if so, what type of weapon).

Table 22.1 Suspect and officer data. Number in parentheses represents the number of forms in which this datum was recorded

Suspect characteristics	*N*	Percentage (%)
Gender (804)		
Male	711	88.4
Female	93	11.6
Race (791)		
African-American	434	54.9
White	128	16.1
Hispanic	213	26.9
Asian/other	16	2.1
Mean age (751)		33.8 years
Suspect emotionally disturbed (752)		
No	116	15.4
Yes	636	84.6
Suspect intoxicated (745)		
No	640	85.9
Yes, drugs	43	5.8
Yes, alcohol	47	6.3
Yes, both	15	2.0
Suspect armed (735)		
No	518	70.5
Yes	217	29.5
Suspect violent (740)		
No	25	3.4
Yes, toward self	103	13.9
Yes, toward officer	443	59.9
Yes, toward others	23	3.1
Yes, multiple	146	19.7

Table 22.1 (continued)

Suspect characteristics	*N*	Percentage (%)
Officer characteristics		
Rank (748)		
Patrol officer	185	24.7
Detective	324	43.3
Supervisor	239	32.0
Command (729)		
Emergency Service Unit	470	64.5
Other	193	35.5
Back-up present (716)		
No	76	10.6
Yes	640	89.4
Supervisor present (707)		
No	57	8.1
Yes	650	91.9

22.3.2.1 Demographics

Suspects in the TASER CEW incidents were primarily male (88.4%); just over half were described as African-American (54.9%), 16.1% were white, 26.9% were Hispanic, and 2.1% were Asian or another ethnic group. Suspects involved in TASER incidents tended to be older than the typical crime suspect, with a mean age of 33.8. Several incidents involving use of the CEW by officers in other police jurisdictions have received national attention because the suspect was either a minor or a senior citizen. Review of the age distribution in this study indicates that 23 TASER incidents involved subjects under the age of 18 (ages 13–17). In each of these cases, the juvenile subject was described as exhibiting violent behavior (and was armed in 1/2 of the cases). Alternatively, nine incidents involved suspects age 65 or older (ages 65–72). In 7/9 cases, the suspect was armed with a weapon and in all of the incidents the suspect was engaging in physical violence.

22.3.2.2 Suspects' Emotional and Physical State

Table 22.1 indicates that most of the suspects involved in the TASER CEW incidents did *not* appear to be under the influence of drugs or alcohol (85.9%). (This variable is based on the police officer's assessment of the suspect at the time of the incident. It is not based on more definitive tests such as a urinalysis or blood/hair analysis.) Of the 14.1% reported to be intoxicated, 5.8% were reported to be under the influence of drugs, 6.3% were under the influence of alcohol, and 2.0% were under the influence of both drugs and alcohol. Overall, it appears that the TASER CEW is not used frequently by NYPD officers on suspects who are intoxicated.

Table 22.1 also shows that a large majority of suspects involved in CEW incidents were considered emotionally disturbed persons (EDPs). (This is also the officer's assessment based on available evidence.) Only 15.4% of suspects were not classified as emotionally disturbed. Although this suggests that police officers in the NYPD use the TASER disproportionately against the mentally ill in crisis, this finding must be interpreted within the context of departmental policy relating to use of the CEW. As noted earlier, rank-and-file officers do not have access to a TASER CEW; only Emergency Service personnel and supervisors are authorized to use the weapon. In fact, Table 22.1 indicates that nearly 2/3 of the officers who used the TASER CEW were assigned to the Emergency Service Unit. Per department policy, the Emergency Service Unit is dispatched when the patrol officers or supervisors on the scene determine that the situation involves an EDP who is behaving in a manner that could result in physical injury or death to the EDP or others. Thus, the findings are a reflection of the types of suspects that the specialized Emergency Service Unit is typically called to handle – not the suspects typically handled by line officers. Moreover, since the analysis does not include data for all Emergency Service Unit cases (regardless of TASER CEW use), it is not clear what proportion of ESU incidents involve police use of the TASER CEW as compared to incidents involving police use of empty hand force and other low lethal weapons.

22.3.2.3 Possession of Weapons

Table 22.1 indicates that just under one-third of the suspects were armed with a weapon (29.5%), while 70.5% were unarmed (information was missing in 85 cases). Of the 217 cases in which the suspect was armed with a weapon, 159 involved a knife or cutting instrument (73% of armed suspects; 22% of all cases). In 6 incidents the suspect was armed with a handgun. Of the remaining cases involving an armed suspect, the most common weapon was a blunt object, such as a metal pipe, baseball bat, or a large stick.

22.3.2.4 Physical Violence

In addition to exhibiting signs of mental illness, the majority of suspects – 96.6% – were engaging in violent conduct. This violent behavior was directed at a police officer(s) in more than 1/2 of the cases (59.9%). Approximately 1/6 of the cases involved a threat of suicide or self-harm (13.9%), 3.1% entailed violence toward another citizen, and 19.7% involved violence toward multiple individuals at the scene. As discussed earlier, some of the controversy surrounding the TASER CEW has focused on when it should be used; specifically, is it appropriate to use the TASER CEW against noncombative, unarmed suspects who fail to follow verbal instructions (i.e., passive resistance)? The NYPD has avoided much of this controversy by restricting use of the device to suspects who have posed a violent (and often armed) threat to themselves, a police officer or other citizens.

22.3.2.5 Officer Characteristics

Unfortunately, the CEW/Stun Device reporting form captures limited information concerning the officer who deployed the weapon. Table 22.1 describes the rank of the deploying officer, as well as his or her command assignment. Just under 1/2 of the officers were detectives (43.3%), 24.7% were patrol officers, and 32.0% were supervisors. As noted above, 64.5% of the responding officers were assigned to ESU.

In the majority of cases, the officer who deployed the TASER CEW was not working alone. Back-up officers were present during 89.4% of the incidents. A supervisor was present during 91.9% of the incidents. [In fact, there were only 10 cases (less than 3%) where the officer was alone when deploying the TASER CEW (no backup and no supervisor present)]. The high frequency of back-up officers and supervisors is likely influenced by the fact that most of these cases involve ESU officers who are typically called to the scene by the first responding officer, and often a supervisor will also respond.

22.3.3 Incidents

22.3.3.1 Types of Encounters

Table 22.2 displays a number of incident-related characteristics that illustrate the types of encounters in which the TASER CEW is typically used. Nearly 3/4 of the cases occurred indoors; approximately 29% occurred outdoors. Interestingly, 54 cases occurred in a precinct house, usually in the holding cell area. In all of these stationhouse cases, the suspect was already in custody but was engaging in physical violence toward himself or others. As a result, the Emergency Service Unit was dispatched to resolve the situation. Table 22.2 also shows that most suspects (65.2%) were NOT arrested after the incident. Approximately 20% had already been arrested and were in custody, and an additional 14.8% were arrested as a result of the incident that led to the TASER CEW deployment. (A review of charges indicates that among those arrested, 28% was charged with a serious person offense such as assault, robbery and attempted murder.)

One possible explanation for most of the suspects not being arrested stems from their mental illness. Although scholars have argued that mental illness is being increasingly criminalized [9], conventional wisdom currently suggests that diverting those suffering from mental illness out of the criminal justice system is a preferable approach [10]. According to department policy, a large majority of suspects (93.2%) were transported to a hospital for a physical examination (with 1/4 subsequently arrested for criminal offenses). However, these data do not indicate how often the encounter led to civil commitment or long-term hospitalization.

Table 22.2 Incident data. Number in parentheses represents the number of forms in which this datum was recorded

	N	Percentage (%)
Location(747)		
Indoors	533	71.4
Outdoors	214	28.6
Suspect arrested (733)		
No	478	65.2
Yes	255	34.8
Suspect transported to hospital (734)		
No	50	6.8
Yes	684	93.2
Number of CEW deployments (695)		
1	582	83.7
>1	113	16.3
Mean distance between officer and suspect (707)		4.76 ft (1.45 m)
Contacts on target (602)		
2	531	88.2
1 (1 contact missed)	38	6.3
0 (Both contacts missed)	13	2.2
Contacts hit but fell from clothing	20	3.3
Was Suspect Incapacitated? (598)		
No	76	11.3
Yes	598	88.7
Mean time to incapacitation	598	7.97 s
Did suspect continue resistance? (724)		
No	513	70.9
Yes	211	29.1
Officer satisfied with CEW? (718)		
No	150	20.9
Yes	568	79.1
Use within department policy (691)		
No	0	0.0
Yes	691	84.3

22.3.3.2 TASER CEW Deployment

The field reports also contain information relating to the actual deployment of the TASER. As seen in Table 22.2, the average distance between the suspect and the officer at the time of deployment was 4.76 feet (1.45 meter). In nearly 2/3 of the incidents (63.2%), the officer and suspect were 5 feet (1.5 meter) or less from one another at the time of the deployment. In 83.7% of the incidents, the TASER CEW was only used once by the officer; in 1/6 of the cases the device was deployed multiple times.

In 88.2% of the incidents, both contacts hit the suspect as intended (see Table 22.2). In 6.3% of cases, 1 of the contacts missed, and in just 2.2%, both contacts missed the suspect. Finally, in 19.5% of the cases officers also used another nonlethal device, most typically a stun device (11.9%) or mace (3.4%).

22.3.4 Outcomes

Aside from characterizing the suspects and officers involved in CEW encounters, a central question addressed by this research involves whether or not the TASER achieved it primary objective: to successfully incapacitate a combative suspect without serious injury to the officer or suspect. Table 22.2 shows that 88.7% of suspects were subdued by the TASER CEW and subsequently taken into custody. The average time to incapacitation (from CEW deployment) was 7.97 seconds; although of those who were incapacitated, 61% were subdued in 5 seconds or less. [The median time to incapacitation was 5.0 seconds, indicating that the mean is skewed a bit by a small number of cases with excessive times. There are five cases with the following times to incapacitation: 60 seconds (n = 2), 90 seconds (n = 2), 120 seconds (n = 11), and 180 seconds (n = 1).]

In 29.1% of the incidents the suspect continued to resist against the officer after being struck by the TASER CEW. These 211 cases can be broken down into two categories based on when the resistance occurred. In 76 cases, the resistance continued immediately following TASER CEW deployment because the suspect was not incapacitated by the weapon. In the remaining cases (n = 135), the suspect was initially incapacitated by the TASER CEW, the officer(s) gained control of the individual, but the suspect began resisting again at a subsequent point in time. Two important findings emerge from this distinction regarding continued resistance. First, since the objective of using a TASER CEW is to gain immediate control over a combative suspect, the 76 cases where this did not occur may be seen as representing the failure rate. As a proportion of the total number of cases, this represents a 10.5% failure rate (information is missing for 96 cases), or alternatively, an 89.5% success rate. Second, the issue of continued resistance at a later point in the encounter highlights the fact that the TASER CEW is intended to temporarily incapacitate a suspect; the involuntary loss of muscle control is not long-term and the suspect will regain full functioning in a relatively short period of time.

Another measure of TASER CEW effectiveness is officer satisfaction with the device. In 79.1% of the incidents, the TASER was reported to have performed satisfactorily. Not surprisingly, the 20.9% of cases where officers rated the TASER CEW as performing poorly include all of the cases where the suspect was not immediately subdued.

The reporting form also includes a section for a supervisor's assessment of whether deployment of the device conformed to departmental policy. In 84.3% of the cases, a supervisor indicated that use of the CEW was consistent with departmental policy. In the remaining 15.7% of the incidents (n = 129), the form was not signed and contained no information about whether the deployment was consistent with departmental policy. However, the narratives of those cases suggest that they conformed to department policy on use of the TASER CEW.

22.3.4.1 Reducing the Potential for a Violent Outcome

The primary objective of the CEW is to gain control of a combative suspect while reducing the likelihood of serious injury or death to both suspects and police officers. The descriptive information above has partially addressed this issue, but in order to more fully investigate the violence reduction question, we created a preliminary *violence escalation scale* and classified all incidents according to that scale. The scale ranges from 0 (least potential for a violent outcome) to 9 (highest potential for a violent outcome) with points assigned as shown in Table 22.3.

The rationale for assigning *violence escalation points* for armed and violent suspects is fairly straightforward. Points are added for intoxicated and mentally ill (in crisis) suspects because research demonstrates that individuals in those mental and physical states are more likely to resort to violence [11,12,13]. Additionally, if the officer is alone it is reasonable to expect an increased likelihood that the suspect will actively resist. Finally, analysis indicates that the mean weight of suspects who continued to resist after being struck with the TASER CEW was nearly 10 pounds (4.5 kilograms) heavier than who did not continue resistance (202 vs. 193 pounds or 92 vs. 88 kilograms).

Table 22.4 illustrates how the cases were scored on the *violence escalation scale*, ranging from values of 0 through 8. [Note that scores were not calculated for 216 cases because information on at least one of the variables was missing (i.e., to be scored the case must have valid information on all scoring variables)]. These scores were then collapsed into a three-level escalation risk classification: 0–3 = low; 4–5 = medium; and 6–8 = high. Table 22.5 shows the percentage of cases in each violence risk level: 48.8% were low risk, 30.1% were medium risk, and 21.1% were high risk. Within each risk level, we then examined officer satisfaction (a measure of TASER CEW effectiveness): low risk = 80.0%,

Table 22.3 Violence escalation scale

Points	Item
+3	Armed with a gun
+3	Armed with an edged weapon <20 ft (7 m)
+1	Armed with an edged weapon ≥20 ft (7 m)
+2	Violent behavior (toward officer, self, others, or multiple)
+1	Intoxicated (drugs, alcohol, or both)
+1	Exhibiting signs of being mentally ill, in crisis
+1	Sole officer (no backup or supervisor present)
+1	Suspect weight >210 lbs (95 kg)

Table 22.4 Violence escalation score distribution

Score	N	Percentage (%)
0	3	0.5
1	8	1.3
2	62	10.3
3	222	36.8
4	152	25.2
5	30	5.0
6	87	14.4
7	37	6.1
8	3	0.5
Total	604	100.0

Table 22.5 Distribution of low, medium, and high violence escalation scores

Risk of violence escalation	N	Percentage satisfactory use	Percentage of total
Low (0–3)	295	80.0	48.8
Medium (4–5)	182	74.0	30.1
High (6–8)	127	81.5	21.1
Total	604		100.0

medium risk = 74.0%, and high risk = 81.5%. While this is a rudimentary analysis with a preliminary scaling measure, the results taken with earlier findings suggest that the TASER CEW performs satisfactorily in most cases, and its effectiveness does not appear to vary substantially by the potential risk for injury or death.

22.3.4.2 Fatalities

We have had two incidents in which a suspect died after being subjected to the CEW. Both cases involved EDPs and the Emergency Services Unit, and a supervisor and back-up officers were present. In the 2005 incident, the suspect, Terrence Thomas, reportedly swallowed crack cocaine prior to his arrest to hide evidence. Several hours later, he became ill in the holding cell at the precinct. When medics attempted to examine him, the suspect became violent and ESU was called. The suspect continued to act violently, and ESU used the TASER CEW. While he was being transported to the hospital by ambulance, the suspect went into cardiac arrest and died. The medical examiner concluded that the death was due to acute cocaine intoxication. In the 2007 incident, still under investigation at the time this chapter was written, suspect Blondel Lassegue reportedly stopped taking his medication. When police officers arrived, he attacked one of the officers, punching his face repeatedly. After he was subdued with a TASER CEW, paramedics examined him on the scene and declared him "OK" for transport. Shortly thereafter, the individual became nonresponsive and paramedics were unable to revive him. In sum, we can conclude that during a 6.5-year period two deaths occurred after TASER deployments in 820 cases – approximately 0.24% of all cases.

22.3.5 Discussion

Using data from one major police department over a 6.5-year period, we sought to characterize the prevalence, use and outcomes of the TASER CEW through a descriptive analysis of 820 cases. The background and context for this research centers on limited independent empirical research on police use of the TASER and the serious questions that have emerged as a result. Given the limitations of our research stated earlier, the findings presented here are generally positive regarding the use of the TASER CEW. Key findings include:

- TASER CEW use has increased considerably during the study period, but still these cases represent a very small proportion of all NYPD/citizen encounters.
- Few suspects were under the influence of alcohol or drugs, but a large majority of the individuals were classified as exhibiting signs of mental illness (85%). (Again, these are not clinical judgments. Rather, they are conclusions drawn by the officers on-scene based on available evidence.)
- Nearly all suspects were engaging in physically violent behavior (97%).
- Just under 1/3 of suspects were armed, and among armed suspects, the majority possessed a knife or cutting instrument (22% of all cases).
- The majority of police officers using the TASER CEW were assigned to the Emergency Service Unit (65%).
- Back-up officers and supervisors were present in nearly all cases;
- A large majority of suspects were incapacitated by the TASER CEW (89%), and most were incapacitated within 5 seconds.
- Although the suspects in these incidents are disproportionately from a vulnerable population that many argue are at higher risk for suffering serious physiological side effects, findings indicate the TASER CEW was highly effective in these cases.
- In fewer than 1/3 of the cases, the suspect continued to resist after being subjected to the TASER CEW, but the resistance in the majority of these cases occurred at a later point in time – highlighting the temporary effect of the weapon.
- In more than 3/4 of the cases, the officer reported that the TASER CEW performed satisfactorily.
- In a preliminary violence risk analysis, the TASER CEW's effectiveness (i.e., officer satisfaction) does not vary by risk potential;

22.4 Conclusions

The favorable findings presented here may be largely a consequence of how the NYPD issues, monitors, and controls use of the weapon. Beyond the health risk issue, the controversy surrounding the TASER CEW has focused on when it should be used – where along the force continuum – how should it be used, and who it should be used against. More specifically, some police departments have approved the use of the TASER CEW in response to nonphysical resistance, such as not following verbal commands, allowing officers to use the TASER CEW in place of empty hand force. Due to the NYPD success the deployment was greatly expanded in June 2008 to include all seargents.

References

1. International Association of Chiefs of Police (IACP) (2005). *Electro-Muscular Disruption Technology (EMDT). A Nine-Step Strategy for Effective Deployment*. IACP: Alexandria, VA, USA.
2. Kalamazoo Gazette (2004). *Michigan Officers Find Tasers Stunningly Effective*. March 7, 2004.
3. U.S. Government Accountability Office (2005). *Taser Weapons: Use of Tasers by Selected Law Enforcement Agencies*. Report to the Chairman, Subcommittee on National Security, Emerging Threats and International relations, Committee on Government reform, House of Representatives.
4. *New York Times* (2004). *Claims Over Tasers' Safety are Challenged*. November 26, 2004.
5. TASER International (2002). *Advanced Taser M26 Field Report Analysis*. November 7, 2002.
6. Seattle Police Department Special Report. 2002.
7. Heck, J. (2003). Kinetic impact weapons: the potential for injury. *Tactical Edge*, 62–63.
8. New York Police Department (2000). *Study Police Department Patrol Guide*. Study City: Study Police Department.
9. Teplin, L.A. (2000). Keeping the peace: Police discretion and mentally ill persons. *National Institute of Justice Journal*, 8–15.
10. Lurigio, A.J., Fallon, J.R., & Dincin, J. (2000). Helping the mentally ill in jails adapt to community life: A description of a postrelease ACT program and its clients. *International Journal of Offender Therapy and Comparative Criminology*, 44(5), 532–548.
11. Swanson, J.W., Holzer, C.D., Ganju, V.K., & Jono, R.T. (1990). Violence and psychiatric disorder in the community: Evidence from the Epidemiologic Catchment Area Surveys. *Hospital and Community Psychiatry*, 41, 761–770.
12. Newhill, C.E. & Mulvey, E.P. (2002). Emotional dysregulation: The key to a treatment approach for violent mentally ill individuals. *Clinical Social Work Journal*, 30(2), 157–171.
13. Mulvey, E.P. & Fardella, J. (2000). Are the mentally ill really violent? *Psychology Today*, 33(6), 39–50.

Chapter 23
Impact of CEW and Other Types of Force and Resistance on Officer and Suspect Injuries

Michael R. Smith, Robert J. Kaminski, Jeffrey Rojek, Geoffrey P. Alpert, and Jason Mathis

The use of force by police has been the subject of empirical inquiry for more than 40 years. In that time, much has been learned about the nature and extent of the force used by police and the conditions and correlates that affect its application. Among the most important issues that have received attention from use-of-force researchers over the years are those involving injuries to officers and suspects. Almost half a century later, however, much of the research on injuries remains descriptive in nature or contains substantial data and analytic limitations that prevent the research from being used optimally to make policy or training decisions at the agency level. Furthermore, with the proliferation in recent years of conducted electrical weapons (CEWs) such as those of the Taser® and Stinger® brands, questions have arisen regarding the safety of such weapons and what their impact has been on injuries and in-custody deaths [1]. The lack of cross over research on CEWs and injuries has again left law enforcement agencies without the information they need to make sound policy decisions or to respond to inquiries from citizens, special interest groups, and policy-makers, some of whom question whether CEWs are an appropriate nonlethal alternative for general police use.

In the early to mid-1990s, police found themselves in a similar position with respect to oleoresin capsicum (OC) or pepper spray. In those days, OC was spreading rapidly among American police forces and concerns were being raised concerning its misuse and safety by the same special interest groups [2]. The National Institute of Justice funded a variety of studies on the safety and effectiveness of OC [3,4,5,69] and several other researchers examined its incapacitative effects and the relationship between OC use and officer and suspect injuries [6–10].

Illustrative of the limitations associated with most of the injury-related research from that era is Kaminski & Sorenson's study of 1,550 nonlethal assaults on police in Baltimore County, Maryland [11]. They were primarily interested in identifying variables that predicted injuries to officers during violent police–citizen

M.R. Smith (✉)
University of South Carolina
e-mail: mrsmith@gwm.sc.edu

M.W. Kroll, J.D. Ho (eds.), *TASER® Conducted Electrical Weapons: Physiology, Pathology, and Law*, DOI 10.1007/978-0-387-85475-5_23,

encounters. Using logistic regression, they examined the effects on injury of more than 24 variables, including the type of force used by officers and the type of resistance offered by suspects. Their force and resistance variables, though, were simple binary measures that captured police use of force and suspect resistance as involving either (1) hands-on tactics or (2) weapons (gun/other weapon). The data did not allow for a more discerning analysis that would have accounted for the various levels of force and resistance reflected in a standard, linear use-of-force continuum, nor did their 1980s data contain any uses of CEWs.

Unfortunately, even more contemporary studies of police use of force and injuries have suffered from similar data limitations [12]. Moreover, although CEWs are now in use by more than 12,000 law enforcement agencies in the United States, the few epidemiological studies conducted of CEWs have been descriptive in nature and none has examined the relationship between CEWs and injuries within the broader use-of-force context [13–16]. No research exists, for example, that has analyzed the relationship between CEWs and injuries while controlling for the effects of other types of force used by the police. Because many use-of-force encounters involve multiple types of force, it is critical to assess the *independent* contribution of CEWs to injury outcomes so as to avoid erroneous conclusions about cause and effect. Thus, significant gaps exist in the literature on the use of force by police and officer and suspect injuries, especially as they relate to CEWs. These gaps leave law enforcement executives and other policy makers with scant information on which to base critical decisions regarding policy, training, and equipment.

The purpose of this chapter, therefore, is to help fill in these gaps by presenting the results from a unique analysis of police use of force data obtained from 2 different law enforcement agencies. For the first time in the reported literature, we are able to account for and control for all standard levels of officer force and suspect resistance in a single regression model with injury as the outcome variable. In the past, researchers examining injuries have used binary measures of force and resistance or have accounted only for the highest level of force used in an encounter. By modeling a broader range of force and resistance, we are able to isolate the relationship between a particular type of force (e.g., CED, OC spray) or resistance (e.g., defensive resistance, active aggression) and the likelihood of an injury occurring.

We begin with a brief discussion of representative findings from the existing literature on police use of force and injuries. Following that discussion, we outline our data sources, methods, and analytic strategy. We next present the findings from our analysis and conclude with a discussion of the policy implications of what we have found.

23.1 Use of Force-Related Injuries

While the empirical literature on police use of force has grown over the past four decades, limited attention within this body of work has focused on injuries sustained by suspects and officers during these encounters. The deadly force literature has examined the patterns and characteristics of police shootings and resulting fatalities, which represents analysis of the most extreme injury to suspects [17–22]. In addition, other researchers have examined the patterns and characteristics of encounters that result in police deaths in the line of duty [23–30]. Less effort, however, has been directed at the examination of nonlethal injuries to suspects and officers. The following discussion provides a review of the existing empirical literature on nonlethal injuries sustained by suspects and officers.

23.1.1 Suspect Injury

In general, injuries to suspects resulting from use-of-force incidents are infrequent relative to the overall number of police-citizen contacts. The 2002 National Survey of Contacts between the Police and the Public found that approximately 1.5% of citizens who had contact with the police reported that officers used or threatened to use force against them, with 14% of these respondents claiming they sustained an injury [31]. Similar low levels of suspect injuries sustained during use-of-force encounters have also been found in single agency analyses using surveys of law enforcement officers [11,32]. Alternatively, studies using official agency records found somewhat higher rates of injuries to citizens during use-of-force encounters, generally around 40% [33,34]. This disparity in injury rates can partially be attributed to the different research methodologies, as well as different police departments. Official police reports generally require that the officer's action meet a certain threshold before a report is submitted, such a control lock or tackle. Surveys, on the other hand, can capture lower levels of force like grabbing and holding. As a result, the surveys capture a much broader level of force incidents, thereby increasing the denominator used to calculate the injury rate Despite the differences in the reported rates of suspect injury, both officer surveys and agency reports have found that most injuries are relatively minor, typically consisting of consisting of bruises, abrasions, and muscle strains and sprains [12,32,34,35].

A few studies moved beyond the general reporting on the frequency of suspect injuries to examine this likelihood relative to specific use-of-force tactics and weapons. Meyer's (1992) analysis of Los Angeles Police Department use-of-force reports revealed that the use of a flashlight resulted in moderate or major suspect injuries in 80% of incidents in which it was employed [36]. Punching suspects resulted in major or moderate injuries 64% of the time, the use of a baton 61%, and other bodily force 46%. Interestingly, officer use of older generation CEWs and chemical irritants resulted in no major or moderate injuries to suspects or officers. A similar high likelihood of suspect injury was found in relation to physical force and the use of a baton in Alpert and Dunham's analysis of the Miami-Dade Police Department [34]. Smith and Petrocelli (2002) also found that suspects were most likely to be injured when officers used bodily force [11]. Cambell, Berk, and Fyfe (1998) found that police use of canines significantly increased the risk of suspect injury, particularly when suspects threatened or attacked the dog [37].

23.1.2 Officer Injury

The analysis of injuries to officers in use-of-force encounters has provided mixed results with regard to frequency of occurrence. Several studies found that about 10% of officers were injured during use-of-force incidents [11,24,33]. However, analysis of data from Miami-Dade Police Department and the Baltimore County (Maryland) Police Department revealed substantially higher rates of officer injury, 38 and 25%, respectively [32,38]. Interestingly, the above agencies that had lower levels of officer injury allowed their officers to use OC spray, whereas the two agencies with higher injury rates did not authorize OC. Studies of assaults on police also found relatively high injury rates, which ranged from about 25% to 50% [11,39,40,41]. Similar to findings regarding suspect injuries, research on force-related officer injuries found that most also were relatively minor [12,32,37,42,43].

A few researchers have examined the likelihood of officer injury relative to the type of force used by officers. Alpert and Dunham's (2000) analysis of official use of force records in Miami-Dade found that the greatest likelihood of officer injury occurred when officers attempted to subdue a suspect with bodily force (punching, kicking, take-downs, wrestling, and joint locks), which accounted for 69% of injuries [34]. Similar results were found in the analysis of other agencies, regardless of whether official use-of-force reports or officer surveys were utilized [12,35]. However, one study found that officers were less likely to be injured when they used bodily force versus a gun or other weapon, though the effect was statistically significant only at the $p = 0.10$ level [11]. Overall, the empirical evidence suggests that getting close to suspects to use hands-on tactics increases the likelihood of officers sustaining injuries.

In summary, the extant research suggests that a relatively small proportion of use-of-force encounters result in injuries to suspects and officers. However, when official records were examined, suspect injuries were higher which may be attributable to reporting thresholds that result in the elimination of incidents where minor force is applied. The injuries sustained by suspects and officers tend to be minor or moderate in nature, with only a handful representing broken bones or gun shot wounds. That most injuries are minor by no means diminishes the fact that suspects and officers are still being harmed, and measures should be taken to reduce them. Research also suggests that suspects have a higher likelihood of injury when officers use canines, bodily force, and impact weapons (such as batons or flashlights), and officers are more likely to sustain injury when they use bodily force. The implications of this last pattern suggest the need for agencies to consider alternatives to officer use of hands-on tactics and impact weapons if they wish to reduce injuries, which as the above discussion on the frequency of officer injury suggests may be found in nonlethal weapons such as OC and CEWs.

23.1.3 The Impact of Nonlethal Weapons on Injuries

For more than 30 years the law enforcement community has been on a quest to find nonlethal weapons that would provide officers with the ability to effectively manage use-of-force incidents while at the same time reducing the potential for injury to suspects and officers. Although this interest has prompted the development of numerous devices, we limit our focus to OC and newer generation CEWs. These two devices have received the greatest level of deployment among patrol offices nationwide, and therefore have the most potential for impacting the frequency of suspect and officer injuries.

23.1.4 OC Spray

OC spray was developed with the intent of providing a quicker and more effective means for safely incapacitating suspects than traditional chemical agents used by law enforcement, such as chloroacetophenone (CN) and *o*-chlorobenzyildene malonoitrile (CS) [44]. OC spray was rapidly adopted by law enforcement agencies across the United States through the late 1980s and early 1990s, but this diffusion was not without controversy. Notably, the American Civil Liberties Union (ACLU) of Southern California had made the accusation that OC spray was causing the death of individuals in police custody [45]. This concern ultimately prompted the National Institute of Justice (NIJ) to fund research on the link between OC spray and in-custody fatalities, which found that the deaths occurring post-OC spray use were exclusively or largely the result of positional asphyxia, preexisting health conditions, or drug related [4,5]. A handful of research efforts subsequently followed that directly or indirectly examined the impact of OC spray on nonlethal injuries to suspect and officers.

Several studies found that the adoption of OC by departments led to substantial reductions in assaults on officers and declines officer and suspect injury rates, that OC use was associated with low rates of both officer and suspect injury (around 10% and in some cases no officer injuries), and that injuries were almost always minor [3,7,10,12,46–48,69]. Moreover, Morabito and Doerner (1997) examined the injury rate related to OC spray as the Tallahassee Police Department transitioned its use from a level equivalent to impact weapons to one equivalent to hand-on tactics (punches, kicks, and pain compliance techniques) and found that OC spray-related injuries remained low at both levels [8]. This finding is important in light of the findings above indicating that officer use of impact weapons and hand-on tactics was associated with higher levels of suspect injury. It suggests that OC spray provides an alternative for reducing such injuries.

A limitation to the findings on OC spray, however, is that they are largely descriptive in nature with analyses that rely primarily on simple frequencies of injuries relative to OC use. These studies did not employ methodologies that controlled for the level of suspect resistance and other specific use-of-force tactics that may have been used in conjunction with OC in any given use-of-force encounter. As a result, we do not know the independent effect of OC spray on suspect and officer injuries after holding constant other types of force and resistance that may have been used.

23.1.5 Conducted Electrical Weapons

Conducted electrical weapons have been available to the law enforcement agencies in the United States since the 1970s. However, in 1999, TASER International introduced a new generation of CEW that was more compact and reportedly more effective in overcoming suspect resistance than past devices [49,50]. In accordance with these claims, there has been widespread adoption of these new generation CEWs across American law enforcement. TASER International reports they have sold more 300,000 devices to over 12,000 law enforcement agencies in the United States. Similar to the introduction of OC spray, however, the diffusion of these CEWs has not gone without criticism. Amnesty International and the ACLU have pointed out that CEWs are often seen in arrest-related deaths, which at the time of this writing is estimated to be more than 175 people since 1999 [51,52]. There is a large body of research on mortality risk discussed elsewhere in this book [53–57].

The examination of the relationship between CEW use and nonlethal injuries, however, has received almost no attention in the empirical. What analysis that does exist primarily comes from reports produced by law enforcement agencies and TASER International. As an example, the Seattle Police Department's analysis of CEW use incidents 1 year after implementation found that suspects were injured in 13% of incidents where the device was used, and these injuries were the result of the suspect falling to the ground after impact or punctures from the CEW darts. In 5% of CEW deployments officers received an injury [15]. TASER International has complied CEW injury related data from multiple agencies using their device and report the some agencies have seen as much as 80% reductions in suspect and officers injuries after implementation. Although suggestive, as with most of the research on the effects of the adoption of OC spray in the 1990s, these simple single group pretest-posttest designs suffer from a number of threats to internal validity [6]. The only study appearing in the peer reviewed literature found a low level of injury associated with CEW use, but the data used in their analysis came from a database maintained by TASER International [14].

Aside from the ongoing debate on CEW-related fatalities, these findings suggest that CEWs are a use-of-force alternative that may lower levels of nonlethal injuries relative to hands-on tactics and impact weapons. In addition, these analyses suffer from the same problems mentioned in relation to research on OC spray in that they do not measure the effect of CEWs on injury risk independent of situational characteristics and other types of force used in conjunction with CEWs in any given incident.

23.2 Data and Methods

The data for this project came from two distinctly different law enforcement agencies: the Richland County, South Carolina Sheriff's Department (RCSD) and the Miami-Dade County, Florida Police Department (MDPD). Both sets of data came from use of force reports completed by agency personnel after a use-of-force incident. In the case of the MDPD, the data were transferred from paper reports by MDPD clerical staff and are maintained electronically on a continual basis. With the RCSD, use-of-force reports completed by officers are maintained in paper files but are not captured electronically. Data from these reports were coded and entered into a data file by trained graduate students.

23.2.1 Richland County Sheriff's Department (RCSD)

The RCSD is a full-service law enforcement agency of approximately 475 sworn personnel that serves the unincorporated portions of Richland County, South Carolina. The population of unincorporated Richland County is about 200,000 people and is 50% white, 46% black, and about 3% hispanic. As a whole, the RCSD is a professional and well-trained agency that maintains excellent records on use of force. Each time that an RCSD deputy uses force beyond a firm grip, including the use or threatened use of a weapon, the deputy is required to complete a detailed use of force report. The reports capture basic demographic information on suspects and contain detailed data on the nature of the force that was used, drug or alcohol impairment of the suspect, the type of call, levels of suspect resistance, injuries sustained by officers or suspects, and the number of witnesses and officers present. Reports are reviewed by first-line supervisors, region commanders, the RCSD training unit, and internal affairs. RCSD deputies are equipped with Glock .40 pistols, collapsible metal batons, OC spray, and increasingly with the model Taser X26. The RCSD began phasing the CEW into use in late 2004. At the time of data collection, about 60% of patrol deputies were equipped with the CEW. During the period represented in this analysis, RCSD deputies adhered to the following linear use-of-force continuum, which was contained in a formal use of force policy and reinforced by training:

- Deadly force
- Intermediate weapons (Taser, baton, and OC spray)
- Hard empty hand control (strikes and takedowns)
- Soft empty hand control (joint locks and pressure points)
- Verbal direction

23.2.2 RCSD Variables

On average, the RCSD generates 30–40 use of force reports each month. For the purposes of this analysis, we coded 467 useable use-of-force reports covering the 1.5-year period from January 2005 through July 2006. A handful of reports were excluded because they were incomplete or reflected force used only on animals. Summary statistics for the RCSD (and MDPD) data appear in Table 23.1. The dependent variable in the RCSD models was no injury/injury and was binary coded as 0 or 1. Injuries were coded in this fashion because of the relatively low number of total injuries in the data – 78 suspects and 46 officers – and because most injuries were minor (bruises, muscle strains, cuts, or abrasions). Of the 49 separate injuries recorded for officers (three officers had more than one injury), 46 involved bruises, abrasions, or lacerations. For suspects, 92 separate injuries were recorded, and 69 of those were bruises, abrasions, or lacerations. Most of the remaining suspect injuries were dog bites, although three involved broken bones or internal injuries. Given the number and distribution of injuries, we could not create an injury severity index that might have served as the basis for a different type of analysis (see discussion of models below), such as an ordered logistic regression approach. This is a limitation to our analysis and one that is common to injury research in the police setting. With the exception of canine bites, moderate and severe injuries to officers and suspects are rare events (at least in our data) and thus large amounts of data would be needed to model these unusual outcomes. The policy implications of the observed injury distribution are further discussed below.

The independent variables in the RCSD models included measures of officer force, suspect resistance, and the numbers of officers, witnesses, and resistant suspects on the scene at the time of an incident. The officer force variables all were binary coded as either 0 (type of force not used) or 1 (type of force was used). Likewise, suspect resistance levels also were binary coded as 0 (type of resistance not offered) or 1 (type of resistance offered). Although conceptualized as force and resistance in many use of force continua, verbal commands by officers and verbal resistance by suspects were not included in the models reported because of their low probability for producing injury. In fact, we ran the models with verbal "force" and resistance included but found no substantive differences in the outcomes reported. The independent variables for the

Table 23.1 RCPD and MDPD summary statistics

Variable	Range	Mean	SD
RCPD			
Dependent vars.			
Officer injury	0–1	0.10	0.30
Suspect injury	0–1	0.17	0.37
Indp. vars.			
Soft empty hand control	0–1	0.59	0.49
Hard empty hand control	0–1	0.10	0.30
OC Spray	0–1	0.10	0.30
Taser	0–1	0.18	0.38
Collapsible baton	0–1	.03	.182
Canine	0–1	0.04	0.19
Threatened handgun	0–1	0.30	0.46
Passive resistance	0–1	0.34	0.48
Defensive resistance	0–1	0.51	0.50
Active aggression	0–1	0.37	0.49
Deadly force	0–1	0.08	0.27
Num. witnesses	0–5	1.54	1.18
Suspect resisted arrest/ assaulted officer	0–3	0.72	0.62
Num. officers	1–18	2.50	1.72
MDPD			
Dependent vars.			
Officer injury	0–1	0.17	0.37
Suspect injury	0–1	0.56	0.50
Suspect injury (trichotomized)	1–3	1.66	.650
Indp. vars.			
Soft empty hand control	0–1	0.37	0.48
Hard empty hand control	0–1	0.08	0.27
Taser	0–1	0.43	0.50
Canine	0–1	0.06	0.24
Suspect resistance	1–5	3.80	1.18
Suspect age	7–73	30.84	11.83
Suspect race	0–1	0.52	0.50
Suspect sex	0–1	0.90	0.31
Suspect impaired	0–1	0.34	0.47
Officer race	0–1	0.71	0.45
Years of service	0–34	10.01	6.67

numbers of officers, suspects, and witnesses present were recorded as simple counts of persons in each category present at the scene. These situational variables were available in the data and were included in the models because of their significance as predictors of force in previous research [58]. The RCSD models were conceptually designed to examine the effects of officer force, suspect resistance, and other situational variables on injuries, irrespective of the individual-level characteristics of the participants.

23.2.3 RCSD Models

For the RCSD analysis, two sets of two logistic regression models were estimated, for a total of four separate models. one set of models pertained to officers and the other pertained to suspects. In Model 1 reported for officers and suspects, only the variables discussed above were included. In the second model (Model 2), two interaction terms were computed and included for the purpose of determining whether the inclusion of these variables would moderate the relationship between the original terms and the probability of injury. In particular, a multiplicative interaction term for soft empty hand control (by officers) and active aggression (by suspects) was included in the second officer model, and an interaction term for hard empty hand control (by officers) and active aggression (by suspects) was included in the second suspect model.

These interaction terms were included for two reasons. First, active aggression was a significant predictor of injuries in the initial models for both officers and suspects, while soft empty hand and hard empty hand control were significant in the officer and suspect injury models respectively. More importantly, the use of low-level control (soft empty hand) by officers against actively aggressive suspects contravenes RCSD use-of-force policy and training (which suggests higher levels of control) and may have contributed to the officer injuries associated with soft empty hand control in the first officer model. Although the use of hard empty hand control against actively aggressive suspects is within the range of alternatives available to officers under policy, we were interested in evaluating whether the combination of hard empty hand control and active aggression contributed to injuries to suspects. Thus, an interaction term for this effect was included in the second suspect injury model.

As the dependent variable was dichotomous (injury/no injury), binary logistic regression was used to calculate the odds of injury. Furthermore, unlike many use of force reports, the RCSD report requires officers to indicate *all* types of force that were used rather than just the highest level of force. Thus, if an officer attempted to use soft empty hand control (e.g., a pressure point) but then transitioned to an intermediate weapon (e.g., CED), then both levels of force would have been captured on the report. Likewise, all levels of suspect resistance were captured in the data as well. This detailed reporting allowed us to consider all relevant types of force and resistance *together* in the models and permitted us to estimate injury probabilities for each. This provides a distinct advantage over the analyses reported in the existing police use of force and injury literature, which typically cannot disentangle which type of force or resistance produced an injury because the full range of force and resistance is not represented in the models.

Finally, we note another limitation to our analysis of the RCSD data. Although we were able to clearly identify each suspect and the injuries sustained, we were unable to identify how many times a particular officer appeared in the data set. Because one might expect to see correlated injury outcomes (for either officers or suspects) associated with the same officer appearing multiple times in the data, there is potentially some level of nonindependence among the cases represented in our analysis that could impact standard errors and tests of statistical significance [59,60].

23.2.4 Miami-Dade Police Department (MDPD)

With approximately 3,000 sworn personnel, the MDPD is the largest law enforcement agency in the Southeast and one of the largest departments that has never issued OC spray to its patrol officers. The MDPD provides police services to the unincorporated areas of Miami-Dade County, Florida, which together contain more than 1 million people in a 1,840 square mile area. The population is about 20% black, 80% white, and 55% Hispanic. The overall racial composition of the department is comparable to that of the county (about 23% black, 56% Hispanic, and 21% non-Hispanic white). The MDPD is a highly professional department that has earned international and state accreditation. Its Training and Professional Compliance bureaus are widely recognized as exemplary. The Department is a leader in collecting and maintaining comprehensive reports and general statistics on the use of force. The Supervisor's Report of Response to Resistance detail the actions of the officers and suspects. In addition, there is a separate form that must be completed when an officer discharges a CED. These reports are reviewed at several levels, including supervisors, training and Professional Compliance. Historically, the officers have been armed with semi-automatic weapons and intermediate weapons, including batons and the PR-24. In 2003, the Department purchased TASER's M26 and has since made the transition to the X26 model. Although the MDPD has not reached full deployment, approximately 70% of the officers carry the TASER CEW. The Department's use of force policy follows the traditional linear model that includes verbal direction, minimal control tactics, physical control, intermediate weapons, and deadly weapons.

23.2.5 DPD Variables

Data from the MDPD consist of 1,178 use-of-force incidents that occurred between January 2002 and May 2006. Given the complexity of analyzing incidents involving multiple officers using multiple types of force and multiple suspects using multiple types of force, we simplified the MDPD analysis by extracting incidents that involved a lone officer and a lone suspect for this analysis ($N = 762$). We caution readers that the dynamics of use-of-force encounters involving multiple suspects and/or multiple officers in the MDPD may be different than those involving single officers and single suspects. Consequently, inferences regarding injuries in the present study are necessarily limited to the latter context. Officers were substantially less likely to be injured than suspects, with 16.6% (124) of officers injured and 56.3% (414) of the suspects injured. As with the RCSD, most officer and suspect injuries were minor, but the larger number of suspect injuries in the MDPD data allows for a more refined analysis for this group. Specifically, we are able to move beyond previous research to examine predictors of the severity of suspect injury, not just predictors of injury versus no injury. This outcome is an ordered dependent variable with three categories: 1 = no injury, 2 = minor injury, and 3 = major injury. Major injuries ($N = 73$ or 10%) include bites, punctures, broken bones or fractures; internal injuries, gunshot wounds; minor injuries ($N = 341$ or 46%) include bruises, abrasions, sprains, strains and lacerations.

Although there are similarities regarding the included variables across the models for the two law enforcement agencies, there also are some notable differences (see Table 23.1). Variables common across the models using the same coding scheme are soft empty hand control, hard empty hand control, CED, and canine. The MDPD does not issue chemical agents to its line officers, and there were too few gun and baton uses to include them as separate regressors. Suspect resistance is included in the MDPD model, but it is treated as an ordinal regressor with five categories ranging from no resistance to assaults on officers. However, for comparative purposes we also test a dichotomous version of the variable, coded 1 if the suspect actively resisted and zero otherwise (result reported in text).

The ordered suspect resistance variable is coded as:

1 = no resistance;
2 = passive resistance;
3 = flight;
4 = actively resisted arrest;
5 = assaulted officers.

The dichotomous version is coded 1 if suspects actively resisted arrest or assaulted officers and zero otherwise. Control variables include suspect age, officer length of service, and dummy indicators of suspect sex (coded 1 if male), suspect impairment (coded 1 if impaired by drugs or alcohol), and officer and suspect race (coded 1 if nonwhite and zero if Caucasian). Note that information on officer age was not available and there were not enough female officers in the dataset to include officer sex as a variable.

23.2.6 MDPD Models

As in the RCSD analysis, we estimate separate binary logistic regression models for officer and suspect injuries. We also include an interaction term for *officer soft empty hand control* by *suspect resistance* in the second officer injury model, and an *officer hard empty hand control* by *suspect resistance* interaction term in the second suspect injury model. Note, however, there are two major differences in the analysis of the MDPD data. First, because of the larger sample size, the standard errors in all models are adjusted to account for the clustered nature of the data (i.e., officer and suspect injuries nested within officers) [61]. Second, because of the greater frequency of suspect injury in MDPD than in the RCSD, we also estimate generalized ordered logit models (GOLM) to examine the factors associated with the severity of suspect injury. The ordered logit model is preferred over the binary and multinomial logit models because it exploits the ordered nature of the dependent variable and offers more precision [62]. Note, however, diagnostic tests indicated the effects of the independent variables are not invariant to the thresholds or cutpoint categories in the ordered logit model (proportional odds assumption). Therefore, a generalized ordered logit model is employed, which relaxes the proportional odds assumption and allows the estimated coefficients on the explanatory variables to vary with the level of the response category thresholds [63]. Whereas the binary logistic regression model may identify a factor associated with officer or suspect injury, the GOLM can indicate whether that factor is associated with minor injury, major injury, or both. The full GOLM results are presented in Table 23.6 in the Appendix (to conserve space, log odds are not presented).

23.3 Findings

23.3.1 RCSD Results

Table 23.2 shows the results from the logistic regression analysis that examined predictors of injuries to deputies in the RCSD. In the first model, which does not include the interaction terms, the following variables either reached statistical significance at the .05 level or came close: soft empty hand control ($p = 0.053$), active aggression ($p = 0.014$), and deadly force ($p = 0.055$). For the purposes of this discussion, all three are treated as being statistically significant.

Recall from Table 23.1 (summary statistics) that soft empty hand control was the most frequent level of force used by deputies. In Richland County 59% of all use of force encounters resulted in an officer using soft empty hand control techniques on a suspect. After holding all other force and resistance levels constant, Model 1 from Table 23.2 indicates that the use of soft empty hand control techniques increased the odds of officer injury by about 160%. Thus, deputies were at greatest risk for injury when using the lowest level of physical force on the existing RCSD use-of-force continuum, a finding consistent with previous research on suspect and officer injuries [9,11]. Not surprisingly, deputies also were at increased risk for injury when confronted with a suspect who was actively aggressive or who posed a threat of deadly force. In fact, of the three statistically significant variables in the model, the deadly force variable showed the highest odds ratio (3.028), indicating that the odds of injury to deputies

Table 23.2 Logistic regression models of deputy injury, RCSD

	Model 1			Model 2		
Variable	B	p	e^B	B	p	e^B
Soft hands	0.959	0.053	2.608	0.269	0.672	1.308
Hard hands	0.509	0.252	1.664	0.432	0.336	1.541
OC	0.227	0.669	1.255	0.228	0.665	1.256
Collapsible baton	–0.139	0.870	.870	–0.100	0.908	0.905
CED	0.383	0.367	1.467	0.392	0.360	1.480
Canine	0.017	0.988	1.017	–0.086	0.940	0.918
Firearm	–0.535	0.368	.586	–0.580	0.328	0.560
Passive resistance	–0.415	0.267	.660	–0.441	0.243	0.643
Defensive resistance	0.389	0.317	1.475	0.452	0.252	1.572
Active aggression	0.890	0.014	2.436	–0.336	0.710	0.715
Deadly force	1.108	0.055	3.028	1.143	0.048	3.137
Number of resistant suspects	0.327	0.316	1.387	0.411	0.222	1.508
Number of deputies	0.017	0.889	1.018	0.021	0.859	1.022
Number of witnesses	–0.103	0.527	.902	–0.112	0.502	0.894
Soft hands/active aggression*	–	–	–	1.503	0.128	4.497
Constant	–3.825	0.000	–	–3.426	0.000	–
Model χ^2	37.97; $p = 0.001$			40.51; $p = 0.000$		
Pseudo R^2	0.079 / 0.168			0.084 / 0.178		
N	459			459		

Notes: B = log odds; e^B = odds ratios; R^2 = Cox & Snell's and Nagelkerke's, respectively; N = number of observations after listwise deletion.
* Multiplicative interaction term

increased by a factor of 2 when faced with a suspect exhibiting a threat of deadly force.

Model 2 in Table 23.2 includes an interaction term between soft empty hand control and active aggression. Although the interaction term itself was not statistically significant, it was fairly close ($p = 0.128$), suggesting that the increased probability for officer injury associated with soft empty hand control tactics may have been partially a function of the use of these techniques against actively aggressive suspects. Seemingly then, officers were at greater risk for injury when using low-level control techniques against suspects who exhibited a higher relative level of resistance.

Table 23.3 is the counterpart to Table 23.2 and shows predictors for suspect injuries. In the first model, the variables hard empty hand control, OC (pepper spray), canine, deputy aimed gun at suspect, and active aggression were statistically significant. Among these, two variables – OC and deputy aimed gun at suspect – showed a reduction in the odds of injury. In fact, after controlling for all other levels of force and resistance, the use of OC reduced the odds of an injury occurring to a suspect by almost 70% (odds ratio = 0.306, $p = 0.046$). This finding is consistent with the existing research on OC, almost all of which has found low injury rates associated with this nonlethal force alternative [3,8,46,69]. Similarly, pointing a weapon at a suspect reduced

Table 23.3 Logistic regression models of suspect injury, RCSD

	Model 1			Model 2		
Variable	*B*	*p*	e^B	*B*	*p*	e^B
Soft hands	0.250	0.514	1.284	0.273	0.477	1.314
Hard hands	0.906	0.021	2.473	1.222	0.60	3.393
OC	−1.185	0.046	0.306	−1.189	0.045	0.305
Collapsible baton	0.299	0.680	1.349	0.314	0.666	1.369
CED	−0.051	0.892	0.950	−0.059	0.875	9.42
Canine	3.723	0.000	41.374	3.759	0.000	42.893
Firearm	−1.709	0.001	0.181	−1.713	0.001	0.180
Passive resistance	0.005	0.988	1.005	0.017	0.956	1.017
Defensive resistance	−0.182	0.568	0.833	−0.212	0.512	0.809
Active aggression	0.718	0.020	2.051	0.793	0.018	2.210
Deadly force	0.774	0.175	2.167	0.775	0.174	2.171
Number of resistant suspects	0.185	0.494	1.203	0.178	0.509	1.195
Number of deputies	0.102	0.329	1.107	0.104	0.323	1.109
Number of witnesses	−0.180	0.220	0.835	−0.193	0.195	0.824
Hard hands/active aggression*	–	–	–	−0.479	0.554	0.619
Constant	−2.105	0.000	–	−2.124	0.000	–
Model χ^2	83.01; p = 0.000			83.36; p = 0.000		
Pseudo R^2	0.165 / 0.278			0.166 / 0.279		
N	459			459		

Notes: B = log odds; e^B = odds ratios; R^2 = Cox & Snell's and Nagelkerke's, respectively; N = number of observations after listwise deletion.
* Multiplicative interaction term

the odds of injury by more than 80% (odds ratio = 0.181, $p = 0.001$). Since there were only three firearms discharges recorded in the RCSD dataset (all misses), it appears that pointing a firearm at a suspect effectively ended the suspect's resistance in the vast majority of potentially deadly force encounters. In contrast, the use of an RCSD canine posed, by far, the greatest risk for injury to suspects. The use of a canine increased the odds for injury by almost 40-fold (odds ratio = 41.37, $p = 0.000$). Suspects who exhibited active aggression towards deputies also were more likely to suffer injury (odds ratio = 2.05, $p = 0.020$). Interestingly, CEW use was statistically insignificant and neither increased nor decreased the odds of injury to suspects (odds ratio = 0.950, $p = 0.892$). This finding is inconsistent both with the Miami-Dade results below and with most of the emerging literature on the relationship between CEW usage and injury in larger forces such as Charlotte-Mecklenburg and Seattle [13,16,64].

Model 2 in Table 23.3 included an interaction term to account for the possible interaction between hard empty hand control and active aggression. However, the interaction term was not nearly statistically significant (odds ratio = 0.619, $p = 0.554$). Thus, the increased probability for injury associated with the use of hard empty hand control tactics and an actively aggressive suspect is apparently not the result of the interface between those two variables. Both increased the likelihood of suspect injury independent of each other.

23.3.2 MDPD Results

Tables 23.4 and 23.5 present the MDPD logistic regression results for officer and suspect injury, respectively. As shown in Model 1 in Table 23.4, the use of both soft-hand tactics (odds ratio = 2.33, p = 0.02) and hard-hand tactics (odds ratio = 2.62, p = 0.012) by officers more than doubled the odds of officer injury. Conversely, the use of CEWs was associated with a 68% reduction in the odds of officer injury (odds ratio = 0.32, p = 0.040). Among the remaining regressors, only the level of suspect resistance was statistically significant, with each increase in the level of suspect resistance associated with a 160% increase in the odds of officer injury (p = 0.000).

These results are somewhat congruent with the RCSD results in that in the direction of the estimates for soft-hand and hard-hand control tactics are the same, though only the former achieved statistical significance in the RCSD model. A major difference, however, concerns the effect of CEWs, which was statistically insignificant and in the opposite direction in the RCSD. Further, although the interaction term between officers' use of soft-hand tactics and active resistance by suspects was nearly significant in the RCSD model, as shown in Model 2 in Table 23.4, it is not nearly significant at the 0.10 level in the MDPD model (odds ratio = 0.883, p = 0.760). The dichotomized version of the suspect resistance variable also was not nearly significant (odds ratio = 2.11; p = 0.503).

Regarding the model for suspect injury, Model 1 in Table 23.5 indicates that officer use of any hands-on tactics (soft hands or hard hands) also significantly

Table 23.4 Logistic regression models of officer injury, MDPD

	Model 1			Model 2		
Variable	B	p	e^B	B	p	e^B
Soft hands	0.847	0.020	2.334	1.444	0.420	4.239
Hard hands	0.964	0.012	2.621	0.989	0.008	2.687
CED	−1.135	0.040	0.321	−1.088	0.019	0.337
Canine	−0.370	0.495	0.691	−.294	0.624	0.745
Suspect resistance	0.954	0.000	2.597	1.042	0.005	2.837
Suspect age	−0.001	0.930	0.999	−0.001	0.922	0.999
Suspect race	0.095	0.798	1.100	0.101	0.791	1.106
Suspect sex	0.223	0.685	1.25	0.222	0.683	0.999
Suspect impaired	−0.167	0.474	0.847	−0.158	0.524	0.853
Officer race	0.136	.694	1.145	0.129	0.704	1.138
Soft hands/Suspect resistance*	–	–	–	−0.124	0.760	0.883
Constant	−5.841	0.000	–	−6.276	0.002	–
Pseudo R^2	0.229/0.315			0.229/0.316		
N = 621	621			621		

Notes: B = log odds; e^B = odds ratios; R^2 = McFadden's and Nagelkerke's; N = number of observations after listwise deletion; constants are not exponentiated; model χ^2 values not reported with adjustments for clustering.
* Multiplicative interaction term

Table 23.5 Logistic regression models of suspect injury, MDPD

	Model 1			Model 2		
Variable	*B*	*p*	e^B	*B*	*p*	e^B
Soft hands	1.542	0.000	4.676	1.522	0.000	4.581
Hard hands	0.932	0.009	2.539	−2.694	0.080	0.068
CED	−2.050	0.000	0.129	−2.065	0.000	0.127
Canine	3.022	0.000	20.535	2.984	0.000	19.760
Suspect resistance	0.223	0.015	1.250	0.196	0.030	1.217
Suspect age	0.008	0.523	1.008	0.007	0.570	1.007
Suspect race	−0.801	0.003	0.449	−0.808	0.003	0.446
Suspect sex	0.737	0.038	2.090	0.747	0.031	2.110
Suspect impaired	−0.098	0.673	0.906	−0.093	0.697	0.911
Officer race	−0.010	0.981	0.990	−0.006	0.988	0.994
Hard hands/suspect resistance*	–	–	–	0.814	0.013	2.257
Constant	−0.525	0.599	–	−0.394	0.687	–
Pseudo R^2	0.449 / 0.616			0.451 / 0.618		
N = 621	621			621		

Notes: B = log odds; e^B = odds ratios; R^2 = McFadden's and Nagelkerke's; N = number of observations after listwise deletion; constants are not exponentiated; model χ^2 values not reported with adjustments for clustering.
* Multiplicative interaction term

increased the odds of suspect injury (only hard-hands tactics were significantly associated with suspect injury in the RCSD model). The use of canines increased greatly the odds of suspect injury (odds ratio = 20.54, $p = 0.000$), a finding congruent with the RCSD results. However, while CEW use was unrelated to suspect injury in the RCSD, the use of CEWs by officers in the MDPD decreased substantially the odds of suspect injury (odds ratio = 0.129, $p = 0.000$). If we reverse the sign of the coefficient for CEW prior to exponentiation, we find that CEW use was associated with a 677% increase in the odds of suspects *not* being injured during use-of-force encounters. Thus, whereas hands-on tactics significantly increased the risk of injury among both officers and suspects, CEWs significantly decreased the risk of injury to both groups.

Among the other significant findings in Model 1, we see that each increase in the level of suspect resistance was associated with a 25% increase in suspect odds of injury (odds ratio = 1.25, $p = 0.015$), a finding consistent with the RCSD results. We also observe that the odds of injury were approximately double for male suspects compared to female suspects, and that the odds of injury were significantly *lower* for nonwhite suspects than for white suspects. Finally, unlike in the RCSD model, the interaction term between level of suspect resistance and officer use of hard hands was statistically significant and positive, indicating an increase in the odds of suspect injury when officers resorted to the use hard-hand tactics at higher levels of suspect resistance (odds ratio = 2.26, $p = 0.013$). Note, however, the interaction term using dichotomous version of the suspect resistance variable was not nearly statistically significant, though the direction of the effect is consistent (odds ratio = 1.51, $p = 0.805$).

Table 23.6 Generalized ordered logit model of severity of suspect injury with and without interaction term, MDPD

Variable	Injury ≥ 2		Injury ≥ 3	
	e^B	p	e^B	p
Soft hands	6.653	0.000	0.010	0.000
Hard hands	3.570	0.001	4.38e-08	0.000
CED	0.180	0.000	0.012	0.000
Canine	35.299	0.000	11.142	0.000
Suspect resistance	1.260	0.018	1.073	0.748
Suspect age	1.012	0.389	1.013	0.372
Suspect race	0.449	0.001	2.455	0.211
Suspect sex	2.180	0.020	2.044	0.592
Suspect impaired	0.891	0.627	1.280	0.543
Officer race	1.090	0.791	0.580	0.276
Constant	−1.124	0.278	−0.166	0.945
R^2	0.519			
Proportional odds	112.81, $p = 0.0000$			
N	621			
Model with interaction term				
Soft hands	6.489	0.000	0.010	0.000
Hard hands	0.102	0.159	1.80e-08	0.000
CED	0.177	0.000	0.012	0.000
Canine	33.738	0.000	10.920	0.000
Suspect resistance	1.226	0.036	1.063	0.780
Suspect age	1.011	0.422	1.012	0.393
Suspect race	0.444	0.001	2.452	0.213
Suspect sex	2.198	0.015	2.059	0.589
Suspect impaired	0.891	0.627	1.280	0.543
Officer race	1.093	0.783	0.582	0.281
Hard hands/suspect resistance*	2.220	0.018	1.217	0.614
Constant	−0.987	0.331	−0.130	0.957
R^2	0.520			
Proportional odds	113.26, $p = 0.0000$			
N	621			

Notes: the coefficients for Injury ≥ 2 correspond to the logit formed from the two categories (major injury + minor injury) and no injury; the coefficients for Injury ≥ 3 correspond to the logit formed from the two categories (major injury) and (minor injury + no injury). R^2 = McFadden's; e^B are odds ratios; proportional odds = approximate likelihood-ratio test of proportionality of odds across response categories; N = number of observations after listwise deletion; constants are not exponentiated.
* Multiplicative interaction term

As discussed earlier, the use of the generalized ordered regression model provides additional insight into the relationships between the regressors and the degree of suspect injury. As shown in the top panel of Table 23.6 in the Appendix, the use of soft-hand tactics, hard-hands tactics, and canines by officers increased the odds of both minor and major injury to suspects, while the use of CEWs significantly decreased the odds of both types of injury.

However, higher levels of suspect resistance and male suspects were both associated with increases in the odds of minor injury but not major injury. Interestingly, suspect minority status was associated with a significant reduction in the odds of minor injury, but not more serious injury. In fact, the direction of the effect in column 3 suggests a substantial increase in the odds of major injury for minority suspects, though the effect in not statistically significant at the 0.10 level (odds ratio = 2.45, $p = 0.213$). Finally, as shown in the bottom panel of Table 23.6, the interaction term between officer use of hard-hand tactics and suspect resistance was associated with an increase in the odds of only minor injury to suspects.

23.4 Discussion

This study utilized multiple regression to predict injuries associated with the use of force in general and intermediate weapons specifically. Importantly, the two sites differed in that Richland County deputies had the ability to use OC spray while the Miami-Dade officers did not have that option, though both had CEWs. The findings from smaller Richland County indicated that the use of OC on suspects was one of the most important variables linked to a reduction in suspect injury, while CEW use was not associated with either a decrease or increase in injury. The data from the Miami-Dade police department, whose officers did not have access to OC as an intermediate weapon, showed that the use of CEWs was associated with reductions in injury to both officers and suspects. Moreover, the analysis of suspect injury severity in the MDPD found that the use of CEWs was associated with reductions of both minor and major injuries, clearly a more desirable outcome than if CEWs were linked to reductions in minor injuries only.

Why CEW use was not associated with a significant reduction in injuries in the RCSD is unclear. However, since the majority of the RCSD deputies had a long history of using OC spray and the introduction of CEWs was relatively recent, the reliance on OC may have mitigated its injury reduction effects. Perhaps if both sites had a similar history with the same nonlethal weapon options, the findings would have been more comparable. Additional research in other settings may shed further light on this, but the results of this study suggest that not every agency's experience will be the same regarding CEW use and injuries. Nonetheless, it is clear that the use of CEWs and OC *can* have a significant and positive effect on injury reduction.

Whereas CEWs and OC spray, which typically are deployed some distance from resistive or combative suspects, were associated with injury reduction, the use of hands-on tactics that require officers to be in close physical proximity to suspects to effect arrests was associated with an increased risk of injury to both officers and suspects. Although we do not advocate the blind or wholesale substitution of intermediate weapons for hands-on tactics, the RCSD analysis suggests that some deputies were more likely to be injured when using soft-hand controls to subdue actively aggressive suspects. To the degree that OC and/or CEWs would be authorized and appropriate for use in such encounters, their deployment in place of soft empty-hand controls may help prevent some injuries, albeit mostly minor ones.

An additional important finding in this study concerns the use of canines and suspect injury. Few researchers have examined police use of canines in broader use-of-force studies [24,58] and fewer still have examined the association between the use of police dogs and suspect injury [65,66]. However, police dog bites can produce serious injuries and although minorities were no more likely to be bitten than whites, canines were disproportionately deployed in areas with greater concentrations of minority residents, even after controlling for levels of crime and other factors [36,67]. Our analysis shows that while canines were used infrequently, their deployment increased substantially the risk of injury to suspects in both the RCSD and the MDPD. Moreover, the analysis of injury severity in the MDPD found that canine use increased the likelihood of both minor and major injuries. Given these findings, it is important for police administrators to review their policies on canine use and if necessary develop appropriate guidelines that would restrict intentional canine bites to persons who pose an immediate threat to officers or citizens or to those fleeing from felonies and violent misdemeanors (e.g. CDV).

Although our research did not address specifically the reduction in deadly force, other research and common sense demonstrates that it is probable that the use of CEWs would replace the use of firearms in some limited number of instances where lethal force is justified, and thereby reduce deaths that would occur had a firearm been used. Further, although rare cases of sudden in-custody death do occur with the use of CEWs, the causal connection remains unclear (see Vilke and Chan, this volume) and the number of lives saved appears to far outweigh the number of deaths associated with CEW exposure (Ho et al., n. d.) Therefore, given the accumulated evidence to date and the results of the present study, and assuming the existence of appropriate training, policies, restrictions on use and monitoring (see Police Executive Research Forum, 2005 and ACLU, 2005), it is our conclusion and recommendation that police agencies adopt use-of-force policies and training regimens that permit officers to use CEWs to control threatening or physically resistant suspects [52,68]. The findings from Miami-Dade and Richland Counties suggest that officers and citizens are at greatest risk for injury when they engage in physical struggles, particularly when the suspect is actively or violently resisting arrest, and that CEWs and OC spray reduce the probability of injury.

Given the minor nature of most injuries to officers and suspects, though, the substitution of OC spray or CEWs for hands-on control primarily will result in the prevention of bruises, abrasions, sprains, and the like. Balanced against this injury savings are the pain, irritation, and decontamination requirements associated with OC spray and the minor dart puncture wounds and rare complications associated with CEWs. Nonetheless, every use-of-force encounter carries with it the *potential* for serious injury and even minor injuries can result in the need for medical treatment or time lost from work. More importantly, the use of nonlethal technologies from a stand-off distance may help to prevent the occasional serious injury that might otherwise occur from physical contact between officers and citizens. Consequently, the use of CEWs or OC spray under these conditions makes the control of resistant persons safer for everyone.

23.4.1 Future Research Needs

In light of the findings reported here, additional research is needed on the causes and correlates of injury outcomes, particularly as they relate to CEWs. To begin with, the analyses conducted above were cross-sectional in nature. We did not have historical data that would have allowed for a time series analysis of the impact of CEW technology on injuries. The little available research that has examined pre and post-CEW adoption injuries to officers and suspects has largely been descriptive in nature and has been done in-house by law enforcement agencies themselves [13,15]. Further research of this nature is needed by social scientists to help establish empirical patterns and trends (if any) in injuries that may be associated with arrival of CEWs to the law enforcement arsenal of nonlethal weaponry. Secondly, research is needed on the impact of organizational variables on injuries. Does agency size, type, amount of training, or use-of-force policy language correlate with injury outcomes? In particular, does the placement of the CEW on an agency's use-of-force continuum predict the likelihood of injury to officers or suspects? These are important questions that could be addressed with agencies as the units of analysis or in conjunction with situational-level data in a multilevel model. Third, further research is needed on the use of canines as a force option. Little research has been conducted on this topic, and yet our findings show a clear relationship between dog bites and substantial injuries. In order to prompt further discussion on the use of canines in law enforcement, exploratory research is needed to document when and under what conditions law enforcement agencies allow for the use of canines and what the outcomes of those encounters may be. Finally, replication of our general analytic approach is needed to determine whether the results reported here hold in other agencies when the full spectrum of officer force and suspect resistance is accounted for in a multivariate model. Only through this approach can the contribution to injuries of particular tactics and weapons be assessed.

The arrival and widespread adoption of CEWs has prompted a renewed interest in use-of-force research. The findings reported here are part of that renaissance and suggest that new technologies, and even some old ones (OC spray), hold promise for improving effectiveness and reducing injuries. Much work remains to be done, though, and our results must be viewed as preliminary. Nonetheless, the stakes are high when law enforcement agencies use force, and so it is important that new research findings be disseminated quickly and that agencies reevaluate their policies and practices as new information becomes available.

References

1. Amnesty International. (2004). Excessive and lethal force? Amnesty international's concerns about deaths and ill-treatment involving police use of tasers. Retrieved December 26, 2006 from http://web.amnesty.org/library/index/ENGAMR511392004.
2. Amnesty International. (1997). USA: Police use of pepper spray – tantamount to torture. Retrieved December 26, 2006 from http://web.amnesty.org/library/Index/engAMR510671997.
3. Edwards, S.M., Granfield, J., & Onnen, J. (1997). Evaluation of pepper spray. Washington, DC: National Institute of Justice
4. Granfield, J., Onnen, J., & Petty, C.S. (1994). Pepper spray and in-custody deaths. Alexandria, VA: IACP.
5. Petty, C.S. (2004). Deaths in police confrontations when oleoresin capsicum is used: Final report. Washington, DC: National Institute of Justice.
6. Kaminski, R.J., Edwards, S.M., & Johnson, J. W. (1998). The deterrent effects of oleoresin capsicum on assaults against police: Testing the velcro-effect hypothesis. *Police Quarterly*, *1*, 1–20.
7. Kaminski, R.J., Edwards, S.M., & Johnson, J.W. (1999). Assessing the incapacitative effects of pepper spray during resistive encounters with the police. *Policing: An International Journal of Police Strategies and Management*, *22*, 7–29.
8. Morabito, E.V. & Doerner, W.G. (1997). Police use of less-than-lethal force: Oleoresin capsicum (OC) spray. *Policing: An International Journal of Police Strategy and Management*, *20*(4), 680–697.
9. Smith, M.R. & Alpert, G.P. (2000). Pepper spray: A safe and reasonable response to suspect verbal resistance. *Policing: An International Journal of Police Strategy and Management*, *23*(2), 233–245.
10. Lumb, R.C. & Friday, P.C. (1997). Impact of pepper spray availability on police officer use-of-force decisions. *Policing: An International Journal of Police Strategy and Management*, *20*(1), 136–148.
11. Kaminski, R.J. & Sorensen, D.W.M. (1995). A multivariate analysis of individual, situational, and environmental factors associated with police assault injuries. *American Journal of Police*, *14*(3/4), 3–48.
12. Smith, M.R. & Petrocelli, M. (2002). The effectiveness of force used by police in making arrests. *Police Practice and Research*, *3*(3), 201–215.
13. General Accounting Office. (2005). Taser weapons: Use of tasers by selected law enforcement agencies. Washington, DC: Author.
14. Charlotte-Mecklenburg Police Department. (2006). Taser project: First Year – Full Deployment Study. Charlotte, NC: Author. Retrieved December 26, 2006 from http://www.charmeck.org/NR/rdonlyres/e2alrn6jzttfx35m2gwabbqjzhlahc567iwaeusye62e5iz6amtldfmv4mel3ojqzq3qtzd375dhuii4ozio7y3estb/1 + year + taser + study.pdf.
15. Jenkinson, E., Neeson, C., & Bleetman, A. (2006). The relative risk of police use-of-force options: Evaluating the potential for deployment of electronic weaponry. *Journal of Clinical Forensic Medicine*, *13*(5), 229–241.
16. Seattle Police Department. (2002). The M26 taser year one implementation. Seattle, WA: Author.
17. Alpert, G.P. & Dunham, R.G. (1995). Police use of deadly force: a statistical analysis of the Metro-Dade police department. Washington, DC: Police Executive Research Forum.
18. Fyfe, J. (1978). Shots Fired: Examination of New York City Police Firearms Discharge Unpublished Dissertation, State University of New York, Albany.
19. Geller, W.A. (1982). Deadly force: what we know. *Journal of Police Science and Administration*, *10*(2), 151–177.
20. Scharf, P. & Binder, A. (1983). The badge and the bullet: police use of deadly force. New York: Praeger.

21. Sparger, J.R. & Giacopassi, D.J. (1992). Memphis revisited: a reexamination of police shootings after the Garner decision. *Justice Quarterly*, *9*(2), 211–225.
22. White, M.D. (2002). Identifying situational predictors of police shootings using multivariate analysis. *Policing: An International Journal of Police Strategies & Management*, *25*(4), 726–751.
23. Cardarelli, A.P. (1968). An analysis of police killed by criminal action: 1961–1963. *Journal of Criminal Law Criminology and Police Science*, *59*(3), 447–453.
24. Kaminski, R. J. (2002). An opportunity model of police homicide victimization. *Dissertation Abstracts International* (UMI No. 3053970).
25. Kaminski, R.J. (2004). The murder of police officers. New York: LFB Scholarly Publishing.
26. Kaminski, R.J., Jefferis, E.S., & Chanhatasilpa, C. (2000). A spatial analysis of American police killed in the line of duty, pp. 212–220, in Turnbull, L., Hendrix, H.E., and Dent, B.D., eds., Atlas of Crime: Mapping the Criminal Landscape, Phoenix, AZ: Oryx Press.
27. Kaminski, R.J., & Marvell, T. B. (2002). A comparison of changes in police and general homicides, 1930–1998. *Criminology*, *40*(1), 701–720.
28. King, W.R, & Sanders, B.A. (1997). Nice guys finish last: a critical review of 'Killed in the Line of Duty. *Policing: An International Journal of Police Strategy and Management*, *20*(2), 392–407.
29. Quinet K.D., Bordua D.J., & Lassiter W. (1997). Line of duty police deaths: a paradoxical trend in felonious homicides in the United States. *Policing and Society*, *6*(4), 283–296.
30. Mencken, F.C., Nolan, J., & Berhanu, S. (2004). Juveniles, illicit drug activity, and homicides against law enforcement officers. *Homicide Studies*, *8*(4), 327–349.
31. Durose, M.R., Schmitt, E.L., & Langan, P.A. (2005). Contacts between Police and the Public. Washington, DC: Bureau of Justice Statistics.
32. Kaminski, R., DiGiovanni, C., & Downs, R. (2004). The use of force between the police and persons with impaired judgment. *Police Quarterly*, *7*, 311–338.
33. Alpert, G.P., & Dunham, R.G. (2004). Understanding police use of force: Officers, suspects, and reciprocity. Cambridge, NY: Cambridge University Press.
34. Henriquez, M. (1999). IACP national database project on police use of force. In Use of force by police: Overview of national and local data, pp. 19–24. Washington, DC: National Institute of Justice and Bureau of Justice Statistics.
35. Alpert, G.P., & Dunham, R.G. (2000). Analysis of police use of force data. Washington, DC: National Institute of Justice.
36. Meyer, G. (1992). Nonlethal weapons vs. conventional police tactics: Assessing injuries and liabilities. *The Police Chief*, *59*, 10–17.
37. Campbell, A., Berk, R.A., & Fyfe, J.J. (1998). Deployment of violence: The Los Angeles Police Department's use of dogs. *Evaluation Review*, *22*(4), 535–561.
38. Alpert, G.P., & Dunham, R.G. (2000). Analysis of police use of force data. Washington, DC: National Institute of Justice.
39. Hirschel, D.J., Dean, C.W., & Lumb, R.C. (1994). The relative contribution of domestic violence to assault and injury of police officers. *Justice Quarterly*, *11*, 99–116.
40. Uchida, C.D., Brooks, L.W., & Koper, C.S. (1987). Danger to police during domestic encounters: Assaults on Baltimore county police. *Criminal Justice Policy Review*, *2*, 357–371.
41. U.S. Department of Justice. (2006). Law enforcement officers killed and assaulted 2005. Washington, DC: Federal Bureau of Investigation. Retrieved January 20, 2007, from http://www.fbi.gov/ucr/killed/2005/table68.htm
42. Brandl, S. (1996). In the line of duty: A descriptive analysis of police assaults and accidents. *Journal of Criminal Justice*, *24*, 255–264.
43. Brandl, S.G., & Stroshine, M.S. (2003). Toward an understanding of the physical hazards of police work. *Police Quarterly*, *6*, 172–191.

44. Chan, T.C., Vilke, G.M., Clausen, J., Clark, R., Schmidt, P., Snowden, T., & Neuman, T. (2001). Pepper spray's effects on a suspect's ability to breathe. National Institute of Justice: Research in Brief. Washington, DC: National Institute of Justice.
45. American Civil Liberties Union of South California. (1995). Pepper spray update: more fatalities, more questions.
46. Gauvin, R. (1995). Oleoresin capsicum spray: A progress report. *The ASLET Journal, May/June*, 29–32.
47. National Institute of Justice. (2003). The effectiveness and safety of pepper spray. Research for Practice. Washington, DC: National Institute of Justice.
48. Nowicki, E. (1993). Oleoresin Capsicum: A non-lethal force alternative. *Law Enforcement Technology*, *20*, 24–27.
49. Laur, D. (2000). Taser Technology Research Paper. Ottawa, Canada. Retrieved January 9, 2007 from http://www.cprc.org/tr/tr-2000-01.pdf.
50. TASER International. (2006). Deadly Rhetoric: How the ACLU of Northern California's fight against law enforcement control tools endangers communities. Phoenix, AZ: Author. Retrieved January 9, 2007 from http://www.taser.com/savinglives/documents/Deadly%20 Rhetoric%20V12.pdf
51. Amnesty International. 2006. USA Amnesty International's continuing concerns about Taser use. AI Index AMR 51/030/2006, http://www.amnestyusa.org/countries/usa/docu ment.do ?id = ENGAMR510302006. London: Amnesty International.
52. American Civil Liberties Union of Northern California. (2005). Stun Gun Fallacy: How the lack of taser regulation endanger lives. San Francisco, CA: Author.
53. Ho, J.D., Miner, J.R., Lakireddy, D.R., Bultman, L.L., & Heegaard, W.G. (2006). Cardiovascular and physiologic effects of conducted electrical weapon discharge in resting adults. *Academic Emergency Medicine*, *13*, 589–595.
54. Jauchen, J.R., Sherry, C.J., Fines, D.A., & Cook, M.C. (2005). Acidosis, lactate, electrolytes, muscle enzymes, and other factors in the blood of Sus scrofa following repeated TASER exposures. *Forensic Science International*, *161*(1), 20–30.
55. Lakkireddy, D., Wallick, D., Ryschon, K., Chung, M.K., Butany, J., Martin, D., Saliba, W., Kowalewski, W., Natale, A., & Tchou, P.J. (2006). Effects of cocaine intoxication on the threshold of stun gun induction of ventricular fibrillation. *Journal of the American College of Cardiology*, *48*(4), 805–811.
56. Levine, S.D., Sloane, C., Dunford, J., Chan, T., Vilke, G., & Dunford, J. (2005). Cardiac Monitoring of Subjects Exposed to the Taser, 12, Suppl 1, 71.
57. Nanthakumar, K., Billingsley, I.M., Masse, S., Dorian, P., Cameron, D., Chauhan, V.S., Downar, E., & Sevaptsidis, E. (2006). *Journal of the American College of Cardiology*, *48*(4), 798–804.
58. Garner, J., Maxwell, C., & Heraux, C.G. (2003). Characteristics associated with the prevalence and amount of force used by the police. *Justice Quarterly*, *19*(4), 705–746.
59. Bliese, P.D., & Hanges, P.J. (2004). Being both too liberal and too conservative: The perils of treating grouped data as though they were independent. *Organizational Research Methods*, *7*(4), 400–417.
60. Moerbeek, M. (2004). The consequence of ignoring a level of nesting in multilevel analysis. *Multivariate Behavioral Research*, *39*(1), 129–149.
61. Long, J.S., & Freese, J. (2001). Regression models for categorical dependents variables using Stata. College Station, TX: Stata Press.
62. Long, J.S. (1997). Regression models for categorical and limited dependent variables. Thousand Oaks, CA: SAGE Publications.
63. Fu, V.K. (1998). Estimating generalized ordered logit models. *Stata Technical Bulletin*, *44*, 27–30.
64. Hougland, S., Mesloh, C., & Henych, M. Use of force, civil litigation, and the Taser. *FBI Law Enforcement Bulletin*, *74*, 24–30.

65. Hickey, R.H., & Hoffman, P.B. (2003). To bite or not to bite: Canine apprehensions in a large, suburban police department. *Journal of Criminal Justice*, *31*, 147–154.
66. Mesloh, C. (2006). The impact of training on police canine force outcomes. *Police Practice and Research*, *7*(4), 323–335.
67. Dill, L.P. (1992). Police dog attacks: A dogmatic approach to crime control. *Whittier Law Review*, *13*, 515.
68. Police Executive Research Forum. (2005). Conducted Energy Device Policy and Training Guidelines for Consideration. PERF Center on Force & Accountability. Washington DC, USA.
69. Garner, J.H., Maxwell, C.D., & Heraux, C.G. (2002). Characteristics associated with the prevalence and severity of force used by the police. *Justice Quarterly*, *19*(4), 705–746.

Chapter 24
Field Statistics Overview

James E. Brewer and Mark W. Kroll*

We now have enough years of experience and enough CEW deployments to answer many of the common statistical questions that arise. This chapter will deal with the following questions:

1. How many human beings have experienced a CEW exposure?
2. What is the net impact on officer and suspect injuries?
3. Is there any truth to the common perception that multiple CEW exposures are more dangerous?
4. How often is the CEW blamed as a cause of death in an arrest-related death?

24.1 Total Human Exposures

24.1.1 Field Usage Exposures

Previous publications have reported the field usage for various law enforcement agencies [1,2]. We sought to calculate the overall usage from local usage rates and detailed CEW sales data.

Reports of law enforcement CEW usage were gathered from web searches. These covered 187 reports from 118 unique agencies for the years 1986–2008. A total of 156 reports were from the United States (83%) with the remainder coming from Canada. The reporting period duration for an agency ranged from 6 weeks to 6 years with a mean of 1.03 ± 0.53 years. The reports covered departments with 10–40,000 officers with 2–3847 TASER CEWs deployed in a department.

**Mark W. Kroll reports serving on the Corporate and Scientific/Medical Advisory Board of TASER International, Inc. and receives compensation in both roles. No other potential conflict of interest relevant to this chapter was reported. University of Minnesota*

J.E. Brewer (✉)
e-mail: jamesbrewer@wcta.net

M.W. Kroll, J.D. Ho (eds.), *TASER® Conducted Electrical Weapons: Physiology, Pathology, and Law*, DOI 10.1007/978-0-387-85475-5_24,

A CEW field use was defined as either a drive-stun or barb-launched application or attempted application. Brandishing, arcing, or laser painting were not counted for this analysis. A total of 22,160 field uses were reported. The usage rate per CEW varied significantly with departments (1.2 ± 2.1) with the pattern of use following a log normal distribution (m = 1.2, μ = 1.08). Most of that variation encountered was due to different deployment levels with an impressive correlation of $r^2 = 0.72$ as seen in Fig. 24.1.

The deployment level is the number of CEWs divided by the number of sworn officers. Full patrol officer deployment is typically found at a deployment level >0.75. This is explained by the fact that not every sworn officer is issued a CEW as they are not all active patrol officers.

$$\mathrm{DL(deployment\_level)} = \frac{\mathrm{number\_of\_CEWs}}{\mathrm{number\_of\_sworn\_officers}}$$

If DL was >75% it was then set to 0.75 (75%).

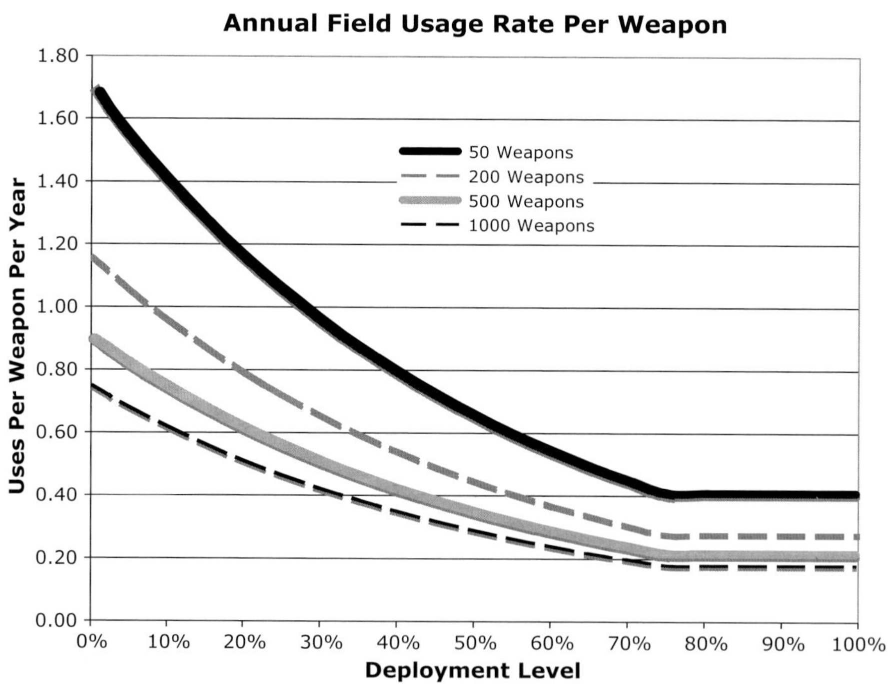

Fig. 24.1 Low deployment departments have much higher usage rates per CEW as officers in these departments are called out to many different situations while officers in full deployment departments have individual CEWs

The predicted annual usage rate per CEW was then given by:

$$\text{Log}_e(\text{usage rate}) = 1.6009 - 1.9047\ \text{DL} - 0.2746\ \log_e(\text{number of CEWs})$$

For example (see Fig. 24.1), in a city that has few CEWs and issued only to the SWAT unit, each of those devices will be called on often and they will be used, on average, about 1.7 times per year. The more officers equipped with CEWs the lower the usage rate, which drops to a rate of 0.2–0.4 per year per CEW depending on the number of CEWs issued.

There was no difference between usage rates in the United States versus Canada nor was there any apparent correlation between the year reported and the usage rate.

This model of usage rate versus deployment rate was then utilized on a TASER CEW sales database of 7617 agencies covering 219,970 CEWs. This database included department size, number of CEWs owned, and deployment level. The model gave an average annual usage rate for each CEW of 0.550 ± 0.008. Each CEW in law enforcement hands was used once every two years, on average, across all departments.

The TASER sales history was then integrated over time from Q1 2000 to 30 June 2008 and yielded 1,102,254 "CEW years" with 606,395 field uses. These numbers did not include those of the 150,000 CEWs in civilian hands (since 1993). They also do not include the large number of noncontact uses including brandishing, arcing, and laser "painting."

24.1.2 Training Exposures

In summer 2007 a survey was sent to all TASER CEW certified instructors. A total of 2082 surveys were completed which covered 106,637 TASER CEWs. This is about 30% of the devices then fielded with law enforcement and thus the survey has unusually high statistical confidence. Instructors were asked about their training policies and the number of human training exposures per CEW.

As can be seen in Fig. 24.2, the most common law enforcement department policy decision was to encourage the law enforcement officer to receive a training CEW exposure. The second most common policy was to make the training exposure mandatory. The *least* common policy was to forbid an exposure (5.4%). Note that the percentages were calculated from the individual instructor responses. Since larger departments had multiple instructors, this method gave an estimate of the number of officers subject to a given policy.

Using the responses from the responding 2,082 certified instructors, we found a weighted mean of 2 (1.98) human training exposures per CEW. The reason this number was >1 was that in low deployment departments many officers share a CEW. A statistical "bootstrap" technique with 1,500 samples of 1,581 data each was used to estimate the confidence limits on this training exposure rate. The 95% confidence limits were 1.85–2.12.

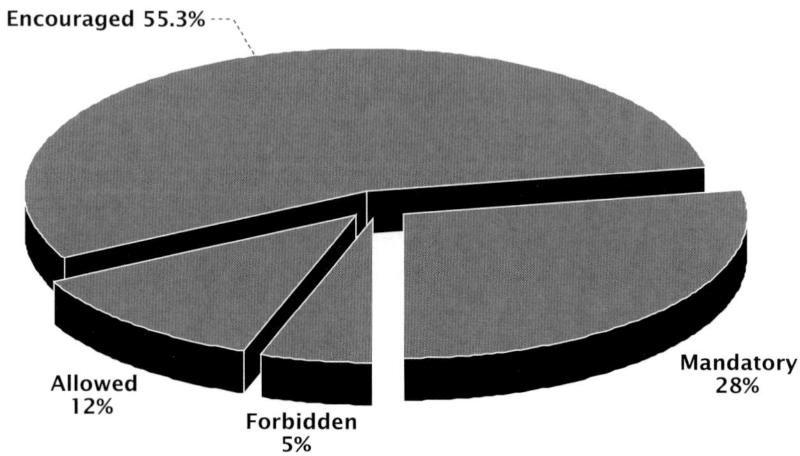

Fig. 24.2 Distribution of officer training policies

The mean response date to the survey was September 18, 2007. The mean calculated training hits given for officer turnover was 4.62% per year. This was 1.17% per quarter.

An estimated 378,731 CEWs were fielded by June 30, 2008. This number multiplied time 1.98 gives 749,175 CEW training exposures. Adding in 1.17% per quarter for turnover exposures gave an estimate of 758,385 training exposures with confidence limits of ±53,000.

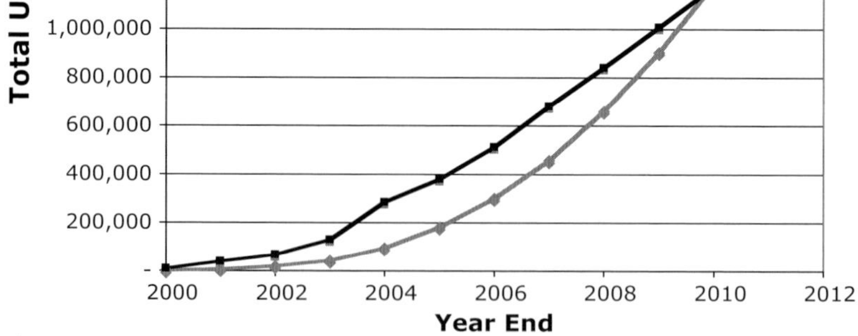

Fig. 24.3 Total field and training exposures. The numbers up to mid-2008 were based on actual CEW deployments while the remaining were based on conservative sales estimates

The total human exposure estimates are shown in Fig. 24.3. Note that the field uses are expected to exceed the training uses by 2010. The reason for this is that training exposures are primarily driven by new deployments and thus are approximately proportional to new sales. However, field uses are driven by the total number of CEWs in the field and thus are predicted by the aggregate number of deployed units. Since this, in turn, is the aggregate number of sales we would hypothesize a quadratic fit to the year, which was the case ($r^2 = 0.9988$). One factor that could influence this analysis, in the future, is the actual need for field usage. There is substantial evidence that the behavioral abnormalities that lead to CEW usage are a function of the illegal street-based drug supply. If this drug supply should expand or contract, field usage could increase or decrease accordingly.

24.1.3 Impact on Officer and Suspect Injuries

Recent publications have demonstrated the low rates of injury from CEW usage [1,3,4]. To establish the average agency results we performed a broad search for reports relating CEW introduction to officer and suspect injuries.

24.1.4 Officer Injuries

The results are shown in Table 24.1. There were 25 law enforcement agencies reporting data. The year of maximum deployment was compared to a baseline year. The baseline year was typically the year before. However, in cases of gradual deployment the baseline year was the latest year with no CEW deployment. The postdeployment year ranged between 2002 and 2007. The number of CEWs in the departments was 54–1444 (mean 456 $\pm$ 446).

The reported officer injury rate reduction ranged from 20% to 100%. The injury reduction statistics were weighted by the number of CEWs. The weighted mean injury reduction was 63%. The 95% confidence bounds were 55–72%. There was no univariate or multivariate correlation between the injury rate reduction and the year or number of CEWs in the department.

24.1.5 Suspect Injuries

The results are shown in Table 24.2. There were data from nine agencies. Both the postdeployment (comparison) year and the number of CEWs are shown. The comparison years were 2004–2005 and the number of CEWs was 205–2569.

The injury reduction ranged from 24% to 82%. These were weighted by the number of CEWs. The weighted mean injury rate reduction was 64%. The 95% confidence bounds were 52–75%. There was no univariate or multivariate

Table 24.1 Officer injury reductions with number of CEWs and post deployment year. The weighted mean injury rate reduction was 63% (55–72% confidence limits)

Location	CEWs	Postdeployment year	Injury reduction (%)
Austin, TX	1144	2004	50
Cape Coral, FL	243	2004	93
Charlotte, NC	1444	2004	59
Cincinnati, OH	1221	2004	56
Columbus, OH	205	2005	23
Concord, CA	71	2006	65
El Paso, TX	869	2007	86
Garner, NC	56	2004	20
Glenn County, CA	54	2006	100
Leon County, FL	203	2004	65
Long Beach, CA	1108	2005	25
Maui, HI	413	2007	77
Minneapolis, MN	128	2006	75
Oakland County, MI	410	2004	100
Omaha, NE	96	2005	47
Orange County, FL	1344	2002	80
Peel Regional, OT	64	2004	37
Putnam County, FL	129	2005	86
Sarasota, FL	220	2006	65
South Bend, IN	275	2004	66
Topeka, KS	147	2003	46
Toronto, ON	630	2006	100
Ventura County, CA	538	2007	72
Queensland, Australia	493	2007	40
Wichita, KS	308	2006	46

correlation between the injury rate reduction and the year or number of CEWs in the department.

It is interesting to note that CEW deployment appeared to help suspects (64 $\pm$ 11% injury rate reduction) and law enforcement officers (63 $\pm$ 8% reduction) equally.

Table 24.2 Suspect injury reductions. The baseline for Phoenix was August 2001–August 2002. Mean suspect injury rate reduction was 64% (52–75%)

Location	TASER CEWs	Post deployment year	Injury reduction (%)
Austin, TX	1144	2004	82
Cape Coral, FL	243	2004	68
Charlotte, NC	1444	2004	79
Cincinnati, OH	1221	2004	35
Columbus, OH	205	2005	24
Lynchburg, VA	40	2007	58
Maui, HI	413	2007	48
Peel Regional, OT	205	2005	47
Phoenix, AZ	2569	2004	67

There were several limitations to this analysis. The data were self-reported (often without independent quality control) and covered varying deployment years.

24.2 Are Multiple Exposures More Dangerous?

A commonly heard hypothesis is that multiple or prolonged CEW exposures may be more dangerous. A limit of three CEW "hits" was proposed as a safe limit [5]. A large body of 292 media-linked death cases, in which the number of exposures was ascertainable, were analyzed to see if there was any statistical support for this hypothesis.

A total of 267 autopsies were obtained, and police records or media accounts were analyzed for the remaining 25 cases. The results are shown in Fig. 24.4. It can be seen that 85% of fatalities were preceded by three exposures or less. Over 75% of the deaths involved only one or two exposures. The distribution of the number of CEW exposures was then compared to the exposure distribution for 3200 CEW exposures of the Royal Canadian Mounted Police (RCMP) [6]. These distributions were fitted to a Gumbel-Gompertz model and then were compared. Main and secondary distribution lobes, including the tail, showed no differences (log-rank $p = 0.48$). We concluded that there appeared to be no correlation between the number of exposures and the mortality rate.

Thus, there does not appear to be any basis for the baseball-reminiscent "3-strike" rule of no more than three CEW discharges. Even if there was an identifiable pathology associated with a fourth exposure, enforcing this "baseball" rule would affect only 15% of cases. In these 15% of cases, officers would

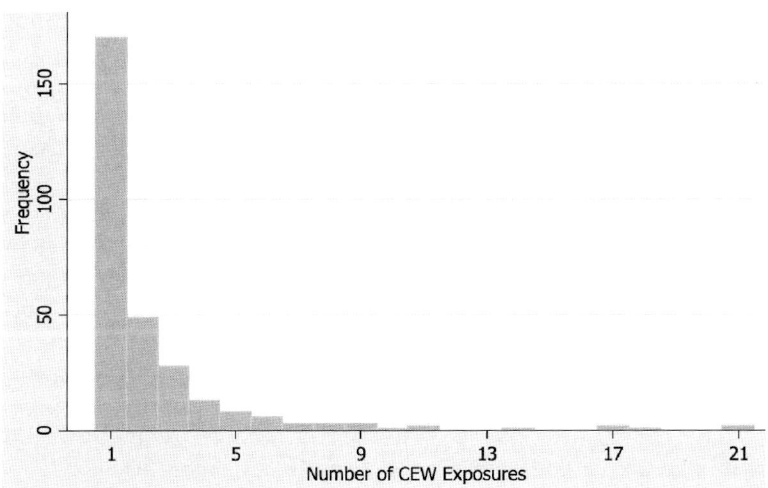

Fig. 24.4 Frequency of various numbers of exposures in 292 reported deaths

then be forced to switch to alternatives such as OC spray, baton strikes, prolonged physical struggle, or firearm discharge. Based on the demonstrated increased injury rates with these alternatives overall injury rates would likely increase.

These conclusions are supported by the recent human data with exposures out to 45 seconds [7] and animal data with exposures out to 30 minutes [8].

24.3 How Often Is the CEW Blamed as a Cause of Death?

The arrest-related death (ARD) is an phenomenon that occurs about 800 times per year in North America. This number is estimated from a population-based adjustment from the 700 annual ARDs found in the 47 reporting states of the USA [9]. These deaths include both criminal suspect arrests and attempts to control someone in order to render medical assistance. In the 700 annual ARDs a CEW was used in 1.8% (12) of the cases. The results of smaller studies (eliminating firearm cases) have demonstrate that CEWs, such as the TASER® X26, had been used during approximately 30% of ARDs in the United States [10,11]. As more law enforcement agencies adopt these devices the percentage will increase.

The medical examiner (ME) is under great pressure when investigating an ARD. They must be impartial in spite of great media and advocacy pressure (especially in the case of an ethnic minority death). Adding to the pressure has been the controversy and paucity of scientific literature regarding these tools. This was also true with chemical irritant aerosols, "hog-tying," carotid neck holds, and restraint "asphyxia." Finally, advocacy groups have always been slow to acknowledge exculpatory scientific evidence, even after it has been published in the peer-reviewed literature. For example, both Amnesty International and the American Civil Liberties Union (ACLU) have still not withdrawn their previous "concerns" regarding alleged deaths following the use of oleresin capsicum (OC) spray [12,13].

Although the use of CEWs was often temporally associated with the occurrence of ARD, medical examiners cited the device as the primary cause of death in five cases (This is now down to four cases as a judge ordered a medical examiner to correct her autopsy in one of these cases.) [14] If the time frame was expanded going back to 1983, and included cases where the CEW was listed as one of several causes of death, the total rises to 12. Twelve cases (out of nearly 1.4 million uses) give the devices a death rate of less than one in 100,000.

Since electrical current does not linger or accumulate in the body, some medical examiners have, in the past, erred on the side of including the CEW as a contributory cause of death, even though they had no explanation for how it could have caused or contributed to the death. Since electrical current does not accumulate in the body, we hypothesized that, as more peer-reviewed data were published, medical examiners would be able to make more accurate judgments about the causes of death.

ME autopsy report accuracy has been previously explored [15–18] as well as clinical autopsy accuracy [19–21]. We decided to explore the rate of ME errors with ARDs with usage of CEWs.

24.3.1 Possible Areas for Confusion

24.3.1.1 Electrocution

"Electrocution" is the term first coined to describe the government's execution of a convicted criminal by use of electricity. Today the term "electrocution" is more broadly used to describe the induction of a cardiac arrest by the application or exposure to electrical shock. This has been theorized as a mechanism by which a TASER CEW, could, allegedly, kill a person. If electrocution – by electrical stimulation – does occur, death is immediate and occurs within seconds. This is distinct, for example, from the results of a high power shock such as lightning strike, which may will cause long term damage including myocardial necrosis [22,23]. Electrical stimulation effects do not linger or build up in the body like a poison beyond the first few seconds [24–32].

The electrical induction of ventricular fibrillation (VF) has recently become one of the best scientifically researched causes of death. Paradoxically, this has been due to the surgical implantation of lifesaving implantable cardioverter defibrillators (ICDs). About 500 times per day a cardiac electrophysiologist will intentionally use electrical current to induce a cardiac arrest to test an ICD device immediately following its insertion [33,34].

From this experience with over 1,000,000 such intentionally induced cases of cardiac arrest in the cardiac cathertization laboratory, certain facts have been medically and scientifically established beyond question:

a. VF is either induced or not induced within 1–5 seconds of current application [30,32,35].
b. Asystole or PEA (pulseless electrical activity) is never induced [36].
c. The cardiac pulse disappears immediately [37].
d. The patient loses consciousness within 5–15 seconds [37].
e. A sufficiently strong defibrillation shock within the first one minute following VF – either internal or external – restores a cardiac sinus rhythm 99.9% of the time [38].

24.3.1.2 Long Duration Shocks

The regulations of both the International Electrotechnical Commission (IEC) [39,40] and Underwriters Laboratories (UL) regulations recognize that electrocution either happens in the first few seconds or does not occur [41]. Currents that will not induce VF in a few seconds will not induce VF in 1 minute as shown in Fig. 24.5 taken from Chilbert [41].

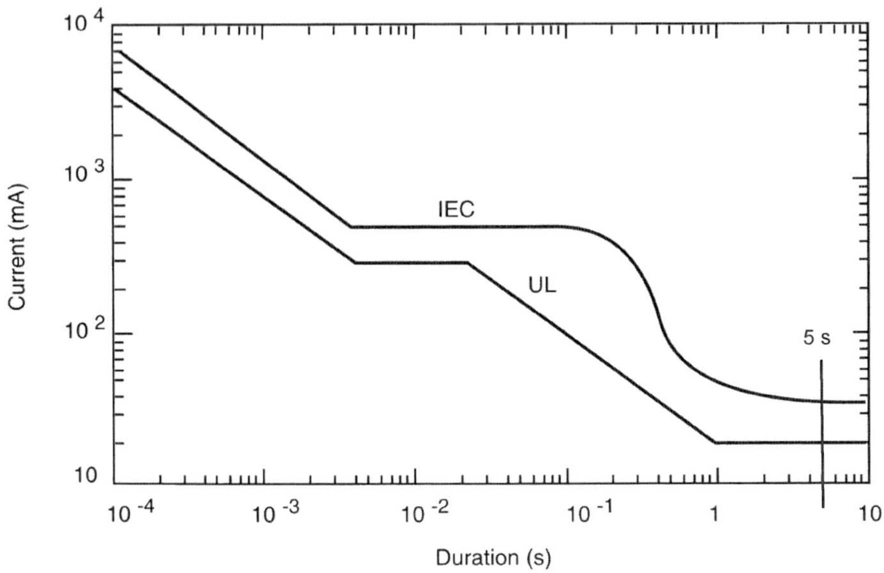

Fig. 24.5 The level of electrical current to induce ventricular fibrillation does not decline after a few seconds

Animal studies going back to the 1930s showed that the risk of inducing VF did not build up (increase) after a critical exposure time of a few seconds. These studies have found that the critical exposure time ranged over 0.8–5.0 seconds [24,26–29,42]. Based on these animal results above, Beigelmeier and Lee calculated that this critical time to induce fibrillation ranged from 2 to 5 seconds for humans due to the fact that humans have lower heart rates than experimental animals [26,27].

When TASER probes were buried under the skin of small pigs (50 kilograms), with a barb over the most sensitive part of the heart [43], experimenters found no difference in the ability of either a 5-second or a 15-second X26 application to induce ventricular capture (24/25 vs. 28/28, p = NS by Yates-corrected χ^2). Due to the differences in thoracic geometry, bilateral passage of current through the heart, as occurred in the pigs, would almost certainly be impossible in humans because of insulation provided by the lungs. Also, any vulnerable side-to-side orientation would be very rare since the CEW barbs are launched in a vertical plane. See Chapter 8 for further discussion of the swine–human differences.

One human study found that connecting a 9-volt cell directly to the inside of the heart induced VF within 3 seconds in the majority of patients [32]. An intracardiac human study found that the current duration required to cause fibrillation (at a 96% success rate), with a small steady direct current (DC), was 3.8 ± 1.4 s [30].

24.3.1.3 Effects of Electrical Current on Breathing

Due to the routing of the phrenic nerves it is extremely difficult to electrically induce respiratory paralysis in the human [44]. The phrenic nerve derives from the C3–C5 cervical plexus and the point of closest passage of these nerves to the skin is just above the clavicle, near the sternocleidomastoid muscle. The left and right phrenic nerves travel through the center of the thorax passing just on the margins of the heart on the way to enervate the left and right hemidiaphragm muscles. The nerves are surrounded by the highly insulative lungs throughout this passage, thus making them very insensitive to external electrical currents. Indeed, when electrical devices are used to stimulate the phrenic nerve (as in a paraplegic), surgical insertion of the electrodes is required, and they must be wrapped directly around the nerve to have any positive effect. As discussed in the chapter by Dawes on breathing effects, it does not appear that CEWs interfere with human breathing. This is true for applications across the chest [45–47] and for "drive-stun" applications focused over the trapezius muscle near the phrenic nerves [48].

24.3.1.4 Drug Dysynergies

As discussed in the chapters of Karch, Tchou, and Evans, chronic abuse of stimulants – such as cocaine – can do permanent damage to the heart and lead to an arrhythmic death without any electrical stimulation. Thus, there has been speculation that the acute usage of stimulants may also exacerbate the risk of electrocution. In fact, the opposite has been found to be typically true [49,50]. For example, cocaine intoxication is a strong sodium channel blocker and actually makes it more difficult to induce VF electrically [51–53]. This has recently been confirmed with actual TASER X26 waveforms [54]. With cocaine intoxication the safety margin rose significantly and was almost doubled for barbs near the heart. The occasional cocaine abuser with the syndrome variously referred to as excited delirium (a subgroup of the population likely to receive a Taser discharge) were usually found to be in asystole [11,36]. The appearance of this rhythm disorder remains unexplained, but it may be of central origin. Thus the induction of VF was irrelevant in these cases.

24.3.1.5 Autopsy Analysis

We performed extensive searches for the years 2001–2006 to find cases of an ARD with any mention of a CEW usage. Once they were identified, written requests were made for the associated autopsy reports.

Any material failure to appreciate the scientific facts regarding electrocution was scored as an error. This included ignoring any of the following: (1) a delayed collapse, (2) failure of immediate defibrillation, or (3) a non-VF rhythm. Other errors were counted if the report reflected hypotheses not supported by known literature. These included: (1) blaming the CEW for

Table 24.3 Scored autopsy errors in decreasing order of frequency [55]

Scored error	Rate of finding
Delayed collapse ignored	16
Nebulous equivocal comment such as "could not be ruled out"	16
Non-VF rhythm ignored	13
Failure of defibrillation ignored (includes cases where a non-VF rhythm was noted by paramedics)	9
Discharge duration or parity stressed	9
Drive-stun mode ignored	8
Assumed drug-CEW dysynergy	6
"Straw" comment	6
Cardiac damage blamed on CEW	4
Impaired respiration assumed	2

myocardial physical changes [23], (2) inclusion of a unsupported innuendo or a nebulous equivocal comment (e.g. "we were unable to eliminate the role" of the CEW), (3) assuming prolonged CEW applications were more dangerous than other restraint techniques, (4) speculating that CEWs impaired breathing, (5) presumption of a lethal synergy between stimulant drug intoxication and the CEW, and (6) use of the CEW only in the "drive stun" mode since this involved current passing between 2 very close electrodes and did not create any major organ involvement. Finally, the use of an unscientific lay term such as the metaphoric "last straw" or "pushed over the edge" was scored as an error (Table 24.3).

24.4 Results

We obtained 301 autopsies and summaries covering deaths occurring between January 1, 2001 and December 31, 2006. Between the autopsies and the summaries obtained, we believe that we were able to analyze the findings of almost ARDs in which a ME might have cited the CEW.

There were 39 cases (9.4%) where the autopsy report listed the CEW as a possibly contributory or as an "unknown" factor. The listing rate declined from 33% in 2001 to 3.3% in 2006 ($r^2 = 0.73$, $p = 0.031$ for linear fit) as seen in Fig. 24.6. Autopsy reports were reviewed for these cases and errors were tabulated. The decedents were primarily male (38 M/1 F) with mean age 35.0 ± 10.9 years (median = 32) which was consistent with other reported ARD data [9,11,56].

We found a mean of 2.9 ± 1.3 scored errors per autopsy report with a range of 1–6. This error rate declined steadily over time as seen in Fig. 24.7 ($p = 0.002$, $r^2 = 0.33$). We performed a multivariate analysis for predictors of the error rate. Suspect age, date of death, body mass index, heart mass, suspect

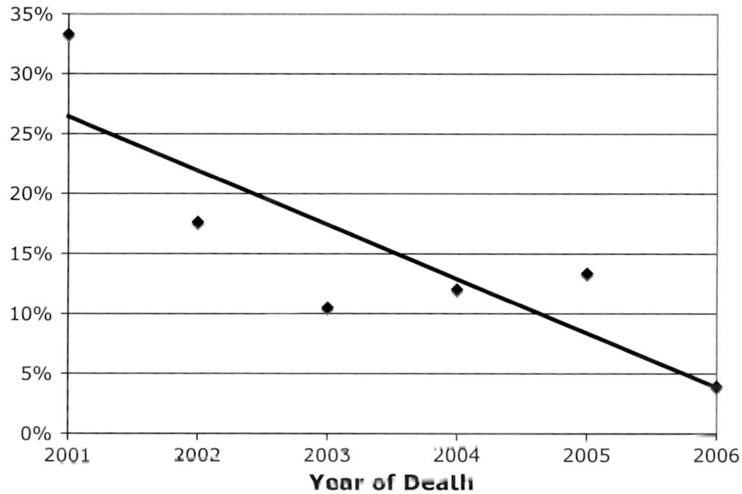

Fig. 24.6 The rate at which medical examiners mentioned a CEW as possibly contributing to a CRD declined significantly over the study period

mass and height, and race were analyzed. The only multivariate predictors found to be significant covariates were date-of-death ($p = 0.034$) and race of subject ($p = 0.028$). Black subject autopsies had slightly higher error rates averaging an additional 0.80 ± 0.30 (SEM) errors per report. Hispanic subject autopsies had slightly lower error rates averaging 0.98 ± 0.38 (SEM) less errors

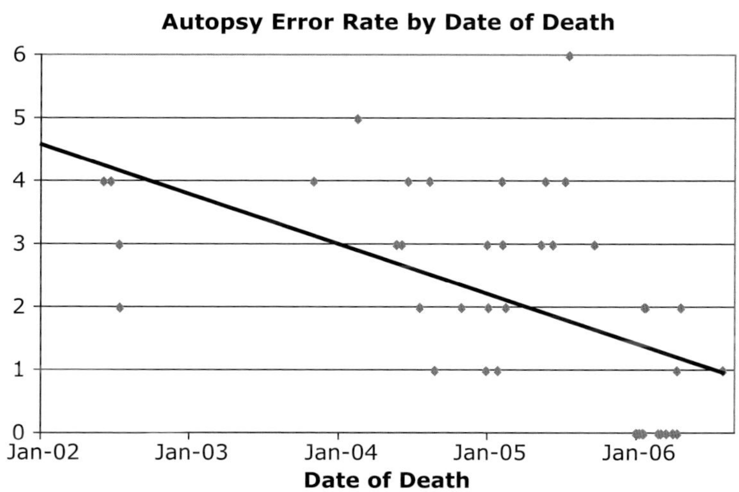

Fig. 24.7 The scored error rate fell significantly from 3.5 per autopsy in 2002–2004 to a rate of 1–2 in 2006

per report. This effect of race was also significant as a univariate predictor of error rate by ANOVA ($p = 0.047$).

The rate at which medical examiners mentioned CEWs as possibly contributing to a ARD went from 33% in 2001 to 10% in 2003. It appeared that the rapid acceptance of these devices by law enforcement agencies began to increase their temporal association with ARDs in the media. This appeared to affect a few medical examiners as the rate of mention began to level off somewhat (p = NS) to 13% in 2005. During 2005 and 2006, numerous peer-reviewed publications and conference presentations addressed the speculated safety concerns regarding these devices [1,2,45–47,57–81]. In addition, published books addressed the critical issue of excited delirium [82,83]. These factors may explain the fall in the mentions of CEW contribution to a ARD. The rate that the CEW was mentioned (as possibly contributory) fell to 3.3% by 2006.

24.5 Conclusions

About 1,400,000 human beings have received CEW exposures as of July 2008. Statistical analysis showed that many of the urban myths surrounding the use of CEW were false. The adoption of these devices has demonstrated a reduction in both suspect and officer injuries. There was no evidence that longer exposures were more dangerous. Presently, medical examiners rarely suggest a link between a CEW exposure and the death of a suspect.

References

1. Smith R. *TASER® Non-LETHAL WEAPONS: Safety Data and Field Results*. Paper presented at: American Academy of Forensic Sciences, Seattle, WA, USA; 2006.
2. White M, Ready J. The TASER as a less lethal force alternative. findings on use and effectiveness in a large metropolitan police agency. *Police Quarterly*. March 2006.
3. Bozeman W, Winslow J, Hauda W, et al. Injury Profile of TASER® electrical Conducted Energy Weapons (CEWs). *Annals of Emergency Medicine*. 2007;50:S65.
4. Eastman AL, Metzger JC, Pepe PE, et al. Conductive electrical devices: a prospective, population-based study of the medical safety of law enforcement use. *The Journal of Trauma*. Jun 2008;64(6):1567–1572.
5. Czarnecki F. *Recommendations for the Use of the TASER by Law Enforcement Officers*. 2005 International Association of Chiefs of Police Conference, Legal Officers Section. Miami Beach, FL; 2005.
6. Anonymous. *Multiple RCMP Taser Zaps on Rise Despite Warning: Canadian Press-CBC Analysis*. Canadian Press. June 11 2008.
7. Dawes D, Johnson M, Lundin E, et al. Breathing Parameters, Venous Blood Gases, and Serum Chemistries With Exposure to a New Wireless Projectile Conducted Electrical Weapon in Human Volunteers. *Annals of Emergency Medicine* 2007;50:S133.
8. Hughes E, Kennett M, Murray W, et al. *Electro-Muscular Disruption (EMD) Bioeffects: A Study on the Effects a Continuous Application of the TASER@X26 Waveform on Swine*. Philadelphia, PA: Penn State University; 30 November 2007.

9. Mumola C. *Arrest-Related Deaths in the United States, 2003–2005*. Bureau of Justice Statistics Special Report October 2007(NCJ 219534).
10. Ho J, Reardon R, Heegaard W. Deaths in police custody: an 8 month surveillance study. *Annals of Emergency Medicine*, 2005;46(suppl) abstract:S94.
11. Stratton SJ, Rogers C, Brickett K, et al. Factors associated with sudden death of individuals requiring restraint for excited delirium. *Am J Emerg Med*. May 2001;19(3):187–191.
12. Amnesty I. *United States of America: Excessive and Lethal Force?* Amnesty International's concerns about deaths and ill-treatment involving police use of TASERs. 2004:https://web.amnsety.org/library/index/ENGAMR511392004.
13. ACLU. *Pepper Spray: A Magic Bullet Under Scrutiny*. ACLU of Southern California; 1993.
14. TASER International & City of Akron v Chief Medical Examiner of Sunnit County, Ohio. Schneiderman: Court of Common Pleas, Summit County, Ohio; 2008.
15. Comstock RD, Mallonee S, Jordan F. A comparison of two surveillance systems for deaths related to violent injury. *Inj Prev*. Feb 2005;11(1):58–63.
16. deJong JL, Hanzlick R. Level of agreement between opinions of medical examiner investigators and forensic pathologist medical examiners regarding the manner of death. *Am J Forensic Med Pathol*. Mar 2000;21(1):11–20.
17. Hanzlick R. Quality assurance review of death certificates: a pilot study. *Am J Forensic Med Pathol*. Mar 2005;26(1):63–65.
18. O'Carroll PW. A consideration of the validity and reliability of suicide mortality data. *Suicide Life Threat Behav*. Spring 1989;19(1):1–16.
19. Association of Directors of Anatomic and Surgical Pathology, Nakhleh R, Coffin C, et al. Recommendations for quality assurance and improvement in surgical and autopsy pathology. *Hum Pathol*. Aug 2006;37(8):985–988.
20. Chandramohan D, Setel P, Quigley M. Effect of misclassification of causes of death in verbal autopsy: can it be adjusted? *Int J Epidemiol*. Jun 2001;30(3):509–514.
21. Mollo F, Bertoldo E, Grandi G, et al. Reliability of death certifications for different types of cancer. An autopsy survey. *Pathol Res Pract*. Aug 1986;181(4):442–447.
22. Colonna M, Caruso G, Nardulli F, et al. Myocardial haemorrhagic necrosis in delayed death from electrocution. *Acta Med Leg Soc (Liege)*. 1989;39(1):145–147.
23. Lichtenberg R, Dries D, Ward K, et al. Cardiovascular effects of lightning strikes. *J Am Coll Cardiol*. Feb 1993;21(2):531–536.
24. Antoni H. Pathophysiological basis of ventricular fibrillation. In: Bridges JF, Ford GL, Sherman IA, et al., eds. *Electrical Shock Safety Criteria*. New York: Pergammon Press; 1985:33–43.
25. Antoni H. Cardiac sensitivity to electrical stimulation. In: Reilly J, ed. *Applied Bioelectricity: From Electrical Stimulation to Electrical Pathology*. Springer; 1998:194–239.
26. Biegelmeier. *Effect of Current Passing Through the Human Body and the Electrical Impedance of the Human Body: A guide to IEC-Report 469*. VDE,-Verlag, Berlin: ETZ; 1987. 20.
27. Biegelmeier G, Lee WR. New considerations on the threshold of ventricular fibrillation for a.c.shocks at 50∼60 Hz. *IEE Proc*. 1980;127(2):Pt. A: 103–110.
28. Jacobsen J, Buntenkotter S, Reinhard HJ. [Experimental studies in pigs on mortality due to sinusoidal and phase-controlled alternating and rectified currents (author's transl)]. *Biomed Tech (Berl)*. Jun 1975;20(3):99–107.
29. Roy OZ, Park GC, Scott JR. Intracardiac catheter fibrillation thresholds as a function of the duration of 60 Hz current and electrode area. *IEEE Trans Biomed Eng*. 1977;BME-24(5):430–435.
30. Sharma AD, Fain E, O'Neill PG, et al. Shock on T versus direct current voltage for induction of ventricular fibrillation: a randomized prospective comparison. *Pacing Clin Electrophysiol*. Jan 2004;27(1):89–94.

31. Swerdlow CD, Olson WH, O'Connor ME, et al. Cardiovascular collapse caused by electrocardiographically silent 60-Hz intracardiac leakage current. Implications for electrical safety. *Circulation*. May 18 1999;99(19):2559–2564.
32. Weismuller P, Richter P, Binner L, et al. Direct current application: easy induction of ventricular fibrillation for the determination of the defibrillation threshold in patients with implantable cardioverter defibrillators. *Pacing Clin Electrophysiol*. 1992;15(8):1137–1143.
33. Kroll M, Tchou P. Testing of implantable defibrillator functions at implantation. In: Ellenbogen K, Kay G, Lau C, et al., eds. *Clinical Cardiac Pacing, Defibrillation and Resynchronization Therapy*. 3rd ed. Philadelphia: W.B. Saunders Company; 2006:531–557.
34. Singer I, Lang D. The defibrillation threshold. In: Kroll M, Lehmann M, eds. *Implantable Cardioverter-Defibrillator Therapy: The Engineering-Clinical Interface*. Boston: Kluwer; 1996.
35. Frame R, Brodman R, Furman S, et al. Clinical evaluation of the safety of repetitive intraoperative defibrillation threshold testing. *Pacing Clin Electrophysiol*. Jun 1992;15(6):870–877.
36. Swerdlow C, Kroll M, Williams H, et al. Presenting Rhythm in Sudden Custodial Deaths After Use of TASER® Electronic Control Device. *Heart Rhythm* May 2008;5(5):S44.
37. Schipke JD, Heusch G, Sanii AP, et al. Static filling pressure in patients during induced ventricular fibrillation. *Am J Physiol Heart Circ Physiol*. Dec 2003;285(6):H2510–H2515.
38. Frame R, Brodman R, Furman S, et al. Clinical evaluation of the safety of repetitive intraoperative defibrillation threshold testing. *Pacing Clin Electrophysiol*. 1992;15(6):870–877.
39. IEC. *Effects of Current on Human Beings and Livestock, CEI/IEC 479-2: Effects of currents passing through the human body*, 2nd ed: IEC, Geneva, Switzerland; 1987.
40. IEC. *Effects of Current on Human Beings and Livestock, CEI/IEC 479-1: General Aspects*, 3nd ed: IEC, Geneva, Switzerland; 1994.
41. Chilbert M. Standards and rationale. In: Reilly J, ed. *Applied Bioelectricity: From Electrical Stimulation to Electrical Pathology*. New York: Springer; 1998:454–501.
42. Ferris LP, King BG, Spence PW, et al. Effect of electric shock on the heart. *Electrical Eng*. 1936;55:498–515.
43. Nanthakumar K, Billingsley IM, Masse S, et al. Cardiac electrophysiological consequences of neuromuscular incapacitating device discharges. *J Am Coll Cardiol*. Aug 15 2006;48(4):798–804.
44. Geddes LA, Voorhees WD, Lagler R, et al. Electrically produced artificial ventilation. *Med Instrum*. Oct 1988;22(5):263–271.
45. Vilke GM, Sloane CM, Bouton KD, et al. Physiological effects of a conducted electrical weapon on human subjects. *Ann Emerg Med*. Nov 2007;50(5):569–575.
46. Ho JD, Dawes DM, Bultman LL, et al. Respiratory effect of prolonged electrical weapon application on human volunteers. *Acad Emerg Med*. Feb 5 2007;14:197–201.
47. Chan T, Sloane C, Neuman T, et al. The impact of the taser weapon on respiratory and ventilatory function in human subjects. *Acad Emerg Med* 2007;14:191–192.
48. Ho J, Lapine A, Joing S. Confirmation of respiration during trapezial conducted electrical weapon application. *Acad Emerg Med*. 2008;15:398.
49. Mitrani RD, Miles WM, Klein LS, et al. Phenylephrine increases T wave shock energy required to induce ventricular fibrillation. *J Cardiovasc Electrophysiol*. Jan 1998;9(1):34–40.
50. Inoue H, Saihara S, Toda I, et al. Summation and inhibition by ultrarapid train pulses in dogs: effects of frequency and duration of trains, lidocaine, and beta blockade. *Pacing Clin Electrophysiol*. 1989;12(11):1777–1786.
51. Schwartz AB, Boyle W, Janzen D, et al. Acute effects of cocaine on catecholamines and cardiac electrophysiology in the conscious dog. *Can J Cardiol*. May 1988;4(4):188–192.
52. Schwartz AB, Janzen D, Jones RT. Electrophysiologic effects of cocaine on the canine ventricle. *J Cardiovasc Pharmacol*. Feb 1989;13(2):253–257.
53. Tisdale JE, Shimoyama H, Sabbah HN, et al. The effect of cocaine on Ventricular fibrillation threshold in the normal canine heart. *Pharmacotherapy*. May–Jun 1996;16(3):429–437.

54. Lakkireddy D, Wallick D, Ryschon K, et al. Effects of cocaine intoxication on the threshold for stun gun induction of ventricular fibrillation. *J Am Coll Cardiol.* Aug 15 2006;48(4):805–811.
55. Kroll M, Panescu D, Ho J, et al. Potential Errors in Autopsy Reports of Custodial Deaths Temporally Associated With Electronic Control Devices: A Cardiovascular Prospective. Proceedings of American Academy of Forensic Sciences. San Antonio, TX, 2007:284–285.
56. Ho J, Reardon R, Heegaard W. *Deaths in Police Custody: a 12 Month Surveillance Study.* 2006.
57. Vilke GM, Sloane C, Levine S, et al. Twelve-lead electrocardiogram monitoring of subjects before and after voluntary exposure to the Taser X26. *Am J Emerg Med.* Jan 2008;26(1):1–4.
58. Mesloh C, Wolf R, Henych M, et al. Less lethal weapons for law enforcement: A performance-based analysis. *Law Enforcement Executive Forum.* 2008;8(1):133–149.
59. Tchou P, Lakkireddy D, Wallick D. *Effects of Torso Dart Position and Cocaine Intoxication on TASER® Induction of Ventricular Fibrillation.* Paper presented at: American Academy of Forensic Sciences. Annual Meeting 2007; San Antonio, TX, USA.
60. Sloane C, Vilke G, Chan T, et al. Serum Troponin I measurement of subjects exposed to the Taser X-26. *J Acad Emerg Med.* 2008;35:29–32.
61. Panescu D. Design and medical safety of neuromuscular incapacitation devices. *IEEE Eng Med Biol Mag.* Sep–Oct 2007;26(5):57–67.
62. Panescu D. Less-than-lethal weapons: Design and Medical Safety of Neuromuscular Incapacitation Devices. *IEEE Eng Med Biol Mag.* July/August 2007;26(4).
63. Moscati R, Ho J, Dawes D, et al. Physiologic Effects of Prolonged Conducted Electrical Weapon Discharge on Intoxicated Adults. Society of Academic Emergency Medicine abstract issue. 2007.
64. Levine SD, Sloane CM, Chan TC, et al. Cardiac monitoring of human subjects exposed to the Taser. *J Emerg Med.* Aug 2007;33(2):113–117.
65. Lakkireddy D, Khasnis A, Antenacci J, et al. Do electrical stun guns (TASER-X26(R)) affect the functional integrity of implantable pacemakers and defibrillators? *Europace.* Jul 2007;9(7):551–556.
66. Kroll MW, Calkins H, Luceri RM. Electronic control devices and the clinical milieu. *J Am Coll Cardiol.* Feb 13 2007;49(6):732; author reply 732–733.
67. Kroll M, Luceri RM, Calkins H. A very interesting case study involving a TASER Conducted Electrical Weapon (CEW) used on a patient with a pacemaker. *J Cardiovasc Electrophysiol.* Dec 2007;18(12):E29–E30; author reply E31.
68. Kroll M. Potential Autopsy Errors with In-Custody-Deaths: The Ronald Hasse Case Study Institute for the Prevention of In-Custody-Death. http://www.ipicd.com/docs/Hasse_Case_Study.pdf. 2007.
69. Kroll M. Crafting the Perfect Shock. *IEEE Spectrum.* Dec 2007;44(12):27–30.
70. Ideker RE, Dosdall DJ. Can the direct cardiac effects of the electric pulses generated by the TASER X26 cause immediate or delayed sudden cardiac arrest in normal adults? *Am J Forensic Med Pathol.* Sep 2007;28(3):195–201.
71. Holden SJ, Sheridan RD, Coffey TJ, et al. Electromagnetic modelling of current flow in the heart from TASER devices and the risk of cardiac dysrhythmias. *Phys Med Biol.* Dec 21 2007;52(24):7193–7209.
72. Ho JD, Dawes DM, Reardon RF, et al. Echocardiographic Evaluation of a TASER-X26 Application in the Ideal Human Cardiac Axis. *Acad Emerg Med* 2008.
73. Ho J, Dawes D, Johnson M, et al. The Neuroendocrine Effects of the TASER X26 Conducted Electrical Weapon as Compared to Oleoresin Capsicum. American College of Emergency Physicians Annual Meeting. Oct 2007.
74. Ho J. Physiologic effects of prolonged conducted electrical weapon discharge on acidotic adults. *SAEM.* 2007(abstract).
75. Ho J. Absence of electrocardiographic change following prolonged application of a conducted electrical weapon in physically exhausted adults. *SAEM.* 2007(abstract).

76. Dawes DM, Ho JD, Johnson MA, et al. 15-Second conducted electrical weapon exposure does not cause core temperature elevation in non-environmentally stressed resting adults. *Forensic Sci Int*. 2008;176:253–257.
77. Cao M, Shinbane JS, Gillberg JM, et al. Taser-induced rapid ventricular myocardial capture demonstrated by pacemaker intracardiac electrograms. *J Cardiovasc Electrophysiol*. Aug 2007;18(8):876–879.
78. Bouton KD, Vilke GM, Chan TC, et al. Physiological Effects of a Five Second Taser Exposure: 1897: Board #185 May 31 8:00 AM–9:30 AM. *Med Sci Sports Exerc*. May 2007;39(5 Suppl):S323.
79. Sweeney J, Kroll M, Panescu D. Analysis of Electrical Activation of Nerve and Muscle by TASERs. Paper presented at: American Academy of Forensic Sciences, 2006; Seattle, WA, USA.
80. Stratbucker RA, Kroll MW, McDaniel W, et al. Cardiac current density distribution by electrical pulses from TASER devices. *Conf Proc IEEE Eng Med Biol Soc*. 2006;1:6305–6307.
81. Panescu D, Kroll MW, Efimov IR, et al. Finite element modeling of electric field effects of TASER devices on nerve and muscle. *Conf Proc IEEE Eng Med Biol Soc*. 2006;1:1277–1279.
82. Wetli C. Excited delirium. In: Chan R, ed. *Sudden Deaths in Custody*. Totawa: Humana Press; 2006:99–112.
83. DiMaio T, VJM D. *Excited Delirium Syndrome Cause of Death and Prevention*. Boca Raton: Taylor & Francis; 2006.

Chapter 25
Sudden In-Custody Death

Samuel J. Stratton

"Sudden in-custody death" does not have a formal medical definition. But the general term "sudden death" is used medically to describe the rapid, unexpected death of an individual within 24 hours of the onset of symptoms. It follows then that sudden in-custody death refers to rapid, unexpected death during detention of individuals by law enforcement or public safety personnel. This term comprises the vast majority of "arrest-related" deaths (the preferred term in criminology literature) which include the fairly rare fatal police shootings [1].

Sudden death can occur as a result of a number of medical conditions as listed in Table 25.1. Sudden death can often be anticipated because of obvious symptoms or injury that present prior to the event. The conditions that result in sudden death can usually be determined by a thorough autopsy. Although there are varied causes for sudden death, the initial medical management of the situation is resuscitation and is based on standard medical maneuvers. The initial step in these maneuvers is called cardiopulmonary resuscitation (CPR). After initial resuscitation, the medical management of sudden death victims is determined by the underlying condition that caused the physical collapse.

Different from the general term sudden death, sudden in-custody death most often refers to death that occurs while a person is being detained and for which there is no obvious cause for death determined after a thorough autopsy. While sudden cardiac and respiratory death have been explored and researched to a great extent, sudden in-custody death has only recently been a subject of medical research.

Sudden in-custody death occurs to individuals usually during physical restraint – that often first appear physiologically stable. The individual being restrained and detained is usually highly agitated and resisting restraint efforts.

S.J. Stratton (✉)
Professor, University of California, Los Angeles School of Public Health and the David Geffen School of Medicine at the University of California, Los Angeles; Medical Director, Health Disaster Management/Emergency Medical Services Orange County Health Care Agency
e-mail: strattos@ucla.edu

M.W. Kroll, J.D. Ho (eds.), *TASER® Conducted Electrical Weapons: Physiology, Pathology, and Law*, DOI 10.1007/978-0-387-85475-5_25,

Table 25.1 Common medical conditions associated with sudden death

1. Cardiac disease
   a. Myocardial infarction (blockage of coronary arteries)
   b. Acute heart failure
   c. Dysrhythmia (sudden abnormal beating of the heart)
2. Neurological catastrophe
   a. Intracranial hemorrhage (acute bleeding into the brain or cranium)
   b. Stroke (blockage of major arteries supplying blood to the brain)
3. Vascular accident
   a. Dissection of the aorta (leakage into the tissue spaces of the wall of the artery)
   b. Dissection of carotid artery or coronary artery
   c. Rupture of the aorta
   d. Blockage (usually by clot) of the aorta or other major artery such as a carotid artery)
4. Pulmonary disease
   a. Pulmonary embolism (blood clot in lung arteries)
   b. Pneumothorax (collapse of one or both lungs)
   c. Acute pulmonary edema (fluid accumulation in lung air spaces)
   d. Toxic pulmonary reactions to chemicals, allergens, or hazardous materials
5. Occult hemorrhage
   a. Gastrointestinal bleeding (ulcer or erosion of the gastrointestinal tract lining)
   b. Spontaneous hemorrhage of internal organ (liver or spleen)
6. Hypoxia
   a. Oxygen deprivation
   b. Carbon monoxide poisoning
   c. Hydrogen sulfide and Cyanide poisoning
7. Acute drug or toxin reaction
   a. Anaphylactic reaction to drug or toxin
   b. Overdose of drug or toxin
   c. Cardiac, neurological, respiratory side effects of drug or toxin
8. Trauma
   a. Blunt injury with subsequent hemorrhage, rupture or dysfunction of a vital organ
   b. Penetrating injury of a vital organ
   c. Spinal injury causing inability to breathe
   d. Head or neck injury causing blockage of the airway
   e. Blunt cardiac contusion (bruising) causing dysrhythmia
   f. Electrocution

At the time of detention, it is not uncommon for sudden in-custody death victims to be confused, and disoriented with exaggerated body motions. This confusion and agitation is called excited or agitated delirium.

Sudden in-custody death occurs without conventional medical warnings. (For example, chest pains are usually noted before the death from a myocardial infarction.) The usual presentation is of an extremely agitated person vigorously resisting restraint, who then becomes calm and is subsequently noted to have ceased breathing or have no pulse. The unexpected nature of sudden in-custody death and the subtle, rapid presentation of death make effective intervention difficult. However, bizarre behavior hours before police involvement is typical.

25.1 Historical Perspective

Sudden death occurring with agitated psychiatric states had been recognized and reported in the literature as far back as the mid-1800s, but these types of deaths rarely involved law enforcement personnel and had become infrequent in the 1960s after the advent of effective antipsychotic medications [2–6].

Sudden in-custody death was first described in the modern medical literature during the 1980s [7–13]. Coincident with the recognition of sudden in-custody death was the rise in popularity in abuse of amphetamines and cocaine. As stimulants, these drugs are associated with agitation and confusion.

Initial medical theories of the cause for sudden in-custody death centered on investigation and discussion of drug toxicity, cardiac dysfunction, or positional asphyxia (discussed below) as the primary cause for the event [7–28]. Developing medical experiments to test theories for the cause of sudden in-custody death are essentially impossible because such studies would place human subjects at risk for physical harm. Therefore, the medical investigation of sudden in-custody death has been based on organized observational studies, simulations, and animal model studies of such incidents.

Current opinion regarding the causes for sudden in-custody death have centered on a number of conditions and situations shown to have an association with the event. Multiple factors have been shown to have an association with sudden in-custody death [29]. Many of the factors are outside the control of those tasked with protecting a potential victim of sudden in-custody death who is being detained for their own safety. Therefore, an emphasis on the knowledge that certain situations place one at risk for sudden in-custody death and prompt management of collapse are now the area of emphasis in daily practice [29,30].

25.2 Associated and Unassociated Factors

While a number of factors are known to have an association with sudden in-custody death, there are others that may or may not have an association. As noted above, a true causal relationship between a factor and sudden in-custody death has not been determined and experiments to identify such a factor are not feasible because human subjects would be put at risk. Table 25.2 lists factors that have been found to have an association with sudden in-custody death. There are, most certainly, other associated factors that have not been identified. Discussed below are factors that have both been and not been shown to have an association with sudden in-custody death.

25.2.1 Conducted Electrical Weapons

Sudden in-custody death is associated with severe agitation and physical resistance by the individual being detained. Because conducted electrical weapons

Table 25.2 Factors associated with sudden in-custody death

1. Excited delirium requiring physical restraint
a. Chronic stimulant abuse
b. Schizophrenia, especially with paranoid features
2. Forceful irrational struggle against restraints
3. Acute stimulant drug use (positive toxicology test):
a. Cocaine
b. Amphetamines
c. Ethanol
4. Autopsy evidence of chronic heart disease
5. Clinical medical abnormalities, including:
a. Dehydration
b. Hyperthermia
c. Metabolic acidosis

(CEWs), often called TASERs, are often deployed to physically detain such individuals, it is inevitable that sudden in-custody death can occur in the setting where CEWs are used. Because an electric current is used to disable the person being detained by means of CEWs, there has been strong interest in the risks of the device.

Di Maio and Di Maio describe a report from Amnesty International of 70 deaths following the use of CEWs during the time period of 2001–2004 [31]. The concern with these cases is cardiac death due to ventricular fibrillation or spasm of the heart caused by the CEW electric impulse delivered. The Di Maios' report reviewing the Amnesty International reports and concluding that most of those dying when the CEW was deployed did not immediately collapse as would be expected with a primary cardiac event such as ventricular fibrillation from electric shock [31].

In 2006, Strote and Hutson reported a case series of 75 deaths in which CEW was employed. They secured 37 autopsies among the 75 identified cases and report that all were male with ages ranging from 18 to 50 years. Chronic cardiovascular disease was found in 54.1% and stimulant drug use in 86.2% [32]. The cases reported in this series were similar to those reported in another study published in 2001 describing 18 sudden in-custody deaths for which the CEW was used in five of the cases [29]. In both papers, the final conclusion was that the deaths due to excited delirium (essentially sudden in-custody death) were the result of multiple factors without a strong association to use of CEWs [29,32]. To date, there has not been a study that shows a strong association between the CEW as a single factor for sudden in-custody death. In fact, autopsy reviews by cardiac electrophysiologists suggest that the presenting rhythms and time sequence would tend to eliminate the CEW as contributing to the death since they are not the type inducible electrically [33,34].

25.2.2 Pepper (Capsicum) Spray

Capsicum-derived ingredients (pepper spray) irritate mucous membranes and skin when applied externally and are used both by law enforcement and the general public as a nonlethal deterrent or defense device. Capsicum is generally recognized as safe as a food component by the US Food and Drug Administration [35]. Capsicum derived ingredients are found widely, and used as skin-conditioning agents, external analgesics, flavoring agents, and as fragrance components in comestics [35].

When used as a deterrent, capsicum functions primarily as an irritant to the corneal and scleral membranes (the thin outer layers of the eye). Capsicum can also irritate the mucous membranes of the throat and upper airway when inhaled, potentially resulting in an allergic or inflammatory reaction that can cause swelling and obstruction of the upper airway. Because capsicum is a nonlethal deterrent used at times to subdue agitated confused persons who subsequently suffer sudden in-custody death, it has been considered by some to have a direct relationship to the fatal event, but such a relationship has not been shown in scientific studies.

The primary medical concern with capsicum spray is eye irritation and development of corneal abrasions (superficial erosion of the membrane covering the pupil and iris). Usual medical management of a person exposed to capsicum spray is vigorous irrigation of the external eye to wash out the capsicum. Rarely are other medical issues found with capsicum spray. Brown and coauthors described 100 cases of capsicum spray exposure among persons presenting to a jail emergency medical treatment area and found seven cases of corneal abrasion [36]. Although significant eye irritation and corneal abrasion occur with capsicum spray, these conditions are rarely complicated by further progression of the problem and are self-limiting.

Case reports have noted use of capsicum spray in sudden in-custody death cases, but no direct association with the fatal outcome [11,29]. In his Canadian study, Pollanen reported use of capsicum spray in four of 21 cases of unexpected death related to excited delirium [11]. This low ratio does not show an association of the capsicum spray with the deaths because capsicum spray is commonly used in both nonfatal and fatal events in which a deterrent is required. One issue with use of capsicum spray is the escalation of agitation that can occur when the spray irritates the eyes of an already agitated and confused person. It has not been shown that the additional stress and agitation of having been sprayed with capsicum increases the risk for sudden in-custody death.

Other than self-limiting eye irritation, capsicum spray has not been associated with serious medical complications. Capsicum is a widely used nonlethal deterrent with occasional use in situations where sudden in-custody death occurs. There is no scientific evidence that establishes a direct association between sudden in-custody death and use of capsicum spray.

25.2.3 Stimulant Drugs

There is a strong association for sudden in-custody death and the chronic and acute use of stimulant drugs. In 1981 and 1985, Wetli and Fishbain authored papers that described sudden death in individuals who were suffering excited delirium associated with cocaine use [37,38]. In the second paper, the authors presented a case series of seven sudden deaths of which five occurred while the individuals were in police custody [38]. Prior to these reports, little had been reported in the medical literature for twenty years with regard to sudden death occurring in association with agitation. The temporal relationship of the increased use of recreational cocaine and amphetamine in the US population and the appearance of sudden death in association with agitation have led many to consider stimulant drugs of abuse as having a high relationship with sudden in-custody death. Law enforcement is often called and required to detain and restrain someone who is experiencing agitation or excited delirium.

Stimulant drug use has been shown to have a high association with sudden in-custody death [11,15–17, 25,27,32,38]. But, it is important to note that not all victims of sudden in-custody death show evidence for use of illegal drugs, including cocaine and amphetamines [29,32].

In addition to stimulant drugs, alcohol is often associated with sudden in-custody death. Alcohol has an association with increased stimulation of the nervous system and as a single drug or in combination with other stimulants accentuate the effects of norepinephrine [39]. Alcohol in the blood stream and tissues can combine with cocaine metabolites to form a more potent stimulant (cocaethylene) than either drug alone [40,41]. It is common for cocaine or amphetamine abusers to include alcohol in their patterns of drug dependence. Many chronically abuse alcohol as well as stimulant drugs that predispose one to physiologic changes of the heart, brain, liver, and vascular system that generally contributes to a higher risk for sudden death.

Another drug of abuse to consider in sudden in-custody death is phencyclidine (PCP or angel dust) that causes changes in brain metabolism that can mimic the acute psychosis seen in true psychotics. This psychosis can lead to agitation and delirium, placing an individual at risk for confrontation with law enforcement and potential sudden in-custody death. As with cocaine and amphetamines, PCP acts as a stimulant for the human central nervous system [42]. The thought disorder effects of PCP can persist or recur over many years, even with cessation of use of the drug [42]. As of this time, PCP is rarely seen in most US emergency departments, but a closely related drug, ketamine, used as a sedative and anesthetic, has become a popular drug of abuse among older adolescents and young adults [43]. Ketamine can produce similar effects as PCP when taken in large doses and on a chronic basis. Interestingly, the two drugs are molecularly similar and ketamine toxicity is detected in the clinical medical setting by the drug cross reacting positively to some urine toxicological tests for PCP. Ironically, ketamine, when used in medical settings, has one of the best studied safety profiles of all of the sedatives used [44].

25.2.4 Underlying Cardiac Disease

Abnormal heart findings are not uncommon when autopsy results of sudden in-custody death are reported. Enlargement of the heart, scarring of the heart muscle, and structural abnormalities were reported in 54.1% and 61.1% of cases reported in two papers [29,32]. The average age in these series of cases was 35.6 and 32.1 years, an extremely young group for more than 1/2 to manifest chronic heart disease findings [29,32]. Enlargement and scarring (fibrosis) of the heart can occur for a variety of reasons, most commonly hypertension, coronary artery disease, congenital etiologies, viral infection, and chronic drug and alcohol abuse.

The high rate of chronic heart disease noted in persons succumbing to sudden in-custody death is important because there is an association of sudden death with these cardiac conditions. Different cardiac conditions can lead to sudden death with dysrhythmia being an important one in a young adult population. Normally, the heart beats in a rhythmic, sequential pattern to effect circulation of blood through the vascular system. Lack of oxygen, excessive stimulation, biochemical and metabolic disruption, and fibrosis (scarring) can lead to disruption of the cardiac circuit resulting in unorganized, useless chaotic quivering of the heart termed ventricular fibrillation. Unless immediately reversed, usually by supplying a quick, phased electrical shock (defibrillation) to the heart, ventricular fibrillation is immediately fatal.

The risk of spontaneous ventricular fibrillation increases when the heart is exposed to sympathetic stimulation. This stimulation occurs with use of cocaine, amphetamines, and alcohol [45,46]. Surprisingly – at first blush – the risk of electrically induced fibrillation is only increased for the first few minutes of high dose sympathetic stimulants [47]. Further, the physiologic stress of excited delirium alone can strain the heart and lead to dysrhythmia. To further support the importance of the heart as a factor in sudden in-custody death is the fact that cocaine and amphetamines are associated with coronary artery spasm, causing a decrease or cessation of oxygen delivery to the cardiac muscle [45]. This coronary artery spasm may be undetectable after the death of an individual because all blood vessel tone is lost with death and the vessels return to a normal, relaxed diameter. Finally, alcohol is a direct myocardial (heart muscle) toxin.

In addition to dysrhythmia as a potential cardiac cause for sudden in-custody death, acute heart failure can occur. Heart failure can be rapid in onset (flash pulmonary edema) and triggered by extreme blood pressure elevation, coronary artery occlusion or spasm with a lack of oxygen supply to the myocardium, and chronic enlargement of the heart with ineffective myocardial contraction to pump blood. In these forms of heart failure, there is generally build up of fluid in the lungs that can be detected on autopsy. But, an increasingly common form of heart failure called diastolic dysfunction has been described which is difficult to detect on autopsy and can lead to a conclusion that someone died of sudden in-custody death (death with negative autopsy findings) [48,49]. In diastolic dysfunction, heart failure occurs because the myocardium (muscle) of the left ventricle or high pressure pumping chamber for the heart becomes stiff and ineffective in contraction. Diastolic dysfunction is more common in persons with highly elevated blood pressure and cardiac enlargement. The recognized cardiac abnormalities associated with sudden in-custody death, cardiac enlargement and fibrosis of the heart muscle, have a high association with diastolic dysfunction. Add to this predisposition, markedly elevated blood pressures seen with stimulant drug abuse and a substantial risk for diastolic dysfunction heart failure can be appreciated in the setting associated with sudden in-custody death.

25.2.5 Restraint Procedures and Positional Asphyxia

Positional asphyxia occurs when a person is held or trapped in a position that inhibits the effective motion of the diaphragm and chest or can obstruct the upper airway. Such positioning can lead to asphyxia by inhibiting the ability to effectively inhale and exhale [8,50]. A common illustrative example of positional asphyxia is the death by suffocation that occurs when a worker is caught in a ditch cave-in where dirt collapses around the chest and abdomen causing an inability to move the muscles needed to continue breathing even when the face is still above ground. Another form of positional asphyxia is restricted positioning of the head and neck such that the upper airway which leads to the lungs is blocked, not allowing the free passage of air into the lungs. Blockage of the airway can occur when someone is unconscious with loss of the ability to properly position the head and neck for effective breathing (this is well recognized and opening the airway is emphasized as a first step of cardiopulmonary resuscitation) [51].

In the early 1990s, interest developed within the medical field concerning the potential association of positional asphyxia due to restraint procedures and sudden in-custody death [8]. This hypothesis generated intense controversy among those researching and publishing on the concept of restraint related death due to positional asphyxia. This controversy was driven by multiple wrongful death legal cases that drew researchers into the court room to give opinion and support their research findings. As the courts argued the foundations for the hypothesis of sudden in-custody death in relation to restraint and positional asphyxia, the scientific exploration and explanations became more controversial [78–12].

Aggressive restraint is often required because of agitation and delirium in a setting that can result to sudden in-custody death. It is difficult to separate the potential association of restraint from the event because restraint almost always occurs and cannot be compared to nonuse of restraint for control of those experiencing excited delirium.

To address potential adverse outcomes related to restraining agitated persons, current education of the public safety and medical community has emphasized minimal restriction of normal body movement while still accomplishing restraint and frequent assessment of the vital physiologic functions (breathing in particular) to insure the restrained person is fairing well. Further, restraint complications including potential positional asphyxia is minimized with modern restraint training of law and medical personnel and the development of devices that minimize potential airway obstruction and respiratory compromise.

25.2.6 Dehydration, Hyperthermia, and Metabolic Acidosis

Agitated persons often fail to pay heed to their own thirst demands for fluids. Agitation and excited delirium causes a hyperkinetic physical state which results in increased loss of body fluids through sweating and loss of water vapors through the lungs with associated hyperventilation. Without adequate oral hydration, these body water losses result in dehydration.

Dehydration places demands on the heart, causes electrolyte in-balance and a build up of metabolic acid byproducts in the tissues. Dehydration and metabolic derangement add further to a person's agitation because of release of epinephrine and other stress hormones as a normal physiologic reaction to the abnormalities. Finally, with progression of dehydration and metabolic derangement, vital organ systems such as the heart, kidneys, liver, and particularly the brain begin to fail, adding further to the physiologic crisis that develops. As dehydration develops, medical shock develops (lack of blood flow to the vital organs) and the condition known as Systemic Inflammatory Response Syndrome develops with release of inflammatory response proteins that cause internal disruption and death of cells. Those who are resuscitated from severe shock or cardiopulmonary arrest associated with agitation and excited delirium require large amounts of intravenous solutions to replace lost body fluid.

Hyperthermia has been associated with sudden death from agitation and delirium. Hyperthermia leads to dehydration and vascular collapse. Persons abusing stimulant drugs are at particular risk for hyperthermia because the usual thermal regulatory systems of the brain are altered. Additionally, stimulants increase the metabolic rate mimicking the effects of excessive heat. The state of hyperthermia raises the risk for sudden in-custody death by stressing the heart, increasing the demand for oxygen from the respiratory system and raising the risk of acute brain changes. As with dehydration, hyperthermia is clinically recognized to be commonly associated with excited delirium. Resuscitation of a person experiencing excited delirium often requires prompt application of cooling measures when hyperthermia is present.

Along with the clinical recognition of dehydration and hyperthermia associated with sudden in-custody death, metabolic acidosis has been described [24]. Metabolic acidosis, the build up acidic products of metabolism within the body tissues, can be the result of vigorous physical activity as seen with excited delirium and resistance of detention. Metabolic acidosis is also associated with dehydration and hyperthermia. An acidotic state within the tissues of the body leads to poor function of vital organs and eventual vascular collapse. In addition, acidosis raises the risk for fatal cardiac dysrhythmias – typically asystole and pulseless electrical activity – which are not treatable with defibrillation shocks. Clinically, metabolic acidosis in the setting of excited delirium requires correction of underlying physiologic disorders such as dehydration, hyperthermia, and potential drug toxicity. For persons who exhibit extreme agitation with muscle hyperactivity pharmacologic sedation (often called chemical restraint) is indicated. It is also common for victims of severe agitation who have required restraint and are stimulant drug toxic to have seizures. Prompt pharmacologic control of seizures is required to minimize increasing the risk of metabolic acidosis and subsequent complications.

25.2.7 Gender, Weight, and Ethnicity

Males are overwhelmingly reported as those who have sudden in-custody death. A female sudden in-custody death has been reported with the woman presenting with typical excited delirium and stimulant drug toxicity found on autopsy [29]. Whether gender places one at higher risk for sudden in-custody death has not been formally studied.

Early theory regarding sudden in-custody death assumed that obesity was a contributing factor with the event. Case series reports have not supported this theory with the reported frequency of obesity noted to be the same as the frequency of obesity in the general population [29,32]. The same similarity in height has been found, with those undergoing sudden in-custody death having heights that reflect that found in the general population.

Ethnicity is not associated with a higher risk for sudden in-custody death. As with obesity and height, the ethnic distribution of those who have sudden in-custody death reflects the ethnicity of the typical arrestee population [29,32].

25.3 Conclusions

Any encounter between law enforcement or public safety personnel and agitated, confused individuals with hyperactivity is a risk for potential sudden in-custody death. There are a number of conditions known to be associated with sudden in-custody death (Table 25.2). Most of these conditions are not in the control of those tasked with detaining and protecting an agitated person who is a danger to himself and others [52–56]. This is often not recognized after the fact and is often the basis for numerous litigation complaints.

Persons exhibiting signs of excited or agitated delirium should always be brought to medical attention. Restraint of these agitated individuals is required to manage the scene and protect them from self-harm and to facilitate moving them to medical evaluation. Standard practice is to use physical restraints that allow for adequate ventilation and maintenance of an open airway. Some Emergency Medical Service systems are now employing the field use of midazolam (Versed®) as a chemical restraint and using aggressive intravenous fluid hydration to correct acidosis. Frequent assessment of vital signs, breathing, pulse, and consciousness is required to quickly respond with cardiopulmonary resuscitation if sudden death occurs.

The use of CEWs has not been shown to have an association with sudden in-custody death. CEWs used by trained law enforcement and public safety personnel can quickly and effectively control a highly agitated person so that he will do no harm to himself or others and can be transported for medical evaluation.

References

1. Mumola C: Arrest-related deaths in the United States, 2003–2005. Bureau of Justice Statistics Special Report 2007.
2. Bell LV: On a form of disease resembling some advanced stages of mania and fever. American Journal of Insanity 1849; 6:97–127.
3. Derby I: Manic-Depressive "Exhaustion' Deaths". Psychiatry Quarterly 1933; 7:436–449.
4. Kraines SH: Bell's Mania (acute delirium). American Journal of Psychiatry 1934; 91:29–40.
5. Lehmann HE, Ban TA: The history of the psychopharmacology of schizophrenia. Canadian Journal of Psychiatry 1997; 42:152–162.
6. Cancro R: The introduction of neuroleptics: A psychiatric revolution. Psychiatry Services 2000; 51:333–335.
7. Bell MD, Rao VJ, Wetli CV, Rodriguez RN: Positional asphyxiation in adults: a series of 30 cases from the Dade and Broward County Florida Medical Examiners Offices from 1982 to 1990. American Journal of Forensic and Medical Pathology 1992; 13:185–192.
8. Reay DT, Flinger CL, Stilwell AD, Arnold J: Positional asphyxia during law enforcement transport. American Journal of Forensic and Medical Pathology 1992; 13:90–97.
9. Bell MD, Rao VJ, Wetli CV, Rodriguez RN: Positional asphyxiation in adults: a series of 30 cases from the Dade and Broward County Florida Medical Examiners Offices from 1982 to 1990. American Journal of Forensic and Medical Pathology 1992; 13:185–192.
10. O'Halloran RL, Lewman LV: Restraint asphyxiation in excited delirium. American Journal of Forensic and Medical Pathology 1993; 14:289–295.
11. Pollanen MS, Chiasson DA, Cairns JT, Young JG: Unexpected death related to restraint for excited delirium: a retrospective study of deaths in police custody and in the community. Canadian Medical Association Journal 1998; 158:1603–1607.
12. Stratton SJ, Rogers C, Green K: Sudden death in individuals in hobble restraints during paramedic transport. Annals of Emergency Medicine 1995; 25:710–712.
13. Reay DT, Howard JD, Flinger CL, Ward RJ: Effect of positional restraint on oxygen saturation and heart rate following exercise. American Journal of Forensic and Medical Pathology 1988; 9:16–18.

14. Karch SB, Green GS, et al.: Myocardial hypertrophy and coronary artery disease in male cocaine users. Journal of Forensic Science 1995; 40:591–595.
15. Karch SB, Stephens BG: Acute excited states and sudden death; acute excited states are not caused by high blood concentration of cocaine. British Medical Journal 1998; 316:1171.
16. Karch SB, Stephens BG, Ho CH: Methamphetamine related deaths in San Francisco: demographic, pathologic and toxicologic profiles. Journal of Forensic Science 1999; 44:359–368.
17. Logan BK Flinger CL, Haddix T: Cause and manner of death in fatalities involving methamphetamine. Journal of Forensic Science 1998; 43:28–24.
18. Wetli CV, Mash D, Karch SB: Cocaine-associated agitated delirium and the neuroleptic malignant syndrome. American Journal of Emergency Medicine 1996; 14:425–428.
19. Chan TC, Vilke GM, Neuman T, Clausen JL: Restraint position and positional asphyxia. Annals of Emergency Medicine 1997; 30:578–586.
20. Schmidt P, Snowden T: The effects of positional restraint on heart rate and oxygen saturation. Journal of Emergency Medicine 1999; 17:777–782.
21. Roeggla M, Wagner A, Muellner M, Bur A, Roeggla H, Hirschl MM, Laggner An, Roeggla G: Cardiorespiratory consequences to hobble restraint. Wien Klin Wochenschr 1997; 109:359–361.
22. Reay DT: Death in custody. Clinical Laboratory Medicine 1998; 18:1–22.
23. Ross DL: Factors associated with excited delirium deaths in police custody. Modern Pathology 1998; 11:1127–1137.
24. Hick JL, Smith SW, Lynch MT: Metabolic acidosis in restraint-associated cardiac arrest: a case series. Academic Emergency Medicine 1999; 6:239–243.
25. Karch SB: Drug abusers who die during arrest or in custody. Journal of the Royal Society of Medicine 1999; 92:110–113.
26. Chan TC, Vilke GM, Neuman T: Reexamination of custody restraint position and positional asphyxia. American Journal of Forensic and Medical Pathology 1998; 19:201–205.
27. Mirchandani HG, Rorke LB, Sekula-Perlman A, Hood IC: Cocaine-induced agitated delirium, forceful struggle, and minor head injury: a further definition of sudden death during restraint. American Journal of Forensic and Medical Pathology 1994; 15:95–99.
28. Chan TC, Neuman T, Clausen J, Eisele J, Vilke GM: Weight force during prone restraint and respiratory function. American Journal of Forensic and Medical Pathology. 2004; 25:185–189.
29. Stratton SJ, Rogers C, Brickett K, Gruzinski G: Factors associated with sudden death of individuals requiring restraint for excited delirium. American Journal of Emergency Medicine 2001; 19:187–191.
30. Robinson L: "Prevention of Excited Delirium Syndrome: The Police and First Responders" In: Excited Delirium Syndrome, Di Maio TG, Di Maio VJM (Eds), Boca Raton, FL, 2006, CRC press, pp. 97–113.
31. Di Maio TG, Di Maio VJM: "Tasers" In: Excited Delirium Syndrome, Di Maio TG, Di Maio VJM (Eds), Boca Raton, FL, 2006, CRC press, p. 42.
32. Strote J, Hutson R: Taser use in restraint-related deaths. Prehospital Emergency Care 2006; 10:447–450.
33. Kroll M, Panescu D, Ho J, Luceri R, IR E, Calkins H, Tchou P: Potential Errors in Autopsy Reports of Custodial Deaths Temporally Associated With Electronic Control Devices: A Cardiovascular Prospective American Academy of Forensic Sciences. San Antonio, TX, 2007:284–285.
34. Swerdlow C, Kroll M, Williams H, Biria M, Lakkireddy D, Tchou P. Presenting Rhythm in Sudden Custodial Deaths After Use of TASER® Electronic Control Device. Heart Rhythm 2008; 5:S44.
35. Final report on the safety assessment of capsicum annuum extract, capsicum annuum fruit extract, capsicum annuum resin, capsicum annuum fruit powder, capsicum

frutescens fruit, capsicum frutescens fruit extract, capsicum frutescens resin and capsaicin. International Journal of Toxicology 2007;26 Suppl 1:3–106.
36. Brown L, Takeuchi D, Challoner K: Corneal abrasions associated with pepper spray exposure. American Journal of Emergency Medicine 2000; 18:271–272.
37. Fishbain DA, Wetli CV: Cocaine intoxication, delirium and death in a body packer. Annals of Emergency Medicine 1981; 10:531–532.
38. Wetli CV, Fishbain DA: Cocaine-induced psychosis and sudden death in recreational cocaine users. Journal of Forensic Science 1985; 30:873–880.
39. Harris DS, Everhart ET, Mendelson J, Jones RT: The pharmacology of cocaethylene in humans following cocaine and ethanol ingestion. Drug and Alcohol Dependence 2003; 24:169–182.
40. Pennings EJ, Leccese AP, Wolff FA, Effects of concurrent use of alcohol and cocaine. Addiction 2002; 97:773–783.
41. Newsome HH, Ethanol modulation of plasma norepinephrine response to trauma and hemorrhage. Journal of Trauma-Injury Infection and Critical Care 1988; 28:1–9.
42. Jentsch JD, Roth RH: The neuropsychopharmacology of phencyclidine: from NMDA receptor hypofunction to the dopamine hypothesis of schizophrenia. Neuropsychopharmacology 1999; 20:201–225.
43. Rimsza ME, Moses KS: Substance abuse on the college campus. Pediatric Clinics of North America 2005; 52:307–319.
44. Smally AJ, Nowicki TA: Sedation in the emergency department. Current Opinion in Anaesthesiology 2007; 20:379–383.
45. Lange RA, Hillus LD: Cardiovascular complications of cocaine use. New England Journal of Medicine 2001; 345:351–358.
46. Billman GE, Cocaine: a review of the toxic actions of cardiac function. Crticical Review of Toxicology 1995; 25:113.
47. Han J, Garciadejalon P, Moe GK. Adrenergic Effects on Ventricular Vulnerability. Circulation Research 1964; 14:516–524.
48. Aurigemma GP, Gaasch WH: Clinical practice. Diastolic heart failure. New England Journal of Medicine 2004; 351:1097–1105.
49. Sanderson JE: Heart failure with a normal ejection fraction. Heart 2007;93:155–158.
50. Marks W: Physical restraints in the practice of medicine: current concepts. Archives of Internal Medicine 1992;152:2203–2206
51. 2005 American Heart Association Guidelines for Cardiopulmonary Resuscitation and Emergency Cardiovascular Care, Part 3: Overview of CPR. Circulation 2005;112(suppl IV):IV-12-IV-18.
52. Chan TC, Neuman T, Clausen J, Eisele J, Vilke GM: Weight force during prone restraint and respiratory function. The American Journal of Forensic Medicine and Pathology 2004;25:185–189.
53. Chan TC, Neuman T, Vilke GM, Clausen J, Clark RF: Metabolic acidosis in restraint-associated cardiac arrest. Academic Emergency Medicine 1999;6:1075–6; author reply 1076–7.
54. Chan TC, Vilke GM, Neuman T: Reexamination of custody restraint position and positional asphyxia. The American Journal of Forensic Medicine and Pathology 1998;19:201–205.
55. Chan TC, Vilke GM, Neuman T: Restraint position and positional asphyxia. The American Journal of Forensic Medicine and Pathology 2000;21:93.
56. Chan TC, Vilke GM, Neuman T, Clausen JL: Restraint position and positional asphyxia. Annals of Emergency Medicine 1997;30:578–586.

Chapter 26
Stimulant Abuse and Sudden Cardiac Death

Steven B. Karch

Chronic illegal stimulant usage alters the structure of the heart. Most of these alterations are well recognized, but the underlying etiology of many still remains obscure. Cocaine was the first local anesthetic and, like all local anesthetics, it blocks sodium channels in the heart [1], but it also has the potential ability to block potassium channels as well. Blockade of potassium channels disrupts the normal dispersion of electrical signals in the heart. Methamphetamine produces many of the same changes, but does not directly interact with any of the heart's ion channels.

Cocaine interacts with one particular type of potassium ion channel called hERG [2], and it does so in a unique way. Other drugs that interact with cardiac conduction channels can do so only when the channel is in its open or closed configuration. Cocaine can bind to hERG no matter whether it is open or closed. To complicate matters further, the hERG gene is polymorphic. That is to say, the structure of the hERG channel is different in different people, which may or may not contribute to the electrical perturbations that result in sudden cardiac death (SCD).

Finally, cocaine and methamphetamine also change the actual physical structure and internal chemistry of heart cells [3,4] causing the heart to undergo a process known as remodeling [5]. Except for the occasional unfortunate who has undiagnosed coronary artery disease who tries cocaine or methamphetamine and immediately dies of myocardial infarction, most deaths in stimulant abusers can be explained by a mechanism referred to by cardiac electrophysiologists as the process of "multiple hits," referring to the fact that multiple abnormalities and [2] chemical interactions are required before SCD can occur [6].

S.B. Karch (✉)
Consultant Pathologist and Toxicologist, Berkeley, California
e-mail: skarch@sonic.net

M.W. Kroll, J.D. Ho (eds.), *TASER® Conducted Electrical Weapons: Physiology, Pathology, and Law*, DOI 10.1007/978-0-387-85475-5_26,

26.1 Myocardial Remodeling

Remodeling is the heart's response to insult or injury. It involves both individual heart cells as well as the structural network of fibrous tissue that supports them. Remodeling allows the heart to change shape so that it can adapt to new working conditions. Remodeling can be initiated by many different events. High blood pressure is probably the most common cause in our society because so many cases of high blood pressure go untreated. In the face of increased peripheral vasoconstriction, the heart must pump harder to supply blood to the body because it is pumping against increased resistance. The end result is concentric myocardial hypertrophy, where the heart becomes rounded, enlarged, and thickened [7]. A totally different sort of hypertrophy is seen in athletes whose hearts must pump increased amounts of blood, but do so against normal resistance. This leads to a condition called "eccentric hypertrophy," where the heart is increased in size, but the wall dimensions remain normal and there is no microscopic remodeling [8]. Common to both conditions is the occurrence of cell death (by apoptosis). When the cell dies, it is replaced by collagen in the form of fibrous scar tissue. The greater the injury, the more myocardial collagen content increases and the more scared (fibrotic) the ventricle becomes [9,10].

The remodeling process involves more than just changes in the gross structure of the heart. The process also involves the ion channels that control the depolarization–repolarization process [11]. Cocaine causes both types of remodeling. It activates certain genes in the cell nucleus causing changes in the type and quantity of enzymes produced, both in the ion channels themselves, and, ultimately, in the gross structure of the heart. Activation of one gene in particular, the gene for the enzyme CaMKII (calmodulin kinase II), leads to the enlargement of individual cardiac muscle cells [12].

As the muscle cells enlarge, their internal structure changes. The predominant form of myosin (the fibers that do the actual contraction each time the heart beats) changes, as does production of the hormone known as atrial naturetic factor. The hormonal change leads to the activation of the rennin-angiotensin system, the basic mechanism that underlies hypertension. It also causes a decrease in the number of calcium ions stored within the cardiomyocyte endoplasmic reticulum [13]. The force of muscle contraction depends on calcium concentrations within the cell, and those are largely dependent on the amount of calcium stored within the endoplasmic reticulum. Thus, as the amount of calcium in the endoplasmic reticulum falls, so does cardiac output, and cardiomyopathy may result [14,15]. While all of these disparate processes are occurring, other changes are occurring in the number and function of potassium channels that control the process of myocardial depolarization. Specifically, they become less efficient [11,16].

Contraction of the heart depends on the depolarization of individual heart cells, which is initiated when potassium ions pass outward through specific channels. Figure 26.1 show a depolarization–repolarization cycle, and where in the cycle each sodium and potassium channel opens. Potassium flow occurs in

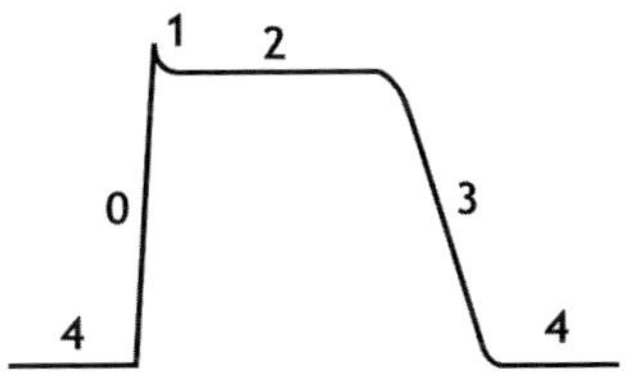

Fig. 26.1 This diagram shows the action potential for a normal heart beat. The numbers indicated the order in which pores open and close, allowing ion movement to produce the action potential. At time 0 sodium enters the cell through fast sodium channels; at time 1, fast sodium channels close; at time 2, calcium and additional sodium ions enter the cell through slow channels; at time 3, potassium exits the cell, and resting membrane potential is reestablished. At time 4, equilibration of sodium and potassium occurs. The hERG channel opens and closes during phase 3. Drawing taken from Wikipedia, with permission

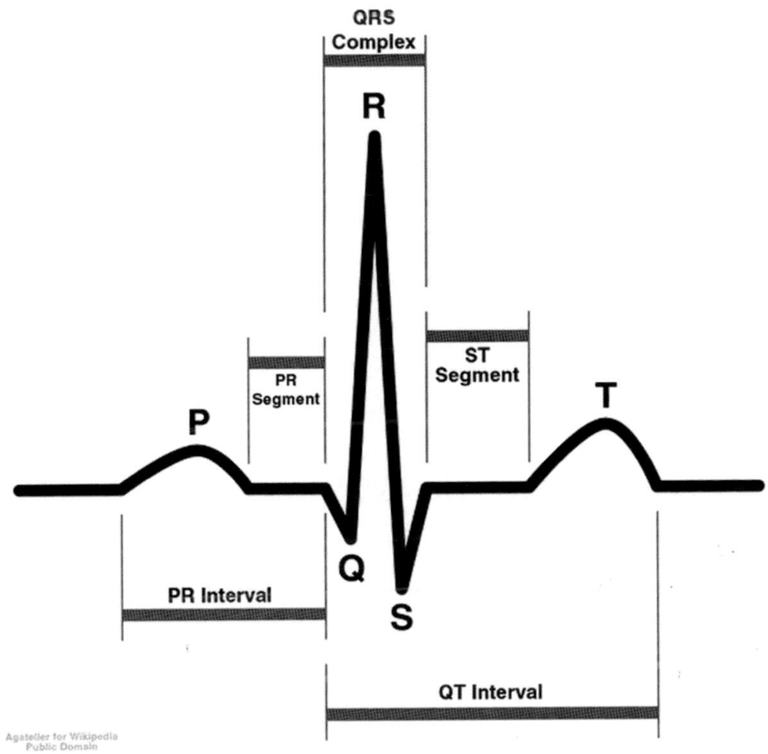

Fig. 26.2 The different intervals measured in the electrocardiogram during one normal depolarization. The QT interval measures the time it takes to repolarize after contraction. The longer the QT interval, the greater the chance of cardiac arrhythmia, especially a type of ventricular tachycardia known as torsades des pointes

2 different phases – the first rapid and the second slow. If these channels do not function normally, the individual heart cells take longer to repolarize, and a process known as QT prolongation may occur. QT prolongation favors the occurrence of sudden cardiac death, especially if there are other drugs present that also interact with the hERG potassium chain to prolong depolarization [17]. Figure 26.2 shows the different, distinct, electrical intervals that occur during one depolarization of the heart.

Two separate genes (KCNQ1 and KCNE1) must function together to produce a potassium channel. The channel itself is actually composed of many different molecular complexes, each with different functions. Some really do behave as pores, while other molecules hold the pore in place, or in some way modify pore behavior. One of the vital molecules involved in channel alteration and control is calcium-dependent phosphatase calmodulin; this molecule binds to L-type calcium channels, sodium channels, and most of the potassium family of channels [18]. It also binds to KCNQ1. When calmodulin binds to the KCNQ1 portion of the channel, flow within that channel is modified. If a mutation occurs, and they occur with some frequency, then flow through the channel can be exaggerated or, alternatively, stopped completely.

Cardiac enlargement only occurs when a molecule called calmodulin interacts with CaMKII. The existence of this complex is responsible not only for cardiac enlargement but also arrhythmias in patients (and experimental animals) with structural heart disease [19]. How this occurs in not known with certainty, but what is known for certain is that both cocaine and methamphetamine directly activate the gene that produces CaMKII [20], and that activation of that gene leads to myocardial hypertrophy and myocardial remodeling [21]. The only important difference between cocaine and methamphetamine is that methamphetamine only causes hypertrophy – it does not interact with any known conduction channel. It has been known for many years that myocardial hypertrophy, no matter the cause, is a strong predictor for sudden cardiac death [22], even in those who do not use drugs.

26.2 Myocardial Hypertrophy and Ischemic Sudden Death

If CaMKII production is activated in a healthy runner, the outcome is called "runners" heart (see above). Enlargement occurs, but the heart retains a normal shape and blood supply [7]. If the process is set off by cocaine or methamphetamine abuse, hypertrophy and fibrosis occur simultaneously, resulting in "concentric hypertrophy." Unlike "runner's heart," which is a benign condition, concentric hypertrophy is potentially lethal [3,23–25]. The pattern of enlargement differs very little from the pattern seen in untreated patients with hypertension. In fact, the modern treatment for hypertension is the administration of drugs that prevent myocardial hypertrophy and remodeling.

Why does concentric hypertrophy increase the risk for SCD? The most likely explanation is that concentrically hypertrophied myocardium is, even at rest, ischemic. The process of new blood vessel formation (angiogenesis) seems to lag the process of hypertrophy. Morphometric (quantitative measurements of microscopic distances measured within and between cells) have shown that the distance between myocytes and the nearest blood vessel are increased in various forms of cardiomyopathy, including ischemic; the distance is much greater than in a normal heart [26–28]. The result is that less blood is available to supply more muscle.

When the blood flow to coronary muscle is reduced, the "coronary flow reserve" (CFR) is diminished. CFR is defined as the ratio of maximum flow to resting flow for any given epicardial coronary artery. The greater the blockage in coronary artery flow, the greater the reduction in CFR [29]. CFR is decreased in cocaine abusers, even if their major coronary vessels do not show signs of significant obstruction. This is partly due to disease of the very small arteries that traverse the heart, and partly because of the mismatch between muscles and blood vessels as described above.

As a consequence, stimulant abusers with left ventricular hypertrophy have, by definition, ischemic myocardium, even when they are at rest. This may explain some cases of SCD in stimulant abusers, especially those that occur during extreme exertion, as might occur in a law enforcement confrontation or emergency medical assistance intervention; the catecholamine surge generated by an altercation will make the heart work harder, and increase its oxygen requirements, which cannot be met because of diminished CFR. Unfortunately, myocardial hypertrophy often goes undiagnosed during life, particularly in patients who are significantly overweight (body mass index >30). A 10% increase in heart weight is likely to go unrecognized at autopsy, even though it is enough to increase an individual's chances of dying [30].

26.3 Myocardial Hypertrophy and QT dispersion

Resting ischemia is not the only reason that stimulant abusers are likely to die suddenly. They are also prone to SCD because their hearts do not propagate the electrical signals normally. The reason for that is two fold: hypertrophic left ventricular cardiomyocytes do not propagate signals normally. The degree of risk from cardiac enlargement can be quantitated. Abnormalities of electrical conduction cause the portion of the electrocardiogram that measures repolarization time (QT interval) to become prolonged. The degree of prolongation is different in each lead of the electrocardiogram (EKG) because each lead reflects electrical activity in a different portion of the heart. The term QT dispersion is used to describe the difference between the longest and shortest QT interval measured in each lead of a 12 lead EKG.

The greater the degree of myocardial dispersion (i.e., the difference between the longest and shortest QT interval in each lead of the cardiogram), the greater the probability that torsades des pointes, a lethal form of ventricular tachycardia, will occur (see below) [31,32]. Under normal circumstances the QT interval should be less than 440 milliseconds. Prolongation of more than 520 milliseconds is associated with 3.5x the normal mortality rate [33]. The simplest explanation for why QT dispersion occurs, and why some parts of the heart repolarize more quickly than others, is that it takes longer for an electrical depolarization front to traverse a thicker ventricular wall than a wall of normal thickness. The heart is thicker in some places than in others, and the process of repolarization begins at different times in different cells [34,35]. If electrical homogeneity does not exist, the probability of arrhythmia increases [36].

26.4 The Theory of "Multiple Hits"

If millions of people use cocaine every month – and they do – and if cocaine interacts with the hERG channel – and it does – why do so few people die of cocaine induced arrhythmia? The simple answer is that more than hERG channel blockade is required; there must also be abnormal signal dispersion, and that generally requires the existence of acquired structural heart disease (hypertrophy and scarring), which is itself the result of complex gene–environment interactions. The most common form of structural disease to be encountered in the general population, cocaine users or not, is likely to be an area of healed myocardial infarction.

The scar resulting from a myocardial infarction leads to ventricular dysfunction (there is less heart muscle to do the pumping), electrical and structural remodeling, and abnormal impulse propagation. All of these abnormalities occur more or less at the same time. Should there also be an electrolyte imbalance (for example, hypokalemia secondary to diuretic therapy for hypertension), or neurohumoral activation [37], for example, elevated levels of norepinephrine from stimulant abuse or an altercation, then either might serve as a trigger for generating an arrhythmia [6]. Of course it also helps if the hERG channel itself is, in some way abnormal, which is not uncommon [38].

Stimulant induced sudden death is a disease with an incubation period. Cocaine has the ability to interfere with sodium and potassium channel function as soon as it is taken, but that seldom leads to illness, let alone lethal arrhythmias. Structural alterations are required before arrhythmias can occur; months, or perhaps years, of chronic use are required before these alterations occur. Deaths that occur after the first or second use (barring substantial overdose) are almost always due to preexisting structural abnormalities, such as coronary artery disease, hypertension induced cardiac hypertrophy, and small vessel disease, as might be seen in a diabetic or poorly treated hypertensive.

26.5 Excited Delirium and Sudden Death

Cocaine abusers have elevated circulating catecholamine (epinephrine and norepinephrine) concentrations. This effect has been demonstrated both in experimental animals and in humans [39–43] and this elevation occurs even when cocaine users are not exercising. Muscle contraction depends on calcium concentrations within the cell; calcium levels must rise roughly 100-fold before a contraction cycle can be initiated. All drugs that are given to help the failing heart (ionotropes) work by causing a slight increase in the amount of calcium within the cell. High catecholamine concentrations cause calcium to enter cardiomyoctes, which is one of the reasons that catecholamine concentrations rise during exercise, but there is a downside to increasing intracellular calcium; it briefly lowers the fibrillation threshold [44,45].

Some have theorized that this is why an individual delirious from drug use, with an enlarged and scarred heart, might die suddenly during a police confrontation. Catecholamines do lower the defibrillation threshold – but only for the first few minutes – after which time the threshold actually increases [46].

The other drawback to this catecholamine theory is the fairly universal observation that those succumbing to excited delirium, even though their hearts are almost always enlarged and fibrotic, are almost never found in ventricular fibrillation. Rather, those succumbing to excited delirium are almost always found to be in asystole or PEA (pulseless electrical activity) [47,48]. This finding

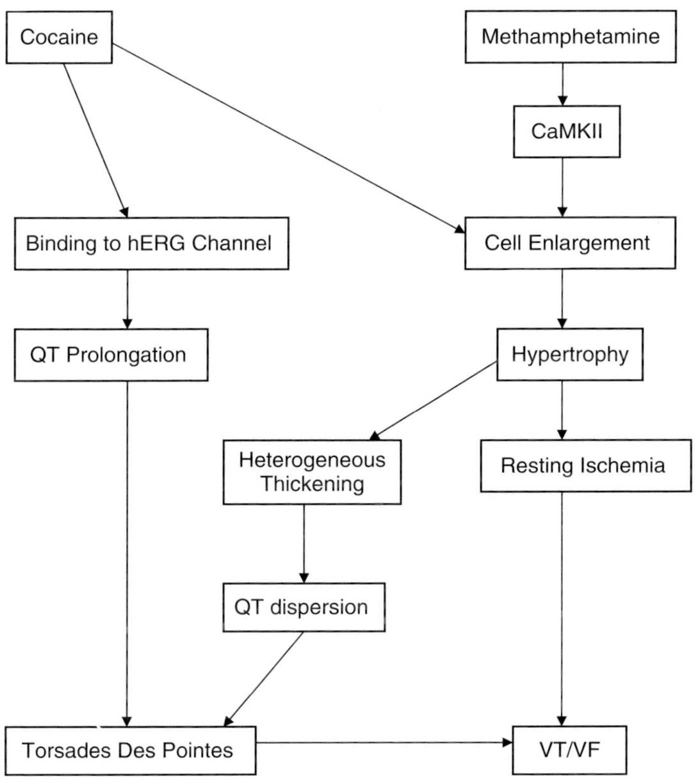

Fig. 26.3 The effects of cocaine and methamphetamine on the heart

suggests, but does not prove, that cardiac arrest in this unfortunate group of individuals, in spite of their obvious heart disease, is centrally mediated. There appears to be some abnormality in the way the brain regulates the heartbeat. What this abnormality might be is not known, but we do know that chronic stimulant abuse disrupts numerous transmitter systems in the brain [49–51] and there is no reason to rule out the possibility that primary brain abnormalities, rather than heart disease, is the proximate cause of death.

An emerging theory for the death in excited delirium cases is that acidosis leads directly to asystole. Only time, and more research will tell what the exact mechanisms of death are in the cases.

26.6 Conclusions

The chronic abuse of stimulants such as cocaine and methamphetamine has multiple and devastating effects on the heart (Fig. 26.3). These effects put stimulant abusers at a highly elevated risk for sudden cardiac death.

References

1. Ma YL, Peters NS, Henry JA. Alpha 1-acid glycoprotein reverses cocaine-induced sodium channel blockade in cardiac myocytes. *Toxicology* 2006; 220:46–50.
2. Karle CA, Kiehn J. An ion channel 'addicted' to ether, alcohol and cocaine: the HERG potassium channel. *Cardiovasc Res* 2002; 53:6–8.
3. Karch SB, Green GS, Young S. Myocardial hypertrophy and coronary artery disease in male cocaine users. *J Forensic Sci* 1995; 40:591–5.
4. Karch SB, Stephens BG, Ho CH. Methamphetamine-related deaths in San Francisco: demographic, pathologic, and toxicologic profiles. *J Forensic Sci* 1999; 44:359–68.
5. Opie LH, Commerford PJ, Gersh BJ, Pfeffer MA. Controversies in ventricular remodelling. *Lancet* 2006; 367:356–67.
6. Shah M, Akar FG, Tomaselli GF. Molecular basis of arrhythmias. *Circulation* 2005; 112:2517–29.
7. de Simone G. Concentric or eccentric hypertrophy: how clinically relevant is the difference? *Hypertension* 2004; 43:714–5.
8. Maron BJ, Towbin JA, Thiene G, Antzelevitch C, Corrado D, Arnett D, Moss AJ, Seidman CE, Young JB. Contemporary definitions and classification of the cardiomyopathies: an American heart association scientific statement from the council on Clinical Cardiology, Heart Failure and Transplantation Committee; Quality of Care and Outcomes Research and Functional Genomics and Translational Biology Interdisciplinary Working Groups; and Council on Epidemiology and Prevention. *Circulation* 2006; 113:1807–16.
9. Abbate A, Scarpa S, Santini D, Palleiro J, Vasaturo F, Miller J, Morales C, Vetrovec GW, Baldi A. Myocardial expression of survivin, an apoptosis inhibitor, in aging and heart failure. An experimental study in the spontaneously hypertensive rat. *Int J Cardiol* 2006; 111:371–6.
10. Lafontant PJ, Field LJ. The cardiomyocyte cell cycle. *Novartis Found Symp* 2006; 274:196–207; discussion 208–13, 272–6.

11. Swynghedauw B. Molecular mechanisms of myocardial remodeling. *Physiol Rev* 1999; 79:215-62.
12. Henning RJ, Cuevas J. Cocaine activates calcium/calmodulin kinase II and causes cardiomyocyte hypertrophy. *J Cardiovasc Pharmacol* 2006; 48:802–13.
13. Henning RJ, Silva J, Reddy V, Kamat S, Morgan MB, Li YX, Chiou S. Cocaine increases beta-myosin heavy-chain protein expression in cardiac myocytes. *J Cardiovasc Pharmacol Ther* 2000; 5:313–22.
14. Wijetunga M, Seto T, Lindsay J, Schatz I. Crystal methamphetamine-associated cardiomyopathy: tip of the iceberg? *J Toxicol Clin Toxicol* 2003; 41:981–6.
15. Arora S, Alfayoumi F, Srinivasan V. Transient left ventricular apical ballooning after cocaine use: is catecholamine cardiotoxicity the pathologic link? *Mayo Clin Proc* 2006; 81:829–32.
16. Furukawa T, Kurokawa J. Potassium channel remodeling in cardiac hypertrophy. *J Mol Cell Cardiol* 2006; 41:753–61.
17. Guo J, Gang H, Zhang S. Molecular determinants of cocaine block of human ether-a-go-go-related gene potassium channels. *J Pharmacol Exp Ther* 2006; 317:865–74.
18. Wu Y, Temple J, Zhang R, Dzhura I, Zhang W, Trimble R, Roden DM, Passier R, Olson EN, Colbran RJ, Anderson ME. Calmodulin kinase II and arrhythmias in a mouse model of cardiac hypertrophy. *Circulation* 2002; 106:1288–93.
19. Kirchhof P, Fabritz L, Kilic A, Begrow F, Breithardt G, Kuhn M. Ventricular arrhythmias, increased cardiac calmodulin kinase II expression, and altered repolarization kinetics in ANP receptor deficient mice. *J Mol Cell Cardiol* 2004; 36:691–700.
20. Ouchi Y, Kubota Y, Ito C. Serial analysis of gene expression in methamphetamine- and phencyclidine-treated rodent cerebral cortices: are there common mechanisms? *Ann N Y Acad Sci* 2004; 1025:57–61.
21. McKinsey TA. Derepression of pathological cardiac genes by members of the CaM kinase superfamily. *Cardiovasc Res* 2007; 73:667–77.
22. Kannel WB, Abbott RD. A prognostic comparison of asymptomatic left ventricular hypertrophy and unrecognized myocardial infarction: the Framingham Study. *Am Heart J* 1986; 111:391–7.
23. Haider AW, Larson MG, Benjamin EJ, Levy D. Increased left ventricular mass and hypertrophy are associated with increased risk for sudden death. *J Am Coll Cardiol* 1998; 32:1454–9.
24. Karch SB, Stephens BG. Drug abusers who die during arrest or in custody. *J R Soc Med* 1999; 92:110–3.
25. Patel MM, Belson MG, Wright D, Lu H, Heninger M, Miller MA. Methylenedioxymethamphetamine (ecstasy)-related myocardial hypertrophy: an autopsy study. *Resuscitation* 2005; 66:197–202.
26. Yarom R, Levy E, Horowitz M. Myocardial pathology in rats exposed to prolonged environmental heat. *Cardiovasc Res* 1990; 24:982–6.
27. Mosseri M, Yarom R, Gotsman MS, Hasin Y. Histologic evidence for small-vessel coronary artery disease in patients with angina pectoris and patent large coronary arteries. *Circulation* 1986; 74:964–72.
28. Mosseri M, Schaper J, Admon D, Hasin Y, Gotsman MS, Sapoznikov D, Pickering JG, Yarom R. Coronary capillaries in patients with congestive cardiomyopathy or angina pectoris with patent main coronary arteries. Ultrastructural morphometry of endomyocardial biopsy samples. *Circulation* 1991; 84:203–10.
29. Bishop AH, Samady H. Fractional flow reserve: critical review of an important physiologic adjunct to angiography. *Am Heart J* 2004; 147:792–802.
30. Karch SB, Wetli CV. Agitated delirium versus positional asphyxia. *Ann Emerg Med* 1995; 26:760–1.
31. Anderson KP. Sympathetic nervous system activity and ventricular tachyarrhythmias: recent advances. *Ann Noninvasive Electrocardiol* 2003; 8:75–89.

32. Anderson ME. QT interval prolongation and arrhythmia: an unbreakable connection? *J Intern Med* 2006; 259:81–90.
33. Antzelevitch C. Cardiac repolarization. The long and short of it. *Europace* 2005;7 Suppl 2:3–9.
34. Kang J, Reynolds WP, Chen XL, Ji J, Wang H, Rampe DE. Mechanisms underlying the QT interval-prolonging effects of sevoflurane and its interactions with other QT-prolonging drugs. *Anesthesiology* 2006; 104:1015–22.
35. Schillaci G, Pirro M, Ronti T, Gemelli F, Pucci G, Innocente S, Porcellati C, Mannarino E. Prognostic impact of prolonged ventricular repolarization in hypertension. *Arch Intern Med* 2006; 166:909–13.
36. Schmidt A, Azevedo CF, Cheng A, Gupta SN, Bluemke DA, Foo TK, Gerstenblith G, Weiss RG, Marban E, Tomaselli GF, Lima JA, Wu KC. Infarct tissue heterogeneity by magnetic resonance imaging identifies enhanced cardiac arrhythmia susceptibility in patients with left ventricular dysfunction. *Circulation* 2007; 115:2006–14.
37. Burns J, Sivananthan MU, Ball SG, Mackintosh AF, Mary DA, Greenwood JP. Relationship between central sympathetic drive and magnetic resonance imaging-determined left ventricular mass in essential hypertension. *Circulation* 2007; 115: 1999–2005.
38. Witchel HJ. The hERG potassium channel as a therapeutic target. *Expert Opin Ther Targets* 2007; 11:321–36.
39. Gunne LM, Jonsson J. Effects of cocaine administration on brain, adrenal and urinary adrenaline and noradrenaline in rats. *Psychopharmacologia* 1964; 6:125–9.
40. Chiueh CC, Kopin IJ. Radioenzymatic paper-chromatographic assay for dopamine and norepinephrine in cerebroventricular cisternal perfusate of cat following administration of cocaine or d-amphetamine. *J Neurochem* 1978; 31:561–4.
41. Dixon WR, Chang AP, Machado J, Lau B, Thompson A, Gallagher S, Sanders W. Effect of intravenous infusion and oral self-administration of cocaine on plasma and adrenal catecholamine levels and cardiovascular parameters in the conscious rat. *NIDA Res Monogr* 1989; 95:335–6.
42. Kiritsy-Roy JA, Halter JB, Gordon SM, Smith MJ, Terry LC. Role of the central nervous system in hemodynamic and sympathoadrenal responses to cocaine in rats. *J Pharmacol Exp Ther* 1990; 255:154–60.
43. Mahlakaarto J, Ruskoaho H, Huttunen P, MacDonald E, Pasanen M. Norcocaine is a potent modulator of haemodynamic responses, plasma catecholamines and cardiac hormone release in conscious rats. *Toxicology* 1998; 128:101–11.
44. Hong K, Kusano KF, Morita H, Fujimoto Y, Nakamura K, Yamanari H, Ohe T. Involvement of Ca(2+) in antiarrhythmic effect of ischemic preconditioning in isolated rat heart. *Jpn J Physiol* 2000; 50:207–13.
45. Pak HN, Kim YH, Lim HE, Chou CC, Miyauchi Y, Fang YH, Sun K, Hwang C, Chen PS. Role of the posterior papillary muscle and purkinje potentials in the mechanism of ventricular fibrillation in open chest dogs and Swine: effects of catheter ablation. *J Cardiovasc Electrophysiol* 2006; 17:777–83.
46. Han J, Garciadejalon P, Moe GK. Adrenergic effects on ventricular vulnerability. *Circ Res* 1964; 14:516–24.
47. Stratton SJ, Rogers C, Brickett K, Gruzinski G. Factors associated with sudden death of individuals requiring restraint for excited delirium. *Am J Emerg Med* 2001; 19:187–91.
48. Swerdlow C, Kroll M, Williams H, Biria M, Lakkireddy D, Tchou. P. Presenting rhythm in sudden custodial deaths after use of TASER® electronic control device. *Europace* 2008;8 submitted.
49. Ruttenber AJ, Lawler-Heavner J, Yin M, Wetli CV, Hearn WL, Mash DC. Fatal excited delirium following cocaine use: epidemiologic findings provide new evidence for mechanisms of cocaine toxicity. *J Forensic Sci* 1997; 42:25–31.

50. Staley JK, Talbot JZ, Ciliax BJ, Miller GW, Levey AI, Kung MP, Kung HF, Mash DC. Radioligand binding and immunoautoradiographic evidence for a lack of toxicity to dopaminergic nerve terminals in human cocaine overdose victims. *Brain Res* 1997; 747:219–29.
51. Qin Y, Ouyang Q, Pablo J, Mash DC. Cocaine abuse elevates alpha-synuclein and dopamine transporter levels in the human striatum. *Neuroreport* 2005; 16:1489–93.

Chapter 27
The Systemic Role of Illicit Drugs and Their Toxicology

Joshua Gunn, Michael A. Evans, and M. Scott Kriger

Illicit drug use is commonly associated with sudden death attributed to excited delirium syndrome. Many of these deaths occur during or shortly after a physical struggle. Excited delirium syndrome sequentially passes through four stages including hyperthermia (typically), delirium with agitation, respiratory arrest, and death. Illegal street drugs such as cocaine, methamphetamine and phencyclidine (PCP) have long been known to cause episodes of excited delirium due to their ability to interact with certain brain systems resulting in markedly elevated levels of catecholamines. Interactions of such drugs with the central nervous system (CNS) have also been linked to the acute onset of mania and violent behaviors, including aggression, combativeness, hyperactivity, extreme paranoia, hallucinations, superhuman strength, or incoherent shouting. Subjects exhibiting such violent behaviors often encounter a physical struggle with law enforcement or emergency medical personnel attempting to restrain them. In such cases, multiple personnel are often required to restrain the individual by using increased levels of physical force or multiple restraint techniques due to the elevated level of physical strength displayed by the subject as a result of chemical impairment. Following successful restraint, agitation ceases and subjects become tranquil and unresponsive. In addition to being responsible for the initial onset of symptoms associated with excited delirium, illicit substances such as cocaine, methamphetamine and PCP also appear to play a significant role in the sudden death of subjects experiencing this condition. Although the exact mechanism of these sudden deaths is still the topic of some debate, it appears that the physiological effects of the physical struggle combined with the sympathomimetic effects of these drugs causes a toxic catecholamine assault on the cardiovascular system. Compounding these effects even further are the sudden fluctuations in electrolyte concentrations that are possible following strenuous exercise such as a physical struggle. Increased myocardial oxygen demand, combined with poststruggle

J. Gunn (✉)
AIT Laboratories
e-mail: JGunn@aitlabs.com

M.W. Kroll, J.D. Ho (eds.), *TASER® Conducted Electrical Weapons: Physiology, Pathology, and Law*, DOI 10.1007/978-0-387-85475-5_27,

hypokalemia is thought to lead to prolongation of the QT interval, development of ventricular tachycardia, and fatal arrhythmias causing sudden cardiac death. While sudden fluctuations in electrolyte concentrations are consistent with fatal cardiac arrhythmias, kalemic cardiac arrest does not necessarily explain why almost all of these sudden arrest-related deaths proceed through an initial stage of respiratory arrest. Until the mechanism of this initial period of quiescence is identified, several possible causations including kalemic cardiac arrest remain largely theories rather than proven etiologies as the mechanism of death due to excited delirium syndrome.

27.1 Cocaine

Cocaine is obtained from the leaves of the plant *Erythroxylon coca* and is included in the tropane alkaloid family. Although the leaves of *E. coca* and other related species have been used by the Peruvian Indians for centuries to increase endurance and improve well-being, the active ingredient cocaine, was not isolated until the mid-nineteenth century. Albert Niemann, a graduate student at the University of Gottingen was the first person to devise and report a technique for the isolation of cocaine in 1860.

In 1884, 24 years after the first reported isolation of cocaine, Dr. Karl Koller discovered that cocaine was an effective local anesthetic. In the years following Koller's discovery, physicians around the world were employing cocaine as an anesthetic in ophthalmologic, dental, and general surgical procedures. Although analogs of cocaine such as procaine and lidocaine are still employed as anesthetic agents and antiarrhythmic agents, the free distribution of cocaine was banned by the Harrison Narcotics Act in 1914. By the mid-1900s the recreational use of cocaine had become a significant concern across socioeconomic lines.

Cocaine is widely available on the street in either the base form (street name "*crack*") or as the hydrochloride salt. Although both forms of the drug are available in high purity at a similar street value, the free-base "crack" cocaine is predominately used for smoking while cocaine hydrochloride is mainly used for intravenous injection and nasal insufflations ("snorting").

27.1.1 Mechanism of Action

Cocaine is a naturally occurring central nervous system (CNS) stimulant that interferes with the actions of dopamine, norepinephrine, and serotonin in functioning nerves. Cocaine is classed as a sympathomimetic agent due to its ability to activate the sympathetic nervous system both centrally and peripherally [1]. Stimulation of the sympathetic nervous system results from cocaine's ability to selectively bind dopamine reuptake transporters (DATs) in the brain [2]. Clearance of dopamine at the synapse and subsequent termination of dopaminergic neurotransmission is achieved through reuptake into the presynaptic neuron which is mediated by DATs [3]. By binding to DATs, cocaine impairs the reuptake of dopamine into the presynaptic neuron resulting in elevated dopamine levels in the central nervous system. Direct stimulation of the central nervous system by cocaine and other sympathomimetic agents also results in increased norepinephrine release from peripheral synapses. Cocaine not only facilitates increased release of norepinephrine peripherally, it also acts to inhibit its reuptake at the synapses causing it to remain in the synaptic cleft for a prolonged period of time. As a result of cocaine's ability to inhibit the reuptake of dopamine centrally and norepinephrine peripherally, the natural effect of these neurotransmitters is amplified. Excessive levels of CNS dopamine and peripheral norepinephrine account for the feelings of euphoria and increased alertness associated with cocaine use [1].

When self-administered, cocaine doses range from 10 to 120 milligrams. Cocaine's ability to release norepinephrine and other catecholamines results in almost immediate cardiovascular effects such as elevated blood pressure and heart rate. Vasoconstrictor actions of cocaine and its major metabolite, benzoylecgonine cause the coronary arteries to constrict, reducing blood supply to the heart [4]. At higher doses, cocaine can produce marked cardiac arrhythmias and sudden death through its ability to slow impulse conduction by blocking the movement of sodium (Na) ion in cardiac channels. This not only explains cocaine's effectiveness as an anesthetic at lower doses but also provides a possible cellular explanation for its cardiotoxic effects [5]. However, recent studies indicate that cocaine may also act to stabilize membranes and reduce cardiac vulnerability to electrically induced ventricular fibrillation. The ability of cocaine to increase the safety margin of CEW induced ventricular fibrillation suggests that its arrhythmogenicity may not increase vulnerability to ventricular fibrillation unless in the presence of compounding hemodynamic or metabolic disturbances [6].

27.2 Methamphetamine

Methamphetamine is a sympathomimetic amine meaning its effects mimic those of a stimulated sympathetic nervous system. Methamphetamine was first synthesized by Ogata, a Japanese chemist in the 1920s who was attempting to synthesize ephedrine to meet the needs of asthma sufferers [2]. Methamphetamine is a member of the amphetamine drug class which is chemically and pharmacologically related to the hormone epinephrine and the neurotransmitter norepinephrine. The amphetamine drug family was first introduced in the 1920s when amphetamine was marketed as a nasal decongestant replacement for ephedrine. Clandestine synthesis of amphetamine-like drugs began shortly thereafter using a variety of chemical processes and employing a number of chemical precursor molecules. By the time World War II began in the late 1930s, the abuse potential of the amphetamines was well known, however, this did not deter many armies from using amphetamine to maintain alertness and decrease battle fatigue [4]. In today's society, the amphetamines are classed as a drug of abuse with very limited therapeutic value. Of the amphetamine family, methamphetamine is the most commonly used illicitly.

Clandestine synthesis of methamphetamine produces two forms of the drug. The free base ("meth base") and the hydrochloride salt ("meth hydrochloride"). The free base, which is the initial product of a clandestine synthesis, is a liquid at room temperature. The hydrochloride salt is produced by bubbling hydrogen chloride gas through a solution of the free base [7]. The hydrochloride salt is the solid form found on the streets. Methamphetamine is well absorbed by a variety of methods including smoking, nasal insufflation, oral and intravenous use [4].

27.2.1 Mechanism of Action

Methamphetamine is similar to cocaine in its ability to modify the actions and levels of catecholamines. Methamphetamine acts to stimulate the sympathomimetic nervous system both centrally and peripherally. This again is achieved through increasing levels of dopamine and norepinephrine however, the mechanism by which methamphetamine achieves this differs slightly from other stimulants such as cocaine. Methamphetamine is chemically similar to dopamine and norepinephrine allowing it to enter the presynaptic terminal assisted by protein molecules that would normally transport dopamine and norepinephrine back into the nerve terminal from the synaptic cleft. Once in the presynaptic terminal, methamphetamine acts to release dopamine and norepinephrine from vesicles resulting in increased levels of free catecholamines in the nerve ending. Methamphetamine also inhibits monoamine oxidase (MAO), an enzyme responsible for the deactivation of free catecholamines in the presynaptic terminal. As a result, excess levels of dopamine and norepinephrine are transported out of the presynaptic terminal and into the synapse where they produce feelings of pleasure and euphoria.

27.3 Phencyclidine

Phencyclidine (PCP) is a dissociative anesthetic which was administered to humans until 1963 when it was discovered that patients recovering from the anesthetic activity of phencyclidine often developed an altered state of consciousness leading to hallucinations [4]. The dissociative anesthetic effects of PCP also resulted in patients becoming delirious and unmanageable for several hours following surgical procedures [8]. PCP is administered by a variety of means including smoking, nasal insufflation, oral or intravenous injection. Typical recreational doses of PCP range from 1 to 6 milligrams and result in lethargy, disorientation, hallucinations, and loss of coordination [9]. PCP use was prevalent on the USA west coast during the 1960s and early 1970s but it quickly earned a reputation for causing delirium, aggression, and convulsions causing it to nearly disappear from the illicit drug market [10]. In the mid-1970s, PCP reappeared on the drug market in powder form which came to be known as "*Angel Dust.*" Smoking of angel dust often mixed with low-grade marijuana became the preferred route of administration as it slightly alleviated the powerful mind-altering effects of the drug. PCP is still available on the streets in several forms, however, its illicit use has decreased significantly since the mid-1980s.

27.3.1 Mechanism of Action

PCP belongs to the family of dissociative anesthetics and works primarily as an N-methyl-d-aspartate (NMDA) receptor antagonist in the cortex and limbic regions of the brain. Like other arylcyclohexylamine compounds, PCP interacts with several neurotransmitter systems and as a result, produces a combination of CNS stimulant and depressant effects. Glutamatergic alteration is thought to be the primary mechanism in which PCP interferes with cognitive and other functions of the nervous system but studies have shown that the drug also possesses comparable affinities for the dopamine and serotonin receptors [11]. It is this dopaminergic dysregulation that is thought to be responsible for changes seen in behavior while PCP's interaction with the sigma-opiod system in the hippocampus may be responsible for its dysphoria. PCP is also thought to activate the alpha-adrenergic system while inhibiting GABA and blocking the cholinergic system, resulting in anticholinergic activity.

27.4 Excited Delirium Syndrome

Excited or agitated delirium syndrome (also known as Bell's mania, acute exhaustive mania, lethal catatonia) was first described in 1849 by Dr. Luther Bell, an American physician at the McLean Asylum for the Insane. Bell believed he was witnessing the rise of a new disease after noticing that a number of his patients were dying unexpectedly after experiencing a brief period of mania and fever. In the 12 years and 1700 admissions prior to Bell's description of the new disease, he identified 40 cases of excited delirium, 3/4 of which resulted in death. Bell's initial report of the clinical symptoms included an acute onset of mania, violent behavior, need for restraint, refusal of food, inability to sleep and fatigue deteriorating to exhaustion and eventual circulatory collapse. In the mid-1900s century, reports of new cases involving excited delirium simply disappeared from the literature. The disappearance of new reports was attributed to the new treatment regimes introduced for patients suffering from endogenous mental diseases. The introduction of antipsychotic drugs such as chlorpromazine meant that patients no longer required full-time medical assistance and were released back into the community [3].

Excited delirium syndrome was reintroduced in the early 1980s when Fishbain and Wetli described death due to excited delirium in a subject suffering from acute cocaine intoxication [12]. This was the first modern report of excited delirium syndrome and the first time in which the condition had been linked to the use of illicit stimulants. In the following years, new case reports of cocaine-associated excited delirium began to appear in the literature due to the widespread abuse of cocaine across socioeconomic lines. The introduction and widespread popularity of crack cocaine in the early 1990s saw the number of reported cocaine-associated excited delirium cases continue to rise; and in the mid-1990s it was reported that men with agitated delirium accounted for 10% of cocaine deaths in Miami [13].

Many cocaine-associated excited delirium deaths occurred during or shortly after an intense physical struggle with law enforcement officers. Such struggles often occur when excited; agitated, and delirious subjects resist arrest. Subjects are often hyperthermic at the time of the struggle and have been known to exhibit amazing feats of physical strength in attempts to avoid restraint. Following restraint, agitation ceases and subjects have been known to develop a shallow or labored breathing pattern before suffering cardiopulmonary arrest [14].

In 1993, a review of 11 cases of sudden death following restraint in a prone position by police officers was reported [15]. Of the one subjects, nine of the men were "hogtied," one was tied to a hospital gurney, and one was manually held prone. All subjects were reported to be in a state of excited delirium and eight of the subjects were acutely delirious from drugs. The report represents the first case in which death from excited delirium was associated with methamphetamine use. Police were dispatched to a freeway where the subject, a 41-year-old man was running in traffic lanes, shouting, and waving his arms in an excited manner while continuously shouting "Don't let them get me!" The subject was physically removed from the traffic, placed prone on the ground with wrists and ankles restrained behind his back. After breaking one of the ankle restraints the subject suddenly became quiet and stopped breathing. A friend indicated that the subject had smoked methamphetamine several hours before and had become paranoid and excited, hallucinating that police were chasing him [15].

In 2003, Pestaner and Southall detailed two cases which occurred over a 10-year period and in which the manner of death appeared to be excited delirium syndrome associated with PCP use [16]. Although few reports of phencyclidine-associated excited delirium exist in the literature, it presents the same clinical symptoms as excited delirium associated with cocaine and methamphetamine use. In both reported cases subjects were handcuffed and placed in the prone position. Minutes later both subjects were found unresponsive and cardiopulmonary resuscitation attempts were unsuccessful [16].

Since reappearing in the early 1980s this modern form of excited delirium syndrome is characterized by the acute onset of bizarre or violent behaviors, including aggression, combativeness, hyperactivity, extreme paranoia, hallucinations, superhuman strength, or incoherent shouting. Symptoms are often, but not always accompanied by hyperthermia [17]. There are two distinct differences between the clinical symptoms of excited delirium syndrome as described by Bell some 150 years ago and the modern form of the condition. Firstly, Bell states that the time between onset of symptoms and death was anywhere from a matter of days to several weeks (Bell, 1849). In modern cases of excited delirium syndrome, death occurs minutes or hours after the cessation of a struggle. It is likely that the same physiological mechanism that was responsible for the deaths of mentally ill patients in the 1800s is also causing death in excited delirium syndrome victims today. Due to the permanent housing of the mentally ill in psychiatric facilities, however, episodes of excited delirium were masked by the knowledge that the individual was suffering from an intrinsic mental disease. Many of the symptoms of excited delirium such as episodes of mania, violent behavior, refusal of food, inability to sleep, and fatigue most likely occurred regularly in patients and were therefore not acted upon. Eventually patients most likely died of dehydration, electrolyte imbalances and disturbances, chronic catecholamine insult on the cardiovascular system, and acidosis [3].

Following the introduction of antipsychotic drugs in the 1960s many long-term mental facilities were shut down and mentally ill patients were able to return to society due to the sedating and antipsychotic properties of drugs such as chlorpromazine. At approximately the same time as such patients were self-medicating from home, illicit stimulants such as cocaine, methamphetamine, and PCP found widespread popularity as recreational drugs. The combination of these two factors formed the basis for the reappearance of excited delirium syndrome in the early 1980s as subjects suffering from intrinsic mental disease such as schizophrenia would regularly stop taking prescribed medication, leading to a reoccurrence of hallucinations, aggression, and violent behavior. At the same time, users of cocaine, methamphetamine, and PCP were experiencing episodes of bizarre or violent behaviors, including aggression, combativeness, hyperactivity, extreme paranoia, hallucinations, and incoherent shouting. It is at this point that the two types of subjects become indistinguishable from each other and when law enforcement officers are called to neutralize the violent behaviors they have no choice but to physically restrain the subjects. Following physical struggle both types are experiencing excited delirium and will cease agitation and most likely suffer cardiopulmonary arrest. The use of physical force or restraint occurs almost immediately in these cases and because death normally occurs minutes after the subject is physically restrained, there is minimal time lapse between the onset of symptoms and death. Physical restraint would most likely not occur until much later in the intervention in the case of the mentally ill housed in permanent facilities. This, combined with the knowledge that the person was suffering from an intrinsic disease rather than illicit substance intoxication, probably meant that the subject exhibited excited delirium symptoms for a prolonged period of time before physical restraint was used.

The use of physical restraints is a common denominator in a significant proportion of excited delirium deaths and it is because of this that many people view the restraint as possibly being causative. Although physical restraint often precedes deaths attributed to excited delirium, significant anecdotal evidence documents cases where subjects are found dead, unattended and unwitnessed in a death scene consistent with the person having had an excited delirium condition prior to death. The absence of law enforcement officers and therefore physical restraints in such cases suggests that the exact causation of death is more complex than the use of restraints and/or illicit substances. Another notable difference exists between the clinical symptoms of excited delirium syndrome as described by Bell some 150 years ago and the modern form of the condition. Modern day excited delirium victims generally have a history of stimulant abuse. Since the introduction of cocaine, methamphetamine, and PCP as recreational drugs, many users have experienced excited delirium and the number of deaths attributed to excited delirium syndrome associated with illicit substance abuse continues to rise.

27.4.1 Excited Delirium Syndrome: Mechanism of Death

When the body is in a state of stress or alarm, large portions of the sympathetic nervous system often become stimulated simultaneously resulting in a phenomenon known as mass sympathetic discharge [18]. Mass sympathetic discharge results in increased activity of many functions in the human body. Increases in arterial pressure, blood supply to the tissues, rate of cellular metabolism, blood glucose concentration, and mental activity are commonly seen. Such increases in bodily function allow the person to perform strenuous physical activity to a far greater extent than would normally be possible [18]. Since physical stress is usually responsible for excitation of the sympathetic nervous system, it is believed that the purpose of this response is to provide extra energy for the body when it is in a state of stress. The bodies' response to states of stress or alarm is often termed the "fight or flight" response or the sympathetic stress reaction.

When the body perceives danger or stress, a signal is relayed through the hypothalamus and is integrated in the brain stem. Triggered by hypothalamus stimulation, signals then transmit downward through the brain stem to the autonomic control centers and into the spinal cord where they produce a massive sympathetic discharge [3]. Mass sympathetic discharge is characterized by the release of norepinephrine at the synapses of the noradrenergic neurons and norepinephrine and epinephrine into the circulating blood from the adrenals [19]. Norepinephrine is released from the postganglionic endings of the sympathetic nervous system following certain nerve impulses which allow the hormone to permeate the membrane at the fiber ending. Upon release, norepinephrine causes increased heart rate, increases in blood coagulation and glucose concentration, increased mental activity, and constriction of essentially all blood vessels in the body [19]. Norepinephrine continues to exert these effects on the organs of the body until it is reabsorbed into the sympathetic nerves or until it undergoes enzymatic modification. Cells of the adrenal medullae are excited by the direct stimulation of sympathetic nerves and as a result, release large quantities of both norepinephrine and epinephrine into the circulating blood where they are transported to all tissues of the body [19]. Once released into the blood, norepinephrine has almost the same effects on different organs as those caused by direct sympathetic stimulation; however, these effects occur for a prolonged period of time as removal of norepinephrine from blood occurs slowly. Epinephrine imposes a greater effect on cardiac activity than norepinephrine. As a result, cardiac output is increased considerably when epinephrine is released from the adrenals due to its effect on the heart and veins [19].

Catecholamines such as norepinephrine and epinephrine increase the rate and force of heart contraction and cause hypertension which in turn increases the myocardial oxygen demand [20]. Vasoconstriction resulting from the interaction of norepinephrine with adrenoceptors decreases the amount of oxygenated blood being supplied to the myocardium at a time when increased levels are needed [3]. Oxygen deprivation of the myocardium predisposes the subject to development of a cardiac arrhythmia.

Mass sympathetic discharge involving the release of large quantities of catecholamines occurs when an individual is experiencing stress or is in a state of alarm such as during a physical struggle. It is common for individuals to become tachycardic and hypertensive during a physical struggle due to the release of norepinephrine and epinephrine which places stress on the heart [21]. If such a struggle occurs while the subject is under the influence of a sympathomimetic drug such as cocaine or methamphetamine, cardiac stress will be amplified as such drugs act to further increase levels of epinephrine and norepinephrine either through neurotransmitter release or reuptake inhibition.

27.4.2 Hypokalemia and Sudden Cardiac Death

During intense physical exercise to exhaustion, blood potassium concentrations may increase by as much as 5 milliequivalents per liter [22]. Such elevations in concentration occur when potassium ions are released from contracting muscles and into the bloodstream. Within a 5-minute period following exercise, potassium concentrations in arterial plasma rapidly fall to concentrations lower than the preexercise resting values [23]. Such abrupt, large, and sustained changes in blood potassium concentrations during and upon cessation of exercise have been implicated in the genesis of electrophysiological changes leading to sudden cardiac death. Such implications arise due to suggestions that hyperkalemia associated with strenuous exercise, followed by prolonged hypokalemia upon recovery may result in cardiac arrhythmias leading to sudden cardiac death [23]. Such arrhythmias are thought to arise due to the effect that abrupt and substantial changes in plasma potassium concentrations has on cardiac muscle contractility and excitability [24]. Although sudden fluctuations in electrolyte concentrations can predispose individuals to cardiac arrhythmias, the remarkably low incidence of such abnormalities indicates that electrolyte disturbances due solely to exercise are not sufficient to produce significant cardiac arrhythmias [23].

27.4.2.1 The Role of Illicit Drugs

Cocaine, methamphetamine and PCP all promote increased activity of catecholamines leading to chronic elevations in blood concentrations of epinephrine and norepinephrine. Specifically, cocaine inhibits the reuptake of norepinephrine in the synaptic cleft resulting in elevated blood concentrations which are sustained for a prolonged period of time [2]. In the brain, cocaine acts primarily on dopamine-containing nerves and blocks the actions of dopamine transporter molecules normally responsible for recycling the neurotransmitter back to the nerve terminal. Although cocaine-associated excited delirium has the net effect of dangerously increasing the concentration of catecholamines in the blood, initial excited delirium episodes are thought to arise through cocaine's ability to alter both dopamine receptors and transport molecules in the brain [25]. Cocaine's ability to increase dopamine concentrations at the synapse results in highs that are associated with cocaine use and prolonged use can trigger paranoia, bizarre, erratic or violent behavior or delusions [3]. Direct CNS stimulation from cocaine use through interaction with the dopamine system causes increased release of norepinephrine at the peripheral synapses and epinephrine from the adrenals [3]. Prevention of norepinephrine reuptake at the synapses results in significantly elevated norepinephrine levels in cocaine users [26]. Cocaine's interaction with dopamine receptors and transport molecules is also thought to cause hyperthermia, a condition often seen in subjects experiencing excited delirium. Cocaine users experiencing psychotic episodes show marked reductions in the number of dopamine D2 receptors. These receptors which are located in the temperature regulatory centers of the hypothalamus are known to decrease core body temperature. Decreased numbers of these receptors may therefore explain the occurrence of hyperthermia in subjects experiencing excited delirium [13].

Methamphetamine promotes increased release of norepinephrine from the synaptic cleft into general circulation resulting in feelings of euphoria, increased wakefulness, and increased physical activity [2]. Like cocaine, methamphetamine acts on the brain to cause increased accumulation of the neurotransmitter dopamine by decreasing the number and effectiveness of dopamine transporters responsible for reuptake [27]. Methamphetamine-associated excited delirium has the net effect of significantly elevating levels of blood catecholamines. Episodes of excited delirium also arise through methamphetamine's ability to elevate dopamine levels resulting in agitation, confusion, hallucinations, paranoia, and aggressiveness [3].

Phencyclidine (PCP) stimulates the autonomic system centrally and also possesses sympathomimetic effects due to its ability to prevent the reuptake of norepinephrine and dopamine. PCP's ability to interact with a number of brain systems including NMDA receptors in the cortex and limbic regions, the sigma-opioid receptor, and the dopaminergic system result in irritability, hyperactivity, impaired attention, psychosis and mood lability. PCP is believed to cause psychosis and episodes of excited delirium due to its ability to interact with dopamine receptors and transport molecules in the brain. PCP-associated excited delirium results in significantly elevated levels of catecholamines.

Subjects experiencing excited delirium often die during or shortly after a physical struggle which arises due to the erratic and violent behavior exhibited by the individual. Subjects experiencing excited delirium will often experience hallucinations and episodes of psychosis with violent behavior. When law enforcement officers and/or other individuals attempt to restrain the individual for the safety of everyone involved, the pharmacological effects of the illicit drugs combined with the physiologic effects of the physical struggle can lead to an acute cardiac arrhythmia causing sudden cardiac death [21]. Following cessation of strenuous exercise such as a physical struggle, norepinephrine and epinephrine levels continue to rise. This assault of catecholamines on the cardiovascular system combined with the already elevated levels due to drug use can result in catecholamine toxicity. Such toxicity will likely result in fatal arrhythmia leading to sudden cardiac death which is often seen in subjects experiencing excited delirium during a physical struggle. Arrhythmia is most likely due to a surge of catecholamines released by the stress response during the struggle, superimposed on a myocardium already deprived of oxygenated blood due to the sympathomimetic effects of drugs [13]. Physiological effects of markedly elevated levels of catecholamines due to drug use and strenuous physical exercise are further compounded by rapid fluctuations in electrolyte levels following the cessation of the struggle. Rapid decreases in blood potassium concentrations in the minutes following the physical struggle will cause the subject to experience hypokalemia, placing additional stress on the heart. This combined with the catecholamine assault due to illicit drug use and the body's natural stress response are thought to cause myocardial ischemia leading to fatal arrhythmia [3].

Chronic use of cocaine and methamphetamine has been shown to increase heart weight (cardiac hypertrophy) putting chronic users at a greater risk of suffering myocardial ischemia leading to fatal arrhythmia. Myocardial oxygen demand is greater during times of stress and while under the influence of sympathomimetic drugs. An increase in heart weight due to chronic substance abuse will heighten the oxygen demands of the heart [28].

27.5 Specimen Collection

Proper postmortem specimen collection and storage is of paramount importance for the proper interpretation of toxicology results. There are many factors to consider when interpreting toxicology results, including the exact site of specimen collection. Following death, changes in drug and chemical levels can result as drugs and other toxins normally bound in tissue and major organs are released and as unabsorbed stomach contents continue to diffuse from the stomach. There is extensive literature to support the phenomenon known as postmortem redistribution, wherein drug concentrations can vary depending on the time of specimen collection after death and the exact location of specimen collection. Studies have shown that as much as a tenfold difference between central blood (heart) and peripheral blood (femoral) can exist as a result of postmortem redistribution.

Equally as important as the actual specimen collection, is the proper labeling and storage of specimens. Each specimen should be individually packaged and sealed to avoid cross-contamination. Each specimen container should be clearly labeled with the full name of the deceased, the date of collection, the specimen type, and agency case number or other identifier. It is very important that specimen be collected into the appropriate container for testing and storage. Table 27.1 provides some guidance as to the appropriate container for each specimen type. Additionally, it is important that specimens be stored refrigerated (2–8°C) prior to shipping to the toxicology lab for testing.

Table 27.1 Useful specimens for death investigation

Specimen	Analysis	Quantity	Container	Comments
Postmortem blood	Qualitative drug analysis	Minimum 20 mL	NaF/EDTA tube/bottle (or equivalent)	Collect blood from heart – right atrium, inferior vena cava
	Quantitative drug analysis	Minimum 10 mL	NaF/EDTA tube/bottle (or equivalent)	Collect blood from two distinct peripheral sites – preferably right and left femoral veins
	Heavy metals	Minimum 10 mL	Royal blue Top tube	
Antemortem Blood	Quantitative drug analysis	Minimum 10 mL	NaF/EDTA tube/bottle (or equivalent)	Typically admission blood

Table 27.1 (continued)

Specimen	Analysis	Quantity	Container	Comments
Serum or Plasma	Quantitative drug Analysis	Typically 2–5 mL	Serum tube	Typically from admission
Urine	Qualitative and quantitative drug analysis	Minimum 20 mL	Standard urine cup – no preservatives	Useful for determining recent drug use; commonly used to detect drugs of abuse; can be useful in interpreting blood results
Vitreous humor	Volatiles – to include ethanol and acetone Electrolytes Qualitative and quantitative drug analysis	Minimum 1 mL	Collect into a 10 mL tube that does not contain preservative or anticoagulant	Useful for comparison to blood results when there is uncertainty Relatively unsusceptible to postmortem redistribution and microbial activity
Stomach contents	Trace analysis Qualitative drug analysis	10–20 g	Sterile plastic container	Useful in acute poisoning cases and identification of drug substances not fully dissolved or absorbed
Bile	Quantitative and qualitative drug analysis	20–30 mL	Sterile plastic container	
Liver	Quantitative and qualitative drug analysis	~20 g	Sterile plastic container	Typically more case study data for liver tissue
Brain	Quantitative and qualitative drug analysis	~ 20 g	Sterile plastic container	Useful when investigating deaths related to certain drugs (for example cocaine) and volatiles

Table 27.1 (continued)

Specimen	Analysis	Quantity	Container	Comments
Lung	Quantitative and qualitative drug analysis	~20 g	Sterile plastic container	Can be useful for investigating deaths related to volatiles – for example "huffing"
Hair	Quantitative and qualitative drug analysis Heavy metals	Entire length of hair approximately 0.5 cm in diameter bundle (minimum)	Plastic container or bag – take care to accurately mark the proximal end of the hair	Can be useful when investigating deaths due to chronic drug or chemical abuse (for example heavy metal poisoning)
Bone and nails	Trace Analysis – drugs and chemicals Heavy metals	Representative portion	Sterile plastic container	Can be useful when investigating chronic exposure or poisoning cases (e.g., heavy metal poisoning) May also be useful when investigating cases in which no soft tissue remains
Pills, liquids, syringes, contaminated materials, clothing, chemicals and chemical containers	Quantitative and qualitative drug and chemical analysis	Representative amount	Collect and package each substance separately	Trace substances

27.5.1 Urine

If available, a urine specimen, however small, should be collected for all cases. Urine specimens can be of great value when screening for drugs and poisons. The preferred specimen volume is 20 milliliters in a sterile plastic container. The analysis of urine is useful for determining recent drug use and the results are typically used to corroborate the interpretation of blood results. Analysis of ethanol in urine can be of particular value when the validity of the blood result is called into question.

27.5.2 *Blood*

Blood is the obvious specimen of choice for the majority of toxicological analyses. There are, however, some considerations to be made when collecting blood specimens.

Heart blood is typically the easiest source that contains sufficient volume for screening (qualitative) and confirmation (quantitative) testing. The heart blood, however, is susceptible to postmortem redistribution that can complicate interpretation of results. It is typically recommended that approximately 20 milliliters of blood be collected from the heart (preferably right atrium or inferior vena cava). This specimen should be placed into a 20-milliliter sterile plastic bottle that contains NaF preservative and EDTA or other suitable anticoagulant. The container should be properly labeled, sealed, and stored refrigerated.

Peripheral blood, while not always attainable without contamination from other sources, is the specimen of choice for determining drug levels at the time of death. This specimen is less susceptible to postmortem redistribution and contamination from perhaps gastric contents. Again, the specimen should be collected into a sterile bottle or tube containing NaF preservative and EDTA or other suitable anticoagulant. The container should be properly labeled, sealed, and stored refrigerated.

27.5.3 *Vitreous Humor*

Whenever possible, vitreous humor should be collected. The vitreous humor specimen is less susceptible to microbial activity and can be of significant value when investigating a case in which decomposition is a factor. The vitreous humor can be analyzed for drugs, volatiles (including ethanol and acetone), and other biochemical tests such as urea, glucose, and electrolytes.

27.5.4 *Tissue and Hair*

Tissue specimens should be collected for each case. Typically a 20-gram section of liver from the right lobe is collected into a sterile plastic container. Collection from the right lobe minimizes contamination from the bile and is less affected by diffusion of gastric contents from the stomach. In addition to liver tissue, portions of kidney, brain, and lung are typically collected, stored, and tested as necessary. Tissue can be a very valuable source of information when investigating decomposition cases and cases in which blood could not be collected. Bone and nails are another good source for testing when other specimens are not available. Bone and nail specimens are especially useful when investigating chronic exposure or poisoning cases involving arsenic and lead.

27.5.4.1 Hair

In addition to specimens collected to provide insight into drug use just prior to death, it may also be helpful and insightful to the death investigation to have evidence of historical use. Hair specimens are ideal for assessing past use of drugs. While death related to an acute drug overdose is best established from the blood specimen, the analysis of hair specimens can help assess whether or not the death is related to first time drug use or is consistent with a pattern of drug use over the past 6 months to a year. If the death is related to an acute overdose from first time use of the drug, this will not be revealed through hair testing as sufficient time will not have elapsed for absorption of the drug into the hair follicle. However, hair analysis can provide a timeline of drug use back to approximately 1 year. The drug is deposited into the hair in a linear fashion and the length of the hair is the limiting factor in how far back in time the drug can be detected. Typical hair testing provides a 3–6 month window of past drug use.

27.5.5 Specimen Stability

In general, drugs and chemicals, when properly collected and stored are stable for extended periods of time. There are, however, several drugs that are particularly susceptible to instability, even when properly collected and stored. The most prominent of these drugs are cocaine, heroin, and nitrazepam. Cocaine is a drug of abuse that is relatively unstable due to metabolically active enzymes that are viable even in postmortem blood. Therefore, when cocaine use is suspected, it is recommended that specimens be collected as close to the time of death as possible. Additionally, it is very important to collect these specimens into a tube or bottle containing sodium fluoride ($\geq$1%) and stored refrigerated (2–8°C). Heroin is another drug of abuse that is rapidly metabolized to morphine. In many cases, it is difficult to discern heroin use from morphine use because of the rapid rate of metabolism of heroin to morphine. An intermediate metabolite between heroin and morphine is 6-monacetylmorphine (6-MAM). This marker is used to distinguish heroin use from morphine use. Unfortunately this metabolite has a short half-life and often goes undetected. It appears that 6-MAM is more stable in vitreous humor fluid. Therefore, it is important to collect vitreous humor fluid when heroin use is suspected. Nitrazepam is an example of a benzodiazepine that is unstable in blood. It is imperative that a blood specimen be collected into a sodium fluoride stabilized tube and frozen as soon after death as possible to be able to detect the nitrazepam parent drug.

27.6 Drug Screening

Illicit stimulants such as cocaine, methamphetamine, and PCP are routinely identified and quantified using various analytical techniques. The technique of choice largely depends on the laboratory performing the analysis and whether the testing aims to provide qualitative or quantitative information. Although the specifics of testing procedures may vary slightly from laboratory to laboratory, the process of measurement is usually divided into an initial screening test followed by a confirmation test. Most laboratories will perform an initial screening test when asked to analyze any specimen for the presence of drugs. Screening tests provide the analyst with a relatively quick process to determine whether a drug or drug class is likely to be present in a given specimen. Screening tests may be designed to detect the presence of a certain drug class, such as benzodiazepines, or may be a broad drug screen performed using gas chromatography–mass spectrometry (GC/MS) which enables the analyst to screen for many drugs based on retention times and mass spectral data. Following a positive screen result, the presence of a drug or drug class should be confirmed using a second technique which draws on a different chemical principle. Confirmatory tests should be more specific for the target analyte than the initial screen. Mass spectrometry (MS) has become the gold standard for confirmatory testing and should be employed where possible and practical. Mass spectrometry provides the analyst with a "chemical fingerprint" of the sample components, making the unequivocal identification of individual drugs possible, even in the presence of chemically similar compounds. Mass spectrometers coupled to either gas chromatographic (GC/MS) or liquid chromatographic (LC/MS) systems are the most commonly employed analytical techniques for confirmatory testing in the toxicology laboratory. Gas chromatography coupled with mass spectrometry has long been the most commonly

Table 27.2 Toxicology and pharmacokinetics of abused drugs

Drug	Route of administration	Quantity of administration	Detection window	Toxic blood levels (ng/mL)	Lethal blood levels (ng/mL)
Cocaine	Intravenous injection Nasal Insufflation Smoking	10–120 mg	Urine – 1 to 4 days Blood – 8 hours	>250	Average 5300 (Range 900–21,000)
Phencyclidine (PCP)	Intravenous injection Nasal Insufflation Oral Ingestion Smoking	1–3 mg 1–3 mg 2–6 mg 1–3 mg applied to plant material	Urine – 2–7 days Blood – 1–7 days		Average 4800 (Range 300–25,000)
Methamphetamine	Smoking Oral Ingestion	Typically 30 mg for casual user Up to ~2000 mg daily for chronic user	Urine – 1–4 days Blood – 1–5 days	>300	>1,000

employed confirmation technique in the toxicology laboratory. In recent years however, the coupling of liquid chromatography with mass spectrometry has found widespread use due to its decreased sample preparation requirements and increased sensitivity. Liquid chromatography offers the analyst the advantage of introducing samples in aqueous solvents and eliminates the need for chemical derivatization, thus greatly reducing sample preparation time. Liquid chromatography employing tandem mass spectrometry has since become the preferred technique for toxicological analyses due to its superior selectivity and sensitivity which enables the analyst to quantify lower levels of analyte in more complex matrices with reduced sample preparation.

Due to the complex nature of biological specimens commonly encountered in forensic toxicology, drugs such as cocaine, methamphetamine, and PCP must first be chemically extracted from the sample matrix before analytical methodologies can be employed to selectively determine and quantitate the drugs. Postmortem concentrations of illicit stimulants need not be large to be of significant importance when determining the degree of impairment or contribution to cause of death (Table 27.2) [4,29]. As a result, a thorough understanding of the physicochemical properties of each analyte must be attained before selective, efficient chemical extractions from complex biological matrices can be designed. Knowing certain physical and chemical properties about each analyte such as p*K*a(s), partition coefficients, and solubility characteristics in organic solvents will enable the analyst to design extraction techniques capable of effectively removing the target from various sample matrices (Table 27.1) [4,30].

References

1. Ross, D.L. and T.C. Chan. *Sudden Deaths in Custody*. 2006, Totowa, NJ: Humana Press.
2. Karch, S.B.. *Karch's Pathology of Drug Abuse*. 2002, Boca Raton, FL: CRC Press.
3. Di Maio, T.G. and V.J.M. Di Maio. *Excited Delirium Syndrome Cause of Death and Prevention*. 2006, Boca Raton: Taylor & Francis Group.
4. Drummer, O.H.. *The Forensic Pharmacology of Drugs of Abuse*. 2001, London: Arnold.
5. Crumb, W.J. and C.W. Clarkson. *Characterization of cocaine-induced block of cardiac sodium channels*. Biophys J, 1990. **57**(3):589–599.
6. Lakkireddy, D., et al.. *Effects of cocaine intoxication on the threshold for stun gun induction of ventricular fibrillation*. J Am Coll Cardiol, 2006. **48**(4):805–811.
7. Salocks, C. and K.B. Kaley. *Technical Support Document: Toxicology Clandestine Drug Labs/Methamphetamine*. 2003.
8. Greifenstein, F.E., et al.. *A study of a 1-aryl cyclo hexyl amine for anesthesia*. Anesth Analg, 1958. **37**(5):283–294.
9. Lundberg, G., R. Gupta, and S. Montgomery. *Phencyclidine: patterns seen in street drug analysis*. Clin Toxicol, 1976. **9**(4): 503–511.
10. Poklis, A., et al.. *Phencyclidine and violent deaths in St. Louis, Missouri: a survey of medical examiners' cases from 1977 through 1986*. Am J Drug Alcohol Abuse, 1990. **16**(3–4): 265–274.
11. Kapur, S. and P. Seeman. *NMDA receptor antagonists ketamine and PCP have direct effects on the dopamine D(2) and serotonin 5-HT(2)receptors-implications for models of schizophrenia*. Mol Psychiatry, 2002. **7**(8): 837–844.

12. Fishbain, D.A. and C.V. Wetli. *Cocaine intoxication, delirium and death in a body packer.* Ann Emerg Med, 1981. **10**(10): 531–532.
13. Wetli, C., D. Mash, and S. Karch. *Cocaine-associated agitated delirium and the neuroleptic malignant syndrome.* Am J Emerg Med, 1996. **14**(4): 425–428.
14. Stratton, S., et al.. *Factors associated with sudden death of individuals requiring restraint for excited delirium.* Am J Emerg Med, 2001. **19**(3): 187–191.
15. O'Halloran, R. and L. Lewman. *Restraint asphyxiation in excited delirium.* Am J Forensic Med Pathol, 1993. **14**(4): 289–295.
16. Pestaner, J. and P. Southall. *Sudden death during arrest and phencyclidine intoxication.* Am J Forensic Med Pathol, 2003. **24**(2): 119–122.
17. Ruttenber, A., et al.. *Fatal excited delirium following cocaine use: epidemiologic findings provide new evidence for mechanisms of cocaine toxicity.* J Forensic Sci, 1997. **42**(1): 25–31.
18. Guyton, A.C.. *Structure and function of the nervous system.* 1972, Philadelphia, PA: W.B Saunders Company.
19. Guyton, A.C.. *Textbook of medical physiology.* 3rd ed. 1966, Philadelphia, PA: W.B Saunders Company.
20. Billman, G.E.. *Mechanisms responsible for the cardiotoxic effects of cocaine.* FASEB J, 1990. **4**(8): 2469–2475
21. Dolinak, D., E.W. Matshes, and E.O. Lew. *Forensic pathology principles and practice.* 2005, Elsevier Academic Press.
22. Medbø, J. and O. Sejersted. *Plasma potassium changes with high intensity exercise.* J Physiol, 1990. **421**: 105–122.
23. Lindinger, M.. *Potassium regulation during exercise and recovery in humans: implications for skeletal and cardiac muscle.* J Mol Cell Cardiol, 1995. **27**(4): 1011–1022.
24. Gettes, L.S.. *Electrolyte abnormalities underlying lethal and ventricular arrhythmias.* Circulation, 1992. **85**((1 Suppl)): I70–176.
25. Seeman, P. and H. Van Tol. *Dopamine receptor pharmacology.* Trends Pharmacol Sci, 1994. **15**(7): 264–270.
26. Billman, G.E.. *Cocaine: a review of its toxic actions on cardiac function.* Crit Rev Toxicol, 1995. **25**(2): 113–132.
27. McCann, U. and G. Ricaurte. *Amphetamine neurotoxicity: accomplishments and remaining challenges.* Neurosci Biobehav Rev, 2004. **27**(8): 821–826.
28. Karch, S., G. Green, and S. Young. *Myocardial hypertrophy and coronary artery disease in male cocaine users.* J Forensic Sci, 1995. **40**(4): 591–595.
29. Baselt, R.C.. *Disposition of Toxic Drugs and Chemicals in Man.* 7th ed. 2004, Foster City, CA: Biomedical Publications.
30. Karch, S.B.. *Postmortem Toxicology of Abused Drugs.* 2008, Boca Raton, FL: CRC Press.

Chapter 28
Excited Delirium Syndrome

Vincent J.M. Di Maio and Theresa G. Di Maio

Delirium involves an acute (minutes to hours), transient disturbance in consciousness and cognition. It is manifested by disorientation, disorganized and inconsistent thought processes, inability to distinguish reality from hallucinations, disturbances in speech, disorientation to time and place, and misidentification of individuals [1]. When the delirium involves combative and/or violent behavior, it is termed excited delirium (ED).

Excited Delirium Syndrome (EDS) involves the sudden death of an individual, during or following an episode of excited delirium, in which an autopsy fails to reveal evidence of sufficient trauma or natural disease to explain the death. In virtually all such cases, the episode of excited delirium is terminated by a struggle with police or medical personnel, and the use of physical restraint [2]. (There have been some cases in which an individual was found dead in their self-vandalized apartment with circumstances suggesting an ED death.) The individual may go into cardiopulmonary arrest during or within minutes following cessation of the struggle. Attempts at resuscitation are usually unsuccessful. If a cardiac monitor is available at the time of the arrest, the rhythm noted is usually Pulseless Electrical Activity (PEA) or asystole. If resuscitation is "successful," the individual is found to have suffered irreversible hypoxic encephalopathy and death usually occurs in a matter of days.

The entity that we now know as "excited delirium," though in a chronic form, was first described by Dr. Luther Bell in 1849 [3]. Typically, his patients presented with fever, a rapid pulse, a lack of appetite and sleep. They were agitated and anxious, with increasing confusion. Attempt to approach the patients resulted in a violent struggle. The patients continued to deteriorate over a course of weeks before dying. Deaths due to what became known as Bell's Mania continued to be reported in the medical literature until the early 1950s when they abruptly disappeared with the introduction of phenothiazines for treatment of mental illness [4,5].

V.J.M. Di Maio (✉)
Chief Medical Examiner (retired), Bexar County (San Antonio), Texas
e-mail: vincent_dimaio@yahoo.com

M.W. Kroll, J.D. Ho (eds.), *TASER® Conducted Electrical Weapons: Physiology, Pathology, and Law*, DOI 10.1007/978-0-387-85475-5_28,

While all of Bell's patients had mental disease, and their symptoms progressed over days to weeks prior to death, deaths seen today in association with excited delirium mainly involve abusers of stimulants, for example, cocaine or methamphetamine, with symptoms present for only hours. Deaths do still occur in mental patients usually in individuals with schizophrenia and occasionally bipolar disease. In individuals with intrinsic mental disease, death usually occurs following use of restraint because of an acute psychotic episode. Deaths occurring in psychiatric patients not on illegal stimulants may be associated with underlying natural disease or the presence of psychotropic drugs. Many of these drugs are cardiotoxic with some having effects on the cardiovascular system similar to cocaine. Just as with ED due to use of illegal stimulants, if a patient becomes violent, steps must be taken to protect the safety of both the patient and others. If all nonphysical methods are utilized without avail, then medical personnel have to resort to use of physical restraints. Deaths due to EDS also occur, though rarely, in association with chronic alcoholism, mental retardation and degenerative diseases of the brain.

The mechanism of death in EDS is controversial. Since such deaths almost always occur after restraint is either instituted or attempted, the cause of death is often attributed to the application of a "choke hold" or "positional asphyxia." ("Positional asphyxia" is often referred to as "restraint asphyxia.") Because of the circumstances surrounding deaths due to excited delirium, there are often charges of police or medical misconduct. In some cases, allegations of murder are made. When no physical cause for the death is found at autopsy, this is ascribed to a cover-up. Some groups and individuals find it suspicious that the most deaths due to EDS occur in association with use of restraint by police. What they ignore is the obvious. If a person goes into ED outside a medical institution, the police are called. Even if the first responders are Emergency Medical Service personnel, they will call the police. The police are then expected to either arrest the individual or have them transported to a hospital. If the individual becomes violent, the police have to restrain them.

In deaths due to EDS, an autopsy fails to reveal evidence of sufficient trauma or natural disease to explain the death. In regards to trauma, the usual findings are minor abrasions and contusions explainable by the struggle that preceded death. If during the struggle, the individual was either hit in the neck or an arm placed around it, hemorrhage in the neck may be present. In rare instances fractures of the superior horns of the thyroid cartilage or the hyoid bone occur. This leads some individuals to contend that manual strangulation has occurred. What they fail to realize is that both hemorrhage in the neck as well as the aforementioned fractures do not equate to death due to strangulation. They are only *markers* indicating that pressure or a blow to the neck has occurred. These injuries in themselves are not lethal. Death from manual strangulation involves constant pressure to the neck over a number of minutes – generally more than two [6].

The two stimulants most commonly associated with death due to EDS are cocaine and methamphetamine. Occasionally, the drug of abuse is phencyclidine

(PCP). Even rarer are deaths seen in association with acute and chronic alcoholism and diphenhydramine.

In deaths suspected of being due to EDS, the forensic pathologist should never issue a ruling as to cause of death until he is cognizant of all the facts surrounding the death, has performed a complete autopsy and has completed all toxicological testing. Prior to performing an autopsy, the medical examiner will attempt to obtain a history on the deceased as well as a detailed account of the circumstances surrounding the incident. This includes the actions of all individuals involved and whether any medications were administered before or after the arrest. If the individual dies in a hospital or institution complete medical records should be requested as well as any original blood dating back to the time of admission.

28.1 Restraint-Related Death

When police encounter an individual in ED, a number of courses of action are open to them. Initially, they may attempt to reason with or "talk down" the individual. Since the individual is in a state of delirium, this usually does not work. The police will often then try to incapacitate the individual with chemical sprays such as pepper spray. Unfortunately, many individuals in the throes of excited delirium appear to be resistant to these chemical sprays.

The next option is to put the individual in physical restraint. The police will attempt to grab the individual, and handcuff them with their hands behind their back. This usually elicits a violent struggle. During the struggle, the Officers may inadvertently place an arm around the neck or try to apply a neck hold either to incapacitate the individual or to restrain them while handcuffs are placed on the individual. Almost inevitably, as a consequence of the struggle, the individual is brought to the ground. The struggle will continue on the ground with the individual bucking, twisting, kicking, and trying to bite. Since the individual has to be handcuffed with his hands behind him, he has to be held prone during the struggle. After placing handcuffs, the individual usually continues to struggle, thrashing about and kicking out with his feet. The police may then place restraints on the ankles.

As one attempts to hold the individual down to handcuff them and place ankle restraints, pressure is often placed on the back by use of knees or someone lying on the individual. Since the individual has to be handcuffed with his hands behind him, he has to be held prone during the struggle. After being restrained, the individual may immediately cease struggling or continue to struggle for a short time. Following the cessation of the struggle, the individual is generally ignored until suddenly it is realized that they are not breathing. Resuscitation is attempted and is unsuccessful [2,7]. Most individuals suffer their cardiopulmonary arrest at the end of the struggle or within a few minutes following cessation of their struggling; in others the cardiopulmonary arrest is delayed.

The individual may arrest in the vehicle transporting them to jail or a hospital or on arrival at an emergency department.

In the case of ED in a medical facility, medical personnel will hold the arms and legs down, sometimes partly lying over the shoulders to prevent bucking. Medication will then be injected, usually intramuscularly. We recommend a minimum of six medical personnel to physically restrain an individual in excited delirium. Following the injection, manual restraint is continued until the sedative attributes of the medication become effective. Just as in the case with the police, after holding the individual down manually for a short time, struggling ceases. Usually, a minute or two later, someone realizes that the patient has arrested. Even in a medical environment, resuscitation is almost uniformly unsuccessful.

Traditionally, two explanations have been put forth to explain restraint-related deaths: use of a neck hold or "positional asphyxia." There are two types of neck holds: the choke hold and the carotid sleeper hold [6]. With both holds, the arm and forearm are used to compress the neck and thus the carotid arteries. If too much force is used, there can be fracture of the larynx or hyoid. These injuries are merely "markers" of force applied to the neck. They indicate that either pressure has been applied or a blow delivered to the neck. The injuries present, either hemorrhage or fractures, are not in themselves the cause of death. It is prolonged compression of the neck that can cause death.

Compression of the airway usually does not occur and is not necessary for either one of these holds to be effective. About 70% of the blood supply to the brain is provided by the carotid arteries with the remainder supplied by the vertebral arteries. Compression of the carotid arteries for 10–15 seconds produces cerebral hypoxia and loss of consciousness [6]. After the choke hold is released, the victim should regain consciousness within 20–30 seconds. If the carotid arteries are continuously occluded for 2–3 more minutes, on release of the pressure on the neck, respiration will usually not return spontaneously. The individual, however, should respond to cardiopulmonary resuscitation.

Occasionally, it is claimed that the death of the healthy individual, following transient pressure applied to the neck, is due to a vasovagal reaction from stimulation of the carotid sinuses, that is a reflex cardiac death. In normal individuals, pressure on the carotid sinus produces minimal effects with a mild decrease in heart rate (bradycardia) of less than 6 beats a minute and only an insignificant reduction in blood pressure (less then 10 mmHg) [8]. Some individuals, however, have an extreme reaction to stimulation of the carotid sinuses. In individuals with a hypersensitive carotid sinus (i.e. carotid sinus syndrome), there is an exaggeration of the normal response with syncope and marked hypotension occurring. Occasional, deaths have been referenced [9]. Review of the original literature in regard to the alleged deaths, however, reveal that the individuals dying all had serious underlying cardiovascular disease which in itself could explain death. Additional investigation of these phenomena has revealed that hypersensitive vasovagal reactions appear to be confined to individuals over the age of 55 years and who also have significant cardiovascular disease [10]. In contrast, individuals dying in restraint are young males without any natural disease sufficient to explain death.

In 1988, Reay et al. conducted a series of experiments to determine the effects on peripheral oxygen saturation and heart rate when an individual is hog-tied and placed prone following exercise [11]. Peripheral oxygen saturation's and heart rate were determined using a pulse oximeter. They concluded that hog-tie restraint prolongs recovery from exercise as determined by changes in peripheral oxygen saturation and heart rate. They speculated that restriction of thoracic respiratory movements could be one of the mechanisms for this occurrence and recommended that positional restraint and its effects should be considered in the investigation of individuals restrained in the prone position. In an article published in 1993 by O'Halloran and Lewman, the association of restraint with asphyxiation and hogtying with death were codified in the concept of "restraint asphyxia" or "positional asphyxia" [12]. Even with the elimination of hog-tying, however, the number of deaths continued, if not increased. Almost immediately after the concept of positional asphyxia was offered, the concept was expanded such that whenever anyone was restrained and died, positional or restraint asphyxia was said to be the cause of death whatever the position of the deceased, the method of restraint, or the presence of drugs.

The problem was that Reay et al.'s original findings were wrong. This would not be known, however, until 1997 when Chan et al. published their studies on restraint asphyxia. Chan et al. repeated the experiments using a more systematic approach and more sophisticated technology [13]. Pulmonary function testing (forced vital capacity; forced expiratory volume in 1 second and maximal voluntary ventilation) was performed on 15 individuals, ages 18–40 years, in the sitting, supine, prone and (hog-tying) restraint position. The subjects were then subjected to two exercise periods and two rest periods. Exercise consisted of 4 minutes on an exercise bicycle. During the rest periods, determinations of arterial blood gas, pulse rate, oxygenation by co-oximetry and pulse oximetry, and pulmonary function testing (PFT) were performed. Determinations at the rest periods were made with the subject alternatively in the sitting position and restrained position. Changes in the heart rate occurred with exercise with a maximum of 164 ± 18.9 beats per minute at the beginning of the sitting rest period and 174 ± 15.3 beats per minute at the beginning of the restraint rest period. Placing individuals in the restraint position after exercise resulted in restrictive pulmonary functioning as measured by PFT. However, the PFT changes – while statistically significant – were not clinically significant. Based on arterial PO_2 and co-oximetry, oxygenation of blood increased with exercise, what one would expect and in contrast to Reay et al.'s findings [11]. Most important was the fact that there was no evidence of hypoxia in the restraint position after exercise with no evidence of hypercapnia either during exercise or in restraint. Chan et al. concluded that there was no evidence that body position while in the 'hog-tie' or 'hobble' restraint position, in and of itself, causes hypoventilation or asphyxiation.

In an attempt to counter Chan et al.'s work, some individuals now claim that the restraint death is due to compromise in ventilation occurring when an LEO (Law Enforcement Officer) or medical worker applies body weight to the upper torso of an individual in an attempt to restrain them and prevent further struggle. This is usually accomplished by lying across an individual's back, or applying pressure on the back with a knee or hands. Because of this allegation,- Michalewicz et al. conducted a series of experiments published in 2007 [14]. They investigated ventilatory and metabolic demands in healthy adults when placed in the prone maximal restraint position (PMRP), i.e., hog-tie restraint. Maximal voluntary ventilation (MVV) was measured in seated subjects ($n=30$), in the PMRP, and when prone with 90.1–102.3 kilograms (198–225 pounds) of weight on the back. Twenty-seven (27) subjects were then placed in the PMRP and struggled vigorously for 60 seconds. The authors found no clinically important restriction of ventilatory reserve when subjects were placed in the PMRP or when prone with up to 90.1 or 102.3 kilograms of weight on their back. Likewise, when subjects were maximally struggling for 60 seconds while in the PMRP, there were no clinically important limitations of metabolic or ventilatory functions. They stated: "Based on these findings, as well as previously published studies, we suggest that factors other than ventilatory failure associated with the restraining process may be responsible for the sudden unexpected deaths of restrained individuals."

28.2 Cause of Death in EDS

Death in EDS is due to a combination of the normal physiologic changes seen in a struggle, combined with – depending on the case – illicit drugs, prescribed medications and natural disease. In some individuals, polymorphism of cardiac adrenoreceptors with resultant exacerbation of the normal responses to violent physical activity may also play a role.

The sympathetic nervous system is the controller of the "fight or flight" response. Whenever an individual is exposed to stress, there is a widespread physiological reaction throughout the body. This stress can be physical or psychological. The reaction of the body to stress is integrated in the brain through the hypothalamus. Signals are transmitted downward from the hypothalamus through the brain stem, into the spinal cord and then to organs such as the heart producing massive sympathetic discharge with release of the neurotransmitters norepinephrine (noradrenalin) and epinephrine (adrenalin).

Neurotransmitters are substances that travel through synapses (a space between the end of the nerve fiber and the cell of the organ) to deliver information to other neurons or cells [15–17]. They are produced within neurons (nerve cells), stored in vesicles at the end of the axons and released into the synapse upon nerve stimulation. The neurotransmitters we are concerned with are the catecholamines. The principal catecholamines are epinephrine (E), norepinephrine (NE), and dopamine. The main catecholamines in the brain are NE and dopamine; outside the brain NE and E. Outside the brain, sympathetic neurons release NE and the adrenals both NE and E.

Norepinephrine (NE) is released from nerve fibers (axons) into the synapse. The NE interacts with receptors on the cell known as adrenoceptors. Adrenoceptors are sites on cell membranes through which norepinephrine and epinephrine (E) act as neurotransmitters in the brain, the cardiovascular system and other organs. The sympathetic nervous system influences the cardiovascular system through changes in the release of NE from sympathetic nerve terminals and NE and E from the adrenals with these substances then acting on receptors on the organs or tissue.

Reuptake mechanisms involve specific enzymes or diffusion out of the synapse to rapidly inactivate the neurotransmitters released into the synapse. This controls the degree of excitation. Virtually all neurotransmitters are recaptured by transport systems, located at the nerve terminals of the releasing neurons.

The membrane receptors responsible for mediating responses to catecholamines were initially divided into α and β adrenoceptors [15–21]. The α adrenoceptors are differentiated into α_1 adrenoceptors and α_2 adrenoceptors. The α_1 adrenoceptors are usually postsynaptic and in effector organs while $\alpha_2$adrenoceptors are located principally presynaptically and regulate the release of norepinephrine. The α_2 adenoceptors, however, have been identified in both pre and postsynaptic anatomical locations. The α_1 adrenoceptors were thought to be responsible for excitatory responses; α_2 adrenoceptors for inhibitory responses. While true in general, α_2 adrenoceptors can also mediate excitatory responses. Both α_1 and α_2 adrenoceptors can in turn be divided into three subtypes: α_{1A}, α_{1B}, α_{1D}, α_{2A}, α_{2B}, and α_{2C}adrenoceptors.

In the vascular system, both $\alpha_1$adrenoceptors and α_2 adreptors are responsible for vasoconstriction with the $\alpha_1$adrenoceptors dominant in arteries and the α_2 adrenoceptors in veins [20].

There are three β adrenoceptors: β_1, β_2, and β_3. The β_1 and $\beta_2$adrenoceptors mediate the cardiovascular responses to NE released from nerve terminals and to circulating E and NE from the adrenals. The predominate receptor in heart cells is the β_1 subtype. Release of catecholamines causes an increase in the heart rate, and force of contraction by way of β_1 receptors in heart cells.

The small coronary arteries and arterioles are the principal determinants of coronary artery resistance. Both α_1 and α_2 adrenoceptors mediate coronary vasoconstriction, with α_1 predominating in the larger vessels and α_2 in the microcirculation. β_1 receptors are also present in the coronary arteries [22].

Sympathetic activation of normal coronary arteries by either stress or physical activity results in vasodilation of both epicardial and microvessels. In contrast, in the presence of atherosclerosis or endothelial dysfunction, there is vasoconstriction during exercise [22]. This can be sufficient so as to produce myocardial ischemia. The fact that vasoconstriction occurs in the presence of atherosclerosis or endothelial dysfunction is significant in that accelerated development of atherosclerosis and endothelial injury are produced by chronic use of cocaine and methamphetamine [23]. Individuals dying of EDS are typically chronic uses of these stimulants. In addition, there is evidence of a genetic determination of alpha receptor-mediated coronary constriction, with resultant super sensitivity to α_2 receptor-mediated constriction [22]. This is due to a polymorphism in a gene.

On stimulation of the adrenal glands by the sympathetic nervous system, there is released into the blood both NE and E, with E predominating (80% E and 20% NE). These are carried by the blood to the organs where they have basically the same effect as does direct sympathetic stimulation. The only difference is that the effects last 5–10 times as long because these substances are slowly removed from the blood over a period of 1–3 minutes.

28.2.1 Polymorphism

A substantial degree of polymorphism or variation occurs in genes in the general population [24]. Presynaptic $\alpha_2$receptors ($\alpha_{2A}$and α_{2C}adrenoceptors) inhibit the release of NE from cardiac sympathetic nerves via negative feedback. The β_1 receptor is responsive to circulating epinephrine and to NE released from cardiac sympathetic nerves. Activation results in increased cardiac the rate and strength of muscular contraction.

Small et al. demonstrated that in some individuals a polymorph alpha 2c receptor (alpha 2c Del 322-325) produced a substantial loss of the normal negative feed back such that there was an increase in synaptic NE release [25]. In addition, a variant of the $\beta_1$receptors (beta1 Arg 389) enhances $\beta_1$receptor activity. The combination of the two receptor variants in the same individual, resulting in an increase in NE release and enhanced $\beta_1$receptor function at the cardiac myocyte, appeared to act synergistically to increase the risk of heart failure. The presence of polymorphic $\alpha_2$and $\beta_1$receptors may explain why some individuals die following the physical stress of EDS while the bulk of the population does not.

We can see that stimulation of the sympathetic nervous system causes release of NE at the synapses and NE and E into the blood from the adrenals. The NE works on the $\beta_1$myocytes of the heart to cause it to beat harder and faster. This in turn results in a greater demand for oxygen by the myocardium. If the coronary arteries have either endothelial injury or atherosclerosis, typical complications resulting from chronic use of cocaine and methamphetamine, there will be contraction of the coronary arteries with decreased supply of oxygenated blood to the myocardium at a time when increased amounts are needed. This predisposes to development of a cardiac arrhythmia.

28.2.2 Postexercise Peril

In excited delirium syndrome (EDS), death often occurs immediately after the individual is restrained and struggling ceases. This time frame corresponds to the time of "postexercise peril" described by Dimsdale et al. [26]. It is when an individual is unusually susceptible to developing a fatal cardiac arrhythmia.

Dimsdale et al. found that during exercise an individual's blood NE and E increased, with NE increasing more sharply [26]. Peak levels of these catecholamines did not occur during the struggle, however, but in the 3 minutes immediately following cessation of the exercise. Following cessation of exercise, E and NE continue to rise with NE levels more than tenfold above baseline levels and E threefold.

The catecholamines acting on the heart consist of NE from the postsynaptic neurons of the Sympathetic Nervous System and NE and E released from the adrenals. Norepinephrine's action on the heart is predominantly stimulation of the $\alpha_1$and β_1 receptors. Stimulation of β_1 receptors increases heart rate, contractility and velocity of conduction. The $\alpha_1$receptors are found in the coronary arteries. Norepinephrine interacting with $\alpha_1$receptors may cause vasoconstriction, thus decreasing the amount of oxygenated blood being supplied to the myocardium at a time when an increased amount is needed due to greater demand being put on the heart resulting from the stimulation of the β_1 receptors. Constriction occurs if there is either atherosclerosis or endothelial dysfunction of the coronary vessels, a finding common in chronic abusers of stimulants [22,23]. Epinephrine interacts with all four receptors (α_1,α_2, β_1, and β_2) reinforcing the cardiac and coronary actions of NE.

Compounding the physiological actions of elevated levels of catecholamine are changes in blood potassium levels. Young et al. investigated the relationship between stress and blood catecholamine and potassium levels [27]. Like Dimsdale et al, they found that the highest levels of plasma catecholamines occurred during the 3 minutes postexercise. In their studies, E levels peaked 1 minute after cessation of exercise, at which point the mean value was more than eightfold mean resting level. Elevated E levels were present for several additional minutes, with a mean value of 1.5 times the resting level at 10 minutes postexercise. For NE, 1 minute following cessation of exercise the level was seven times greater than the resting level. At 10 minutes, the NE level was two times the resting level.

Blood potassium concentrations have a very narrow range of safety (3.5–5.1 milliequivalents per liter). Fatal cardiac arrhythmia is associated with both hyperkalemia (>5.1 milliequivalents per liter) and hypokalemia (<3.5 milliequivalents per liter). With mild hypokalemia (serum potassium of 3.0–3.5 millimoles per liter), there are usually no symptoms [28]. Severe hypokalemia (<2.5 millimoles per liter) produces symptoms including muscle necrosis. Both mild and severe hypokalemia increase the incidence of cardiac arrhythmia [28]. Hypokalemia predisposes to prolongation of the QT-interval, development of torsade de pointes and sudden cardiac death [29].

Young et al. found that during exercise, the mean plasma potassium increased slightly more than 1 milliequivalents per liter [27]. Following cessation of exercise, the potassium level fell rapidly returning to approximately normal levels in 5 minutes. The maximum rate of fall occurred within the

1–2 minutes postexercise. In some individuals engaged in strenuous exercise, plasma potassium levels of 9 milliequivalents per liter may occur if the exercise is carried to exhaustion [30]. Peak levels are proportional to the intensity of the exercise, contracting muscle mass and duration of exercise. There is little evidence, however, that these extremely high levels of potassium due to exercise adversely affect cardiac functioning [30,31]. Such levels can only be maintained for 1–2 minutes before exercise has to be stopped due to exhaustion. Studies suggest that exercise induced increases in blood catecholamines have a cardio-protective effect on hyperkalemia [32,33]. It is felt that it is the rapid drop to low levels following cessation of exercise that produces an arrhythmia.

28.2.3 Drugs

In most deaths due to excited delirium syndrome (EDS), drugs, whether illicit or prescribed, play a role. The drugs of abuse most commonly associated with EDS are cocaine, methamphetamine, phencyclidine (PCP) and to a lesser degree alcohol. Cocaine and methamphetamine are cardiotoxic. These effects are a consequence of the elevated levels of catecholamines resulting from their use. The effects are dose related and usually cumulative. Chronic abuse of these drugs is associated with cardiomyopathy with an increase in heart weight [34–36]. Sudden cardiac death due to arrhythmia is a known complication of cardiomyopathy [37,38]. Chronic abuse has also been linked to accelerated atherosclerosis of the coronary arteries [23].

Chronic cocaine abusers, in an apparent neuroadaptive measure, increase the density of dopamine transport binding sites in the limbic striatum of the brain [39,40]. In chronic cocaine abusers dying in ED, the number of dopamine reuptake transporters has been found not to be increased [40]. This possibly explains the occurrence of excited delirium in these individuals.

The Central nervous system stimulation from cocaine causes increased release of NE at the peripheral synapses and E from the adrenals. Peripherally, cocaine acts to produce inhibition of NE reuptake at the synapses [41]. Thus, directly by central nervous stimulation and indirectly by blocking the reuptake of NE, cocaine causes increased concentrations of norepinephrime at the synapses between the nerve terminals and the receptors on the organs.

Presumably, myocardial ischemia due to the effects of markedly elevated catecholamine levels, in association with hypokalemia, is the cause of a fatal arrhythmia in individuals with EDS due to cocaine. The ischemia is due to increases in myocardial oxygen demand resulting from increased heart rate, blood pressure, and myocardial contractility (due to increased levels of E and NE at the postganglionic synapses of the heart) in conjunction with a decrease in the supply of oxygen to the myocardium (due to vasoconstriction of the coronary vasculature resulting from increased alpha-adrenergic stimulation). If an EKG is performed immediately after cardiac arrest, individuals arresting due to EDS typically show either asystole or pulseless electrical activity (PEA). Both these arrhythmias have a grim prognosis with little response to cardiac resuscitation.

Methamphetamine stimulates the sympathetic nervous system both centrally and peripherally by increasing levels of dopamine and norepinephrine. Acting on the brain, it causes the accumulation of high levels of the neurotransmitter dopamine by decreasing the number and activity of dopamine transporters [42,43]. In the brain, it appears to have a direct neurotoxic effect, damaging axons and axon terminals [42,43]. The mechanism of death in EDS due to methamphetamine overdose is the same as with cocaine, myocardial ischemia due to the effects of markedly elevated catecholamine levels, in association with hypokalemia. Similarly to cocaine effects, the ischemia is due to an increase in myocardial oxygen demand resulting from increased heart rate, blood pressure, and myocardial contractility (due to increased levels of E and NE at the postganglionic synapses of the heart) in conjunction with a decrease in the supply of oxygen to the myocardium (due to vasoconstriction of the coronary vasculature resulting from increased alpha-adrenergic stimulation). The EKG findings are the same.

Sudden death during a struggle in an individual with a history of alcohol abuse, and in whom only alcohol may be present, occasionally occurs. Alcohol is a recognized cause of a variety of atrial and ventricular arrhythmias [44,45]. In addition, chronic alcoholics have been found to have a prolonged QT interval. Alcoholism has been associated with increased levels of norepinephrine. All these factors predisposing to arrhythmias can be aggravated by catecholamines released during a violent struggle.

Psychotropic drugs, e.g., antipsychotics, neuroleptics, antidepressants, stimulants, and antianxiety agents may cause cardiac arrhythmias and sudden death. Control of the duration of the action potential of ventricular myocytes (heart cells) is based on an equilibrium between inward and outward currents of ions across the cell membrane. The longer action potential demonstrated by QT prolongation results from an imbalance in the flow. These drugs may block outward movement of K^+ and thus repolarization, or inhibit inward ionic currents especially Na^+ and Ca^{2+} in heart cells. Prolongation of the QT interval is associated with the ventricular tachyarrhythmia *torsade de pointes* and sudden death [46,47]. Thus, these drugs, in conjunction with a hyperadrenergic state caused by the delirium and struggle, and in conjunction with hypokalemia may produce a fatal arrhythmia.

28.2.4 Treatment of Excited Delirium

Identifying individuals at high risk for EDS is the *first step* towards death prevention [2]. Once identified, preventive procedures intended to alter actions that initiate the sequence of events culminating in death can be implemented.

Death occurring from "excited delirium syndrome," whether due to intrinsic mental disease or use of stimulants, is characterized by:

1. Acute onset of symptoms (minutes to hours)
2. Delirium with:
   a. acute, transient disturbance in consciousness and cognition; disorientation;
   b. disorganized and inconsistent though processes;
   c. inability to distinguish reality from hallucinations;
   d. disturbances in speech;
   e. disorientation to time and place;
   f. misidentification of individuals.
3. Combative or violent behavior.
4. Use of physical restraint.
5. Sudden cardiac death within minutes to hours after presentation of signs and symptoms.
6. Lack of response to CPR.

The presentation is the same whether the ED is due to stimulant abuse or endogenous mental disease [2]. Acute mania is to be considered a medical emergency and requires rapid control and urgent treatment to prevent ED from escalating to EDS.

In the hospital setting, the initial use of verbal deescalation as well as the offering of sedative medications is recommended to reduce agitation. When attempting deescalation by verbal intervention, it is vitally important that staff members display a calm and nonthreatening behavior towards the patient. No rapid movements towards the patient should be attempted. Psychotic agitated patients are very difficult and dangerous to deal with. Staff cannot expect them to cognitively interpret, rationalize and respond to requests in a "normal" time or manner. If conservative measures are not successful and/or not possible then the staff must use physical restraint. The use of restraint must only be initiated with a clear understanding that there is a potential for sudden death. If physical restraint is necessary, and a CEW is not available, the use of approximately 6 individuals trained in approved physical restraint techniques should be use. It is necessary to obtain a rapid restraint by using a large show of force. This will reduce the time of struggle, thus reducing the cascading physiological response mechanisms inherent in EDS. As soon as the individual is restrained, medication should be administered to calm the patient. The intra-venous route is the most rapid.

In the case of ED encountered by law enforcement, while the presentation is the same, their resources are limited compared to the hospital setting and thus the responses are typically significantly different. The Officers cannot administer drugs and often do not have the overwhelming staff necessary for immediate restraint. Law enforcement officers are usually called because the situation has escalated to violence or a significant risk of injury. They typically do not have the luxury of time for verbal deescalation attempts.

The dramatic increase in deaths due to EDS in the community is due to two factors: the increased use of illegal stimulants, such as cocaine, since the 1980's and the appearance of large numbers of mentally ill individuals in the community at large due to closure of mental institutions and the inability to commit these individuals for involuntary treatment due to interpretation of their civil rights by the courts. Both factors have increased the number of encounters between the police and individuals with ED, and thus the number of deaths.

The individuals police encounter who are susceptible to EDS constitute a subgroup of the population that has a greater potential for violence. A police officer must make rapid assessments of an individual's mental state as it relates to the potential for violence. Many are chronic users of illicit drugs, and have a criminal history, a prior history of violence and possibly possess weapons. Complicating the situation for the police is the fact that they are usually unaware of an individual's past medical history, mental history, history of violence and whether the individual is on drugs.

To prevent deaths from EDS, the police must:

1. identify individuals in ED;
2. (if time is available) attempt to deescalate the situation and calm subject down;
3. use CEW or overwhelming force if restraint must be used;
4. after being restrained, monitor them at the scene and during transport;
5. immediately transport them to a hospital for treatment or observation.

Environmental options available to police for prevention of violence include scanning the scene to remove potentially hazardous objects; removing bystanders who might escalate the individual's level of distrust and agitation and asking others to move away from the individual to reduce stimulation. The police should attempt to reduce the noise level. If there is loud music playing at the scene, they should have it shut off.

Ideally, the police should initially attempt to deescalate the situation and calm the individual. However, an individual in acute psychosis is not experiencing reality; therefore make simple demands stated in a nonchallenging manner. If time allows, the Officers should remain calm and restate requests and not take any action unless there is an immediate threat to the individual or others. Statements should always be phrased in a positive manner with offers to help and assist. It takes time for a highly agitated individual to calm down.

The above traditional advise must be taken in light of the increasing awareness that an ED subject is often already acidotic. A delay in control allows acidosis to increase which may eventually result in a nonresuscitatable arrhythmia. Thus, some localities are now implementing immediate control followed by sedation.

If a verbal intervention does not work, then one must resort to restraint. The police may want to initially try to immobilize the individual with chemicals, e.g., pepper spray, or by use of a TASER CEW. Unfortunately, there are numerous cases where chemicals and a CEW have had no effect. If the need to physically restrain an individual cannot be avoided by other interventions, restraint must occur rapidly to reduce the time of struggle. Any attempt to restrain or gain physical control of a highly agitated and aggressive individual suffering from ED brings with it the possibility of death. It is recommended that all individuals in ED be taken immediately to a medical facility for treatment. If EMS units are at a scene, once the individual is restrained, they should administer a sedative or tranquilizing agent.

Quick physical control can only be gained by use of a CEW or overwhelming force. By reducing the time of struggle, the effects of the continued physiological catecholamine surge and acidosis inherent in the struggle will be reduced and one may prevent death from occurring. At all times, at the scene or during transport, face to face monitoring of the individuals breathing status must be done by police or EMS personnel until arrival at a hospital facility. Any diminished respiratory status or rapid elevation of core body temperature may signal a danger warning for sudden cardiac arrest from EDS. Police and EMS personnel must be ready to immediately begin cardiac resuscitation procedures and to transport all individuals experiencing ED to a hospital even if they appear to be recovering.

28.3 Terminology Confusion

There has been an excessive concern over the variations in terminology for excited delirium. Some of this is by special interest groups with their own agenda, e.g., plaintiffs' attorneys and the American Civil Liberties Union. The claim is often made that ED cannot exist as there is no universally accepted terminology. It is a fact that this condition has been variously referred to as Bell's mania, acute exhaustive mania, acute delirious mania, delirium grave, typhoma, acute delirium, manic-depressive exhaustion, excited catonia, lethal catatonia, malignant hyperthermia, hyperexertion, and neuroleptic malignant syndrome. If a multiplicity of terms over 150 years was enough to question a condition then we could question the existence of coronary artery disease as it is also referred to as ischemic heart disease, coronary heart disease, atherosclerosis, ateriosclerosis, sclerotic heart disease, and hardening of the arteries.

Surprisingly, some nonmedical critics have questioned the scientific basis, or existence of, excited delirium since it has no numerical code for insurance billing purposes with the exact words, "excited delirium." Obviously, this is not a condition for which an insured productive member of society makes psychiatric appointments. This is an emergency typically associated with law enforcement and EMS involvement.

Some have suggested that "excited delirium" does not exist because the AMA (American Medical Association) does not "approve" the diagnosis. This is a "red herring" for several reasons. The AMA does not approve or disprove of any diagnoses – just billable procedures. The AMA database is the Current Procedural Terminology (CPT) listing of insurance billable procedures.

The group that deals with dead people is called the National Association of Medical Examiners (NAME). They do recognize excited delirium and have for years.

References

1. American Psychiatric Association: Diagnostic and Statistical Manual of Mental Disorders, 4th ed. Text Revision. Washington, DC: American Psychiatric Association, 2000.
2. Di Maio T.G. and Di Maio V.J.M. Excited Delirium Syndrome. Boca Raton, FL: CRC Press Inc., 2006.
3. Bell L.V. On a Form of Disease resembling some advanced stages of mania and fever. Amer. J. Insanity 6:97–127, 1849.
4. Cancro R. The Introduction of Neuroleptics: A Psychiatric Revolution. Psychiatric Services 51(3), 2000.
5. Lieberman J.A., Golden R., Stroup S., and McEnvoy J. Drugs of the psychopharmacological revolution in clinical psychiatry. Psychiatric Services 51(10), 2000.
6. Di Maio V.J.M. and Di Maio, D. Forensic Pathology, 2nd ed. Boca Raton, FL: CRC Press Inc., 2001.
7. Stratton S.J., Rogers C., Brickett, K., and Gruzinski G. Factors associated with sudden death of individuals requiring restraint for excited delirium. Amer. J. Emerg. Med. 19(3):187–191, 2001.
8. Weiss S. and Baker J.P. The carotid sinus reflex in health and disease. Medicine 12:297–354, 1933.
9. Thomas J.E. Hyperactive carotid sinus reflexes and carotid sinus syncope. Mayo Clinic. Proc. 44:127–139,1969.
10. Walter P.F., Crawley I.S., and Dorney E.R. Carotid sinus hyppersensitivity and syncope. Amer. J. Cardiology. 42: 396–403, 1978.
11. Reay D.T., Howard J.D., Fligner C.L., and Ward, R.J. Effects of positional restraint on oxygen saturation and heart rate following exercise. Amer J. Forensic Med. Path. 9(1):16–18, 1988.
12. O'Halloran R.L. and Lewman L.V. Restraint asphyxiation in excited delirium. Am. J. Forensic Med. Pathol. 14(4):289–295, 1993.
13. Chan T.C., Vilke G.M., Neuman T., and Clausen L. Restraint positional asphyxia. Ann. Emerg. Med. 1997;30(5):578–586.
14. Michalewicz B.A., Chan T.C., Vilke G.M., Levy S.S., Neuman T.S., and Kolkhorst F.W. Ventilatory and metabolic demands during aggressive physical restraint in healthy adults. J. Forensic Sci. 52(1):171–175, 2007.
15. Guyton A.C. and Hall J.E. Medical physiology, 10th ed. Philadelphia: WB Saunders, 2000.
16. World Health Organization. Neuroscience of Psychoactive Substance Use and Dependence, 2004.
17. Cooper J.R., Bloom F.E., and Roth R.H. The Biochemical Basis of Neuropharmacology, 8th ed. New York: Oxford University Press, 2003.
18. Katz A.M. Physiology of the heart. Philadelphia, PA: Lippincott Williams & Wilkins, 2001.
19. Insel PA. Seminars in medicine of the Beth Israel Hosp., Boston: Adrenergic receptors – evolving concepts and clinical implications (Review article). New Eng. J. Med. 334(9):580–585, 1996.
20. Calzada B.C. and De Artinano A.A. Alpha-adrenoceptor subtypes. Pharm. Res. 44(3):195–208, 2001.

21. Guimaraes S. and Moura D. Vascular adrenoceptors: an update. Pharm. Reviews. 53:319–356, 2001.
22. Heusch G., Baumgart D., Camici P., et al. [alpha]- Adrenergic coronary vasoconstriction and myocardial ischemia in humans. Circulation 101(6):689–694, 2000.
23. Lange R.A. and Hillis L.D. Cardiovascular complications of cocaine use. New Eng. J. Med. 345(5):351–358, 2001.
24. Hajjar R.J. and MacRae C.A. Editorial"Adrenergic-receptor polymorphism and heart failure. New Eng. J. Med. 347(15):1196–1198, 2002.
25. Small K.M., Wagoner L.E., Levin A.M., Kardia S.L.R., and Liggett S.B. Synergistic polymorphisms of B_1 and alpha $_{2c}$ adrenergic receptors and the risk of congestive heart failure. New Eng. J. Med. 347(15):1135–1142, 2002.
26. Dimsdale J.E., Hartley G.T., Guiney T., Ruskin J.N., and Greenblatt D. Post-exercise peril: Plasma catecholamines and exercise. JAMA 251:630–632, 1984.
27. Young D.B., Srivastava T.N., Fitzovich D.E., Kivlighn S.D., and Hamaguchi M. Potassium and catecholamine concentrations in the immediate post exercise period. Am. J Med. Sci. 304:150–153, 1992.
28. Rastergar A. and Soleimani M. Hypokalaemia and hyperkalaemia. Postgrad Med J. 77:759–764, 2001.
29. Gennar F.J. Current concepts: Hypokalemia (review article). New Eng. J. Med. 339(7):451–458, 1998.
30. Medbo J.I. and Sejersted O.M. Plasma potassium changes with high intensity exercise. J Physiol. 421:105–122, 1990.
31. Lindinger M.I. Potassium regulation during exercise and recovery in humans: implications for skeletal and cardiac muscle. J. Mol. Cell. Cardiol. 27(4):1011–1022, 1995.
32. Paterson D.J., Rogers J., Powell T., and Brown H.F. Effect of catecholamines on the ventricular myocyte action potential in raised extracellular potassium. Acta Physiologica. Scand. 148:177–186, 1993.
33. Leitch S.P. and Paterson D.J. Interactive effects of K^+, acidosis, and catecholamines on isolated rabbit heart: implications for exercise. J. Appl. Physiol. 77(3):1164–1171, 1994.
34. Karch S.B., Green G.S., et al. Myocardial hypertrophy and coronary artery disease in male cocaine users. J. Forensic Sci. 40(4):591–595, 1995.
35. Wiener R.S., Lockhart J.T., and Schwartz R.G. Dilated cardiomyopathy and cocaine abuse: report of two cases. Amer. J. Med. 81:699–701, 1986.
36. Karch S.B. Karch's Pathology of Drug Abuse, 3rd. ed. Boca Raton, FL: CRC Press, 2002.
37. Silver M.D., Gotlieb A.L., and Schoen F.J. Cardiovascular Pathology. Philadelphia, PA: Churchill Livingstone, 2001.
38. Spitito P., Bellone P., et al. Magnitude of left ventricular hypertrophy and risk of sudden death in hypertrophic cardiomyopathy. New Eng. J. Med. 342(24):1778–1785, 2000.
39. Staley J.K., Hearn W.L., Ruttenber A.J., Wetli C.V., and Mash D.C. High affinity cocaine recognition sites on the dopamine transporter are elevated in fatal cocaine overdose victims. J. Pharmacol. Exp. Ther. 271:1678–1685, 1994.
40. Mash D.C., Pablo J., Ouyang Q., Hearn W.L., and Izenwasser S. Dopamine transport function is elevated in cocaine users. J. Neurochem. 81:292–300, 2002.
41. Billman G.E. Cocaine: a review of the toxic actions on cardiac function. Crit Rev. Toxicol. 25:113, 1995.
42. Fleckenstein A.E., Gibb J.W., Hanson G.R. Differential effects of stimulants on monoaminergic transporters: pharmacological consequences and implications for neurotoxicity. Eur. J. Pharmacol. 406(1):1–13, 2000.
43. McCann U.D. and Ricaurte G.A. Amphetamine neurotoxicity: accomplishments and remaining challenges. Neurosci. Biobehavioral Rev. 27(8):821–826, 2004.
44. Singer K. and Lundberg W.B. Ventricular arrythmias associated with the ingestion of alcohol. Ann. Intern. Med. 77:247–248, 1972.

45. Ettinger P.O., Wu C.F., De La Cruz C. Jr., Weisse A.B., Ahmed S.S., and Regan T.J. Arrhythmias and the “Holiday Heart”: alcohol-associated cardiac rhythm disorders. Am. Heart J. 95(5):555–562, 1978.
46. Witchel H.J., Hancox J.C., and Nutt D.J. Psychotropic drugs, cardiac arrhythmia, and sudden death. J. Clin. Psychopharmacol. 23(1):58–77, 2003.
47. Roden D.M. Drug-induced prolongation of the QT interval] New Eng. J. Med. 350(10):1013–1022, 2004.

Chapter 29
Biochemical Brain Markers in Excited Delirium Deaths

Deborah C. Mash

Over the past decade, increased attention has been paid to the sudden deaths of some highly agitated subjects held in police custody, which were restrained or incapacitated by CEWs. Medical examiners often have extreme difficulty in identifying the cause of death, but frequently drug intoxication is considered a contributing factor. The symptoms of excited delirium include bizarre and/or aggressive behavior, shouting, paranoia, panic, violence toward others, unexpected physical strength, and hyperthermia. Throughout the United States and Canada, these cases are frequently associated with cocaine or methamphetamine abuse. Acute exhaustive mania and sudden death presents with behavioral signs and symptoms that are the same as in cases of cocaine excited delirium. However, the forensic autopsies fail to demonstrate a history of illicit drug use or positive toxicologic basis for the diagnosis. The striking parallel in behavioral symptoms between the manner of sudden death involving unusual behavior in these individuals suggests that a genetic brain disorder may be the precipitating cause of delirium and death. While the precise cause and mechanism of lethality remains controversial, we have demonstrated that there are network-level changes in dopaminergic synaptic markers in brain that identify and help to confirm the occurrence of the excited delirium syndrome. Characterization of adaptive and maladaptive functional interactions among brain pathways is a critical step towards development of evidence-based biomarkers of excited delirium.

29.1 Differential Diagnosis of Drug-Induced Psychotic Disorders

Delirium is a condition with specific diagnostic criteria, characterized by acute onset, altered level of consciousness, fluctuating course and increased mortality. Clinical subtypes of delirium with unique and definable phenomenological or physical characteristics are not widely accepted. Psychotic symptoms are not

D.C. Mash (✉)
Department of Neurology, Miller School of Medicine, University of Miami
e-mail: dmash@med.miami.edu

M.W. Kroll, J.D. Ho (eds.), *TASER® Conducted Electrical Weapons: Physiology, Pathology, and Law*, DOI 10.1007/978-0-387-85475-5_29,

uncommon in delirium, but specific psychotic symptoms may have different factors contributing to their development. Delirium is often underrecognized or misdiagnosed as other psychiatric conditions.

Excited delirium is a "state" of mental and physiological arousal, agitation, and hostility. Observers typically emphasize the extreme sweating, bizarre behavior and speech, and the subject's extraordinary strength and endurance when struggling, apparently without fatigue. Such states are commonly associated with high blood concentrations of cocaine or other stimulants, though some cases arise in those with histories of schizophrenia or mania and no evidence of illicit drugs or intoxication.

Psychostimulant abuse is a major complicating factor in psychosis, renders the management of psychotic disorders more difficult, and adverse reactions to psychostimulants may mimic psychosis [1]. The differential diagnosis of psychotic disorders in the young has not had consistent definition and the relationship between drug use and psychotic symptoms is controversial. Adverse psychiatric effects associated with acute cocaine or amphetamine intoxication include extreme agitation, irritability or affective lability, impaired judgment, paranoia, hallucinations (visual or tactile), and sometimes manic excitement. Medical and psychiatric symptoms caused by acute psychostimulant intoxication are a common reason for presentation to the emergency department. Psychiatric symptoms of cocaine intoxication usually subside within 24 hours, but some patients may require benzodiazepines for acute agitation. Neuroleptics are often used for the treatment of unremitting paranoid psychosis, hallucinations and delusions. The transient paranoid state is a common feature of cocaine dependence, with affected persons possessing an obvious predisposition to this drug-induced state [2]. Psychiatric complications of cocaine intoxication include cocaine-induced paranoia, delirium, delusional disorder and the depressed mood and dysphoria associated with abrupt cocaine withdrawal.

Extended behavioral signs of cocaine psychosis usually imply the presence of an underlying major psychopathology in susceptible individuals [1]. Cocaine-induced psychosis typically manifests as an intense hypervigilance (paranoia) accompanied by marked apprehension and fear. Auditory and tactile hallucinations, formal thought disorder, and ideas of reference frequently noted with chronic use of amphetamines, are not prevalent in cocaine abusers. Paranoid experience secondary to cocaine use is limited to a drug episode, which dissipates by the time the user awakens from the "crash," usually about 8–36 hours after the cessation of the cocaine "binge". In a sample of 100 cocaine-dependent males, none reported cocaine paranoia extending beyond the crash phase [1,2].

In contrast to the effects of cocaine, amphetamine has apparently greater and longer-acting psychotogenic properties [3]. Angrist has suggested that very high doses of cocaine use with resulting sustained elevations in plasma cocaine levels may be necessary for the development or kindling of an episode of cocaine psychosis. In keeping with this suggestion, certain effects are known to become progressively more intense after repeated cocaine administration, a phenomenon, which is referred to as sensitization. However, Satel and coworkers [2] have provided data to suggest that instances of cocaine-induced paranoia or psychosis lasting more than several days most likely indicate the presence of an underlying primary psychotic disorder.

Victims of excited delirium display sudden onset of paranoia and alternate between calm behavior and extreme agitation [4]. When confronted by police, who are typically called to the scene, the victim intensifies the violence and paranoia. An intense struggle ensues, when the victim exhibits incredible "superhuman" strength and is impervious to the usual police techniques of pain control, including pepper spray, electric stun guns, and peroneal baton strikes (for review, [5]. Usually, within minutes of being restrained, the victim loses all vital signs. Core body temperatures average 105°C [6,7]. Hyperthermia is considered strong supportive evidence for the diagnosis of excited delirium, but is not an absolute requirement [5]. Hyperthermia is a harbinger of inevitable death for virtually all subjects. While many factors are associated with sudden death in individuals requiring restraint for excited delirium (for review [8–12], these individuals develop a disturbance in thought, behavior and mood, and become agitated and violent, in keeping with a central nervous system (CNS) mechanism or underlying brain disorder as a contributing cause of lethality.

Wetli and colleagues have suggested that there are three related syndromes: (1) acute exhaustive mania, as described by Bell [13] in psychiatric patients, (2) excited delirium, due to psychostimulants (cocaine, methamphetamine, MDMA) and psychiatric illness; and (3) the attenuated variant – neuroleptic malignant syndrome (NMS) [5,14]. This classification agrees with the suggestion of Kosten and Kleber [15], who termed agitated delirium as a possible cocaine variant of NMS, a highly lethal disorder seen in patients taking dopamine antagonists or following abrupt withdrawal from dopaminergic agonists [15–19]. NMS is usually associated with muscle rigidity, while the cocaine variant of the syndrome presents with brief onset of rigidity immediately prior to respiratory collapse [15].

29.2 Cocaine Delirium and Sudden Death

Ruttenber and colleagues have examined excited delirium deaths in a population-based registry of all cocaine-related deaths in Dade County, Florida [7]. This study has led to a clear description of the cocaine delirium syndrome, its pattern of occurrence in cocaine users over time, and has identified a number of important risk factors for the syndrome. Cocaine delirium deaths are defined as an accidental cocaine toxicity that occurred in individuals who experienced an episode of bizarre behavior prior to death [20,21]. The victims are more likely to be male, black and younger than other cocaine toxicity deaths. The most frequent route of administration was injection for the excited delirium victims as compared to inhalation for the other accidental cocaine toxicity deaths. However, the frequency of smoked "crack" cocaine was similar for both groups. Thirty-nine percent of the excited delirium victims died in police custody as compared with only 2% for the comparison group of accidental cocaine toxicity cases [7]. A large proportion of these individuals survive between 1 and 12 hours after the onset of the syndrome.

The most striking feature of the excited delirium syndrome is the extreme hyperthermia. The epidemiological data provide some clues for the etiology of the elevated body temperature [7,10]. Victims of cocaine excited delirium have higher body mass indices. This finding suggests that muscle mass and adiposity may contribute to the generation of body heat. Temporal clustering in summer months supports the hypothesis that abnormal thermoregulation is an important risk factor for death in people who develop the syndrome. Being placed in police custody prior to death can also raise body temperature through increased psychomotor activity if the victim struggles in the process of restraint.

Cocaine-associated rhabdomyolysis and excited delirium share many similar features, suggesting that they may be different stages of the same syndrome [22]. Patients with rhabdomyolysis were similar to victims of fatal excited delirium with regard to age, gender, race, route of cocaine administration, the experiencing of excitement, delirium, and hyperthermia, and the absence of seizures. Cocaine-associated rhabdomyolysis and excited delirium have similar clinical features and risk factors, occur in similar populations of drug users, and can be explained by the same pathophysiologic processes. It appears that this syndrome is caused by changes in dopamine processing induced by chronic and intense use of cocaine rather than by the acute toxic effects of the drug.

29.3 Defective Signaling at the Dopamine Synapse

The overlap between the behavioral symptoms associated with cocaine-induced excited delirium and acute exhaustive mania supports the concept of a final common brain pathology. While various neurotransmitter alterations may converge to result in a delirium syndrome or subtype thereof, what constitutes this may involve certain brain regions or circuits and certain neurotransmitters. We have demonstrated that there is a defect in the regulation of the dopamine transporter, which leads to a dopamine excess in these victims [6,23–25]. The failure to upregulate the dopamine transporter with chronic cocaine abuse leads to a hyperdopaminergic state and this in turn, may underlie the development of psychotic symptoms and hyperthermia.

The mesolimbic dopaminergic system plays a primary role in mediating the rewarding, as well as, the lethal effects of most abused drugs (for review [26]). Chronic cocaine use is associated with an increase in dopamine neurotransmission resulting from the blockade of dopamine uptake and mediated by the activation of pre- and postsynaptic dopamine receptors. Cocaine mediates its powerful reinforcement by binding to specific recognition sites on the dopamine transporter protein [27]. Dopamine transporters function to rapidly control the removal of transmitter molecules from the synaptic cleft. Neurochemical abnormalities have been identified in the striatum of cocaine abusers dying of excited delirium [6,25]. Some of the abnormalities in the dopaminergic system involve alterations in certain types of dopamine receptors and in cocaine's ability to block the reuptake carrier, by which dopamine is recycled back in to the presynaptic nerve terminal. Cocaine users often go on binges, consuming a large amount of the drug over a period of a few days. The neurochemical changes occurring over the "binge" and crash periods involve adaptive alterations of the dopamine transporter and receptors on receiving cells. Functional network-level changes in the basal ganglia circuits in brain are not the same after chronic cocaine exposure.

The dopamine transporter is a critical regulator of dopamine disposition within the brain [28,29]. Psychostimulants alter dopaminergic signaling by affecting the release and/or reuptake of dopamine. Many psychostimulants are classified as releasers (i.e., amphetamine analogs) or uptake blockers (i.e., cocaine-like drugs) based on the mechanism of their acute effects on neurotransmitter flux through the dopamine transporter. Radiolabeled cocaine congeners bind to high and low affinity sites on the cloned and native human dopamine transporter, one of which appears to overlap with the functional state of the carrier protein [25,30]. In cases where death was due to cocaine intoxication, the high affinity cocaine recognition sites on the dopamine transporter were upregulated significantly in the striatum as compared to age-matched and drug-free control subjects (Fig. 29.1). These chronic cocaine abusers died suddenly, without evidence of excited delirium. The increase in high affinity [3H]WIN 35,428 binding sites on the human dopamine transporter reflects an increased ability of the protein to transport dopamine [25].

Since cocaine and other psychostimulants (amphetamine, methamphetamine, methylenedioxymethamphetamine) promote increased extracellular dopamine concentrations, this adaptive response may help to explain the addictive liability of cocaine. As the transporter upregulates its apparent density in the nerve terminal to more efficiently transport dopamine, more cocaine will be needed to experience cocaine's reinforcing effects and euphoria (Fig. 29.2). An acute tolerance to the effects of cocaine resulting from a rapid upregulation in dopamine transporter function may help to explain the human pattern of "binge" cocaine use.

Victims of excited delirium have the same brain concentrations of cocaine and ratio of parent to metabolite as measured in cocaine abusers who died suddenly due to the lethal effects of cocaine (Table 29.1). However, the dopamine transporters measured in the cocaine delirium victims did not show the compensatory increase in striatum seen in chronic cocaine abusers (Fig. 29.2), despite the same brain concentrations of cocaine measured at autopsy. Since the concentration of synaptic dopamine is controlled by the reuptake

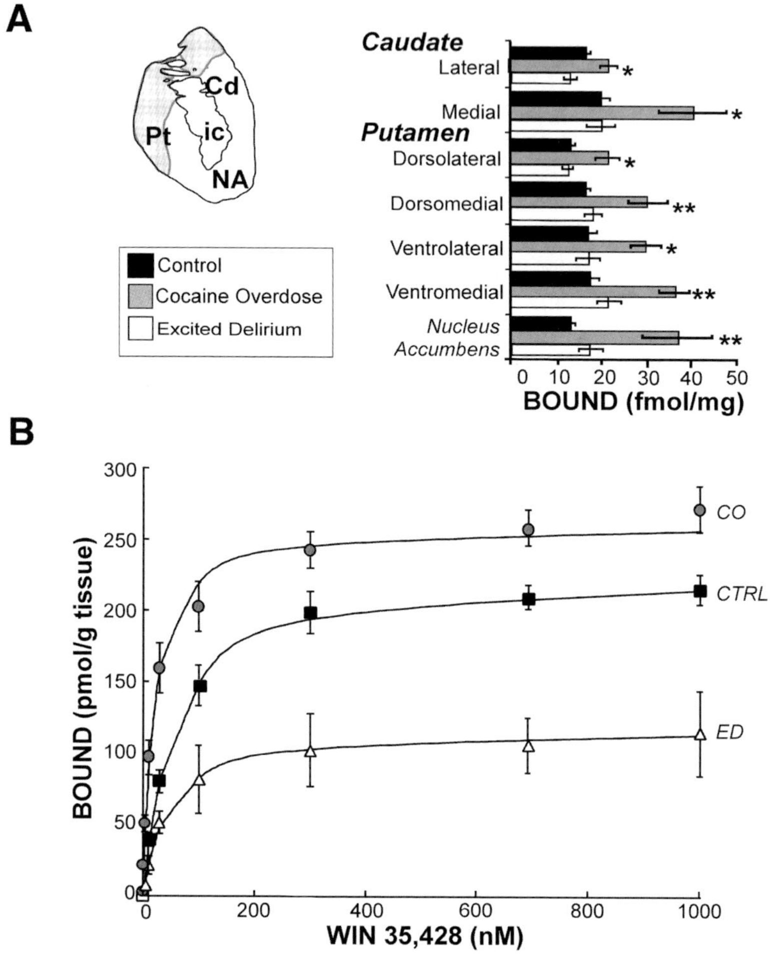

Fig. 29.1 Distribution and density of [3 H]WIN 35,428 labeling of the dopamine transporter in the striatum from victims of cocaine excited delirium. (A) The results shown illustrate autoradiographic measures of the density (mean and SE) for cocaine-excited delirium subgroup (white bars, $N = 8$) and cocaine intoxication deaths (gray bars, $N = 8$) as compared to age-matched control subjects (black bars, $N = 8$). Cocaine abuse leads to an adaptive increase in dopamine transporter density over the striatum in the cocaine overdose victim and the lack of any apparent elevation for the excited delirium victim. Panel at left illustrates striatal regions-of-interest. Abbrevations, Cd, caudate; Pt, putamen; NA, nucleus accumbens. $*p < 0.05$; $**p < 0.001$. (B) Saturation analysis of WIN 35,428 binding to the dopamine transporter in human striatum. Saturation binding analysis was conducted according to the method of Mash et al. (2002). Note the increase in the density of WIN35, 428 binding in the striatum in cocaine intoxication deaths (CO, $N = 25$, $p < 0.001$) compared to excited delirium victims (ED, $N = 30$) and drug-free age-matched control subjects (CTRL, $N = 27$)

Table 29.1 Brain and blood concentrations of cocaine and benzoylecgonine in excited delirium deaths

	COC		*ED*	
	Blood (mg/l)	Brain (mg/kg)	Blood (mg/l)	Brain (mg/kg)
Cocaine	3.83 ± 0.90	6.01 ± 1.15	0.74 ± 0.14*	3.87 ± 1.02
Benzoylecgonine	4.57 ± 0.53	2.40 ± 0.90	3.33 ± 0.45	1.66 ± 0.23

Postmortem blood samples from cocaine intoxication deaths (COC, $N = 100$) and Excited Delirium victims (ED, $N = 74$). Postmortem brain samples (occipital cortex) from COC ($N = 59$) and ED ($N = 52$) subjects. A significant difference was demonstrated for blood cocaine concentrations between COC and ED *($p < 0.0001$, student's t-test). Brain concentrations were not significantly different between the two groups.

mechanism(s), this lack of compensatory increase in dopamine transporters could be the defect in dopaminergic transmission that explains the paranoia and agitation associated with this syndrome. Since cocaine blockade of the dopamine transporter leads to an over abundance of synaptic dopamine, the failure to upregulate dopamine transporter activity as compensatory response to a cocaine binge results in a hyperdopaminergic state (Fig. 29.2).

Paranoia in the context of cocaine abuse is common and several lines of evidence suggest that this phenomenon may be related to the function of the dopamine transporter protein [31–33]. Genetic differences in the makeup of individuals who abuse cocaine or amphetamines may also underlie some of these differences in susceptibility to the development of adverse neuropsychiatric effects with chronic cocaine abuse, that appear to result from a defective regulation of the dopamine transporter protein. As perturbations of dopamine have been linked to the development of psychosis, it follows that abuse of stimulants might be associated with a greater risk of developing psychotic symptoms or triggering delirium.

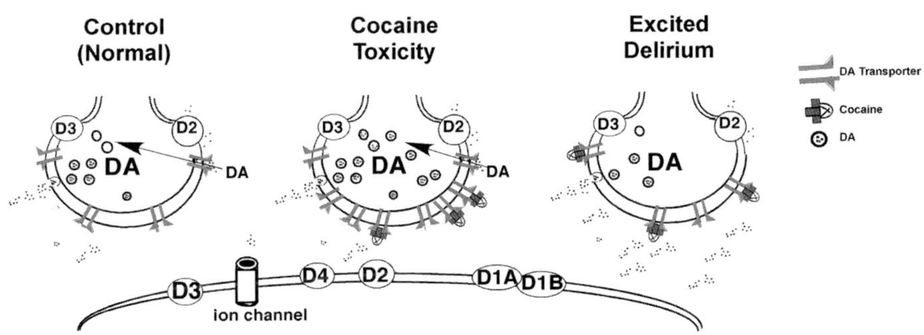

Fig. 29.2 Dysregulated dopamine in excited delirium victims. Cocaine blocks the reuptake of dopamine by the transporter molecule, which upregulates as a compensatory response to repeated exposure of the drug in cocaine abusers. Since the dopamine transporter regulates the synaptic concentration of dopamine, the lack of a compensatory increase in excited delirium victims results in a dopamine overflow following a cocaine "binge." Elevated synaptic dopamine (hyperdopaminergic state) with repeat exposure may kindle the emergence of the excited delirium syndrome due to a loss of homeostatic control in vulnerable subjects

Dopamine receptors are known to play a role in regulating core body temperature by acting in the hypothalamus. Since hyperthermia appears to be a clinical feature of cocaine-related sudden death presenting with excited delirium, Kosten and Kleber [15] have suggested previously that death occurred due to a malfunction in dopaminergic control of thermoregulation. The hypothalamus has two regions for temperature control, including the posterior region for conservation of heat and the preoptic region for heat dissipation. Lesions of the preoptic area in animal models cause hyperthermia [34]. The regulatory sensors of the hypothalamus are a predetermined "set-point", responsive to temperatures above and below the body's normal temperature.

Hypothermia receptors are known to be downregulated by high levels of intrasynaptic dopamine. Direct application of intracerebral dopamine at first lowers body temperature, however, a subsequent "rebound" in body temperature occurs about one hour after discontinuing this stimulation [35,36]. When cocaine is repeatedly administered, dopaminergic receptor numbers are altered [37,38]. The likelihood of hyperthermia may be increased with chronic cocaine abuse if the dopaminergic receptors involved in thermoregulation are undergoing adaptive changes with chronic cocaine exposure.

Excited delirium victims had a different profile of D2 receptor binding within the thermoregulatory centers of the hypothalamus as compared to cocaine overdose deaths [6]. The density of the D2 dopamine receptor subtype in the posterior and preoptic area of the hypothalamus in the cocaine delirium subgroup of cocaine overdose deaths were decreased significantly (Fig. 29.3). These results may be relevant to an understanding of the contribution of selective alterations in D1 and D2 receptor subtypes in central dopaminergic

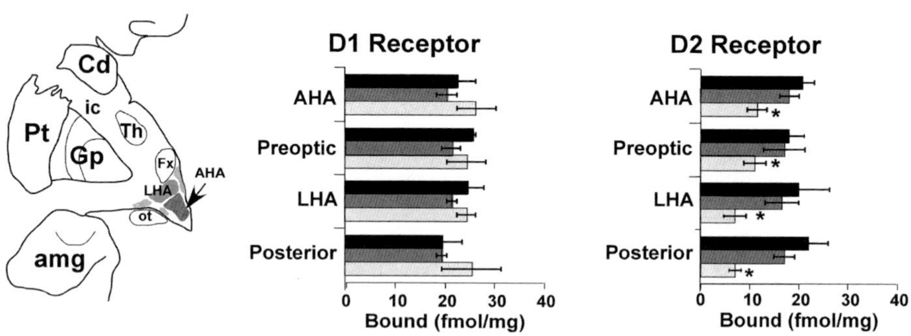

Fig. 29.3 Regional distribution and density of D1 and D2 dopaminergic receptor subtypes in hypothalamic nuclei. The results illustrate the density for the excited delirium subgroup (white bars, $N = 8$) as compared to age-matched control subjects (black bars, $N = 8$) and cocaine overdose cases (*grey bars*, $N = 8$). Values were obtained from densitometric analysis of in vitro autoradiography of specific [^{3}H]SKF-83566 binding for D1 receptors and [^{3}H]raclopride binding for D2 receptors. Abbreviations: AHA anterior hypothalamic area; amg amygdala; Cd caudate; Fx fornix; Gp globus pallidus; ic internal capusule; LHA lateral hypothalamic area; ot optic tract; Pt putamen; Th thalamus

temperature regulation. D1 and D2 receptors mediate opposite effects on thermoregulation, with the D1 receptor mediating a prevailing increase in core body temperature, while the D2 receptor mediates an opposing decrease in temperature [35,39,40]. Thus, the selective downregulation in the density of the D2 dopaminergic receptor subtype within the hypothalamus may explain the loss of temperature regulation in cocaine delirium victims. Cocaine elevations of synaptic dopamine occur in the hypothalamus, resulting in acute downregulation of D2 receptors. The servo-controlled D2 receptor-mediated decrease in core body temperature malfunctions in the hypothalamus, leaving the D1-mediated elevation unopposed.

29.4 Heath Shock Proteins and Hyperthermia

Hyperthermia, a condition of extremely high core body temperatures is associated with an elevation in heat shock proteins. Heat shock protein 70 (Hsp70) is elevated whenever core temperature reaches or exceeds 39°C [41]. In humans, impaired heat dissipation is a major mechanism by which cocaine elevates body temperature. We have demonstrated that heat shock proteins may serve as a biomarker of hyperthermia associated with cocaine-related excited delirium. The lethal effects of cocaine (and perhaps other psychostimulants) may be unique among those of other illicit drugs because they contribute to hypermetabolic states and increased heat production. When healthy, cocaine-naive persons are subjected to passive heating, pretreatment with even a small dose of intranasal cocaine impairs sweating and cutaneous vasodilation, the major autonomic adjustments to thermal stress and heat perception [42].

In cases where excited delirium is a suspected cause of death, a core temperature should be taken as close as possible to the time of death. However, forensic investigators often fail to obtain this information as part of the scene description and history. Core temperatures may not be recorded or they may be measured long after death occurs. The neurochemical pathologic examination of cocaine delirium suggests that measurements of heat shock protein mRNA may be a useful surrogate biomarker of heat stress in cases of suspected excited delirium. The heat-shock response is an immediate transient response to heat, with the level of the 70 kDa protein (Hsp70) the most closely related to the magnitude of the thermal stress [43]. One of the first physiological functions associated with stress-induced accumulation of Hsp70 was acquired thermotolerance, which is defined as the ability of the cell or organism to become resistant to heat stress after a prior sublethal heat exposure (for review [44]). We have observed a greater than five-fold elevation in HSPA1B and HSPA6 transcripts in cocaine delirium cases compared to either drug-free age-matched controls or cocaine intoxication deaths (unpublished data; Fig. 29.4). These transcripts code for Hsp70 protein, which is the most ubiquitous and temperature sensitive member of the heat shock protein family. This result demonstrates

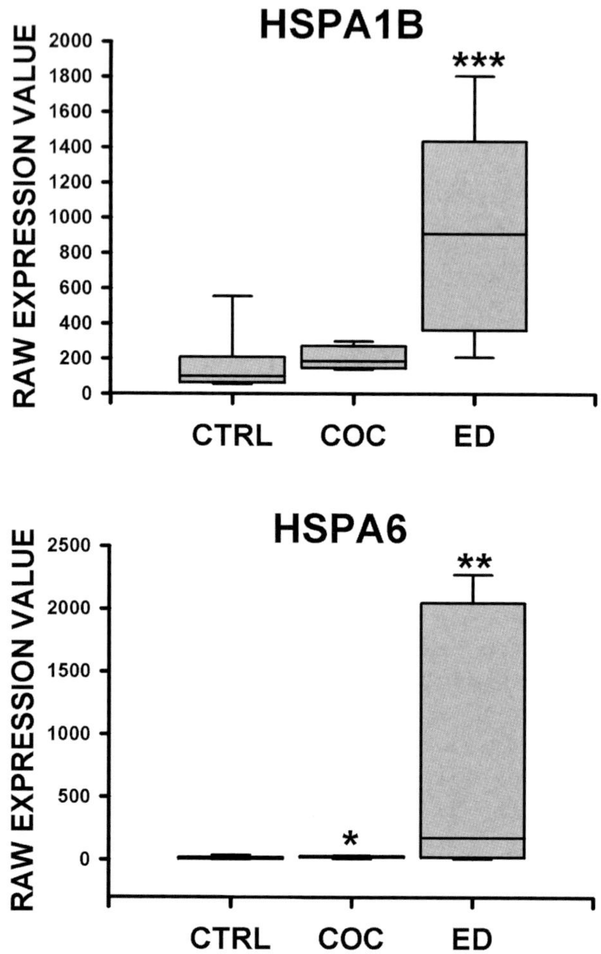

Fig. 29.4 Upregulation of heat shock proteins in excited delirium deaths. Affymetrix gene expression analysis demonstrated increased expression of HSPA1B and HSPA6 transcripts, which encode Hsp70 protein in the amygdala of victims of excited cocaine delirium. Box plot illustrates range and median values ($N = 13$; $^{**}p < 0.05$; $^{***}p < 0.001$). **Abbreviations:** CTRL(age-matched and drug-free controls), COC (cocaine intoxication deaths), ED (excited delirium)

the inducible transcription of heat shock genes occurs in victims of excited delirium as an adaptation to a pathophysiological state of hyperthermia.

29.5 Conclusions

Psychostimulant abuse is associated with neuropsychiatric disorders, including acute psychotic episodes, paranoid states, and delirium. The neurochemical pathology of cocaine-related excited delirium includes dysregulated dopamine transporters and hypothalamic receptors (hyperdopaminergic state). Elevated heat shock proteins are a biomarker for the occurrence of lethal hyperthermia in these victims. Examining validated biomarkers of dopamine dysregulation and heat stress in postmortem brain specimens provides additional forensic biology tools to recognize excited delirium at autopsy. The application of molecular biology and neurochemical approaches to the field of forensic pathology may assist the pathologic and forensic-medical diagnosis of sudden and unexpected excited delirium deaths.

Acknowledgments The author would like to acknowledge the expert technical assistance of Margaret Basile, M.S., Qinjie Ouyang, B.A., and John Pablo, Ph.D. The author is grateful to Charles Wetli, M.D., W. Lee Hearn, Ph.D. and A. James Ruttenber, M.D., Ph.D., for their long-standing collaboration, guidance and helpful discussions on the topic of in custody excited delirium deaths. Julie K. Staley, Ph.D. contributed to the characterization of dopaminergic synaptic markers in postmortem brains of cocaine abusers and to many of the original publications cited here. Charles P. Wetli, M.D. and Steven B. Karch, M.D. have contributed to the author's views on excited delirium and to the application of the research findings in postmortem brain to forensic biology. This work was supported by a USPHS grant from the National Institute on Drug Abuse (DA06627).

References

1. Satel SL, Seibyl JP, Charney DS. Prolonged cocaine psychosis implies underlying major psychopathology. *J Clin Psychiatry* 1991;52:349–50.
2. Satel SL, Southwick SM, Gawin FH. Clinical features of cocaine-induced paranoia. *Am J Psychiatry* 1991;148:495–8.
3. Angrist BM. Cocaine in the context of prior central nervous system stimulant epidemics. In: Volkow N, Swann AC, eds. Cocaine in the Brain (Mind in Medicine Series): Rutgers University Press, 1990:7–24.
4. Wetli CV, Fishbain DA. Cocaine-induced psychosis and sudden death in recreational cocaine users. *J Forensic Sci* 1985;30:873–80.
5. Wetli CV. Excited delirium. In: Payne-James J, et al., eds. Encyclopedia of Forensic and Legal Medicine. Elsevier, 2005:276–281.
6. Wetli CV, Mash D, Karch SB. Cocaine-associated agitated delirium and the neuroleptic malignant syndrome. *Am J Emerg Med* 1996;14:425–8.
7. Ruttenber AJ, Lawler-Heavner J, Yin M, Wetli CV, Hearn WL, Mash DC. Fatal excited delirium following cocaine use: epidemiologic findings provide new evidence for mechanisms of cocaine toxicity. *J Forensic Sci* 1997;42:25–31.
8. Ross DL. Factors associated with excited delirium deaths in police custody. *Mod Pathol* 1998;11:1127–37.
9. Reay DT, Howard JD, Fligner CL, Ward RJ. Effects of positional restraint on oxygen saturation and heart rate following exercise. *Am J Forensic Med Pathol* 1988;9:16–8.
10. Karch SB, Stephens BG. Drug abusers who die during arrest or in custody. *J R Soc Med* 1999;92:110–3.
11. Stratton SJ, Rogers C, Brickett K, Gruzinski G. Factors associated with sudden death of individuals requiring restraint for excited delirium. *Am J Emerg Med* 2001;19:187–91.
12. DiMaio TG, DiMaio VJM. Excited Delirium Syndrome: Cause of Death and Prevention. Taylor and Francis, New York, 2006.
13. Bell LV. On a form of disease resembling some advanced stages of mania and fever. *Am J Insanity* 1849;6:97–127.
14. Wetli CV, Natarajan GA. CUSTODY | Death in, United States of America. In: Payne-James J, Byard R, Corey T, Henderson C, eds. Encyclopedia of Forensic and Legal Medicine. Elsevier, 2005:65–73.

15. Kosten TR, Kleber HD. Rapid death during cocaine abuse: a variant of the neuroleptic malignant syndrome? *Am J Drug Alcohol Abuse* 1988;14:335–46.
16. Friedman JH, Feinberg SS, Feldman RG. A neuroleptic malignant-like syndrome due to levodopa therapy withdrawal. *Jama* 1985;254:2792–5.
17. Levenson JL. Neuroleptic malignant syndrome. *Am J Psychiatry* 1985;142:1137–45.
18. Caroff SN, Campbell EC, Sullivan KA. Neuroleptic malignant syndrome in elderly patients. *Expert Rev Neurother* 2007;7:423–31.
19. Strawn JR, Keck PE, Jr., Caroff SN. Neuroleptic malignant syndrome. *Am J Psychiatry* 2007;164:870–6.
20. Stephens BG, Jentzen JM, Karch S, Mash DC, Wetli CV. Criteria for the interpretation of cocaine levels in human biological samples and their relation to the cause of death. *Am J Forensic Med Pathol* 2004;25:1–10.
21. Stephens BG, Jentzen JM, Karch S, Wetli CV, Mash DC. National Association of Medical Examiners position paper on the certification of cocaine-related deaths. *Am J Forensic Med Pathol* 2004;25:11–3.
22. Ruttenber AJ, McAnally HB, Wetli CV. Cocaine-associated rhabdomyolysis and excited delirium: different stages of the same syndrome. *Am J Forensic Med Pathol* 1999;20:120–7.
23. Staley JK, Hearn WL, Ruttenber AJ, Wetli CV, Mash DC. High affinity cocaine recognition sites on the dopamine transporter are elevated in fatal cocaine overdose victims. *J Pharmacol Exp Ther* 1994;271:1678–85.
24. Staley JK, Boja JW, Carroll FI, Seltzman HH, Wyrick CD, Lewin AH, Abraham P, Mash DC. Mapping dopamine transporters in the human brain with novel selective cocaine analog [125I]RTI-121. *Synapse* 1995;21:364–72.
25. Mash DC, Pablo J, Ouyang Q, Hearn WL, Izenwasser S. Dopamine transport function is elevated in cocaine users. *J Neurochem* 2002;81:292–300.
26. Ritz MC, Lamb RJ, Goldberg SR, Kuhar MJ. Cocaine receptors on dopamine transporters are related to self-administration of cocaine. *Science* 1987;237:1219–23.
27. Mortensen OV, Amara SG. Dynamic regulation of the dopamine transporter. *Eur J Pharmacol* 2003;479:159–70.
28. Uhl GR. Dopamine transporter: basic science and human variation of a key molecule for dopaminergic function, locomotion, and parkinsonism. *Mov Disord* 2003;18 Suppl 7:S71–80.
29. Riddle EL, Fleckenstein AE, Hanson GR. Role of monoamine transporters in mediating psychostimulant effects. *Aaps J* 2005;7:E847–51.
30. Pristupa ZB, Wilson JM, Hoffman BJ, Kish SJ, Niznik HB. Pharmacological heterogeneity of the cloned and native human dopamine transporter: disassociation of [3H]WIN 35,428 and [3H]GBR 12,935 binding. *Mol Pharmacol* 1994;45:125–35.
31. Gelernter J, Kranzler HR, Satel SL, Rao PA. Genetic association between dopamine transporter protein alleles and cocaine-induced paranoia. *Neuropsychopharmacology* 1994;11:195–200.
32. van Dyck CH, Quinlan DM, Cretella LM, Staley JK, Malison RT, Baldwin RM, Seibyl JP, Innis RB. Unaltered dopamine transporter availability in adult attention deficit hyperactivity disorder. *Am J Psychiatry* 2002;159:309–12.
33. van Dyck CH, Malison RT, Jacobsen LK, Seibyl JP, Staley JK, Laruelle M, Baldwin RM, Innis RB, Gelernter J. Increased dopamine transporter availability associated with the 9-repeat allele of the SLC6A3 gene. *J Nucl Med* 2005;46:745–51.
34. Boden AG, Harris MC, Parkes MJ. The preoptic area in the hypothalamus is the source of the additional respiratory drive at raised body temperature in anaesthetised rats. *Exp Physiol* 2000;85:527–37.
35. Costentin J, Duterte-Boucher D, Panissaud C, Michael-Titus A. Dopamine D1 and D2 receptors mediate opposite effects of apomorphine on the body temperature of reserpinized mice. *Neuropharmacology* 1990;29:31–5.
36. Meller E, Hizami R, Kreuter L. Hypothermia in mice: D2 dopamine receptor mediation and absence of spare receptors. *Pharmacol Biochem Behav* 1989;32:141–5.

37. Peris J, Boyson SJ, Cass WA, Curella P, Dwoskin LP, Larson G, Lin LH, Yasuda RP, Zahniser NR. Persistence of neurochemical changes in dopamine systems after repeated cocaine administration. *J Pharmacol Exp Ther* 1990;253:38–44.
38. Kleven MS, Perry BD, Woolverton WL, Seiden LS. Effects of repeated injections of cocaine on D1 and D2 dopamine receptors in rat brain. *Brain Res* 1990;532:265–70.
39. Nunes JL, Sharif NA, Michel AD, Whiting RL. Dopamine D2-receptors mediate hypothermia in mice: ICV and IP effects of agonists and antagonists. *Neurochem Res* 1991;16:1167–74.
40. Barros RC, Branco LG, Carnio EC. Evidence for thermoregulation by dopamine D1 and D2 receptors in the anteroventral preoptic region during normoxia and hypoxia. *Brain Res* 2004;1030:165–71.
41. Wang ZZ, Wang CL, Wu TC, Pan HN, Wang SK, Jiang JD. Autoantibody response to heat shock protein 70 in patients with heatstroke. *Am J Med* 2001;111:654–7.
42. Crandall CG, Vongpatanasin W, Victor RG. Mechanism of cocaine-induced hyperthermia in humans. *Ann Intern Med* 2002;136:785–91.
43. Skidmore R, Gutierrez JA, Guerriero V, Jr., Kregel KC. HSP70 induction during exercise and heat stress in rats: role of internal temperature. *Am J Physiol* 1995;268:R92–7.
44. Kregel KC. Heat shock proteins: modifying factors in physiological stress responses and acquired thermotolerance. *J Appl Physiol* 2002;92:2177–86.

Chapter 30
Sudden Unexpected Death in Custody (SUDIC)

The Sudic Investigative Checklist

Charles V. Wetli

Any death occurring in police custody results in an intense investigation into the actions of the police, their judgment, and whether their use of force was appropriate [1]. In cases of police shooting or in judicial executions, the cause of death is obvious and the investigation focuses on the judgment of the police and whether proper procedures were followed. However, when death occurs suddenly and unexpectedly, speculation becomes rampant and the intensity of the investigation increases. Perhaps in no other circumstance is the failed logic of "*post hoc ergo propter hoc*" ("because of this, therefore this") invoked: the death occurred in police custody, therefore the police must have caused the death. Although the assumption is not valid, it must be acknowledged that the actions of the police may, at times, result in the death of a subject. It is therefore the task of the forensic investigator (viz. Medical Examiner, Coroner, and their investigative agents) to both establish the cause of death to a degree of reasonable medical certainty, and to evaluate the contributory role of the police actions, if any, to the death of that person. Furthermore, the forensic investigator must evaluate whether the application of what is thought to be a nonlethal force (e.g. pepper spray, lateral vascular neck restraint, ligature restraint, CEW, etc.) may have contributed to the death, or perhaps even caused the death. Strict objectivity and attention to detail are paramount since the conclusions will be scrutinized by the public, by the news media, and in criminal, civil and administrative proceedings.

Unfortunately, the sudden unexpected death in police custody may spark political interests that may totally ignore the objectivity of the investigation and the scientific support to attain a "politically correct" outcome [2]. When this occurs, jobs may be lost, careers ruined, and the civil and criminal legal processes circumvented and frustrated. This becomes the unstated occupational hazard of both law enforcement personnel and forensic investigators alike. In

C.V. Wetli (✉)
Chief Medical Examiner and Director of Forensic Sciences, Suffolk, County, NY (retired); Clinical Professor of Pathology, SUNY at Stony Brook (retired)
e-mail: thanatopsis888@yahoo.com

M.W. Kroll, J.D. Ho (eds.), *TASER® Conducted Electrical Weapons: Physiology, Pathology, and Law*, DOI 10.1007/978-0-387-85475-5_30,

some extreme cases this has resulted in very questionable criminal prosecutions of law enforcement officers [3].

Sudden unexpected death in police/correctional custody may be the result of the deliberate actions of the subject (e.g., suicidal hanging) or of another individual (e.g., homicidal strangulation or stabbing). For purposes of this discussion, sudden unexpected death will refer to those deaths that occur suddenly and without warning, and where the reason for the death is not immediately obvious. Interestingly, such deaths involving violent behavior, multiple restraints, and drug intoxication became recognized in the 1980s and coincide with the increase in cocaine abuse seen nationally [4]. Death frequently occurs at the scene but some may occur later while the subject is being transported or even after having been booked into a correctional facility. The forensic death investigation must evaluate the antecedent and terminal events along with autopsy findings demonstrating the presence or absence of disease and injury, drugs, and the actions (or possibly the neglect), of the police and/or correctional officers.

It is axiomatic that "common things happen commonly". Hence cardiac disease would be the primary suspicion for the death in middle aged or older individuals, and the possibility of drug abuse should be suspected in younger individuals. Both factors may, of course be, responsible for a death in police custody regardless of age. Some of the more common considerations to evaluate are listed in Table 30.1.

Table 30.1 Some initial considerations for cause of death

1. Natural disease
a. Coronary artery disease
b. Cardiac
i. cardiomyopathy (including hypertension)
ii. valvular disease
iii. coronary artery disease
c. Aortic dissection
d. Ruptured intracranial aneurysm
e. Intracerebral hypertensive hemorrhage
f. Sickle cell trait
g. Sudden unexpected death in epilepsy (SUDEP)
2. Drugs
a. Cocaine
b. Methamphetamine
c. Phencyclidine
d. Heroin
e. Drug packet ingestion ("mini-packer")
3. Law enforcement actions
a. Bullet wounds
b. Oleoresin capsicum (OC, pepper) spray
c. Conducted electrical weapon (if head injury from fall)
d. Blunt force (ASP or PR-24)
e. Lateral vascular neck restraint ("carotid sleeper hold")
f. Chest or neck compression
g. Restraint ("hog-tying" or maximal restraint)

30.1 Before the Autopsy

30.1.1 Scene Where Vital Signs Were Lost

Because of very efficient and rapid communications, the subject is most often transported from the scene to a local hospital. Often, emergency medical services are already enroute or at the scene when it is discovered that vital signs have been lost, and CPR begins almost immediately. It is also common that an incident begins at one location, and the subject is transported elsewhere or to a correctional facility or hospital, and a renewed struggle ensues followed by loss of vital signs. It is therefore imperative that the Medical Examiner responds to the scene where vital signs were lost even if the body has been transported to a local medical facility. Examination of the other locations where a struggle with the subject took place is also advisable. The purpose of the scene investigation by the Medical Examiner is to gain an appreciation of the environment and obtain at least some information as to the circumstances of the police encounter, and to correlate this information with the subsequent observations of the body. The approach to the autopsy is thus guided by the observations and issues obtained during the scene investigations. Knowledge of the environment may well correlate with observed injuries, and awareness of the details of the encounter may well direct attention to specific areas during the autopsy or suggest specific areas for dissection and documentation. For example, knowing that a police baton or CEW was used during the struggle would correlate with certain pattern injuries. Knowing that the struggle took place on a gravel roadway or that there was broken glass strewn about may also help interpret subsequently observed injuries.

Ideally, the Medical Examiner who will perform the autopsy should be the one to respond to the scene. However, for reasons that are many and varied, this may not be possible or practical in all situations. In such instance it is highly recommended that whoever investigates the scene for the Medical Examiner also attend the autopsy to provide as much information and continuity as possible. Today, it is also possible to provide photographic documentation of the scene to the Medical Examiner prior to the commencement of the autopsy. As with any sudden and unexpected death, knowledge of the scene and terminal event guides the approach to the autopsy and allows the pathologist to anticipate questions and issues that will arise in the future.

30.1.2 Examination at the Hospital

Most often the body has been transported to the hospital where death is officially pronounced after a period of attempted resuscitation. Examination of the subject at this time is obviously preliminary but nonetheless may be quite revealing. Also, it provides the opportunity to obtain a core body temperature that may reveal hyperthermia that often accompanies sudden death in cases of excited delirium. Note that it is the core (rectal or liver) temperature that is important since the skin may feel cool in the presence of hyperthermia. While at the hospital, the clothing can be inspected along with the contents of the pockets. Any fluids taken from the subject (usually blood, urine, and sometimes gastric content) should be obtained for subsequent toxicological testing. Depending on the circumstances of death, trace evidence may be an important consideration in which case further contact by medical personnel or family members should be restricted. If fingernail scrapings for DNA or trace evidence are anticipated, the hands should be covered with paper bags. Photographs of the body may be important in documenting evidence of medical intervention as well as injuries which may become altered in appearance by the time the body is examined in the morgue some hours later. For example, dependent lividity will obscure bruises of the back that were evident at the time the body was examined in the emergency room of the hospital. Also, abrasions may become apparent only after they are dehydrated and thus may not be visible in the hospital but are obvious when examined hours later in the morgue [5].

Efforts should me made to obtain the EMS cardiac rhythm strips or automatic external defibrillator data if in existence. Establishing the presenting rhythm is especially important if a CEW was used since certain rhythms (such as asystole and pulseless electrical activity) are not associated with induction by electrical stimulation.

Generally, Medical Examiners and Coroners are not allowed to become involved in a case until death has actually occurred. Therefore, if the resuscitative attempts were successful there can be no examination of the subject unless

Table 30.2 Preautopsy checklist

1. Prehospital
- a. Visit scene of lost vital signs
  - i. Other struggle locations
- b. Interview officers involved

2. Hospital
- a. Obtain EMS records
  - i. Especially presenting rhythms
- b. Obtain a core body temperature
- c. Examine clothing and contents of pockets, purse, etc.
- d. Consider necessity of preserving trace evidence
- e. Place paper bags on hands if indicated
- f. Obtain all blood, urine and gastric lavage specimens
- g. Photograph the subject, especially the back
- h. Emergency department records
  - i. Toxicology screen
  - ii. Blood gases and pH
  - iii. Myoglobin and CK

there is a court order or permission of the family and hospital personnel is granted. If death is anticipated within a few days (e.g. in excited delirium where rhabdomyolysis, renal failure, and disseminated intravascular coagulation are developing), then a hold should be placed on all blood and urine specimens so these may be toxicologically analyzed once death occurs.

These items are summarized in Table 30.2.

30.2 The Autopsy

The autopsy checklist is summarized in Table 30.3. Once the scene and hospital investigations are completed, the pathologist and other investigators will have formulated some ideas as to the possible cause of death and areas of concern that the autopsy and ancillary studies should address. The high profile nature of in-custody deaths requires extensive diagrammatic and photographic documentation of both what is present and what is not present (e.g., presence or absence of petechiae of the scleral and palpebral conjunctivae). If the terminal event involved a violent struggle, full body x-rays are recommended. Additional

Table 30.3 The autopsy checklist

1. Obtain full body x-rays
2. Photograph all injuries, the hands, mouth, eyes
3. Consider special dissections:
a. Back and extremities (especially wrists and ankles)
b. Layerwise neck dissection
c. Posterior neck dissection
d. Stripping the pleura
e. Facial dissection
4. Consider cardiac and neuropathology consultations
5. Obtain Specimens for Toxicology:
a. Peripheral blood (femoral)
b. Central blood (aorta, pulmonary artery)
c. Vitreous fluid
d. Urine
e. Bile
f. Liver
g. Brain
h. Entire gastric content
6. Consider special testing:
a. Hemoglobin electrophoresis (to detect sickle cell trait)
b. DNA sample for possible paternity testing
c. Hair analysis for drugs
d. Vitreous chemistry
e. Dopamine receptor study of a coronal section of brain

dissections with photographic documentation are also necessary. These include a layerwise neck dissection, dissection of the back and lower extremities to detect contusions not externally visible, a posterior neck dissection, and stripping the pleura to detect rib fractures not readily visible by x-ray. Dissection of the wrists and ankles may demonstrate evidence of continued forceful struggling against the restraints that is often seen in sudden death from excited delirium. If the terminal event suggests the possibility of facial fractures, then an actual facial dissection should be considered since these are frequently not detected radiologically [1,6].

Deaths in police custody mandate extensive toxicological testing. Drugs and their metabolites should be analyzed in a variety of fluids and tissues. Alcohol concentrations should ideally be obtained in peripheral blood, vitreous humor (fluid from the eye), brain, and gastric content. Cocaine and its metabolites should be quantified in peripheral blood and brain. Other drugs should be quantified in peripheral blood and, usually, liver. While peripheral blood drug and alcohol concentrations are preferred, this may not always be possible since the amount of peripheral blood obtainable at autopsy may be limited, in which case blood from the heart and aorta (central blood) must be analyzed. However, some drugs, such as cocaine, may have spurious elevations in central blood due to release of the drug from the tissue into the blood after death. With these considerations in mind, the following specimens should be obtained at the time of the autopsy and submitted to the toxicology laboratory: antemortem (hospital) specimens, peripheral (femoral) blood, blood from the heart or aorta and the pulmonary artery, urine, bile, liver, brain, and the entire gastric content.

Histological sampling should be extensive to document the presence or absence of any natural disease processes, particularly of the cardiovascular and central nervous systems. In some cases, histological documentation of injuries may also be important, especially if some are suspected as being of a different age, or to document the nature of an injury (e.g., an electrothermal burn from a CEW). Vitreous humor should also be submitted for chemistries (electrolytes, glucose, and urea nitrogen), if indicated. Samples of hair (including the hair roots) should be obtained in the event that chronicity of drug use needs to be determined. A blood or tissue sample for subsequent DNA analysis should be preserved in the event that subsequent paternity testing is required. If the subject is of African descent, a blood sample for hemoglobin electrophoresis (to determine the presence of sickle cell trait) should be obtained [2].

If excited delirium is a consideration, a coronal section of brain just anterior to the temporal poles should be frozen and sent for dopamine receptor studies [7]. Prior to the autopsy, the head should be kept cold (e.g., by wrapping in ice) to help prevent deterioration of the receptor sites in the brain.

30.2.1 Some Special Issues

Frequently the family of a person who dies suddenly in police custody will request a second autopsy, or even request that a pathologist of their choice attend the initial autopsy. The latter request is often met with protest and resistance by the police and government attorneys since it is a case under active investigation. However, it might be more prudent to allow an independent observer or pathologist chosen by the family of the decedent to attend the autopsy than to exclude this person and encourage the allegation of a "cover-up" for the police. Second independent autopsies obviously have their limitations since the organs have already been removed and dissected, and there are usually numerous other artifacts (such as embalming) as well. In addition, some tissues (e.g., the neck structures) and organs (e.g., heart and brain) may not be available. Instead of a second autopsy being done at a funeral home where lighting and other facilities are often less than desirable, it is suggested that the Medical Examiner facility be made available for this purpose where the original pathologist can be in attendance to explain various findings and provide tissues and organs that have been retained for further evaluation.

Families may also consent to organ or tissue donation, in which case the Medical Examiner must carefully evaluate what, if any, tissues or organs must be restricted so as not to jeopardize the investigation of the death. Blanket denial of organ and tissue procurement is not indicated, but restrictions of certain organs or tissues may be necessary [8]. Organ donation is usually less problematical since there is time to evaluate the function of the organs and extensive testing in the hospital will have documented the injuries. Tissue donation, however, has time limitations which means the pathologist must either examine and document external findings (e.g., petechiae of the eyes, externally visible contusions and abrasions) before the tissue procurement, or the tissue procurement must take place after the autopsy. In general, communication between the organ and tissue procurement organizations and the Medical Examiner should resolve the issues of the case in allowing the procurement of some tissues and organs while preserving the integrity of the postmortem examination. The option of examining the heart under sterile conditions to allow heart valve procurement and the feasibility of obtaining some tissues after the autopsy should be considered as well.

30.3 Case Study

The following case illustrates the frustration that can arise from the investigation of some deaths in police custody, the problems in resolving issues regarding the cause and manner of death, and the legal resolution that is at times perplexing.

The police were called to a violent domestic dispute involving a 34 year old computer technician and his wife and children. The subject had a history of depression, occasional alcohol consumption, and illicit drug ingestion. He resisted attempts by the police to take him into custody, and during the struggle was placed prone on the floor and handcuffed. Shortly thereafter he stopped breathing. The EMS automatic external defibrillator found no shockable rhythm, and the ECG monitor revealed irregular electrical nonfibrillating cardiac activity. CPR was unsuccessful. It was later learned that when the police arrived at the scene, the subject was on the telephone and leaving a voice-mail message for his mother. On the audiotape he is heard to say "stop choking me, bro." The police denied striking him with an instrument or applying a neck restraining hold. The postmortem examination revealed a variety of contusions, large subgaleal contusions, hepatosplenomegaly, moderate cardiomegaly, and multiple, mostly anterolateral, rib fractures "surrounded by hemorrhage" (not further described). The right sternocleidomastoid muscle had a 3×1.2 centimeters contusion at the sternal insertion and the left mid-sternocleidomastoid muscle had a 4.3×1.8 centimeters contusion. A layerwise dissection of the neck further revealed a fracture of the greater horn of the hyoid bone with "no surrounding hemorrhage," and in the autopsy findings indicated this was "consistent with removal artifact." Toxicological analysis revealed a blood alcohol concentration of 0.20%, a vitreous alcohol concentration of 0.22%, and a blood citalopram concentration of 0.32 milligram per liter. Benzoylecgonine was detected in the urine only. Neurochemical testing of the brain was not consistent with excited delirium.

The cause of death was given as "cardiopulmonary arrest during prone restraint" and the manner of death was recorded as "accident." In the stated rationale for death certification, it was noted that "Sudden death during prone restraint has been recognized for many years," and opined upon theories for the mechanism of death. The manner of death was considered accidental since "the intent of the police officers was to restrain him rather than harm him."

A consultant pathologist for the family opined: there was no evidence for excited delirium; the autopsy did not reveal a cause of death; the toxicological analysis did not reveal a cause of death; a layerwise neck dissection should have precluded an inadvertent fracture of the hyoid bone that would have eluded the notice of the prosecutor; the cause of death was a description of terminal events and not a cause of death supported by medical and scientific literature; that the articles cited from the medical and forensic literature in the stated rationale for the cause of death actually debunked and disproved the theories proposed for the mechanism of death being a mechanical asphyxia; and that the accidental manner of death was illogical since "intent" is not a criteria for homicide and lack of intent is not a criteria for accident. In the opinion of the consultant pathologist, the most likely cause of death was the prolonged or inappropriate application of a lateral vascular neck restraint (carotid sleeper hold). This was based on the location of the sternocleidomastoid contusions and the fracture of the hyoid bone. Also, the photographs, in the opinion of this pathologist, indicated the struggle was more violent than what the autopsy report and the police reports of the incident indicated.

A trial was held in Federal Court and resulted in a hung jury. A second trial was held a few weeks later. The family attorney related that the jury had to determine if (1) there was a violation of Constitutional rights, (2) the police officers were responsible for the legal and the proximate cause of death, and (3) damages. On the second day, the jury asked for a clarification of the second element, implying they found there was a violation of the Constitutional rights and thereby indicating the third day of deliberations was devoted to the calculation for damages. The entire court was therefore shocked when the jury returned a complete defense verdict indicating there was no violation of Constitutional rights.

30.4 Death Certification

Certifying the cause and manner of death is perhaps the most controversial issue in Forensic Pathology. In general, for sudden death occurring in police custody, it should be sufficient to simply list the cause of death since the concepts of proximal and immediate causes are not an issue unless there is a period of survival. However, there is a tendency to incorporate the entire scenario into the cause of death and this actually engenders confusion instead of a meaningful conclusion. This approach to death certification seems to be especially true with deaths due to excited delirium, where the struggle is incorporated into the cause of death and the manner of death then becomes accident or homicide. Curiously, although intent is not a criterion for the classification of homicide, the manner of death is sometimes listed as "accident" because the police did not "intend" to cause the death of the subject. With excited delirium, it is well to remember that a violent struggle is part of the syndrome, and the manner of death is therefore dependent on the cause of the excited delirium. Therefore, if the excited delirium is due to cocaine or methamphetamine, the manner of death is accident. If the excited delirium is due to untreated schizophrenia or bipolar disease, the manner of death is natural.

30.5 Conclusion

A sudden unexpected death in police custody is a high profile event and the investigation will be intensely scrutinized by numerous individuals and agencies. It is imperative to be exceedingly thorough in the investigation, to be objective, and to be neutral. If the police did contribute to the death of the subject, then this should be thoroughly established and duly reflected on the death certificate. However, it must also be remembered that because the death occurred in custody does not necessarily mean the actions of the police were contributory. There is no room for speculation or invoking unproven or debunked theories in determining the cause of death in any case, including deaths in police custody.

References

1. Wetli, C.V. and Natarajan, G.A.: Death in Custody, United States of America. In Encyclopedia of Forensic and Legal Medicine, Vol. 2, pp. 65–73, Payne-James, Byard, Corey and Henderson (eds.), Elsevier, Glasgow, 2005 (ISBN: 0-12-547970-0 {set}).
2. Scheinin, L. and Wetli, C.V.: Sudden Death in Sickle Cell Trait Medicolegal Considerations and Implications. Amer J Forensic Med and Pathol, *in press*.
3. Armon, R.: 5 Summit Deputies Indicted in Jail Death. Akron Beacon Journal. Akron, Ohio, 2007.
4. Grant, J.R., Southall, P.E., Fowler, D.R., Mealey, J., Thomas, E.J., and Kinlock, T.W.: Death in Custody – A Historical Analysis J Forensic Sci, 52:1177–1181, 2007.
5. Wetli, C.V., Mittleman, R.E., and Rao, V.J.: An Atlas of Forensic Pathology American Society of Clinical Pathologists (ASCP Press), Chicago, 1999, pp. 26–27 ISBN 0-89189-430-6.
6. Natarajan, G., M.D., and Fonseca, C., M.D. "Beyond Nose Jobs and Face Lifts: An Illustrated Technique of Facial Dissection". *ASCP Check Sample*, Forensic Pathology, Volume 39, Number 8, 1998, ISSN-1056-5922.*American Society of Clinical Pathologists.*
7. Staley, J.K., Hearn, W.L., Ruttenber, A.J., Wetli, C.V., and Mash, D.C.: High Affinity Cocaine Recognition Sites on the Dopamine Transporter are Elevated in Fatal Cocaine Overdose Victims. J. Pharm. and Exptl. Therapeutics 271:1678–1685, 1994.
8. Pinckard, J.K., Wetli, C.V., and Graham, M.A.: National Association of Medical Examiner Position Paper on the Medical Examiner Release of Organs and Tissues for Transplantation. Amer J Forensic Med and Pathol 28(3):202–207, 2007.

Chapter 31
Legal Basics for the CEW

Michael A. Brave

Since the dawn of human civilization groups of people have banded together with a goal of developing orderly societies. These societies have often created governments. These governments have enacted and enforced limitations on conduct and established procedures for accountability for those who violate these limitations. This chapter provides a basic overview of these limitations that pertain to the possession and use of conducted electrical weapons (CEWs).[1]

This chapter focuses on legal accountability issues in the United States. However, most of the principles – although not the exact legal specifics – pertain to many other countries.

Societies place restrictions on what one member of society may do to another, especially when one person injures or damages another. Some of these restrictions are context specific, for example, a defender of his home may use force to protect his home (the "Castle Doctrine"), or a person may take action against an attacker in self-defense, defense of another, or defense of property that he has a right or obligation to protect. Each of these may have different rules that may vary from jurisdiction to jurisdiction.

Societies also grant some of their members such as law enforcement officers (LEOs) privileges to act against others under governmental power to enforce society's rules. However, these government enforcers have limitations on when, where, and how they can enforce the government's rules.

In the United States there are several forms of government entities, including federal government, state governments, Native American tribal governments, county governments, and city governments. Each government entity is usually subservient to another entity. For example, a city is usually part of a county, a county is part of a state, and a state is part of the federal system. As part of this system, each government can place restrictions on its subparts. For example, a state government could prevent a county or city government from having more restrictive rules than the state enacted rules. Also, in some circumstances, a state

M.A. Brave (✉)
LAAW International, Inc., TASER International, Inc.
e-mail: brave@laaw.com

M.W. Kroll, J.D. Ho (eds.), *TASER® Conducted Electrical Weapons: Physiology, Pathology, and Law*, DOI 10.1007/978-0-387-85475-5_31,

government can have more restrictive accountability standards than the federal government, but, due to the supremacy doctrine, the state generally cannot create a less-restrictive standard that would attempt to circumvent the federal rules.

In this chapter we will take a look at CEWs as they relate to the following:

- General Accountability Themes
- Accountability Sanctions
- CEW Usage by Non-LEOs
- Government Empowered CEW Use
- US Appellate Court CEW Case Law
- US Lower Court CEW Cases

31.1 General Accountability Themes

Our society allows people to be held legally accountable for their misconduct or transgressions upon another. The primary underlying accountability themes include:

- **Do not injure another.** No person has the right to infringe upon (injure, damage, harm) another or his property, without a legally based justification, privilege, or process.
- **If wronged, use the legal system (not self-help).** A person who has been injured by another should use society's legal systems as recourse, unless the infringement is presently taking place to the person or another, or his property. If a thief steals a car (infringes on the owner's property), and the owner later sees the thief driving his car, then society expects the owner to use the established legal systems to attempt to reacquire his property and to hold the thief accountable (under society's procedures) for the thief's breach of society's rules without the owner committing violence or resorting to self-help as a vigilante.
- **Minimize injury to another.** If a person, for whatever lawful justification acts upon another (e.g. a LEO using force on a resisting criminal suspect to accomplish a lawful arrest) the person should use the minimum amount of force necessary to accomplish the lawful objective. Note, this is "not" the legal force standard in the United States, but this is the ideal objective: to cause as little injury or infringement as possible and still be able to get the job done.
- **Different people function under different force paradigms.** There are different basic use-of-force paradigms that apply to people in different circumstances. The most basic paradigms include the following:
    - **LEO "Abuse of Authority" Standard.** While lawfully performing duties, a LEO is held accountable to an "abuse of power" standard. The LEO is restricted from misusing his government-empowered authority.[2] LEOs are empowered to enforce society's laws and thus they are given greater latitude, and protections from sanctions, than others.

- **Private Person (the "Self-Defense" Standard).** A private person is usually held to a "self-defense" standard. Depending upon the particular jurisdiction the person may be entitled to make a citizen's arrest, may be required to retreat prior to using force, or may be required to not re-engage if there is a lull in the encounter.
- **Private Security Guard (or other Special Standard).** Private security guard standards vary greatly from jurisdiction to jurisdiction and the specific circumstances regarding a particular event. For example, a shopping mall security officer may be allowed to use force to detain a shoplifter; a nuclear facility security officer may be allowed to use force to stop and detain a person trespassing at the facility; or a juvenile detention officer may be allowed to use force to control an out-of-control youth under the influence of illegal substances.

There are many legal and situational variables and each person needs to ensure compliance with any applicable restrictions, exemptions, and privileges.

31.2 Accountability Sanctions

There are different penalties and sanctions for different transgressions and degrees of intent. The more intentional the transgression, the greater society's sanctions.

31.2.1 Criminal Sanctions

For some transgressions the legal system allows for the person to be criminally prosecuted and punished (including loss of freedom, and in some circumstances forfeiture of life) for his actions. Usually the burden of proof for the imposition of a criminal sanction requires the wrongdoer to be found guilty beyond a reasonable doubt after a fair adjudication.

In the United States the "general" rule is that a person can only be criminally prosecuted for an offense "one" time. However, this is "sovereign" dependent. If a person robs a federally insured bank on a Native American reservation, then the bank robber can be criminally prosecuted, and receive criminal sanctions from: (1) the federal government, (2) the Tribal government, and (3) the state government. A LEO who misuses his government endowed power by willfully harming someone without legal justification or privilege can be criminally prosecuted by both the federal government and the state government.

31.2.2 Civil Accountability

For some transgressions a wrongdoer can be brought into the legal system and be forced to either financially compensate the wronged party for the transgression (money damages) or to change his ways (equitable actions). Usually in order for a person (plaintiff) to recover civilly against a wrongdoer (defendant) the burden of proof is by a "preponderance" of the evidence or "more likely than not."

If the transgressor's wrongdoing was sufficiently intentional or egregious then the plaintiff may also be able to recover punitive or exemplary damages to punish the transgressor for his intentional misdeeds. In order for a plaintiff to recover punitive damages against a defendant, the plaintiff must usually prove the heightened misconduct of the defendant by clear and convincing evidence, a higher burden of proof than "by a preponderance of the evidence."

31.2.3 Administrative Accountability

If a person (e.g., LEO, private security officer, or correctional officer) has been empowered to transgress on another, the person's employer usually puts restrictions on that privilege. If the empowered person violates or exceeds the employer's standards of conduct, then the employee may be disciplined or terminated from his employment. Employers can usually, but not always, establish force standards that are more restrictive than the applicable federal, state or local legal standards.

If a person has been granted a privilege, for example, a license to possess or carry a handgun, and the person transgresses against another in violation of the rules accompanying that privilege, then, the entity granting the privilege may revoke it.

31.3 CEW Usage by Non-LEOs

CEWs are weapons designed to incapacitate people and animals. As such, the use of a CEW will usually involve the use of force (or the threatened or "coercive" use of force). The degree of a person's privilege to use (or threaten the use of) force upon another person is defined by the rules of the jurisdictions within which the force is used and usually defined by: (1) the person's government endowed authority to act as a representative of the government; (2) a person's privilege to use force based upon his employment (e.g. private security or hospital security officer); or (3) a private person's privilege to use force to make a citizen's arrest or in lawful self-defense, defense of others, or defense of property.

The legal issues regarding CEWs can be categorized as: acquisition, ownership, carrying (including concealed), brandishing (display), use as a coercive device (to display the CEW to gain compliance without deployment), discharge, deployment, use, and reporting. Below we discuss the use of a CEW by private persons and by private security officers (non-LEOs).

31.3.1 Use of a CEW by a Private Person

The US federal government only regulates a person's ability to lawfully acquire and possess a CEW when that CEW is taken onto federal parklands.[3] The ability of a private person to lawfully acquire and possess a CEW is based primarily upon each individual state's laws. Presently, a private person can lawfully acquire and possess a CEW in 43 of the 50 US states, with varying degrees of restrictions (such as age, criminal history, mental health status, license requirement, and concealability).[4] Also, some counties and cities further restrict the acquisition or possession of a CEW. For example, a private person (with some restrictions) in Pennsylvania may possess a CEW; however, the City of Philadelphia completely prohibits possession inside the city limits.[5]

Different states have varying privileges, exceptions, and liability avoidance mechanisms for display, brandishing, and coercive use. "Display" can be defined as simply allowing the device to be seen or disclosed. "Brandishing" is defined as wielding the device in a manner that someone may find threatening or menacing. "Coercive use" can be defined as using the CEW's integrated light or laser or pointing the CEW at a person in a threatening manner in order to gain compliance from the person. To display, brandish, or use for coercive purpose without acceptable legal justification could expose the private CEW user to criminal, civil, or administrative consequences.

Different states also have different rules as to where, when, how, and under what circumstances a CEW user may legally use the CEW. Where a CEW may be possessed may be limited. For example, in Illinois a private person may only possess or carry a CEW in a private home or business.[6] Because CEW possession is regulated by each state, a citizen may not take a CEW that was legally purchased in one state into a state that prohibits the possession of the devices. For example, a citizen who lawfully possesses and carries a CEW in Minnesota may not take the CEW into neighboring Wisconsin, where such possession is a felony criminal offense.

When a CEW may be used is more complicated and varies greatly from state-to-state. Considerations include whether a particular state's legal framework requires a person to attempt to retreat from an encounter rather than engage an attacker, whether a person has a right to use force on a person breaking into the CEW user's home (the Castle Doctrine), and whether the CEW user can lawfully discharge or deploy the CEW if the user is under the influence of drugs or alcohol at the time.

31.3.2 Nonsworn LEOs (Private Security Officers)

A private security officer is also regulated by state specific laws. These security officers may be restricted to simply observing an event, reporting, and summoning authorities; however, private security officers who protect some of the

US nuclear power facilities have much greater authority to use force. Other security officers such as those responsible for a locked mental health hospital ward, have other privileges and restrictions on their possession and deployment of a CEW.

In an extreme case such as Wisconsin – where only a sworn LEO can possess or use a CEW – a nonsworn jail or detention officer would be committing a felony by possessing or using a CEW.[7]

The lesson for private persons is that each jurisdiction may have multiple – and sometimes conflicting and antiquated laws and regulations pertaining to CEW acquisition, possession, display, brandishing, discharging, deployment, use, and reporting. Therefore, private persons must take care to ensure that they are aware of the local laws and regulations pertaining to CEWs before they possess or use a CEW.

31.4 Government Empowered CEW Use

Government employees (including LEOs) are subject to simultaneous force limitations under both federal and state laws. Under federal law, LEOs are prohibited from using force that violates the US Constitution and other applicable statutes. A LEO can be criminally prosecuted under 18 U.S.C. § 242 for willful violation of a person's federal constitutional rights:

> Whoever, under color of any law, statute, ordinance, regulation, or custom, willfully subjects any person in any State, Territory, Commonwealth, Possession, or District to the deprivation of any rights, privileges, or immunities secured or protected by the Constitution or laws of the United States, or to different punishments, pains, or penalties, on account of such person being an alien, or by reason of his color, or race, than are prescribed for the punishment of citizens, shall be fined under this title or imprisoned not more than one year, or both; and if bodily injury results from the acts committed in violation of this section or if such acts include the use, attempted use, or threatened use of a dangerous weapon, explosives, or fire, shall be fined under this title or imprisoned not more than ten years, or both; and if death results from the acts committed in violation of this section or if such acts include kidnapping or an attempt to kidnap, aggravated sexual abuse, or an attempt to commit aggravated sexual abuse, or an attempt to kill, shall be fined under this title, or imprisoned for any term of years or for life, or both, or may be sentenced to death.

The LEO can also be civilly sued pursuant to 42 U.S.C. § 1983. Section 1983 is not itself a source of substantive rights, but merely provides a method for vindicating federal rights elsewhere conferred:[8,9]

> Every person who, under color of any statute, ordinance, regulation, custom, or usage, of any State or Territory or the District of Columbia, subjects, or causes to be subjected, any citizen of the United States or other person within the jurisdiction thereof to the deprivation of any rights, privileges, or immunities secured by the Constitution and laws, shall be liable to the party injured in an action at law, suit in equity, or other proper proceeding for redress, except that in any action brought against a judicial officer for an act or omission taken in such officer's judicial capacity,

> injunctive relief shall not be granted unless a declaratory decree was violated or declaratory relief was unavailable. For the purposes of this section, any Act of Congress applicable exclusively to the District of Columbia shall be considered to be a statute of the District of Columbia.

Qualified immunity shields public officials from civil lawsuits against them in their individual capacities for torts committed while performing discretionary duties unless the tortious act violates a clearly established statutory or constitutional right.[10] Qualified immunity requires a two-step inquiry. First, do the facts (taken in the light most favorable to the party asserting the injury) show that the police officer's conduct violated a constitutional or statutory right and, second, was the constitutional or statutory right clearly established at the time the force was used.

The federal constitutional rights that set out standards for the use of force by a government actor are found in the 4th, 5th, 8th, and 14th Amendments to the US Constitution. In an "excessive force" civil lawsuit brought under 42 U.S.C. § 1983, analysis begins by identifying the specific constitutional right allegedly infringed by the challenged application of force.[11] The first inquiry in any Section 1983 lawsuit is to isolate the precise constitutional violation with which the defendant is charged. In most instances, that will be either the standards in the 4th Amendment's prohibition against unreasonable seizures of the person or the 8th Amendment's ban on cruel and unusual punishments.

LEOs must also comply with their own agency or department standards on the use of CEWs. And, lastly, LEOs, while not necessarily legally obligated to abide by, are guided by numerous advisory standards, such as model and sample policies, advisory standards, and training bulletins promulgated by organizations including the International Association of Chiefs of Police (IACP)[12] and the Police Executive Research Forum (PERF).[13]

31.4.1 Constitutional Use of Force Standards

The first step in analyzing the constitutionality of a LEO's use of force is to determine which of the constitutional force standards applies. Thus, before discussing each standard one must determine which standard will gauge the LEO's use of force (see Fig. 31.1).

31.4.1.1 "Objective Reasonableness" Standard of the 4th Amendment

The 4th Amendment provides:

> The right of the people to be secure in their persons, houses, papers, and effects, against unreasonable searches and seizures, shall not be violated, and no warrants shall issue, but upon probable cause, supported by oath or affirmation, and particularly describing the place to be searched, and the persons or things to be seized.

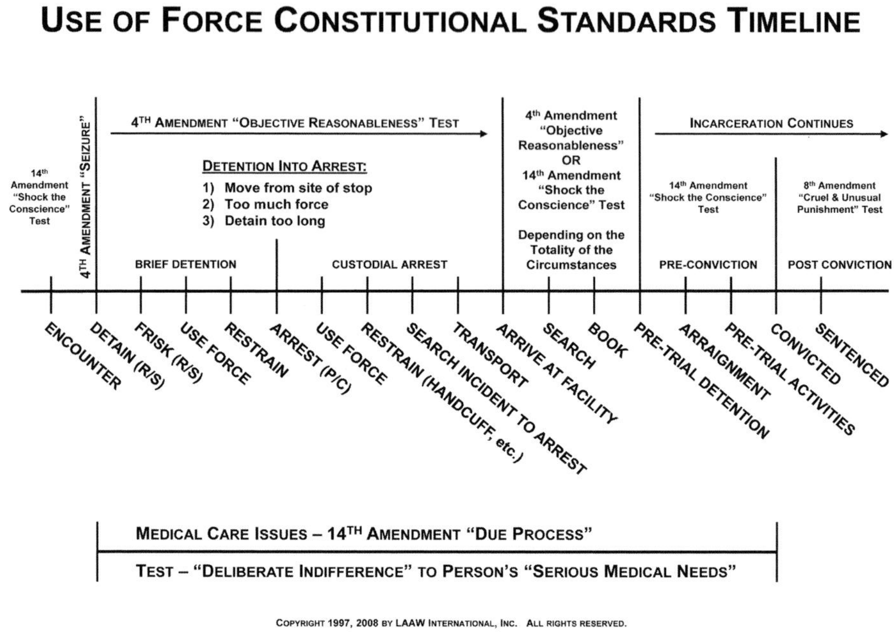

Fig. 31.1 Constitutional timeline

Thus, the question at this stage is whether the LEO's use of force was a "seizure" of a free person? If the LEO's use of force was a seizure of a free person then the 4th Amendment's "objective reasonableness" standard applies. If the LEO's use of force was not a seizure of a free person then the 4th Amendment standard does not apply.

A person is seized, and thus entitled to challenge the government's action, when LEOs, by physical force or a show of authority, terminate or restrain the person's freedom of movement through means intentionally applied.[14] There is no seizure without that person's actual submission.[15] When LEOs' actions do not show an unambiguous intent to restrain or when an individual's submission takes the form of passive acquiescence, the test for telling when a seizure occurs is whether, in light of all the surrounding circumstances, a reasonable person would believe he was not free to leave.[16] But when a person "has no desire to leave" for reasons unrelated to the LEOs' presence, the "coercive effect of the encounter" can be measured better by asking whether "a reasonable person would feel free to decline the officers' requests or otherwise terminate the encounter."[17]

All claims that LEOs have used excessive force – deadly or not – in the course of an arrest, investigatory stop, or other seizure of a free person are properly analyzed under the 4th Amendment's "objective reasonableness" standard. Determining whether the force used to affect a particular seizure is "reasonable" under the 4th Amendment standard requires careful balancing of the nature and quality of the intrusion on the individual's 4th Amendment interests against countervailing governmental interests. "Reasonableness" of a particular use of force must be judged from the perspective of a reasonable LEO on the scene, and the calculus of reasonableness must allow for the fact that LEOs are often forced to make split-second judgments in circumstances that are tense, uncertain and rapidly evolving, about the amount of force that is necessary in a particular situation. The determination of whether force is excessive in violation of the 4th Amendment standard requires careful attention to the facts and circumstances of each particular case, including the severity of the crime at issue, whether the person poses an immediate threat to the safety of the officers or others, and whether he is actively resisting seizure or attempting to flee, and whether the LEOs' actions are "objectively reasonable" in light of facts and circumstances confronting them, without regard to their underlying intent or motivation.

31.4.1.2 "Cruel and Unusual Punishment" Standard of the 8th Amendment

Was the person convicted and incarcerated at the time the force was used? If the answer is "yes" then the analysis is under the 8th Amendment's "cruel and unusual punishment" standard.[18] The 8th Amendment force standard is whether the force was applied in a good faith effort to maintain or restore discipline or maliciously and sadistically applied for the purpose of causing harm:

> It is obduracy and wantonness, not inadvertence or error in good faith, that characterize the conduct prohibited by the Cruel and Unusual Punishments Clause, whether that conduct occurs in connection with establishing conditions of confinement, supplying medical needs, or restoring control over a tumultuous cellblock. The infliction of pain in the course of a prison security measure, therefore, does not amount to cruel and unusual punishment simply because it may appear in retrospect that the degree of force authorized or applied for security purposes was unreasonable, and hence unnecessary in the strict sense. The general requirement that an 8th Amendment claimant establish the unnecessary and wanton infliction of pain should also be applied with due regard for differences in the kind of conduct involved. Thus, where a prison security measure is undertaken to resolve a disturbance that poses significant risks to the safety of inmates and prison staff, the question whether the measure taken inflicted unnecessary and wanton pain and suffering ultimately turns on whether force was applied in a good faith effort to maintain or restore discipline or maliciously and sadistically for the purpose of causing harm.[18]

31.4.1.3 "Shocking to the Conscience" Standard of the 14th Amendment

If at the time of the use of force the person is neither: (1) convicted and incarcerated; or (2) a "seized" free person, then the default standard to be used to analyze the LEO's use of force will be found in the 14th Amendment's Due Process Clause. The 14th Amendment has two force standards distinguished by the degree of urgency in the LEO's decision. The 14th Amendment's standard is that "only a purpose to cause harm unrelated to the legitimate object of arrest will satisfy the element of arbitrary conduct shocking to the conscience, necessary for a due process violation."[19]

Deliberate indifference rests upon the luxury enjoyed of having time to make unhurried judgments, upon the chance for repeated reflection, largely uncomplicated by the pulls of competing obligations. When such extended opportunities to do better are teamed with protracted failure even to care, indifference is truly shocking. But when unforeseen circumstances demand a LEO's instant judgment, even precipitate recklessness fails to inch close enough to harmful purpose to spark the shock that implicates the large concerns of the governors and the governed.[20]

31.5 US Appellate Court CEW Case Law

US appeal court decisions are binding on the states within that appeals court's circuit and are also considered strong precedent nationwide.

31.5.1 4th Amendment "Objective Reasonableness" Standard

There are many legal cases that define and explain the nuances of when the use of a CEW is objectively reasonable under the totality of the circumstances. In *Draper v. Reynolds*, one of the leading CEW cases, the Court states that in a "difficult, tense and uncertain situation" the use of a CEW to subdue a suspect who has repeatedly ignored a LEO's instructions and continues to act belligerently toward police is not excessive force.[21]

In *Ewolski v. City of Brunswick*, LEO's use of CEW nonlethal force to subdue a potentially homicidal individual did not transgress clearly established law. The court held that the use of a CEW nonlethal force against an armed and volatile suspect does not constitute excessive force and concluded that the LEOs were entitled to qualified immunity on the excessive use of force claim. The court further held that in cases in which LEOs must choose among alternative use of force options, a plaintiff must show that the LEOs "knowingly and unreasonably" opted for a course of conduct that entailed a substantially greater total risk than the available alternatives.[22]

31.5.1.1 CEW Use on a Restrained Criminal Suspect

The use of CEWs on handcuffed persons, and a person in a restraint chair have been found objectively reasonable when the person is fighting, struggling, resisting, or attempting to flee. However, if the person is not an immediate threat or attempting to flee such CEW deployments have been found to violate the 4th Amendment's objective reasonableness standard. For current listings of many such cases go to www.ecdlaw.info.

For instance, a LEO did not use excessive force by employing a CEW against a handcuffed arrestee in *Zivojinovich v. Barner*.[23] The Plaintiff was arrested for resisting an officer with violence. A CEW was employed (in drive-stun mode) against the handcuffed arrestee whom the officer reasonably believed was intentionally spraying blood toward the officer through a broken nose.

31.5.1.2 Related Pepper Spray Cases

A LEO who used pepper spray on an arrestee who was already handcuffed and was not a threat is excessive force that is clearly established and is objectively unreasonable according to the decision in *Bultema v. Benzie County*.[24] The LEO who used pepper spray on an arrestee who was already handcuffed, was not a threat, and was not attempting to evade arrest constituted excessive force under the 4th Amendment. The court also found that using pepper spray on a handcuffed person who is not resisting is objectively unreasonable.

In *Champion v. Outlook Nashville, Inc.*, the court held that LEOs used excessive force by pepper spraying a nonresisting mentally challenged arrestee who was immobilized by handcuffs and a hobbling device.[25] Lying on top of the nonresisting mentally challenged arrestee and spraying him with pepper spray after he was in handcuffs and a hobbling device was not objectively reasonable. This use of force violated the arrestee's 4th Amendment rights and thus the LEOs were not entitled to qualified immunity.

31.5.2 8th Amendment "Cruel and Unusual Punishment" Standard

In *Michenfelder v. Sumner*, the appellate court held that the use of CEWs was not cruel and unusual punishment and a policy of allowing the use of CEWs on an inmate who refused to submit to a strip search does not constitute cruel and unusual punishment. The court noted that Nevada's Department of Prison authorities believe the CEW is the preferred method for controlling prisoners because it is the "least confrontational" when compared to the use of physical restraint, billy clubs, mace, or other stun-guns. By disabling the inmate, it prevents further violence. The court in this case held that the CEW is not per se unconstitutional.[26]

However, in the unreported case of *Preston v. Pavlushkin*, LEOs' summary judgment motion was denied in an 8th Amendment excessive force claim for use of a CEW where plaintiff's claim was that the deputies used the CEW to enforce order when the LEOs and the plaintiff inmate were the only occupants of the common area, the other inmates being closed in their cells. The only conduct of the plaintiff was his adamant refusal to comply with a LEO's order. The inmate did not present any threat of physical violence, so the court therefore concluded that the CEW, which was used to shock the inmate three times while he was on the ground and obviously incapacitated, was excessive force.[27]

31.5.3 14th Amendment "Shocking to the Conscience" Standard

A CEW ***cannot*** be used on a nonresisting, fully compliant subject who is not endangering themselves or others. Some courts also factor in whether the person claiming excessive force was warned that a CEW would be deployed against them. If the person claiming excessive force was warned, this would be one factor that weighs ***against*** a finding of excessive force.

A LEO's use of a CEW on an unruly arrestee after she forcefully stated "f*** you" and was not endangering herself or others qualified as wanton and sadistic – *Orem v. Rephann*.[28] In the *Orem* case, a 280 pound LEO's use of a CEW in drive-stun mode, on an unruly handcuffed and hobbled 100 pound female, under the arrestee's breast and on her inner thigh qualified as wanton and sadistic. The LEO deputy used the CED on the arrestee after she forcefully stated the expletive to the deputy. At the time the arrestee used the expletive, she was not endangering herself or others because she was handcuffed and in a hobbling device while locked in the back seat of a police squad car. The court found that the arrestee received more than *de minimus* injuries although the CEW was only applied for 1.5 seconds. The CEW caused pain, electric shock, and the development of a sunburn-like scar on plaintiff's thigh. Ultimately, the court found that the LEO's use of the CEW was not objectively reasonable.

31.6 US Lower Court CEW Cases

The court cases in this section include US Federal and State court decisions which are generally precedent only in the jurisdiction of the respective court issuing the opinion.

31.6.1 4th Amendment "Objective Reasonableness" Standard

Our society's legal standards defining when a CEW can be used are relatively well developed. As a starting point, the *Beaver vs. City of Federal Way*[29] case provides what could be construed as best practices guidance on when a LEO can lawfully deploy a CEW, and what needs to be in a LEO's use-of-force report. In the *Beaver* case, the Federal court in Seattle (WA) found that only the first three of five CEW discharges were objectively reasonable. Each CEW use lasted 5 seconds, and all five CEW uses took place within an 85-second time period. For the first three CEW discharges, a LEO was chasing a known drugged residential burglar alone and the subject was fleeing and then trying to get up. For the last two CEW discharges, a second LEO had arrived on scene and the subject was not attempting to flee, was not an "immediate" "threat," and was not "capable" of following LEOs' conflicting commands. The *Beaver* case outlines the following factors for a court to consider when evaluating the objective reasonableness standard:

- The use of a CEW involves the application of force; therefore, each use of force on a free person that is a seizure is the application of force and must be objectively reasonable under the 4th Amendment standard.
- Each CEW application involves an additional use of force. (This is true of any use of force.)
- Multiple CEW applications cannot be justified solely on the grounds that a suspect fails to comply with a command, absent other indications that the suspect is about to flee or poses an immediate threat[30] to an officer or other person. This is particularly true when more than one officer is present to assist in controlling a situation.
- Any decision to apply multiple CEW applications must take into consideration whether a suspect is capable of complying with LEOs' commands. This would apply to whether a suspect is physically and emotionally capable of complying. In the *Beaver* case, it was alleged that the LEOs gave conflicting commands that reduced the suspect's ability to comply.

31.6.1.1 CEW Use Found Reasonable

Use of CEW on man's neck was found reasonable and summary judgment was granted in favor of officer in *McBride v. Clark*.[31] Here, the CEW was deployed on Plaintiff's neck while he was in a restraint chair. After being placed in the restraint chair Plaintiff yelled, screamed, dislodged his IV causing bleeding, and stated that he had Hepatitis C. Plaintiff did not dispute that the CEW deployment was in the best interest of the officer's and the Plaintiff's safety.

Summary judgment was also granted to officer for use of CEW on handcuffed arrestee, who had kicked detention officer in *Rose v. City of Lafayette*:[32]

> It is undisputed that the [CEW] was not employed until after Plaintiff intentionally or unintentionally kicked Franek and appeared to be grabbing at Franek's weapon. I note that Plaintiff wearing handcuffs at the time, which lessens the threat of harm he may have presented at the time; nonetheless, in light of his previous conduct and the uncertainty presented in the struggle, it was not objectively unreasonable to believe that the gun could be released and cause injury.

Summary judgment was granted to officers who used CEW on a handcuffed, struggling, and resisting 14-year-old male in *Johnson ex rel. Smith v. City of*

Lincoln Park.[33] The use of a CEW on this unarmed 14-year-old male in school who was resisting a search and struggled during arrest after swinging and biting at officer was reasonable under the circumstances.

In the case of *Carroll v. County of Trumbull*, the court granted qualified immunity to the LEO for his use of CEW on a kicking, handcuffed arrestee while observing:

> Defendant Mann asserts ... he [used the CEW on] Plaintiff because he was thrashing about and resisting after officers removed the handcuffs from the front of Plaintiff's person in order to recuff him behind his back. Defendant Mann further attested that when they decided to remove Plaintiff from the back seat in order to cuff him behind his back, Plaintiff continued to resist and officers therefore forced him to the ground. Defendant Mann further stated that when they removed his handcuffs from the front, Plaintiff lifted his left arm and thrashed about again, so he decided to [use the CEW on] Plaintiff in order to gain his compliance so that he could be handcuffed behind his back without injury to himself or to the officers.[34]

In *Willkomm v. Mayer,* the court granted summary judgment to LEOs for 3 CEW uses on a belligerent driver, including 2 uses of the CEW while the arrestee was handcuffed.[35] In this case, after being arrested, the plaintiff – while in the back of the patrol car – was able to reposition his handcuffed hands to the front of his body. The LEOs removed plaintiff from the car and repositioned his handcuffs and secured his legs with flexcuffs. Plaintiff was told to swing his legs into the patrol car, he did not, and after a warning he was shocked with a CEW. Plaintiff again attempted to reposition his handcuffs. Plaintiff was then removed again from the car and a LEO used the CEW in order to reposition the plaintiff's handcuffs. All of the CEW applications were deemed reasonable.

In another case, summary judgment was granted in favor of the LEOs for use of a CEW in drive-stun mode on a handcuffed resisting arrestee in *Devoe v. Rebant.*[36] Plaintiff, while handcuffed, was told to get into the LEO's patrol car. Plaintiff would not comply and was then drive-stunned with a single CEW discharge. The court held that the LEO's action was objectively reasonable.

Additionally, the use of a CEW on a fleeing handcuffed arrestee was found objectively reasonable in *Yarnall v. Mendez.*[37]

31.6.1.2 CEW Use Found Potentially Unreasonable

In a number of cases, courts have found that the use of a CEW on an arrestee is unreasonable.

The District Court for the Eastern District of Michigan in *Sleeman v. Oakland County* found that a nonresisting arrestee has a triable excessive force claim for a CEW being deployed against him after he was handcuffed.[38] In this matter, the plaintiff claimed he was not resisting, had a CEW used against him after he was handcuffed, and that he did not try to assault any of the LEOs or flee. These alleged facts created a question of fact and thus a triable excessive force claim.

In *Parker v. City of South Portland*, the District Court of Maine denied the LEOs' motion for summary judgment because the court found that the plaintiff posed no threat to the police officers and was not resisting arrest.[39] Parker, the plaintiff, was stopped for a moving violation and posed no immediate danger to anyone. Parker claimed he did not use or attempt to use any force against the LEOs nor did he attempt to flee. At the point when one LEO fired his CEW, another LEO was completing his handcuffing of Parker. At the time Parker was being handcuffed he was surrounded by three LEOs, one of whom, was larger than Parker. In this case and under these specific circumstances, the court thus determined that it was a question of fact as to whether the LEOs used excessive force.

The District Court for the Eastern District of Washington in *Richards v. Janis*, found that there was a genuine issue of material fact as to "whether a reasonable officer would know that use of a [CEW] on an unarmed, nonresisting, handcuffed suspect would violate the suspect's constitutional rights."[40] Here, the court was faced with a person not endangering himself or others and not resisting arrest. The plaintiff claimed that he was not resisting arrest when police officers threw him to the ground and banged his head into the concrete sidewalk. Afterwards, another police officer arrived and deployed his CEW on the plaintiff's back. The court stated that there was a genuine issue of material fact in:

> ...whether a reasonable officer would know that [using a CEW on] an unarmed, nonresisting, handcuffed suspect would violate the suspect's constitutional rights. The officer's motion for qualified immunity is denied since the law is "clearly established that a [CEW] must not be used against a nonresisting individual."

Courts have also found that gratuitous force may not be used against a nonresisting handcuffed arrestee who is neither a safety or flight risk. In denying qualified immunity in *Michaels v. City of Vermillion*,[41] the Ohio Court stated that at the time of arrest, the use of force on a subdued suspect who was not a safety or flight risk was excessive under the 4th Amendment. Thus, the police officer who used a CEW multiple times after the arrestee was handcuffed, who was not resisting, and who was already in a patrol car is not entitled to qualified immunity. The use of the CEW in this instance was held to be gratuitous force.

In *Wanbaugh v. Fields*, a court in Arkansas found it unlawful for a law enforcement officer to use a CEW on a handcuffed nonresisting arrestee.[42] In this case, plaintiff was exposed to a CEW after he was in handcuffs and was not resisting the police officers. By the time this case was decided, the state of the law gave the police officers fair warning that repeated deployment of a CEW on a handcuffed arrestee was unconstitutional and the officers were not entitled to qualified immunity.

In *Bareaux v. Taylor*, the Louisiana court found that police officers may have engaged in excessive force where a handcuffed arrestee was allegedly beaten with a flashlight and received a CEW application.[43] Here, plaintiff claimed that after he was handcuffed the police officers sprayed him with an entire can of mace directly in his face, beat him with a flashlight, hit his head on a concrete wall, and used a CEW against him because he was unable to step into the police officer's Ford Expedition SUV due to a knee brace and leg shackles. The court found that these facts, if taken as true, would support a finding of force that was used "maliciously and sadistically for the very purpose of causing harm."

There is an interesting case out of Virginia, *Crihfield v. City of Danville Police Dep.*,[44] where a handcuffed arrestee was drive-stunned multiple times by a CEW. The plaintiff's complaint alleged that he was drive-stunned by a CEW 15–20 times after he was arrested; however, the complaint made no mention as to whether or not the plaintiff had resisted arrest. Thus, the court found that the alleged actions of the police officers were not reasonable despite the fact that the arrestee may have been resisting arrest.

31.6.2 8th and 14th Amendment Cases

Under the 8th and 14th Amendment, several courts have found the use of CEWs unreasonable. In *Batiste v. City of Beaumont*, qualified immunity was denied for police officers during a multi-hour mental health detention transport of a non-resistant and fully compliant handcuffed person.[45] The court found that striking, kicking, dragging, choking, and repeatedly using a CEW against a fully compliant, docile arrestee would constitute "wanton and unnecessary infliction of pain" and thus would be considered excessive and unreasonable force.

An Illinois court, in *DeSalvo v. City of Collinsville*, found that a reasonable police officer would have known that using a CEW against a subject who is not resisting and merely asked why he was being arrested is unlawful.[46] Here, after plaintiff was handcuffed behind his back he asked the police officer why he was being arrested. The police officer did not respond. After plaintiff asked a second time why he was being arrested the police officer threatened he would use his CEW if plaintiff did not get into the squad car. Then – 6 seconds later – the police officer drive-stunned the plaintiff's neck and then placed the CEW against the plaintiff's forehead threatening to use it again. Taking these facts as true, the police officer's use of the CEW constituted excessive force under the 8th and 14th Amendments. In addition, the Court found, "a reasonable [police] officer in Krug's position would have known that it would be unlawful to [use a CEW on] DeSalvo (Plaintiff) under the circumstances of this case." Thus, the Court rejected the police officer's qualified immunity argument.

31.7 Conclusions

While the case law specifically regarding CEW use is rapidly growing, enough themes have already emerged that police officers and agencies can derive useful direction from today's existing case law. As demonstrated in the cases discussed in this chapter, one of the most pressing concerns confronting law enforcement administrators includes implementing strategies to ensure proper and unambiguous reporting of use-of-force incidents. Often there are discrepancies between the LEOs and the criminal suspect or arrestees' versions of what transpired during the CEW incident. Video and audio recordings of the incidents from the officers' perspectives could foreseeably go a long way toward either ferreting out officers who would intentionally misuse their government endowed authority or hold suspects' responsible for their actions that necessitated the officers use of CEWs.

Notes

1. The discussion in this chapter only addresses CEWs that expel probes via compressed gas (not those that use gun powder because the use of gunpowder subjects those particular devices to automatic federal regulation as a firearm).
2. "[T]he 4th Amendment [to the U.S. Constitution] addresses 'misuse of power,' not the accidental effects of otherwise lawful conduct." *Brower v. County of Inyo*, 489 U.S. 593, 596, 109 S.Ct. 1378, 103 L.Ed.2d 628 (1989); *Milstead v. Kibler*, 243 F.3d 157 (4th Cir. 2001).
3. *36 C.F.R. §§ 1.4(a) and 2.4.*
4. State restrictions on CEWs are often changing. As of January 1, 2008, the following states completely prohibit the possession of CEWs by private citizens: Hawaii (*Haw. Rev. Stat. § 134-16);* Massachusetts (*Mass. Gen. Laws Ann. ch. 140, § 131 J);* Michigan (*Mich. Comp. Laws § 750.224a); New Jersey (N.J. Stat. Ann. § 2C:39-3); New York (N.Y. Penal Law § 265.01);* Rhode Island (*R.I. Gen. Laws § 11-47-42); and* Wisconsin (*Wis. Stat. § 941.295).* For a reference list of CEW state specific statutes, visit the Electronic Control Device: Legal Resources website at www.ecdlaw.info.
5. *The Philadelphia Code § 10-825.*
6. *Ill. Comp. Stat. §§ 720 ILCS 5/24-1 and 5/24-1.6*
7. Wis Stat. § 941.295.
8. *Baker v. McCollan*, 443 U.S. 137, 144, n. 3, 99 S.Ct. 2689, 61 L.Ed.2d 433 (1979).
9. *Graham v. Connor*, 490 U.S. 386, 394, 109 S. Ct. 1865, 1870-1871, 104 L. Ed. 2d 443 (1989).
10. *Harlow v. Fitzgerald*, 457 U.S. 800, 818, 102 S. Ct. 2727, 2738, 73 L.Ed.2d 396 (1982).
11. *Baker*, 443 U.S. at 140. ("The first inquiry in any § 1983 suit" is "to isolate the precise constitutional violation with which [the defendant] is charged"). *Graham*, 490 U.S. at 394.
12. http://www.theiacp.org/
13. http://www.policeforum.org/
14. *Florida v. Bostick*, 501 U.S. 429, 434, 111 S.Ct. 2382, 115 L.Ed.2d 389 (1991); *Brower*, 489 U.S. at 597.
15. *See, e.g., California v. Hodari D.*, 499 U.S. 621, 626, n. 2, 111 S.Ct. 1547, 113 L.Ed.2d 690 (1991).
16. *See e.g., United States v. Mendenhall*, 446 U.S. 544, 554, 100 S.Ct. 1870, 64 L.Ed.2d 497 (1980) (principal opinion).
17. *Bostick*, 501 U.S. at 435–436.
18. *Whitley v. Albers*, 475 U.S. 312, 318–326, 106 S.Ct. 1078, 1083–1088, 89 L.Ed.2d 251 (1986) (claim of excessive force to subdue convicted prisoner analyzed under an 8th Amendment standard).

19. *County of Sacramento v. Lewis*, 523 U.S. 833, 836, 118 S.Ct. 1708, 140 L.Ed.2d 1043 (1998).
20. *Lewis*, 523 U.S. at 854.
21. *Draper v. Reynolds*, 369 F.3d 1270, 1278 (11th Cir. 2004).
22. *Ewolski v. City of Brunswick*, 287 F.3d 492 (6th Cir. (Ohio) 2002).
23. *Zivojinovich v. Barner,* 525 F.3d 1059 (11th Cir. 2008).
24. *Bultema v. Benzie County,* 146 Fed. Appx. 28 (6th Cir. 2005).
25. *Champion v. Outlook Nashville, Inc.,* 380 F.3d 893 (6th Cir. 2004).
26. *Michenfelder v. Sumner*, 860 F.2d 328 (9th Cir. (Nev.) 1988).
27. *Preston v. Pavlushkin*, Slip Copy, 2006 WL 686481 (D. Colo., March 16, 2006).
28. *Orem v. Rephann,* 523 F.3d 442 (4th Cir. 2008).
29. *Beaver v. City of Federal Way*, 507 F.Supp.2d 1137 (D. Wash. 2007).
30. "[A] simple statement by an officer that he fears for his safety or the safety of others is not enough; there must be objective factors to justify such a concern." *Deorle v. Rutherford,* 272 F.3rd 1272, 1281 (9th Cir. 2001).
31. *McBride v. Clark*, No. 04-03307-CV-S-REL, 2006 WL 581139 (W.D. Mo., March 8, 2006).
32. *Rose v. City of Lafayette,* No. 05-cv-00311-WDM-MJW, 2007 WL 485228 (D. Colo., Feb. 12, 2007).
33. *Johnson ex rel. Smith v. City of Lincoln Park*, 434 F.Supp.2d 467 (E.D. Mich. 2006).
34. *Carroll v. County of Trumbull*, No. 4:05CV1854, 2006 WL 1134206 (N.D. Ohio, April 25, 2006).
35. *Willkomm v. Mayer*, No. 05 C 523 S, 2006 WL 582044 (W.D. Wis., March 9, 2006).
36. *Devoe v. Rebant*, No. 05-71863, 2006 WL 334297 (E.D. Mich., Feb. 13, 2006).
37. *Yarnall v. Mendez,* 509 F.Supp.2d 421 (D. Del. 2007).
38. *Sleeman v. Oakland County,* No. 06-10953, 2007 WL 1343403 (E.D. Mich., May 7, 2007).
39. *Parker v. City of South Portland,* No. 06-129-P-S, 2007 WL 1468658 (D. Me., May 18, 2007).
40. *Richards v. Janis,* No. CV-06-3064-EFS, 2007 WL 3046252 (E.D. Wash., Oct. 17, 2007).
41. *Michaels v. City of Vermillion,* 539 F.Supp.2d 975 (N.D. Ohio 2008).
42. *Wanbaugh v. Fields,* 508 F.Supp.2d 723 (W.D. Ark. 2007).
43. *Bareaux v. Taylor,* No. 06-1145, 2008 WL 145249 (W.D. La., Jan. 8, 2008).
44. *Crihfield v. City of Danville Police Dept.,* Nos. 4:07CV00010 and 4:07CV00011, 2007 WL 3003279 (W.D. Va., Oct. 11, 2007).
45. *Batiste v. City of Beaumont*, 426 F.Supp.2d 395 (E.D. Tex. 2006).
46. *DeSalvo v. City of Collinsville,* No. 04-CV-0718-MJR, 2005 WL 2487829 (S.D. Ill., Oct. 7, 2005).

Chapter 32
Science and Logic Meet the Law

John G. Peters

Science is technical, making it more than a superficial read for the nonscientist. The legal process can be equally challenging to the nonlawyer, especially when science and law collide in the courtroom.

32.1 Case Study: The Graduation Party

The following actual case study provides an opportunity for applying scientific and legal constructs to a law enforcement incident. Tragically, this incident involved a male teenager, Richard, who died after struggling with the police and after being controlled with a CEW. Richard's estate filed a lawsuit against the governmental entity that employed the law enforcement officers, two of the officers, and the manufacturer of the CEW in federal court. This case will provide the necessary platform to discuss and to demonstrate how science and law can be applied to such cases.

On a May evening, Richard, age 18, attended a high school graduation party for his friends at Adam's house. Richard, a high school dropout who had a long history of deviant behavior and criminal misconduct, also had a history of taking licit and illicit drugs. A friend confirmed that Richard and he "did" lines of methamphetamine and cocaine. Sometime during that evening, his friend testified that he thought Richard ingested two MDMA tablets. MDMA, a neurotoxic drug that is in the amphetamine family, has stimulant and hallucinogenic properties and is also known as "Ecstasy." A female who attended the party revealed that she and Richard each had a shot of whisky and also smoked marijuana at the party. Sometime during the evening, Richard left the party but later returned.

When Richard returned to the party something was visibly wrong with him. Outside the residence where the party was being held, Richard was yelling so loudly that people inside the house heard him. He had removed his shirt,

J.G. Peters (✉)
President and Founder, Institute for the Prevention of In-Custody Deaths, Inc.
e-mail: john@ipicd.com

M.W. Kroll, J.D. Ho (eds.), *TASER® Conducted Electrical Weapons: Physiology, Pathology, and Law*, DOI 10.1007/978-0-387-85475-5_32,

appeared scared, and failed to recognize any of the people who were attending the party. Richard began running around the driveway, acting weird, telling people that he was "God" and "Jesus". He also attempted to kiss people. After about 20 minutes of people chasing Richard to catch him, he ran down the road – fast. One friend told investigators that Richard was making noises like the R2-D2 character of Star Wars® fame.

Richard entered a neighboring property where the owner saw him and called the police. When the first officer arrived at the neighbor's residence in response to her telephone call, she heard and observed Richard screaming and yelling. He was also moving his arms as if he were swinging a baseball bat. During her radio request for back-up officers, the first officer told the dispatcher that she believed the unknown male was on drugs. After giving both coherent and incoherent responses to the officer's questions, he suddenly ran at her. Fearing for her safety, she removed her departmentally-issued CEW, and when Richard went up in the air and lunged at her, she deployed it. When the CEW probes struck Richard's body the device activated, dropping him onto the ground. Because he kept attempting to get up from the ground and backup had not yet arrived, the first officer deployed the CEW at least three more times by pulling the trigger in an effort to capture and to control Richard.

A back-up officer arrived and immediately handcuffed Richard, without putting any weight on Richard's back area. A short time later a sergeant arrived on the scene and witnessed Richard breathing and moaning. Richard, who was now on his side and resting against the back-up officer's leg, suddenly slid down the leg and onto the ground. Richard's eyes began to roll back into his head, and he had a bowel movement.

Emergency Medical Services (EMS) arrived on scene about that time (police had earlier requested EMS) and found Richard nonresponsive. EMS attempted to use an automatic external defibrillator (AED). In the dark field EMS personnel read the AED as ventricular fibrillation (VF). Cardiac electrophysiologists later determined that the EMS personnel erred and the Richard's cardiac rhythm was pulseless electrical activity (PEA). Richard was taken to City Hospital, where he was pronounced dead. After an autopsy, the Medical Examiner's Office issued a media release claiming the TASER CEW was the "straw that broke the camel's back," in essence, the ultimate *cause* of Richard's death and "but for the drug intoxication the CEW would not have resulted in death." The causes of death were listed as: cardiac arrhythmia, due to drug induced psychosis, methamphetamine and MDMA intoxication, acute, with electrical pulse incapacitation as a contributory condition, and was classified as a homicide [1].

32.1.1 What Killed Richard?

The case study's abbreviated summary of facts is often used in law enforcement training presentations prior to discussing science, cause and effect, and other related issues regarding this and similar sudden death events. Some attendees who are unfamiliar with the toxic effects of chronic drug abuse or CEW effects, quickly form a variety of opinions including how CEW electrical shocks might have been the *cause* of Richard's death. However, when these audience members are asked to scientifically support their opinions, the room gets very quiet. They, like many people, have answers (really only unsupported suppositions) to one or more "questions" (Do you think the CEW killed him?) without knowing what the exact question is or what the science says about the device or the inherent physiological effects of a violent struggle.

To better understand the outcome of the lawsuit about Richard's tragic death, legal and scientific criteria used by courts to guide medical and expert witness testimony about causation and related issues are discussed. These criteria will then be applied to the facts surrounding Richard's death.

32.1.2 Facts and Evidence for the Juror

Judges and jurors are often required to sort through a constellation of facts and testimony that are often confusing, controversial, and based on a mixture of nonscience, pseudoscience, and science, and then make a decision about innocence or assigning and apportioning blame. In reaching a verdict at the end of a jury trial jurors must evaluate the credibility of information and evidence that was presented during the trial. Instructions to jurors from the presiding judge are key ingredients that help to guide the jurors' decision-making process.

Jurors are often instructed:

> You must not read anything or listen to anything or watch anything with regard to this trial. It would be a violation of your oath as jurors to decide this case on anything other than the evidence presented at trial and your common sense. You must decide the case solely and exclusively on the evidence that will be received here in court [2].

After hearing the opening argument of the Plaintiff's lawyer in a case such as Richard's, it is reasonable for jurors to wonder or ask themselves: Did the CEW kill Richard? Could multiple CEW applications kill a person? In some cases, jurors may initially come to a subjective conclusion that CEWs are deadly because they discharge the mysterious and invisible electricity into the body. Expert witnesses are typically used to assist jurors in the clarifying and understanding of scientific and other technical information, including evidence.

32.1.3 Experts and Their Testimonies

Federal Rule of Evidence 702 defines an expert witness and also speaks to the admissibility of expert testimony. Rule 702 provides:

> If scientific, technical, or other specialized knowledge will assist the trier of fact to understand the evidence or determine a fact in issue, a witness qualified as an expert by knowledge, skill, experience, training, or education, may testify thereto in the form of an opinion or otherwise, if (1) the testimony is based upon sufficient facts or data, (2) the testimony is the product of reliable principles and methods, and (3) the witness has applied the principles and methods reasonably to the case [3,4].

An expert witness' skillful blending of nonscience, pseudoscience, and science has often confused jurors and courts. Fortunately, over the last two decades the United States Supreme Court has provided guidance regarding the admissibility of expert witness opinion testimony during federal legal proceedings. Jurors and jurists must remember that any *argument* made by an expert or other witness is intended to push the listener toward a conclusion, where, in contrast, an *explanation* is used to justify why a particular claim, declaration, or conclusion is true [5].

The courts have become arbiters of science and also of separating those witnesses who *appear* credible from those who are truly credible. In those cases where science plays an important role, decisions about guilt or innocence are often made by individuals with little scientific knowledge, so the outcome of a particular case does not mean that the "winner" is scientifically accurate. It could be interpreted that one side managed to convince that case's judge or jury.

32.2 The Daubert Trilogy

To help determine whether the methodology used by the expert witness is sufficiently reliable to be considered admissible, it will be reviewed against the following Court-determined criteria: the degree of acceptance within the relevant community; the existence of standards designed to ensure the credibility or accuracy of a particular technique; the likelihood that a particular technique may result in error; and, whether the methodology stands up to the standards applied within the relevant field. The United States Supreme Court has provided guidance to Federal courts regarding expert testimony in a series of three cases, known as the *Daubert* Trilogy: *Daubert v. Merrill Dow Pharmaceuticals, Inc*, *General Electric Co. v. Joiner*, and *Kuhmo Tire, Co., Ltd. v. Carmichael* (Table 32.1).

Table 32.1 The Daubert trilogy

	Rule	Source
1	The theory and technique has been subjected to empirical testing	Daubert
2	It has been subjected to peer review and publication	Daubert
3	It has a known or potential error rate	Daubert
4	There are standards controlling the technique's operations	Daubert
5	It is generally accepted by a relevant scientific community	Daubert
6	The opinion in the specific case is supported by empirical testing	General Electric
7	Above rules are applied to experience-based opinions	Kuhmo Tire

The Court must engage in a preliminary assessment of whether the reasoning or methodology underlying the expert testimony is scientifically valid and whether that reasoning or methodology can properly be applied to the facts in issue [6]. The assessment's goal is to ensure the reliability and relevancy of expert testimony [7]. Federal judges are required to become the gatekeepers preventing unreliable and irrelevant professional and expert testimony from being presented in their courtrooms. Unreliable or irrelevant testimony may create confusion, lack probative value, and fail to be of assistance to the trier of fact (judge or jury). State court judges often follow their Federal brethren in this gate keeping function, but are usually not required to follow Rule 702 of the Federal Rules of Evidence.

When the testimony of an expert witness is called into question, the scientific basis for such testimony (principles, methods, data, and their application to the case) must be evaluated by the trial judge to ascertain if it has a reliable basis in the knowledge and experience of the relevant discipline [7]. In short, the Court's inquiry focuses not on whether the professional or expert witness is correct, but rather on whether his or her testimony has established – by a preponderance of the evidence – that the testimony is reliable in the context of the methodologies or techniques applied within the appropriate field. A person's experience can be argued as a methodology.

32.2.1 *Daubert v. Merrill Dow Pharmaceuticals, Inc*

In 1993, the United States Supreme Court held in *Daubert v. Merrill Dow Pharmaceuticals, Inc.* that scientific testimony will not be admissible unless such testimony was scientifically reliable and relevant [6]. In determining scientific reliability, the Court offered "general observations" of whether proffered evidence was based on the scientific method, although the list is not intended to be used as an exacting checklist. The issues are whether or not the scientific theory and technique: (1) had empirical testing; (2) was subjected to peer review and publication; (3) had a known or potential error rate; (4) has standards controlling the technique's operations, and (5) is generally accepted by a relevant scientific community [6]. In short, the scientific method must be used by the subject matter expert witness; however, a subject matter expert may still qualify by knowledge, skill, experience, training, or education [4].

The Supreme Court defined the scientific method as:

> Scientific methodology today is based on generating hypotheses [research proposition] and testing them to see if they can be falsified; indeed this methodology is what distinguishes science from other fields of human inquiry [6]. It is also equally important to note that the professional or expert witness testimony must not be based on what might be proven at a future time, or what future knowledge may surface, but rather the current state of scientific knowledge [8,9].

32.2.2 *General Electric Co. v. Joiner*

Simply adhering to the scientific method is not enough. In 1997 the Court held that the specific opinion in a given case had to be sufficiently supported by *empirical testing* that validates the conclusions reached by the professional or expert witness [10]. This important topic will be discussed under the scientific method later in this chapter.

32.2.3 *Kuhmo Tire, Co., Ltd. v. Carmichael*

The Court again addressed this issue in 1999 in *Kuhmo Tire*, and held that the reliability and relevance requirements outlined in *Daubert* applied to all expert testimony, including testimony that was experience-based [7]. Expert witnesses could no longer offer opinions such as "I would not have done X that way", without providing a scientific basis or rigorous methodology for the opinion.

32.2.4 *Federal Rules of Evidence*

In 2000, the Supreme Court approved amendments to the Federal Rules of Evidence with respect to opinion evidence and expert testimony that conformed to the *Daubert* trilogy [11]. Rules 701, 702, and 703 now include additional provisions that the testimony of experts be based upon sufficient facts or data, and that the individual has applied the principles and methods reliably to the facts of the case [11]. The Court's objective is to make sure that an expert "employs in the courtroom the same level of intellectual rigor that characterizes the practice of an expert in the relevant field." [6] However, courts enjoy considerable leeway in deciding in a particular case how to go about determining whether particular testimony is reliable [6]. There does not appear to be a rigid checklist for a court to follow, but rather the gate keeping inquiry must be tailored to the facts of the case and the type of professional or expert testimony at issue. When expert testimony is reliable, it still may be excluded if no logical relationship exists between the testimony and a fact or issue in the case [6].

> Specifically, when expert 'testimony's factual basis, data, principles, methods or their application are called sufficiently into question, ... the trial judge must determine whether the testimony has a reliable basis in the knowledge and experience of [the relevant] discipline [3].

A federal judge noted that "the Court's inquiry focuses not on whether the expert is correct, but whether the proponent of expert testimony has established by a preponderance of the evidence that the testimony is reliable in the context of the methodologies or techniques applied within the appropriate field." [3]

32.2.5 Education of the Jury

Most jurors are in need of education about scientific causation principles so they will have a fundamental understanding about what is the scientific method and how it can be used to establish causation. Knowledge can be acquired through several methods listed in Table 32.2. These are: *tenacity*, *intuition*, *authority*, *rationalism*, *mysticism*, *empiricism*, and *science* [12].

The last category, science, focuses upon a systematic or structured process of using empirical observation and logic, for example, to develop one or more theories or hypotheses that can be tested using the scientific method. It is absurd to suggest that science is the only way a person acquires knowledge about their job, nature, life, electricity, world events, local events, and law enforcement activities. While people may obtain knowledge through observation, personal, or others' experiences, they must be careful about how they use this knowledge to formulate conclusions about simple or complex issues. Relying solely upon observation, common sense, intuition, or personal experience, can, and often does serve as the foundation for illogical conclusions and opinions. These nonscientific conclusions and opinions may also be negatively exacerbated if they are fueled by biases, phobias, or negative preconceptions.

The most common example with direct relevance to this book is the intuition that electricity must build up like poison does. This leads to the intuitive (but false) conclusion that there is an increased danger with longer duration CEW applications.

32.2.5.1 CSI Effect

One preconceived idea held by many people (including jurors) is about crime scene evidence collection and processing. Many people who watch the *CSI*® (crime scene investigation) television program erroneously form beliefs about how forensic investigations *should* be conducted and what types of evidence *should* be gathered and analyzed. Known as the *CSI effect*, it is very real and must often be dealt with in the courtroom by legal counsel for both sides,

Table 32.2 How knowledge is acquired

Method	Definition
Tenacity	Partial truths or mistruths are repeated for an extended period of time and accepted as being factual
Intuition	One accepts a position as true because it "feels" true without critical thinking
Authority	A respected source such as a medical doctor, or politician, makes a claim
Rationalism	A person using common sense or principles of logic develops a premise or hypothesis that is however tested in a nonscientific manner
Mysticism	Knowledge is sought from prophets or astrologers
Empiricism	Knowledge developed through observation such as sticking a finger into an electrical outlet
Science	Integration of empiricism and rationalism to develop and test theories

including the Court [13]. Many jurors "learn" from this fast-paced television show and often erroneously conclude that fingerprints, DNA, and other such evidence *must* be collected to find the defendant guilty.

32.3 Scientific Research

32.3.1 Literature Review

People often say they are going to use the Internet or the library to conduct *research*. Although they say "research," what they really are referring to is a literature review. A literature review involves the identification of *primary* and *secondary* literature sources that focus upon the topic about which the individual seeks knowledge. However, this is not *scientific research*.

Primary sources include peer-reviewed journal articles and dissertations [12]. Primary sources identify the researcher's hypothesis or research questions, methodology, statistical analysis, discussion, and conclusion. In contrast, secondary sources include Internet sites, articles in trade publications that are not peer reviewed, and similar documents [14,15,16].

A literature review is the transportation of information from one document to another document, the latter usually being the one an individual is preparing [17,18]. A literature review is typically one of the early steps performed by the researcher when using the scientific method.

32.3.2 Quantitative Research

There are generally two categories of research: *quantitative* and *qualitative* [18,19,20,21]. Quantitative research examines scientific, social, or human problems, and usually involves the testing of a theory (ontological assumption) through the manipulation of an *independent variable* (e.g., CEW application length) to see how it impacts the *dependent variable* (e.g., lactic acid in the blood). The independent variable in this case, CEW applications of a 15-second versus a 5-second duration, is then compared to the outcome, or dependent variable (lactic acid) [19].

This is often referred to as *experimental research* when the *experimental* group receives the treatment (15-second duration) with the outcome data then compared to the *control* group that only received the more common 5- second application. The magnitude of the effect between the results of the two groups is then analyzed to determine if there is a *statistically significant* difference between the experimental and control groups. A statistically significant result means that the *null hypothesis* has been rejected. (The *null hypothesis* of the breathing study would state there is no difference in lactic acid between the 15-second and 5-second duration of CEW application.)

In a quantitative research project, the researcher should remain distant and independent from the topic under investigation, and generally uses *deductive reasoning* (methodological assumption), that is formulating a conclusion about the topic under investigation based upon its findings (i.e., specific to the general). Quantitative research studies may include experimental, quasi-experimental, official statistics, surveys, and correlation [19,21].

32.3.3 Qualitative Research

Quantitative research does not always give meaningful results. The classic example would be a study that shows that the average human being has 1.008 testicles. (There are slightly more males than females worldwide.) In contrast, qualitative research examines topics that may not produce quantifiable data. For example, a researcher interviews volunteers who have implanted pacemakers to ascertain how they felt after receiving a 5-second CEW application to their chest area. The volunteers' answers would be transcribed and reviewed for the identification of possible categories of responses. The qualitative researcher generally uses *inductive reasoning*: formulating a conclusion about the topic under investigation going from the specific to the general. This type of reasoning is often used to formulate theories, which are often based upon observations [12].

Qualitative research has many different types: *case studies* (or anecdotes) where the researcher examines one entity, be it an individual or an organization, where the study is bound by activity and time, *ethnographic studies* (the researcher studies an intact cultural group, such as the Amish or law enforcement officers), *phenomenological studies* (the researcher attempts to understand "lived experiences"), and *grounded theory* (the researcher attempts to develop a theory through data collection, etc.) [19,22]. One drawback to qualitative studies is that the researcher often gets involved with the people being studied, and thus may bring bias to the study. Researcher bias may unintentionally influence data collection and analysis; hence, the researcher's inquiry must be kept as free from bias as possible. However, qualitative studies do enable the researcher to investigate some topics in more depth (watching and listening to jurors discuss and then learning firsthand how they reach a verdict during sequestered deliberations) versus attempting to learn how a verdict is reached using quantitative studies (jurors completing a self-reporting questionnaire about how they reached a verdict).

32.4 Theories, Laws, and Models

32.4.1 Theories

Theories help to define, predict, or explain phenomena and are most often based upon observations [12,22]. Suppose a person observes that it usually rains when the sky gets very dark on a summer afternoon. Based upon one or more observations, that person has developed a theory that when a summer afternoon sky gets dark, it often rains. Although theories help to predict and explain phenomena, there needs to be a significant amount of observation or research conducted before a theory is developed, because scientific theory must have a significant degree of probability. After all, a scientific theory is one that permits testing [12].

In the social sciences, when a theory is tested and disproven, it will be kept for further testing. For example, if a researcher disproved a theory that poverty caused criminal activity in Washington, DC, it would be kept so that theory could again be tested by researchers in other geographic locations. The hard sciences (e.g., physics or chemistry), in contrast, generally do not keep theories that have been disproven. Some researchers have argued that theories are never proven, only probed [23].

32.4.2 Models

Models, in contrast, are not as strong as theories. A model is technically a theory, but is less developed and only *represents* reality, but does not duplicate it [12]. Models are often useful in the development of theories, and according to some writers and researchers, they are often referred to as *minitheories* [12, p. 38].

Models have both strengths and limitations. For example, the *swine model* is often used when researchers conduct CEW research because the swine's heart rhythm is more sensitive to electrical shock, but limitations include that swine hearts are not physiologically identical to those of humans [24]. This is one reason why CEW research using the human model has more validity and reliability.

32.4.3 Laws

The highest level of theories that appear to be true are scientific *laws*, but unlike conventional theories, they do not change [25]. For example, the *Law of Gravity* states that when an object is freely dropped anywhere on planet Earth it will fall toward the earth. Hence, scientific laws such as the *Law of Gravity* or the *Second Law of Thermodynamics* (i.e., when a system uses energy, such as a human system, there is a reduction of usable or free energy for that system) are universal and have stood the tests of time, scientific testing, and are believed to be true. Models, theories, and laws are used when developing a research study using the scientific method.

32.5 The Scientific Method

The etymology of the word *science* is from the Latin word *scientia* meaning *to know* [26]. However, science is not easily defined, because it can mean different things to different people. Regardless of an individual's definition of science,

Table 32.3 The steps of the scientific method

	Step	Example
1	Identification of a question	Why do some people die after struggling with law enforcement officers and receiving a CEW application?
2	Development of one or more hypotheses	There is no difference in human acidosis between a 5- and 15-second CEW application
3	Study design	Manipulation of an independent variable such as the application duration
4	Collection of data	Blood acid is measured
5	Data analysis	Acid levels in both groups are compared using appropriate statistical tests
6	Data interpretation	Interpreting the data after analyzing it
7	Communication	A report is written of the findings, and their generalizability. Future research needs are articulated

there is a systematic and structured scientific approach for conducting scientific research: the *scientific method.*

The scientific method consists of seven phases as enumerated in Table 32.3. These are: *identification of a question* (i.e., why do some people die after struggling with law enforcement officers and receiving an CEW application?); *development of one or more hypotheses* (specific unproven statement, such as: There is no difference in human acidosis between a 5- and 15-second CEW application; *methodology design* (manipulation of an independent variable such as some people receive a 15-second CEW application (experimental group), others receive a 5-second CEW application (control group), sample selection and size, experimental or case study, etc.); *collection of data* (e.g., blood is captured and acid levels measured); *data analysis* (e.g., evaluating lactic acid levels of both groups using appropriate statistical tests or qualitative analysis methods); *data interpretation* (interpreting the data after analyzing it); and, *communication* (written or oral, a report of the study's findings, generalizability of its findings, and future research needs are articulated) [12,17,27].

Some newspaper articles demonstrate the shocking level of ignorance in the scientific method. A study of CEW effects was performed by the US military and closed with the usual suggestions for follow-up research. A headline in the *New York Times* then stated that the "conclusion" of the study was that there was insufficient research on the effects of CEWs! [28].

A scientific study's *methodology* is a key element for a successful study. In short, the study's methodology is the roadmap a researcher follows. If the methodology is ill conceived, the findings of the study will be called into question. Likewise, if the identified statistical tests for data analysis are inappropriate or misapplied, the study's findings and conclusions may be jeopardized [29]. Hence, the importance of a study's methodology cannot be ignored. It is here that critical decisions about the target population, sample size, confidence levels and intervals, etc. are developed and outlined. Many of these decisions are impacted by human subject research guidelines.

32.5.1 Human Subject Research Guidelines

The American Psychological Association was one of the earliest groups to establish human subject guidelines and ethical standards [30]. Many hospitals, universities, and other groups have established similar guidelines, often requiring that any human subject research be submitted to an internal review board (IRB) for approval [31]. If the institution or researcher is the recipient of federally funded or sponsored research, the researcher or IRB is subject to the strict guidelines for the evaluation of research on human subjects according to the Code of Federal Regulations [32]. There are similar research protocols for the ethical and humane treatment of animals.

These research guidelines often limit the type of human testing a researcher can ethically and legally conduct, such as giving research subjects illicit drugs. Ignorance of this fact appears to be widespread as evidenced by the common suggestion that CEWs should be tested on subjects with high levels of illegal drugs.

32.5.2 Population and Sample Size

A scientific study's *sample* (statistical notation, "n") is a portion of a *population* (statistical notation, "N") [33,34].

Sample size is also important to a scientific study. If it is too small, the study's findings cannot be generalized to the target population. Known as the *theory of generalizability*, it is another statistical approach that focuses upon *reliability* of the measurement [35,18]. To properly determine a sample size, the researcher should know the *standard deviation*, *magnitude of error*, and *confidence level* [18,33]. For example, if a researcher used a porcine (swine) model to determine if a CEW caused the heart to go into VF and only used one pig ($n = 1$), the sample size is too small to generalize the study's findings, or there may be plausible rival hypotheses [23,36,37]. There are well-accepted formulae for determining the minimum sample size for a given study design. As a very rough rule of thumb, an n of ≥ 30 is considered a minimum sample size for the generalizability of a study's findings to a target population [38].

In addition to sample size, how the sample was selected (i.e., simple random, systematic, stratified, convenience, etc.) may also have an impact on the reliability and validity of a study's findings [17,18]. The classic example is seen in the marketing of cholesterol drugs to females based on studies of males. It is well known that middle-aged males with high total serum cholesterol levels have an increased mortality. However, women with elevated cholesterol levels actually live longer than women with lower cholesterol levels! [39]. Sample size, coupled with research design, often impacts a study's reliability and validity of findings.

32.5.3 Reliability and Validity

Reliability focuses upon how reliable the findings are, in other words, how free they are from errors [12,18]. In short, if another study were conducted, would it yield consistent findings? For example, the seminal research about the safety of using CEWs on humans conducted by Ho et al. [40,41,42,43], has been validated by other researchers that have replicated these findings [44,45,46,47].

Validity focuses upon how well the study's scale or instrument is capable in assessing what the study intended to measure [12,18]. Remember that a core goal of scientific research is to systematically study a paradigm or theory, and then to replicate and check the scientific findings [27]. The researcher wants to achieve *external validity* or the extent to which his or her study results can be generalized [52]. These concepts are important in research findings, and also in the proffering of professional or expert witness opinions or testimonies.

For example, the seminal research about the restraint method of hogtying causing positional asphyxia [48] was found not to be reliable or valid during subsequent scientific investigation [49]. Although the theory that hogtying caused positional asphyxia and ultimately death has been scientifically disproven (in healthy males), many critics argue that the sample sizes of the studies were too small to generalize their findings to the whole population of people who were arrested and then hogtied; that they were conducted on people who were not on illicit substances and that the test subjects did not have a large body mass index, etc.

These issues are often successfully raised in the courts. In *Cruz v. City of Laramie* the court held that the proffered research study was not "persuasive" and did not permit its findings [50], whereas in *Price v. San Diego* the court permitted a study's findings that scientifically refuted prior scientific findings [51].

32.5.4 Statistical Versus Clinical Significance

Statistical significance is said to be achieved when a study's findings are the unlikely result of solely chance [12]. Statistical significance is usually expressed as a probability that a result was due to chance, (e.g., $p = 0.05$) [34]. This is often referred to as the *confidence* or *alpha* (α) *level* [34].

When a study result has a low probability of being due to chance (e.g., $p < 0.05$) of the results finding is above set limits, the findings are said to be *statistically significant* [17] but *chance has not been totally eliminated.* We are just 95% (100%–5%) confident that it is not due to chance.

32.5.5 Clinical (Practical) Significance

There is also the paradigm of *clinical significance*, or practical importance. The clinical significance of a study's findings is subjective, and is established by the researcher or clinician as there are no objective, set measures or tables. One proposed rule is that of Cohen which suggests that any change of <20% is not *clinically significant* [53]. In other words, some differences are just too small to have any practical impact [52]. As a general rule, the larger the sample size (which reduces the error rate), the easier it is to produce findings that have statistical significance from small differences between groups. This is most commonly seen when very large studies are done of effects with a small percentage of change. The change is not necessarily important. However, due to the large sample size the result can be shown to not be due to chance.

For example, imagine that a large study was done of the difference between left and right-handed people. The study found that left-handers had – on average – higher intelligence by 2 IQ points. This was *statistically significant* with $p < 0.01$ so we are 99% confident that this result was not due to chance. However, since this change (2 IQ points) is only about 2% of the average IQ, this would not be considered *clinically significant* by Cohen's rule.

32.5.6 Rate of Error

There are errors that can surface during the sampling process or during the analysis of a study's findings. These include, but are not limited to: *sampling error* (i.e., differences in samples taken from the same target population), *Type I error* (finding an effect when there was really no effect), and *Type II error* (finding no effect when there really was an effect) [12]. The most common Type 1 error comes from failing to perform statistical analyses which would have shown that the imagined effect was really not statistically significant. For example, two friends have a remission of a cancer and note that they both drank a certain brand of beer. Since the cancer-curing effects of that beer would probably not be shown in a larger study, we can safely conclude that the conclusion of the two friends was the result of a Type I error. The most common source of a Type II error is failing to perform a large enough study to find statistical significance.

Rather than accepting a study's findings at face value, the study should be analyzed to identify if any of these errors are present.

32.5.7 Fallacies

Fallacies are, in short, errors in reasoning, with more than 40 fallacies being reported in the literature [54,55,56]. More than a mistake, fallacies are arguments where the explanations for their conclusions do not provide the needed degree of support. The *post hoc ergo proctor hoc* ("A" caused "B" because "B" happened after "A") [5] is a well-known fallacy that is often found in newspaper articles and argument when implying causation, especially when a person has died after receiving a CEW deployment. Media often help to promote this fallacy by immediately reporting a news story that "TASER kills man."

"If it's in print, it must be true," is another fallacy that research studies confirm most people believe [57]. This "belief" has been extended to "[i]f it is in the paper it must be true," [58] and "[i]f it's on the Internet, it must be true." [59] Unless an exhaustive literature review is conducted to identify scientific studies that confirm or refute the information contained in articles that people read or sound bites that they hear, the information gets passed along or retained and often times is wrongly accepted as "true."

An increasing and troubling trend is the blurring of the reliability of newspapers versus the Internet. Traditionally, professional journalism relied on a code of ethics, a fact-checking function, and editorial supervision, which is often absent on the Internet. However, the Internet is now helping to drive newspapers toward bankruptcy and the solution seen by many newspapers appears to be to match the sensationalism and undisciplined nature of the Internet.

The same is often true for the fallacies that focus on what caused the death of an individual who died after a CEW was deployed. To determine if a person has put forth a fallacy, one must examine the causation conclusion to determine if it has been adequately supported with science or other evidence. Fallacies are often identified when *causation theories* about death are offered and then scientifically challenged.

32.5.8 Causation

Cause (independent variable) is what produces an effect on a condition (dependent variable). There are at least three categories of causation: scientific, medical, and legal. Scientific studies that attempt to identify causation often use experimental research designs, and are often called cause-and-effect studies. Medical causation theories are often identified in an autopsy report, after a medical examiner has attempted to identify the *cause*, *manner*, or *mechanism* of death. Legal causation theories focus on who or what precipitated (direct or proximately caused) the breach, injury or death (effect). Another way to view legal causation is to think of whom or what is *responsible* for the effect, which is a *nonsequitur* in scientific causation.

32.5.9 Temporality as Causation

Temporality focuses upon time and is often improperly applied in causation analysis. For example, a person walked under a ladder that was extended over a sidewalk (A) and then was struck by an out-of-control car that jumped the curb (B). The person told the paramedics that if he had not walked under the ladder, he would not have been hit by the car. Or, a black cat ran across the sidewalk in front of a person (A), and later in the day the person received an email from her stockbroker that showed her stock investments lost money (B). The person blamed the black cat for causing her financial loss. Did "A" cause "B" in either of these two instances? While there is a temporal relationship in both of these events ("A" happened before "B") to claim that "A" caused "B" is a causation mistake.

The fact that "A" precedes "B" does not mean that "A" caused "B." The researcher must examine causality to determine if there is any "cause and effect" that can be ruled out when comparing and interpreting the causal inference. Cook and Campbell noted that causal inference depends upon three factors: First, "A" (cause) must precede "B" (effect); second, "A" and "B" must be related; and third, other explanations of the "A" and "B" relationship must to be eliminated [21]. One group of researchers retrospectively studied arrest-related deaths, finding that all of the individuals were handcuffed. Therefore, applying the illogic of correlation or the *post hoc ergo proctor hoc* fallacy to causation, handcuffs must have killed the individuals, as all were handcuffed (A), and this presumably happened prior to death (B) [60].

Temporality is not an absolute basis for legal causation. Consider the following court holdings,

> Timing may be an important clue to causation, see *Dey v. Colt Construction & Development Co*, 28 F. 3d 2446 (7th Cir.1994) but does not eliminate the need to show causation—and Wilson really has nothing but the *post hoc ergo proctor hoc* 'argument' to stand on [61].
>
> "Post hoc ergo propter hoc is not a good way to establish causation." [62]
>
> "It is well settled that a causation opinion based solely on a temporal relationship is not derived from the scientific method and is therefore insufficient to satisfy the requirements of Fed. R. Evid. 207." [63]
>
> "[A] temporal relationship by itself provides no evidence of causation." [64]

32.5.9.1 Felony Murder Rule Exception

An exception to the temporality-does-not-equal-causation paradigm is the *felony murder* rule [65] that is available in some states. This rule of law holds "that if a killing occurs during the commission or attempted commission of a felony (a major crime), the person or persons responsible for the felony can be charged with murder." [65] Generally, *intent to kill* is a necessary element to prove felony murder, but the *felony murder rule* "becomes operative when there is a killing during or a death soon after the felony, and there is ***some causal connection*** between the felony and the killing [emphasis added]." [65] In those jurisdictions that have and use the felony murder rule, Medical Examiners are often *forced* by state law and the local jurisdictional practice to find the person's death a *homicide* even when there is little or no scientific basis for such a ruling. For example, a person who is standing in a teller's line at a bank collapses and dies from a cardiac arrest after a bank robber points a gun and announces "This is a bank robbery." The bank robber had no intention of anyone dying, but using the *felony murder rule* a Medical Examiner may be forced to conclude that the *temporal* issue of pointing the gun and announcing the bank robbery *caused* the person's death. The local prosecutor may charge the bank robber with felony murder. The *felony murder rule* blurs the line between legal blameworthiness (causality) and scientific causation.

32.5.9.2 Correlation as Causation

Correlation is a statistical technique that is used to measure the strength of two or more variables and does not equal causation [34,52]. The relational strength or "correlation coefficient" (which ranges between 1 and –1) can be strong (0.9 or –0.9), weak (0.1 or –0.1), or in between. Regardless of the calculated strength, correlation does not necessarily imply a cause-and-effect relationship between the variables studied. For example, ice cream sales and shark attack frequency are strongly correlated. This is not because sharks start attacking in response to ice cream, but because the two variables exhibit a common response to the warm season. Another example is the strong correlation between the number of tooth cavities found in elementary school children and their vocabulary size. No one advocates eating more candy to increase knowledge though; these variables are both tied with age.

It is therefore important to differentiate between correlation and causation, especially in death cases, as temporality or a strong correlation does not equate to causation [52].

32.5.9.3 Association as Causation

Association is different than correlation, as it focuses upon how the independent variable provides information about the dependent variable [25]. The independent variable may explain why the association of one geographic area has more suspects dying after struggling with the law enforcement (effect). It may explain that it is not that the law enforcement officers caused their death, but that in metropolitan areas there may be more individuals who confront law enforcement while on illicit drugs, or who are mentally ill and on neuroleptic medications than in rural areas. In short, association does not equal causation. Even when evidence from epidemiological studies that are highly regarded when it comes to testing an hypothesis about a particular substance and its effects on humans shows an association, it is well understood that association does not equal causation [8,9].

32.5.9.4 Necessary and Sufficient for Causation

When independent and dependent variables are evaluated together, their integration is thought to be *necessary and sufficient* for a cause to occur. In other words, causes can be categorized as two types: necessary and sufficient. If independent variables do not produce the same effect each time, the variables are rejected as the cause. Applying necessary and sufficiency to everyday lives, to be a mother it is *necessary* that one be a female, but it may not be *sufficient*.

32.6 Back to the Case Study

The Medical Examiner in Richard's case study identified the CEW as a contributory factor in the person's death [66]. The Medical Examiner testified that when she could not rule out the CEW, she had to rule it in as a contributing factor in the person's death [66]. The opinion proffered by the Medical Examiner raised questions about the scientific basis for the alleged physiological effects of the delivered electrical charge and whether or not she had used the scientific method, or whether she had arrived at this conclusion based upon bias, lack of knowledge, or inferential leaps that were somehow based upon her clinical reasoning, or was it a pure guess or unsupported speculation. As one California Medical Examiner commented that when something such as a CEW cannot be "ruled out," then "ruling it in" is purely a causation *guess*.

32.6.1 Applying Science and Law to the Case Study

The following analysis is only for experiential purposes, and is not intended to be critical of medical professionals, expert witnesses, or others.

The Chief Medical Examiner found that Richard:

> ...came to death by: cardiac arrhythmia due to drug induced psychosis, due to drug (methamphetamine and MDMA/MDA) intoxication, acute; and, electrical pulse incapacitation as a contributory condition." [Manner of death was] Homicide: Used drugs; Sudden death incurred during restraint [66].

32.6.2 Claims Versus Critical Thinking

The Medical Examiner's Office press release stated:

> To our knowledge, cases in which the TASER equipment has been implicated in deaths are victims who either have used drugs or had some sort of medical condition, 'susceptible populations.... For those people, the TASER was the 'straw that broke the camel's back.... [Richard] died from the effects of methamphetamine and Ecstasy which sensitized his heart to the effects of the TASER equipment that was required to subdue him. But for the drug intoxication, the use of the TASER would not have resulted in his death [66].

The Medical Examiner's Office and at least 2 of its employees made *claims* about the CEW and its alleged role in the cause of Richard's death. A claim is simply a statement that is either true or false [5]. From the Medical Examiner's Office and its employees' perspective, what they have said is a "true" statement. However, when *critical thinking* is applied to their statement, it can be rejected. Critical thinking requires that one deliberately think about a statement and then conclude that it is true, false, or suspend judgment about the claim. Critical thinking is not done in a vacuum, but rather it is done after reviewing the relevant information to the claim, and eventually tracing the consequences to the claim.

32.6.3 The "Straw" Causation Theory

"The TASER [CEW] was the straw that broke the camel's back" sounds simple, yet it is very misleading. What medical theory regarding death is known as "the straw that broke the camel's back"? There is none. Yet, the Medical Examiner's Office issued a media release containing this inaccurate, unsupported, and inflammatory claim. Possibly this phrase was taken from another pathologist who appeared on the *Early Show* and opined about another so-called CEW death. Here is an excerpt from that show:

> This is the straw that broke the camel's back.... The application of the TASER [CEW], I believe, was the trigger factor or the stressful event that caused an elevation of blood pressure and an elevation in heart rate which stressed an already damaged heart to the point where it went into cardiac arrest [66].

This is also an example of a fallacy, where many individuals will believe what was written or said, based upon a person's real or perceived authority.

The "straw" theory is sometimes referred to as "pushed over the edge," the "boo causation," or the "1% factor." This "boo" or "1%" causation is a very tempting one as one does not need to establish any strong effect from the TASER CEW. After all, if the suspect were on the edge of the cliff, a tiny push is all that is required to kill. Or, to paraphrase, he was so close to death that if a law enforcement officer yelled "boo" he would have died. However, reflection shows the limitations of this causation "theory": it admits that the contribution of the final "boo" factor was tiny – hence the "1%" term. Also, if the suspect were so close to death that the smallest stimulus would have killed him, what could law enforcement officers have done differently?

A more fundamental problem with this fallacy is that drug abusers suffering from an excited delirium episode may no longer be "on the edge" of the cliff when law enforcement officers get involved. They are in a possibly lethal medical crisis; they are likely to be already rolling downhill rapidly, or as Michael Curtis, M.D. says, they are already on *the freight train of death* [67].

32.6.4 Temporal or Association Causation Theory

After the Richard lawsuit was filed, the Chief Medical Examiner and her associate's depositions were taken, under oath, regarding their medical findings. The associate testified as follows when asked about "A" (CEW deployment) occurring before "B" (Richard's collapse and death). "Question: And so the temporal proximity is the fact upon which you rely to link the TASER [CEW] to [Richard's] death, correct? Answer: Yes" [68].

Somewhat confusing and contradictory, the medical examiner also testified: "Question: If an event happens first and then a second event occurs some time after the first event, do you agree that that does not necessarily mean that the second event was caused by the first event? Answer: Yes."

Temporality and association do not equal scientific causation. Recall, that this is really correlation, which never equates to causation.

32.6.5 Probabilistic Causation

Recall the Medical Examiner's cause of Richard's death listed the CEW as a contributing factor. Yet, the Medical Examiner could not state by what mechanism the CEW allegedly caused or contributed to any injury. When asked how much the CEW contributed to Richard's death, even though there was no identified mechanism of causation, or degree of proof of mechanism of causation, the Medical Examiner testified it was as low as 0.00000000001% [68]. To get a percentage, there needs to be a numerator and a denominator. In this case, based upon her testimony, the associate Medical Examiner appears to have confused "temporality" with "causation" when she testified that the CEW contributed to Richard's death based solely on the temporal proximity of its use to his collapse. Statistically, she had absolutely no mathematical, statistical, or scientific basis for her opinion that the CEW contributed as little as 0.00000000001% to Richard's death.

This is a good example of how a *post hoc ergo propter hoc* fallacy was applied and how it also demonstrated a nonscientific temporal relationship (i.e., timing), which is insufficient, and by itself provided no scientific basis or evidence of causation. Both Medical Examiners also relied upon third party observation and reporting about CEW usage prior to Richard's collapse, which in isolation, were scientifically *insufficient* but were intuitively *necessary* to develop their nonscientific, probabilistic causation opinion. Probabilistic causation contains an element of "chance" that is predicated upon scientific research methodology and findings. The Medical Examiners could not support their 0.0000000001% assertion, because it was not based upon any scientific or statistical research study or finding.

The associate Medical Examiner also testified that she "believe[s] that [the TASER CEW] was involved" in the death of [Richard] [67]. A "belief" is not a scientific basis for developing an opinion to a reasonable degree of medical certainty or scientific certainty.

32.6.6 *It Is in Print Fallacy*

Other experts were engaged by the Plaintiff on the behalf of Richard's estate with one expert writing:

> I am left with a series of anecdotal cases in which TASER [CEW] usage and rapid death happened in the same person. Attached is the recitation of 167 deaths from the Jan. 5, 2006 *Arizona Sun*, an edited but not peer reviewed publication [67].

The "edited publication" to which the expert refers, is an ordinary newspaper. Unless this expert had personally and thoroughly reviewed and analyzed each of these cases including the medical findings about each person's death, which he had not, he is attempting to persuade others by preying upon their naiveté about scientific research. He is relying on third party lay information from a common newspaper to formulate and support his nonscientific opinion. The expert had used a *proof substitute* to suggest that he had scientific proof, but in actuality he offered no scientific proof [56]. However, when proof substitutes or fictional claims are made, it shifts the burden of proof to the other party.

32.6.7 *Bias and Causation Theory*

At this writing (mid-2008), there is no scientific cause-and-effect research study that has scientifically or statistically proven TASER-brand CEWs to be the direct cause of death in human subjects. (There has been a handful of fatal head-injury cases in which the CEW may have contributed to a fall.) Rather, the scientific studies that have been conducted have shown the relative safety of these devices, given the ethical and legal constraints that are imposed on such studies [69].

The Chief Medical Examiner's colleague testified there was never ever a point in time – from the start of the autopsy until the issuance of the autopsy report – where she did not think that "electrical pulse incapacitation" should be included in the autopsy report as a contributing factor [67].

This is not science, but examples of tenacity, intuition, authority, rationalism, ignorance (lack of knowledge), or bias. It appears unequivocal from her testimony that she began with a nonscientific bias that a CEW can kill a person, and then focused upon converting her bias into a "contributory condition" of Richard's death.

The colleague testified that she had not stayed current in the scientific literature regarding CEWs, had not conducted any research or analysis of CEWs, did not know the power (wattage) of a TASER CEW, did not know the energy of a TASER CEW, did not know the current (in amperes) of TASER CEWs, did not know how CEWs work, had very little knowledge about metabolic acidosis, did not consider herself an expert on sudden death, and did not know whether anything happened to Richard physiologically when the TASER CEW was applied to him [68].

32.6.8 Guessing as a Causation Theory

There have been several medical examiners who have opined that when a CEW is deployed at an individual it causes pain. For some people who receive the CEW shock this may be a true statement, but many people who are "high" on drugs are less sensitive to or insensitive to pain. The alleged pain from the CEW, according to the medical examiners, *could* have caused a release of natural stimulants in the human body known as catecholamines, which *could* rush to the heart, causing it to stop. According to Dr. David A. Fishbain, M.D., "there is no evidence in the literature implicating increased catecholamine release with acute pain induction, [and] no evidence in the literature implicating acute pain as causing sudden death" [70].

Recall that a medical examiner in a prior CEW incident had opined on the *Early Show* that the CEW was the trigger factor or the stressful event that caused an elevation of blood pressure and an elevation in heart rate which stressed an already damaged heart to the point where it went into cardiac arrest. Regardless of whether the Medical Examiner's opinion is an increase in blood pressure, an increase in catecholamine levels, an increase in potassium levels, or a decrease in ph levels, one must know the baseline of each of these items prior to claiming that there was an increase. Thus, guessing is another fallacy being substituted for scientific evidence. The incident involving Richard never went to trial because the case was dismissed, in large part due to the misapplication of scientific causation. Eventually, a judicial review of the autopsy report resulted in the autopsy conclusions being overturned.

32.7 Conclusions

Science has virtually replaced common sense and nonscientific *opinion* in the court room, but one must stay current with the science. Take nothing at face value, and when given a scientific study, review it by analyzing it with the many scientific and statistical points discussed in this chapter, understand the underlying foundations of the study, and avoid taking it out of context. The days of experts walking into court rooms and throwing nonscientific or pseudoscientific opinions about causation "against the courtroom walls" hoping that something will stick are over. Applying science and the Court's criteria to the qualification of experts and their opinions will help to keep nonscientific methodology, pseudoscience, and bias out of the court room.

Acknowledgments The author thanks Judy Melinek, M.D., Gregg J. Gunta, J.D., Kevin Reak, J.D., John Wolfgang, J.D., Wayne Schmidt, J.D., L.L.M., Michael A. Brave, J.D., William D. Steeves, Jr., Ed.D., and David A. Donovan, M.A. for their insightful comments and for their peer review of this work. Thanks to Rebecca Bours for proofreading several drafts. They are very gracious, but are not responsible for any errors.

References

1. Peters JG Jr. *The Graduation Party*. Henderson, NV: Institute for the Prevention of In-Custody Deaths, Inc.; 2007.
2. The Committee on Pattern Civil Jury Instructions of the Seventh Circuit. Federal Civil Jury Instructions of the Seventh Circuit.
3. Ruby Mann, et al. v. Taser International, Inc., et al. Order. United States District Court for the Northern District of Georgia, Rome Division; October 1, 2007.
4. Federal Rules of Evidence: Rule 702.
5. Moore NM, Parker R. *Critical Thinking*. Palo Alto, CA: Mayfield Publishing Company; 1986.
6. Daubert v. Merrell Dow Pharm., Inc., 509 U.S. 579 (1993).
7. Kuhmo Tire Co., Ltd. v. Carmichael, 526 U.S. 137 (1999).
8. Hollingsworth JG, Lasker EG. Testing claims of adverse drug effects in the courtroom. In: Karch SB, ed. *Drug Abuse Handbook*. 2nd ed. Boca Raton, FL: CRC Press; 2007: 1156–1173.
9. Hollingsworth JG, Lasker EG. The case against differential diagnosis: *Daubert* medical causation testimony, and the scientific method. *J Health Law*. 2004; 37(1): 85–111.
10. General Electric Co. v. Joiner, 522 U.S. 136; 1997.
11. Gunta GG. *Daubert Issues in Excited Delirium Cases*. Milwaukee, WI: Gunta & Reak, S.C.; 2007.
12. Graziano AM, Raulin ML. *Research Methods*. 4th ed. Boston: Allyn and Bacon; 2000.
13. Judge Bans Jurors From Watching CSI Shows. American Police Beat Oct. 2007:42.
14. Ho JD. Sudden In-Custody Death. *Police Magazine* 2005; 29(8): 47–55.
15. Peters JG Jr. Sudden death, 'excited' delirium, and issues of force: Part I. *Police & Security News* 2006; 22(1): 104–107.
16. Peters JG Jr. Sudden death, 'excited' delirium, and issues of force: Part II. *Police & Security News* 2006; 22(2).
17. Leedy, PD. *Practical Research*. New York: Macmillan Publishing Co.; 1974.
18. Zikmund, WG. *Business Research Methods*. 7th ed. United States: South-Western; 2003.
19. Cresswell JW. *Research Design: Qualitative & Quantitative Approaches*. Thousand Oaks: Sage Publications; 1994.
20. Breakwell GM, Hammond S, Fife-Schaw C. *Research Methods in Psychology*. 2nd ed. London: Sage Publications; 2000.
21. Cook TD, Campbell DT. *Quasi-Experimentation*. Boston: Houghton Mifflin Company; 1979.
22. Silverman D. *Doing Qualitative Research*. Thousand Oaks, CA: Sage Publications; 2000.
23. Bickman L. ed. *Research Design*. Thousand Oaks, CA: Sage Publications, Inc.; 2000.
24. Dennis AJ, Valentino, AJ, Walter RJ, et al. Acute Effects of TASER X26 Discharges in a Swine Model. *J Trauma*, 2007; 63: 581–590.
25. Rubin HJ. *Applied Social Research*. Columbus, OH: Charles E. Merrill; 1983.
26. Nachmias D, Nachmias C. *Research Methods in the Social Sciences*. New York: St. Martin's Press; 1981.
27. Kerlinger FN. *Foundations of behavioral research*. 2nd ed. New York: Holt, Reinhart and Winston, Inc.; 1964.

28. Berenson A. Claims Over Tasers' Safety are Challenged. New York Times. New York: New York Times, 2004.
29. Reay DT, Howard JD, Fligner CL, Ward RJ. Effects of positional restraint on oxygen saturation and heart rate following exercise. *Am J Forensic Med Pathol.* 1988; 9(1):16–18.
30. American Psychological Association; 2001.
31. Polit DF, Beck C T. *Nursing research: Principles and practices.* 7th ed. Philadelphia: Lippencott Williams & Wilkins; 2004.
32. §46.111. *Code of Federal Regulations*; 1991.
33. Minimum EW, King BM, Bear G. *Statistical Reasoning in Psychology and Education.* 3rd ed. New York: John Wiley & Sons, Inc.; 1993.
34. Gravetter FJ, Wallnau LB. *Essentials of Statistics for the Behavioral Sciences.* 3rd ed. Pacific Grove, CA: Brooks/Cole Publishing Company; 1999.
35. Cascio WF. *Applied Psychology in Human Resource Management.* 5th ed. Upper Saddle River, NJ: Prentice Hall; 1998.
36. Dennis AJ, Valentino DJ, Walter RJ, Nagy KK, Winners J, Bokhari F, Wiley DE, Joseph KT, Roberts RR. Acute effects of TASER X26 discharges in a swine model. *J Trauma* 2007;63:581–90.
37. Walter RJ, Dennis AJ, Valentino DJ, Margeta B, Nagy KK, Bokhari F, Wiley DE, Joseph KT, Roberts RR. TASER X26 discharges in swine produce potentially fatal ventricular arrhythmias. *Acad Emerg Med* 2008;15:66–73.
38. Jaisingh LR. *Statistics for the Utterly Confused.* New York: McGraw-Hill; 2000.
39. Jacobs D, Blackburn H, Higging M, Reed D, Iso H, McMillan G, Neaton J, Nelson J, Potter J, Rifkind B, et al. Report of the conference on low blood cholesterol: Mortality associations. *Circulation* 1992;86:1046–1060.
40. Ho JD, Miner JF, Lakireddy DR, et al. Cardiovascular and physiologic effects of conducted electrical weapon discharge in resting adults. *Acad Emerg Med.* 2006; 13:589–595.
41. Ho JD, Dawes DM, et al. Absence of electrocardiographic change following prolonged application of a conducted electrical weapon in physically exhausted adults. *Acad Emerg Med.* 2007 (Suppl. 1); 14:S128–S129.
42. Ho JD, Dawes DM, et al. Respiratory effect of prolonged electrical weapon application on human volunteers. *Acad Emerg Med.* 2007 Mar; 14(3): 197–201.
43. Dawes D, Ho, J et al. 15-second conducted electrical weapon application does not impair basic respiratory parameters, venous blood gases, or blood chemistries and does not increase core body temperatures. *Ann Emerg Med.* 2007; 50(3):S132.
44. Levine SD, Sloane C, Chan TC, Vilke GM, Dunford J. Cardiac monitoring of subjects exposed to the Taser. *Acad Emerg Med.* 2005; 12(Suppl. 1): 71.
45. Sloane C, Vilke GM, et al. Serum Tronopin I measurement of subjects exposed to the Taser X26. *Acad Emerg Med.* 2007 (Suppl. 1); 14: S103–S104.
46. Vilke, GM, Sloane C, et al. Cardiovascular and metabolic Effects of the TASER on Human Subjects. *Acad Emerg Med.* 2007 (Suppl. 1); 14: S104–S105.
47. Vilke GM, Sloane CM, Levine S, Neuman T, Castillo EM, Chan TC. Twelve lead ECG monitoring of subjects before and after voluntary exposure to the Taser X-26. *Am J Emerg Med.* 2007. In press.
48. Reay DT. Positional asphyxia during law enforcement transport. *Am J Forensic Med Pathol.* 1993; 14(2):170–175.
49. Chan TC, Vilke GM, Neuman T, Clausen JL. Restraint position and positional asphyxia. *Ann Emerg Med Ann Emerg Med*, 1997; 30(5): 578–586.
50. Cruz v. City of Laramie, 239 F.3d 1183 (10th Cir. 2001).
51. Price, et al. v. San Diego, et al, 990 F. Supp. 1230 (S.D. Cal.) (1998).
52. Phillips JL. *How to Think About Statistics.* 6th ed. New York: W. H. Freeman and Company; 2000.

53. Cohen J. *Statistical Power Analysis for the Behavioral Sciences*. Hillsdale, NJ: Lawrence Erlbaum Associates, Inc.;1988.
54. The Nizkor Project. Fallacies. Available at: www.nizkor.org/features/fallacies Accessed March 26, 2006.
55. Aldisert RJ. *Logic for Lawyers*. 3rd ed. South Bend, IN: National Institute for Trial Advocacy; 1997.
56. Epstein RL. *Critical Thinking*. 2nd ed. Belmont, CA: Wadsworth; 2002.
57. Better Business Bureau. If It's In Print, It Must Be True. Available at: www.bbbsilicon.org/topic002p.htm. Accessed March 26, 2006.
58. Card H. If it is in the Paper it must be True. Available at: http://www.nysut.org/html Accessed March 24, 2006.
59. Napoli L. If it's on the Internet, it must be true. Available at: http://www.globetechnology.com/servlet/story/RTGAM.20050310.gtgulliblemar10/BN story/technology. Accessed March 26, 2006.
60. Ho JD, Miner, JR, Heegaard, WG, Reardon, RF. Deaths in American Custody: A 12-month Surveillance Study.
61. Bermudez v. TRC Holdings, 138 F.3d 1176 (7th Cir.1998).
62. See Oest v. Illinois Department of Corrections, 240 F 3e 605, 616 & n 8 (7th Cir. 2001).
63. Schmaltz v. Norfolk & Western Ry. Co., 878 F. Supp. 1119, 1122 (N.D. Ill. 1995). See also, Conde v. Velsicol Chemical Corp., 804 F. Supp. 972, 1023 (S.D. Ohio 1992), aff'd, 24 F.3d 809 (6th Cir. 1994); Porter v. Whitehall laboratories, Inc., 9 F.3d 607, 611 (7th Cir. 1993); Cuevas v. E. I. DuPont De Nemours and Co., 956 F. Supp. 1306, 1310-11 (S.D. Miss. 1997); Cartwright v. Home Depot U.S.A., Inc., 936 F. Supp. 900 906 (M.D. Fla. 1996); Cavallo v. Star Enterprise, 892 F. Supp. 756, 757 (E.D. Va. 1995), affirmed in part, reversed in part on other grounds, 100 F.3d 1150 (4th Cir. 1996), cert denied, 522 U.S. 1044 (1998).
64. In re Breast Implant Litigation, 11 F. Supp.2d 1217, 1238 (D. Colo. 1998).
65. Felony-Murder Rule. Available at: http://law.jrank.org/pages/6835/Felony-Murder-Rule.html Accessed January 24, 2008.
66. Kohler LJ. Report of Investigation: Richard Thomas Holcomb. Akron, OH: Medical Examiner, County of Summit; 2005.
67. Curtis M. Excited Delirium Conference. New Berlin, Wisconsin, 2007.
68. Taser International, Inc. And City Of Akron v. Chief Medical Examiner Of Summit County, Ohio Cv 2006-11-7421: In The Court of Common Pleas: Summit County, Ohio, 2008.
69. Morgan J. Medical Panel Issues Interim Findings on Stun Gun Safety. *National Institute of Justice Journal* 2008; 261: 20–23.
70. Fishbain DA. The Nosology, Identification, Pathophysiology and Treatment of Various Types of Delirium. *3rd Annual Sudden Death, Excited Delirium & In-Custody Death Conference*. Las Vegas, Nevada, 2008.

Appendix A
Excited Delirium Checklist

Excited delirium or excited delirium syndrome is the only one form of potential sudden death that law enforcement officers may encounter. Other potential causes of unexpected arrest-related deaths include, but are not limited to: sudden unexpected death in epilepsy, sickle cell sudden death, and severe heart disease.

Present?	Criterion
911 Call – emergency contact for assistance	
	1. Critical call phrases include, "He just freaked out," "just snapped," "flipped out," or a person is "running around naked." [1]
Law enforcement	
	2. Agitation, screaming, extreme fear response, or panic [2–6]
	3. Violence, assault, or aggression towards others [6–9]
	4. Suspicion of impending death. Typical comments include, "I'm dying," "Please save me," or "Don't kill me" [10]
	5. Incoherence or disorganized speech. Grunting or animal sounds [9,11]
	6. Clothing removal inappropriate for ambient temperature or complete nudity [6,12–14].
	7. Disorientation or hallucinations [6,15–18]
	8. Mania, paranoia, anxiety, or avoidance behavior [2,6,19–22]
	9. Constant motion or hyperactivity [2,18,23–25]
Capture, control, and restraint of subject	
	10. Extreme or "super human" strength [9,21]
	11. High threshold of or imperviousness to pain [11,14]
	12. Extreme stamina [26,11]
	13. Brief quiet period before collapse likely corresponding with respiratory arrest [2,5,11,27]
Emergency medical services contact and intervention	
	14. Presenting rhythm of PEA (pulseless electrical activity) or asystole [26,28–30]. Also documented by "No shock advised" with automatic external defibrillator [30]

(continued)

Present?	Criterion
Emergency department	
	15. High core body temperature [3,4,9,19,31,32]
	16. Acidosis (acidic blood) [11,33,34]
	17. Rhabdomyolysis (if suspect is resuscitated) [3,32,35]
Law enforcement/forensic investigator death investigation	
	18. History of chronic stimulant abuse or mental illness [2,7,15,20,25,28,36–39]. History of violence or drug related arrests, mental health histories and treatments, and drug rehabilitation interventions, etc.
	19. Damage to shiny objects such as glass, mirrors, and lights [11]. Reported behaviors may include attacking a squad car light bar or charging oncoming traffic at night. Occasionally generalized vandalism
Pathologist – medical examiner investigation	
	20. Minor injuries from fighting against restraints (e.g. handcuffs, hobbles)
	21. Positive mash (central nervous system biomarkers) test for dopamine transporter assay and heat shock protein [3,19,20,40–44]
	22. Positive brain and hair toxicology screen for chronic stimulant abuse [40,45–49]. Post-incident drug levels may be low to negative

Contributors: Mark Kroll, PhD; Charles Wetli, MD; Deborah Mash, PhD; Steven Karch, MD; Michael Graham, MD, Jeffrey Ho, MD.

Notes: A syndrome is an aggregate of signs and symptoms that define a medical condition. Not all persons with a certain syndrome have all the same signs and symptoms. Not all cases of a syndrome result from the same cause. For example, some persons with carpal tunnel syndrome will have numbness and tingling, while others will have weakness and pain. Also, some persons with carpal tunnel syndrome will have it because of trauma, while others will have the syndrome because of pregnancy, diabetes, rheumatoid arthritis, or thyroid disease. Persons with the excited delirium syndrome will have various combinations of some of the signs and symptoms listed above. The cause (etiology) of the excited delirium syndrome in any individual may be due to one or more of a number of conditions. The most common conditions are mental illness and illegal stimulant abuse (especially cocaine and methamphetamine) [28].

Because the term "excited delirium syndrome" has not been widely used until recent years, many physicians do not recognize the term even though they may be very familiar with agitation and deaths due to drugs and other conditions. It is important to avoid the distraction of the various terms that have been applied to this syndrome. For example, what is now referred to as excited delirium [2–4,14,20,21,24,26–28,32–35,38,41,42,50–57] or agitated delirium [29,44,58–103] has also been called: Bell's mania [18], acute exhaustive mania [104], acute delirious mania [18], delirium grave [18], typhoma [18], acute delirium [18], manic-depressive exhaustion [12], excited catatonia [77], lethal catatonia [105], and neuroleptic malignant syndrome [7,14,31,60,105].

References

1. Code 1069 Excited Delirium Dispatch Policy. In: Sheriff's_Office, ed. Jacksonville, FL, 2006.
2. Pollanen MS, Chiasson DA, Cairns JT, Young JG. Unexpected death related to restraint for excited delirium: a retrospective study of deaths in police custody and in the community. *CMAJ* 1998;158:1603–7.

3. Ruttenber AJ, Lawler-Heavner J, Yin M, Wetli CV, Hearn WL, Mash DC. Fatal excited delirium following cocaine use: epidemiologic findings provide new evidence for mechanisms of cocaine toxicity. *J Forensic Sci* 1997;42:25–31.
4. Blaho K, Winbery S, Park L, Logan B, Karch SB, Barker LA. Cocaine metabolism in hyperthermic patients with excited delirium. *J Clin Forensic Med* 2000;7:71–6.
5. Stefan H. Sudden death of psychiatric patients following great excitation and exhaustion which has no actual anatomic basis. *Dtsch Med Wehnschr* 1934;60:1550–8.
6. Shulack N. Sudden "exhaustive" death in excited patients. *Psychiatr Q* 1944;18:3–12.
7. Kasantikul D, Kanchanatawan B. Neuroleptic malignant syndrome: a review and report of six cases. *J Med Assoc Thai* 2006;89:2155–60.
8. Bell L. On a form of disease resembling some advanced stages of mania and fever, but so contradistinguished from any ordinary observed or described combination of symptoms as to render it probable that it may be overlooked and hitherto unrecorded malady. *Am J Insanity* 1849;6:97–127.
9. Fishbain DA, Wetli CV. Cocaine intoxication, delirium, and death in a body packer. *Ann Emerg Med* 1981;10:531–2.
10. Karch S, Kroll M. Unpublished observations of 150 excited delirium cases, 2008.
11. Wetli C. Excited delirium. In: Chan R, ed. Sudden Deaths in Custody. Totowa: Humana Press, 2006:99–112.
12. Derby I. Manic-depressive "exhaustion" deaths. *Pyschiatric Q* 1933;7:436–49.
13. Shulack N. Sudden "exhaustive" death in excited patients. *Psychiatr Q* 1938;12:282–93.
14. Ross DL. Factors associated with excited delirium deaths in police custody. *Mod Pathol* 1998;11:1127–37.
15. Stagno D, Gibson C, Breitbart W. The delirium subtypes: a review of prevalence, phenomenology, pathophysiology, and treatment response. *Palliat Support Care* 2004;2:171–9.
16. Ross CA, Peyser CE, Shapiro I, Folstein MF. Delirium: phenomenologic and etiologic subtypes. *Int Psychogeriatr* 1991;3:135–47.
17. Davidson G. Concerning the cause of death in certain cases. *Am J Psychiatry* 1934;91:41–9.
18. Kraines S. Bell's mania (acute delirium). *Am J Psychiatry* 1934;91:29–40.
19. Mash DC, Ouyang Q, Pablo J, Basile M, Izenwasser S, Lieberman A, Perrin RJ. Cocaine abusers have an overexpression of alpha-synuclein in dopamine neurons. *J Neurosci* 2003;23:2564–71.
20. Mash DC, Pablo J, Ouyang Q, Hearn WL, Izenwasser S. Dopamine transport function is elevated in cocaine users. *J Neurochem* 2002;81:292–300.
21. Wetli CV, Fishbain DA. Cocaine-induced psychosis and sudden death in recreational cocaine users. *J Forensic Sci* 1985;30:873–80.
22. Anglin MD, Burke C, Perrochet B, Stamper E, Dawud-Noursi S. History of the methamphetamine problem. *J Psychoactive Drugs* 2000;32:137–41.
23. Schroeder U, Schroeder H, Darius J, Grecksch G, Sabel BA. Simulation of psychosis by continuous delivery of phencyclidine from controlled-release polymer implants. *Behav Brain Res* 1998;97:59–68.
24. Pedal I, Zimmer G, Mattern R, Mittmeyer HJ, Oehmichen M. Fatal incidences during arrest of highly agitated persons. *Arch Kriminol* 1999;203:1–9.
25. Fava M. Psychopharmacologic treatment of pathologic aggression. *Psychiatr Clin North Am* 1997;20:427–51.
26. DiMaio T, VJM D. Excited delirium syndrome cause of death and prevention. Boca Raton: Taylor & Francis, 2006.
27. Stratton SJ, Rogers C, Green K. Sudden death in individuals in hobble restraints during paramedic transport. *Ann Emerg Med* 1995;25:710–2.
28. Stratton SJ, Rogers C, Brickett K, Gruzinski G. Factors associated with sudden death of individuals requiring restraint for excited delirium. *Am J Emerg Med* 2001;19:187–91.
29. Park KS, Korn CS, Henderson SO. Agitated delirium and sudden death: two case reports. *Prehosp Emerg Care* 2001;5:214–6.

30. Swerdlow C, Kroll M, Williams H, Biria M, Lakkireddy D, Tchou P. Presenting rhythm in sudden custodial deaths after use of TASER® electronic control device. *Heart Rhythm* 2008;5:S44.
31. Nielsen J, Bruhn AM. Atypical neuroleptic malignant syndrome caused by olanzapine. *Acta Psychiatr Scand* 2005;112:238–40; discussion 240.
32. Ruttenber AJ, McAnally HB, Wetli CV. Cocaine-associated rhabdomyolysis and excited delirium: different stages of the same syndrome. *Am J Forensic Med Pathol* 1999;20:120–7.
33. Allam S, Noble JS. Cocaine-excited delirium and severe acidosis. *Anaesthesia* 2001;56:385–6.
34. Brice JH, Pirrallo RG, Racht E, Zachariah BS, Krohmer J. Management of the violent patient. *Prehosp Emerg Care* 2003;7:48–55.
35. Karch SB. Karch's Pathology of Drug Abuse, CRC Press, Boca Raton, 2002:541.
36. DiMaio VJ. Forensic Pathology, CRC Press, Boca Raton, 2001:656.
37. Gray SD, Fatovich DM, McCoubrie DL, Daly FF. Amphetamine-related presentations to an inner-city tertiary emergency department: a prospective evaluation. *Med J Aust* 2007;186:336–9.
38. Morrison A, Sadler D. Death of a psychiatric patient during physical restraint. Excited delirium – a case report. *Med Sci Law* 2001;41:46–50.
39. Pearlson GD. Psychiatric and medical syndromes associated with phencyclidine (PCP) abuse. *Johns Hopkins Med J* 1981;148:25–33.
40. Stephens BG, Jentzen JM, Karch S, Mash DC, Wetli CV. Criteria for the interpretation of cocaine levels in human biological samples and their relation to the cause of death. *Am J Forensic Med Pathol* 2004;25:1–10.
41. Mash DC, Staley JK. D3 dopamine and kappa opioid receptor alterations in human brain of cocaine-overdose victims. *Ann N Y Acad Sci* 1999;877:507–22.
42. Mash DC, Staley JK, Izenwasser S, Basile M, Ruttenber AJ. Serotonin transporters upregulate with chronic cocaine use. *J Chem Neuroanat* 2000;20:271–80.
43. Stephens BG, Jentzen JM, Karch S, Wetli CV, Mash DC. National Association of Medical Examiners position paper on the certification of cocaine-related deaths. *Am J Forensic Med Pathol* 2004;25:11–3.
44. Chen L, Segal DM, Moraes CT, Mash DC. Dopamine transporter mRNA in autopsy studies of chronic cocaine users. *Brain Res Mol Brain Res* 1999;73:181–5.
45. Berankova K, Habrdova V, Balikova M, Strejc P. Methamphetamine in hair and interpretation of forensic findings in a fatal case. *Forensic Sci Int* 2005;153:93–7.
46. Kimura H, Mukaida M, Mori A. Detection of stimulants in hair by laser microscopy. *J Anal Toxicol* 1999;23:577–80.
47. Takayama N, Tanaka S, Hayakawa K. Determination of stimulants in a single human hair sample by high-performance liquid chromatographic method with chemiluminescence detection. *Biomed Chromatogr* 1997;11:25–8.
48. Kintz P, Cirimele V, Tracqui A, Mangin P. Simultaneous determination of amphetamine, methamphetamine, 3,4-methylenedioxyamphetamine and 3,4-methylenedioxymethamphetamine in human hair by gas chromatography-mass spectrometry. *J Chromatogr B Biomed Appl* 1995;670:162–6.
49. Nagai T, Kamiyama S, Nagai T. Forensic toxicologic analysis of methamphetamine and amphetamine optical isomers by high performance liquid chromatography. *Z Rechtsmed* 1988;101:151–9.
50. Sztajnkrycer MD, Baez AA. Cocaine, excited delirium and sudden unexpected death. *Emerg Med Serv* 2005;34:77–81.
51. O'Halloran RL, Lewman LV. Restraint asphyxiation in excited delirium. *Am J Forensic Med Pathol* 1993;14:289–95.
52. Strote J, Range Hutson H. Taser use in restraint-related deaths. *Prehosp Emerg Care* 2006;10:447–50.
53. Paquette M. Excited delirium: does it exist? *Perspect Psychiatr Care* 2003;39:93–4.

54. Gowers WR. A Manual of Diseases of the Nervous System, J&A Churchill, London, 1896.
55. Nahas GG, Burks TF, Hollister LE. Drug Abuse in the Decade of the Brain, IOS Press, Amsterdam, 1997:281.
56. Levine B. Principles of Forensic Toxicology, AACC, Washington, DC, 2003:385.
57. Karch SB. Drug Abuse Handbook, CRC Press, Boca Raton, FL, 2006:1267.
58. Mets B, Jamdar S, Landry D. The role of catecholamines in cocaine toxicity: a model for cocaine "sudden death". *Life Sci* 1996;59:2021–31.
59. Karch SB, Wetli CV. Agitated delirium versus positional asphyxia. *Ann Emerg Med* 1995;26:760–1.
60. Wetli CV, Mash D, Karch SB. Cocaine-associated agitated delirium and the neuroleptic malignant syndrome. *Am J Emerg Med* 1996;14:425–8.
61. Mirchandani HG, Rorke LB, Sekula-Perlman A, Hood IC. Cocaine-induced agitated delirium, forceful struggle, and minor head injury. A further definition of sudden death during restraint. *Am J Forensic Med Pathol* 1994;15:95–9.
62. Irwin P, Murray S, Bilinski A, Chern B, Stafford B. Alcohol withdrawal as an underrated cause of agitated delirium and terminal restlessness in patients with advanced malignancy. *J Pain Symptom Manage* 2005;29:104–8
63. Morita T, Tei Y, Inoue S. Agitated terminal delirium and association with partial opioid substitution and hydration. *J Palliat Med* 2003;6:557–63.
64. Morita T, Tei Y, Inoue S. Impaired communication capacity and agitated delirium in the final week of terminally ill cancer patients: prevalence and identification of research focus. *J Pain Symptom Manage* 2003;26:827–34.
65. Vatsavayi V, Malhotra S, Franco K. Agitated delirium with posterior cerebral artery infarction. *J Emerg Med* 2003;24:263–6.
66. Vilke GM, Chan TC. Agitated delirium and sudden death. *Prehosp Emerg Care* 2002;6:259; author reply 259–60.
67. Frye MA, Coudreaut MF, Hakeman SM, Shah BG, Strouse TB, Skotzko CE. Continuous droperidol infusion for management of agitated delirium in an intensive care unit. *Psychosomatics* 1995;36:301–5.
68. Levenson JL. High-dose intravenous haloperidol for agitated delirium following lung transplantation. *Psychosomatics* 1995;36:66–8.
69. Sanders KM, Murray GB, Cassem NH. High-dose intravenous haloperidol for agitated delirium in a cardiac patient on intra-aortic balloon pump. *J Clin Psychopharmacol* 1991;11:146–7.
70. Verslegers W, De Deyn PP, Saerens J, Marien P, Appel B, Pickut BA, Lowenthal A. Slow progressive bilateral posterior artery infarction presenting as agitated delirium, complicated with Anton's syndrome. *Eur Neurol* 1991;31:216–9.
71. Mori E, Yamadori A. Acute confusional state and acute agitated delirium. Occurrence after infarction in the right middle cerebral artery territory. *Arch Neurol* 1987;44:1139–43.
72. Julien J, Vital C, Vallat JM, Bourgoin B. Epilepsy and agitated delirium caused by an astrocytoma of the amygdala. *Eur Neurol* 1979;18:387–90.
73. Medina JL, Chokroverty S, Rubino FA. Syndrome of agitated delirium and visual impairment: a manifestation of medial temporo-occipital infarction. *J Neurol Neurosurg Psychiatry* 1977;40:861–4.
74. Medina JL, Rubino FA, Ross E. Agitated delirium caused by infarctions of the hippocampal formation and fusiform and lingual gyri: a case report. *Neurology* 1974;24:1181–3.
75. Horenstein S, Chamberlin W, Conomy J. Infarction of the fusiform and calcarine regions: agitated delirium and hemianopia. *Trans Am Neurol Assoc* 1967;92:85–9.
76. Spiller JA, Keen JC. Hypoactive delirium: assessing the extent of the problem for inpatient specialist palliative care. *Palliat Med* 2006;20:17–23.
77. Pruett JR, Rizvi ST. A16-year-old girl with excited catatonia treated with low-dose oral Lorazepam. *J Child Adolesc Psychopharmacol* 2005;15:1005–10.

78. Kusne S, Smilack J. Transmission of rabies virus from an organ donor to four transplant recipients. *Liver Transpl* 2005;11:1295–7.
79. Duggal MK, Singh A, Arunabh, Lolis JD, Guzik HJ. Olanzapine-induced vasculitis. *Am J Geriatr Pharmacother* 2005;3:21–4.
80. Ogasawara K, Komoribayashi N, Kobayashi M, Fukuda T, Inoue T, Yamadate K, Ogawa A. Neural damage caused by cerebral hyperperfusion after arterial bypass surgery in a patient with moyamoya disease: case report. *Neurosurgery* 2005;56:E1380; discussion E1380.
81. Srinivasan A, Burton EC, Kuehnert MJ, Rupprecht C, Sutker WL, Ksiazek TG, Paddock CD, Guarner J, Shieh WJ, Goldsmith C, Hanlon CA, Zoretic J, Fischbach B, Niezgoda M, El-Feky WH, Orciari L, Sanchez EQ, Likos A, Klintmalm GB, Cardo D, LeDuc J, Chamberland ME, Jernigan DB, Zaki SR. Transmission of rabies virus from an organ donor to four transplant recipients. *N Engl J Med* 2005;352:1103–11.
82. Cameron D, Bridge D, Blitz-Lindeque J. Use of sedation to relieve refractory symptoms in dying patients. *S Afr Med J* 2004;94:445–9.
83. Muller-Busch HC, Andres I, Jehser T. Sedation in palliative care – a critical analysis of 7 years experience. *BMC Palliat Care* 2003;2:2.
84. Cowan JD, Palmer TW. Practical guide to palliative sedation. *Curr Oncol Rep* 2002;4:242–9.
85. Cheng C, Roemer-Becuwe C, Pereira J. When midazolam fails. *J Pain Symptom Manage* 2002;23:256–65.
86. Morita T, Tsunoda J, Inoue S, Chihara S. Terminal sedation for existential distress. *Am J Hosp Palliat Care* 2000; 17:189–95.
87. Travis SS, Conway J, Daly M, Larsen P. Terminal restlessness in the nursing facility: assessment, palliation, and symptom management. *Geriatr Nurs* 2001;22:308–12.
88. Chiu TY, Hu WY, Lue BH, Cheng SY, Chen CY. Sedation for refractory symptoms of terminal cancer patients in Taiwan. *J Pain Symptom Manage* 2001;21:467–72.
89. Ogasawara K, Ogawa A, Okuguchi T, Kobayashi M, Suzuki M, Yoshimoto T. Postoperative hyperperfusion syndrome in elderly patients with chronic subdural hematoma. *Surg Neurol* 2000;54:155–9.
90. Dunlop RJ, Campbell CW. Cytokines and advanced cancer. *J Pain Symptom Manage* 2000;20:214–32.
91. Wein S. Sedation in the imminently dying patient. *Oncology (Williston Park)* 2000;14:585–92; discussion 592, 597–8, 601.
92. Fainsinger RL, Bruera E. When to treat dehydration in a terminally ill patient? *Support Care Cancer* 1997;5:205–11.
93. Black KJ, Perlmutter JS. Septuagenarian Sydenham's with secondary hypomania. *Neuropsychiatry Neuropsychol Behav Neurol* 1997;10:147–50.
94. Vaphiades MS, Celesia GG, Brigell MG. Positive spontaneous visual phenomena limited to the hemianopic field in lesions of central visual pathways. *Neurology* 1996;47:408–17.
95. Moyle J. The use of propofol in palliative medicine. *J Pain Symptom Manage* 1995;10:643–6.
96. Di Salvo TG, O'Gara PT. Torsade de pointes caused by high-dose intravenous haloperidol in cardiac patients. *Clin Cardiol* 1995;18:285–90.
97. Nicolai A, Lazzarino LG. Acute confusional states secondary to infarctions in the territory of the posterior cerebral artery in elderly patients. *Ital J Neurol Sci* 1994;15:91–6.
98. Sanders KM, Stern TA. Management of delirium associated with use of the intra-aortic balloon pump. *Am J Crit Care* 1993;2:371–7.
99. Fernandez F, Holmes VF, Adams F, Kavanaugh JJ. Treatment of severe, refractory agitation with a haloperidol drip. *J Clin Psychiatry* 1988;49:239–41.
100. Swedlow DB, Schreiner MS. Management of Reye's syndrome. *Crit Care Clin* 1985;1:285–311.
101. Price J, Whitlock FA, Hall RT. The psychiatry of vertebro-basilar insufficiency with the report of a case. *Psychiatr Clin (Basel)* 1983;16:26–44.

102. Steinhart MJ. Treatment of delirium – a reappraisal. *Int J Psychiatry Med* 1978;9:191–7.
103. Brenner WI, Lieberman AN. Acute clonidine withdrawal syndrome following open-heart operation. *Ann Thorac Surg* 1977;24:80–2.
104. Wendkos M. Acute Exhaustive Mania Sudden Death and Psychiatric Illness. New York: Medical & Scientific Books division of Spectrum Publications, 1979:Chapter 10.
105. Mann SC, Caroff SN, Bleier HR, Welz WK, Kling MA, Hayashida M. Lethal catatonia. *Am J Psychiatry* 1986;143:1374–81.

Appendix B
Electrocution Diagnosis Checklist

In considering the possibility that a CEW (conducted electrical weapon) may have induced VF (ventricular fibrillation), it is important to remember that:

- No medical, scientific, electrical, or engineering study has found that a CEW can cause cardiac capture or induce VF in a human.
- In addition to other studies, the latest echocardiograpahic human studies have specifically not found cardiac capture or VF during CEW discharge [1,2].
- In over 500 CEW medical studies of human exposures, many of which occurred across the chest, none induced cardiac capture or VF [1-15].[1]
- No closely monitored medical study of CEW field-use has found any case of CEW induced VF [16,17].[2]
- In the entire history of medical, scientific, electrical, and engineering research, no study has found that the amount of electrical charge as delivered by a battery powered handheld CEW causes cardiac capture or induces VF in humans [18].

All 10 criteria below would have to be present in order to have the possibility that a suspect was electrocuted by the CEW.

Present?	Not Present	Criterion	Rationale
From law enforcement reports			
		1. Probe mode deployment (instead of drive-stun).	Drive-stun mode is not able to induce VF even in small swine [19,20].
		2. Use of TASER® X26™ CEW instead of ADVANCED TASER M26™ CEW (M26).	The M26 is unable to induce VF even in smaller swine while this has occurred in swine with the X26 [21].
		3. Collapse within 15 s of *initiation* of a probe-mode application.[3]	The electrical induction of VF requires 2–5 s [22–26]. Collapse in VF occurs within 10 s of the VF initiation [27].

(continued)

From EMS report		
	4. A defibrillator shock applied to the person within 8–10 min of VF terminates the VF. Law enforcement defibrillator may have been used.	Electrically induced VF is terminated by a defibrillation shock 99.9% of the time [28].
	5. Cardiac rhythm seen within 20 min of the collapse is VF. This is demonstrated either by an external defibrillator announcing "shock advised" or by a paramedic's EKG (electrocardiograph) recording.	Most cardiac rhythms seen with acute or chronic drug effects or excited delirium are not VF [29–32]. After 20 min of a possible VF the rhythm may deteriorate to another type of arrhythmia [33,34].
	6. Initial presenting rhythm is not asystole or PEA (pulseless electrical activity).	Asystole and PEA cannot be induced with electrical stimulation. This criterion is not valid after 20 min from collapse as VF will eventually deteriorate into asystole or PEA [34].
From autopsy report		
	7. One probe was directly over a cardiac ventricle (main chambers of the heart).	When VF has been induced in the small swine, at least one of the probes was over a ventricle [19,35].
	8. The probe directly over the cardiac ventricle penetrated the skin and did not simply penetrate or lodge in the person's clothing.	VF has only been induced even in swine with a fully inserted probe [21,35–37]. Note: typical probe penetration is around 4 mm or approximately half of the standard 9 mm barb length.
	9. The penetrating probe over the heart penetrated straight in and was not at an angle that would negatively affect the dart-to-heart distance.	Webster group predicted, from porcine studies, that VF induction would require a very thin person with a full 9 mm barb penetration [35].
	10. The dart tip-to-heart distance is a maximum of 8 mm.[4]	The most extreme distance found by the Webster group for VF induction – in swine – was 8 mm with an average of 6 mm, and a minimum of 2 mm [35]. Swine are easy to fibrillate and thus present a conservative model [38–40].

Contributors: Mark Kroll, PhD, Richard Luceri, MD; Hugh Calkins, MD, DJ Lakkireddy, MD, Jeffrey Ho, MD.

1. There is a single anecdote of a prisoner with a pacemaker with a possible brief cardiac capture [41,42]. Some cardiac electrophysiologists feel that the internal pacemaker recording reflects cardiac capture while others feel that the recordings only show the artifact of the CEW pulses. Regardless, on careful analysis of this case, pacemaker experts have concluded that capture, if any, could only have been caused by the pacemaker lead carrying some current to the inside of the heart. Thus, this case does not provide evidence that a CEW – absent a pacemaker – could cause cardiac capture. There was no harm to the prisoner or pacemaker.
2. An oft-cited anecdote, of possible electrically induced VF was misreported with material omissions [43]. A violent subject exhibiting most of the signs of excited delirium was briefly subdued with a short CEW discharge. Paramedics were present and found a normal pulse and respiration *after* the CEW discharge. After a 14-min delay, the subject collapsed and probably had an ideoventricular rhythm. After an aggressive therapy of 3 defibrillation shocks along with atropine and epinephrine, the subject finally had the VF strip shown in the published anecdote. A total of 23 min elapsed between the CEW application and the published VF strip.
3. Collapse at the end of a longer CEW application is generally due to the "quiet period" after an agitated (or excited) delirium struggle which necessitated the long CEW application. Even before VF would theoretically be induced, the rapid cardiac capture required to induce VF would have resulted in immediate loss of blood pressure and subsequent collapse [44]. This would have ended the struggle.
4. Such a dart tip-to-heart distance is only possible in a very thin person with a small BMI (body mass index). Skin-to-heart distances of 17 mm or less are rarely found in adults, especially males. Also, most law enforcement involved custody or arrest-related deaths involve males with an above-average BMI [29] and dart-to-heart distance significantly greater than 17 mm.

References

1. Ho JD, Dawes DM, Reardon RF, Lapine AL, Dolan BJ, Lundin EJ, Miner JR. Echocardiographic evaluation of a TASER-X26 application in the ideal human cardiac axis. *Acad Emerg Med* 2008.
2. DawesD, Ho J, Miner J. Echocardiographic evaluation of TASER X26® probe deployment into the chest of human volunteers. *Australas Coll Emerg Med Sci Assem* 2008.
3. Dawes D, Ho J, Miner J. The neuroendocrine effects of the TASER X26: A brief report. *Forensic Sci Int* 2009:in press.
4. Vilke GM, Sloane C, Levine S, Neuman T, Castillo E, Chan TC. Twelve-lead electrocardiogram monitoring of subjects before and after voluntary exposure to the Taser X26. *Am J Emerg Med* 2008;26:1–4.
5. Sloane CM, Chan TC, Levine SD, Dunford JV, Neuman T, Vilke GM. Serum troponin I measurement of subjects exposed to the Taser X-26. *J Emerg Med* 2008;35:29–32.

6. Ho JD, Dawes DM, Bultman LL, Moscati RM, Janchar J, Miner JR. Prolonged TASER® use on exhausted humans does not worsen markers of acidosis. *Am J Emerg Med* 2008:in press.
7. Ho J, Lapine A, Joing S, Reardon R, Dawes D. Confirmation of respiration during trapezial conducted electrical weapon application. *Acad Emerg Med* 2008;15:398.
8. Ho J, Dawes D. The effect of the extended range electronic projectile (XREP) on breathing. Australian College of Emergency Medicine Winter Symposium. Newcastle, NSW, 2008.
9. Dawes DM, Ho JD, Johnson MA, Lundin E, Janchar TA, Miner JR. 15-Second conducted electrical weapon exposure does not cause core temperature elevation in non-environmentally stressed resting adults. *Forensic Sci Int* 2008;176:253–7.
10. Vilke GM, Sloane CM, Bouton KD, Kolkhorst FW, Levine SD, Neuman TS, Castillo EM, Chan TC. Physiological effects of a conducted electrical weapon on human subjects. *Ann Emerg Med* 2007;50:569–75.
11. Vilke G, Sloane C, Levine S, Neuman T, Castillo E, Chan T. Does the Taser cause electrical changes in twelve lead ECG monitoring of human subjects. *J Acad Emerg Med* 2007;Abstracts:257.
12. Vilke G, Sloane C, Bouton K, Levine S, Neuman T, Castillo E, Kolkhorst F, Chan T. Cardiovascular and metabolic effects of the TASER on human subjects. *J Acad Emerg Med* 2007;14:104.
13. Moscati R, Ho J, Dawes D, Miner J, Reardon R, Heegaard W, Janchar T, Johnson M, Bultman L. Physiologic effects of prolonged conducted electrical weapon discharge on intoxicated adults. *Soc* Acad Emerg Med *abstract issue* 2007.
14. Levine SD, Sloane CM, Chan TC, Dunford JV, Vilke GM. Cardiac monitoring of human subjects exposed to the taser. *J Emerg Med* 2007;33:113–7.
15. Chan T, Sloane C, Neuman T, Levine S, Castillo E, Vilke G, Bouton K, Kohokorst F. The impact of the Taser weapon on respiratory and ventilatory function in human subjects. *Acad Emerg Med* 2007;14:191–192.
16. Bozeman W, Winslow J, Hauda W, Graham D, Martin B, Heck J. Injury Profile of TASER® electrical conducted energy weapons (CEWs) *Ann Emerg Med* 2007;50:S65.
17. Eastman AL, Metzger JC, Pepe PE, Benitez FL, Decker J, Rinnert KJ, Field CA, Friese RS. Conductive electrical devices: a prospective, population-based study of the medical safety of law enforcement use. *J Trauma* 2008;64:1567–72.
18. Ideker RE, Dosdall DJ. Can the direct cardiac effects of the electric pulses generated by the TASER X26 cause immediate or delayed sudden cardiac arrest in normal adults? *Am J Forensic Med Pathol* 2007;28:195–201.
19. Lakkireddy D, Wallick D, Verma A, Ryschon K, Kowalewski W, Wazni O, Butany J, Martin D, Tchou PJ. Cardiac effects of electrical stun guns: does position of barbs contact make a difference? *Pacing Clin Electrophysiol* 2008;31:398–408.
20. Valentino DJ, Walter RJ, Dennis AJ, Nagy K, Loor MM, Winners J, Bokhari F, Wiley D, Merchant A, Joseph K, Roberts R. Acute effects of MK63 stun device discharges in miniature swine. *Mil Med* 2008;173:167–73.
21. Nanthakumar K, Billingsley IM, Masse S, Dorian P, Cameron D, Chauhan VS, Downar E, Sevaptsidis E. Cardiac electrophysiological consequences of neuromuscular incapacitating device discharges. *J Am Coll Cardiol* 2006;48:798–804.
22. Biegelmeier G. Effect of current passing through the human body and the electrical impedance of the human body: A guide to IEC-Report 469. VDE,-Verlag, Berlin: ETZ, 1987.
23. Biegelmeier G, Lee WR. New considerations on the threshold of ventricular fibrillation for a.c.shocks at 50∼60 Hz. *IEE Proc.* 1980;127:Pt. A: 103–110.
24. Sharma AD, Fain E, O'Neill PG, Skadsen A, Damle R, Baker J, Chauhan V, Mazuz M, Ross T, Zhang Z. Shock on T versus direct current voltage for induction of ventricular fibrillation: a randomized prospective comparison. *Pacing Clin Electrophysiol* 2004;27:89–94.

25. Weismuller P, Richter P, Binner L, Grossmann G, Hemmer W, Hoher M, Kochs M, Hombach V. Direct current application: easy induction of ventricular fibrillation for the determination of the defibrillation threshold in patients with implantable cardioverter defibrillators. *Pacing Clin Electrophysiol* 1992;15:1137–43.
26. Frame R, Brodman R, Furman S, Kim SG, Rot J, Ferrick K, Hollinger I, Gross J, Fisher JD. Clinical evaluation of the safety of repetitive intraoperative defibrillation threshold testing. *Pacing Clin Electrophysiol* 1992;15:870–7.
27. Schipke JD, Heusch G, Sanii AP, Gams E, Winter J. Static filling pressure in patients during induced ventricular fibrillation. *Am J Physiol Heart Circ Physiol* 2003;285:H2510–5.
28. Frame R, Brodman R, Furman S, Kim SG, Roth J, Ferrick K, Hollinger I, Gross J, Fisher JD. Clinical evaluation of the safety of repetitive intraoperative defibrillation threshold testing. *Pacing Clin Electrophysiol* 1992;15:870–7.
29. Stratton SJ, Rogers C, Brickett K, Gruzinski G. Factors associated with sudden death of individuals requiring restraint for excited delirium. *Am J Emerg Med* 2001;19:187–91.
30. Park KS, Korn CS, Henderson SO. Agitated delirium and sudden death: two case reports. *Prehosp Emerg Care* 2001;5:214–6.
31. DiMaio T, VJM D. Excited Delirium Syndrome Cause of Death and Prevention. Boca Raton: Taylor & Francis, 2006.
32. Swerdlow C, Kroll M, Williams H, Biria M, Lakkireddy D, Tchou P. Presenting rhythm in sudden custodial deaths after use of TASER® electronic control device. *Heart Rhythm* 2008;5:S44.
33. Allison JS, Qin H, Dosdall DJ, Huang J, Newton JC, Allred JD, Smith WM, Ideker RE. The transmural activation sequence in porcine and canine left ventricle is markedly different during long-duration ventricular fibrillation. *J Cardiovasc Electrophysiol* 2007;18:1306–12.
34. Veltmann C, Borggrefe M, Schimpf R, Wolpert C. Fatal inappropriate ICD shock. *J Cardiovasc Electrophysiol* 2007;18:326–8.
35. Wu J, Sun H, O'Rourke A, Huebner S, Rahko P, Will J, Webster J. Taser blunt dart-to-heart distance causing ventricular fibrillation in pigs. *IEEE Trans Biomed Eng* 2009:in press.
36. Walter RJ, Dennis AJ, Valentino DJ, Margeta B, Nagy KK, Bokhari F, Wiley DE, Joseph KT, Roberts RR. TASER X26 discharges in swine produce potentially fatal ventricular arrhythmias. *Acad Emerg Med* 2008;15:66–73.
37. Wu J, Sun H, O'Rourke A, Huebner S, Rahko P, Will J, Webster J. Taser dart-to-heart distance that causes ventricular fibrillation in pigs. *IEEE Trans Biomed Eng,* 2007;54:503–8.
38. Dalziel CF, Lee WR. Reevaluation of lethal electric currents. *IEEE Trans Ind Gen Appl* 1968;IGA-4:467–6.
39. Kroll MW, Calkins H, Luceri RM, Graham M, Heegaard W. Sensitive swine and TASER electronic control devices. *Acad Emerg Med* 2008;15:695–6.
40. Pak HN, Kim YH, Lim HE, Chou CC, Miyauchi Y, Fang YH, Sun K, Hwang C, Chen PS. Role of the posterior papillary muscle and purkinje potentials in the mechanism of ventricular fibrillation in open chest dogs and swine: effects of catheter ablation. *J Cardiovasc Electrophysiol* 2006;17:777–83.
41. Cao M, Shinbane JS, Gillberg JM, Saxon LA. Taser-induced rapid ventricular myocardial capture demonstrated by pacemaker intracardiac electrograms. *J Cardiovasc Electrophysiol* 2007;18:876–9.
42. Kroll M, Luceri RM, Calkins H. A very interesting case study involving a TASER conducted electrical weapon (CEW) used on a patient with a pacemaker. *J Cardiovasc Electrophysiol* 2007;18:E29–30; author reply E31.
43. Kim PJ, Franklin WH. Ventricular fibrillation after stun-gun discharge. *N Engl J Med* 2005;353:958–9.
44. Swerdlow CD, Olson WH, O'Connor ME, Gallik DM, Malkin RA, Laks M. Cardiovascular collapse caused by electrocardiographically silent 60-Hz intracardiac leakage current. Implications for electrical safety. *Circulation* 1999;99:2559–64.

Index

D

Printed in the United States of America